ANIMAL NUTRITION

ANIMAL NUTRITION

Editor

Valentin Flater

scitus
academics

Animal Nutrition

Edited by **Valentin Flater**

This Book is Printed on Acid-Free Paper. ♾

ISBN: 9781681179216

Library of Congress Control Number: 2017948950

SCITUS Academics LLC,

616, Corporate Way, Suite 2, 4766,

Valley Cottage, NY 10989

www.scitusacademics.com

Printed in United States of America.

Preface

Meeting livestock nutritional requirements is enormously significant in maintaining satisfactory performance of neonatal, growing, finishing and breeding animals. From a practical point of view, an optimal nutritional program should ensure sufficient intakes of amino acids (both traditionally classified essential and nonessential), carbohydrates, fatty acids, minerals, and vitamins by animals through a supplementation program that corrects deficiencies in basal diets (e.g., corn- and soybean meal-based diets for swine; milk replacers for calves and lambs; and available forage for ruminants). Also crucial to the nutrition program for animals is water. Modern breeds of dairy animals are able to produce huge amount of milk. In attempt to consume, digest and metabolize enough nutrients to satisfy lactation needs, those animals are exposed to serious stress conditions that can affect their health. Health problems which arise from those conditions are mainly related to impaired ability to metabolize enough nutrients to compensate for those lost in milk. They are known as metabolic or production diseases and may be of great economic importance in milk production systems. Although metabolic diseases have become a common problem on dairy farms, they still require a serious attention to be controlled. The incidences of these disorders can be reduced by proper nutrition of animals. Also, some of the specific strategies in feeding practice offer additional advantages in prevention of nutrition-related metabolic diseases.

This volume 'Animal Nutrition' is concerned with the animal feeds and their feeding describing research on feed for ruminants and non-ruminants, including poultry, horses, companion animals and aquatic animals. It encompasses the full coverage of animal nutritional sciences and reviews including, but not limited to, fundamental aspects of animal nutrition such as nutritional requirements, metabolic studies, body composition, energetics, immunology, genetics and molecular and cell biology related to nutrition, and more applied aspects of animal nutrition, such as raw material evaluation, feed additives, nutritive value of novel ingredients and feed safety. This book will be useful for students, researchers, teaching staff, practicing professionals connected with dairy science, animal science, food science, nutrition, physiology, biochemistry, veterinary medicine and other related fields.

Table of Contents

CHAPTER 1

Nutrition and Health of Dairy Animals

Cazim Crnkic and Aida Hodzic

University of Sarajevo, Veterinary faculty, Bosnia & Herzegovina

1. INTRODUCTION

Modern breeds of dairy animals are able to produce huge amount of milk. In attempt to consume, digest and metabolize enough nutrients to satisfy lactation needs, those animals are exposed to serious stress conditions that can affect their health. Health problems which arise from those conditions are mainly related to impaired ability to metabolize enough nutrients to compensate for those lost in milk. They are known as metabolic or production diseases and may be of great economic importance in milk production systems.

Although metabolic diseases have become a common problem on dairy farms, they still require a serious attention to be controlled. The incidences of these disorders can be reduced by proper nutrition of animals. Also, some of the specific strategies in feeding practice offer additional advantages in prevention of nutrition-related metabolic diseases.

2. NUTRITION-RELATED METABOLIC DISORDERS, THEIR ETIOLOGY AND CONSEQUENCES

The most important metabolic diseases in dairy cows, ewes and goats can be discussed as energy and/or fiber related, lipid related or vitamin and mineral related disorders. However, they are not easily categorized according to their cause, thus the pathogenesis and hallmarks of each disease should be considered for their categorization. Energy related disorders (related to energy density of the diet or feed intake) include fatty liver and ketosis, rumen acidosis, laminitis, displaced abomasums and milk fat depression. As energy density and fiber content of the diets are often inversely related, most of these diseases can be considered as fiber related too. Fatty liver and ketosis also can be categorized as lipid related disorders due to changes in lipid metabolism in affected animals (Grummer, 1993). The most important mineral/vitamin related disorders are hypocalcaemia, hypomagnesaemia, udder edema, retained placenta and metritis. Not all of them are ultimately caused by mineral composition of the diet, but may be prevented by manipulation of minerals or vitamins in the diet. Broken homeostatic mechanisms often become major etiological factors in the development of metabolic disease, and consequences varies from reduced production and impaired reproductive performances to increased risk to develop other diseases.

2.1. Fatty liver

Fatty liver (hepatic lipidosis, fat cow syndrome) is a metabolic disorder characterized by a high content of lipids and triglycerides in the liver (Ingvartsen, 2006). The disease occurs in periparturient period, primarily in the first 4 weeks after calving (Grummer, 1993), and as a secondary disease of other production diseases that depress appetite or increase body fat mobilization. The clinical symptoms comprise depression, lack of appetite and weight loss, and the cows are weak and apathetic (Radostits et al., 2000). Most cows suffer from non-specific clinical signs including reduced rumen motility and decreased milk yield. However, this disease occurs especially in its subclinical form and can be a problem for up to 50% of cows in early lactation (Ingvartsen, 2006).

Risk factors for fatty liver in dairy cows may be nutritional, managerial, and genetic. Prepartum risk factors include obesity, severe feed restriction, feeding excess energy, and long calving interval. Postpartum risk factors are different diseases and infection, fasting and feed restriction, as well as ketogenic diets and sudden feed changes (Bobe et al., 2004). Among the highest risk factors of fatty liver is a high rate of mobilization of body lipids around calving in overconditioned cows which have been overfed in late lactation and the dry period and had low feed intake around calving (Ingvartsen, 2006). Increased prepartum body weight gain or body condition increase liver fat content and indicate increased risk of fatty liver and related disorders. Therefore, the primary nutritional risk factor for fatty liver is obesity. In obese cows, lipolysis of adipose tissue is increased more during peripartal period than in cows with normal body condition.

Obese cows have a greater decrease in feed intake around calving and, therefore, have a more severe negative energy balance (Bobe et al., 2004). During the negative energy balance body fat is mobilized into the bloodstream in the form of non-esterified fatty acids (NEFA). NEFA are taken by the liver in proportion to their supply, but the liver does not have capacity sufficient to oxidize and use all amount of NEFA for energy. Therefore, cows are predisposed to accumulate NEFA as triglycerides within the liver when large amounts of NEFA are released from adipose tissue (Overton & Waldron, 2004). Rate of triglyceride synthesis is proportional to plasma NEFA concentration, thus fatty liver is likely to develop when plasma NEFA are elevated (Grummer, 1993).

Severe negative energy balance caused by low feed intake around calving leads to increase in body fat mobilization and plasma NEFA concentration. Thus, the severity of fatty liver may be reduced if drop in feed intake before calving is prevented (Bertics et al., 1992).

Diets deficient in cobalt (Co) have been shown to cause fatty liver in sheep. It has been termed as ovine white liver disease. Hepatic lipidosis in goats due to feeding low levels of Co in their diet has also been reported (Johnson et al., 2004).

Fatty liver is associated with decreased health status and reproductive performance. In severe cases, milk production and feed intake are also decreased (Bobe et al., 2004). Although fatty liver is often a reversible condition, it predisposes cows to reduced liver function and to a number of other production diseases (Ingvartsen, 2006). The incidence of fatty liver is strongly associated with the incidence of especially ketosis and displaced abomasum because these disorders are related to severe negative energy balance too (Bobe et al., 2004). Inhibition of gluconeogenesis may also occur when triglycerides accumulate in liver (Overton & Waldron, 2004) and additionally increase the risk for ketosis. Different aspects of the immune response are also suppressed in cows with fatty liver (Bobe et al., 2004; Overton & Waldron, 2004) and those animals are more prone to infectious diseases. However, the exact costs of fatty liver are difficult to estimate because condition can be diagnosed only by liver biopsy (Bobe et al., 2004).

2.2. Ketosis

Ketosis (acetonemia) is a metabolic disorder characterized by elevated concentrations of the ketone bodies (acetoacetate, beta-hydroxybutyrate and acetone), and a low to normal concentration of glucose in the blood. This disorder occurs both subclinically and clinically. Prevalence of subclinical ketosis is the highest in the first 2 weeks of lactation ranging from 8,9% to 34% in various studies (Ingvartsen, 2006; Rushen et al.,

2008). Blood glucose in clinically affected cows fall below the level required to support nerve and brain function and cows often exhibit signs of central nervous system dysfunction. Ketotic cows also suffer inappetence which further exacerbates the negative energy balance. Milk production falls seriously which helps the cow to cope with the negative energy balance (Goff, 2006a).

Four main types of ketosis are: primary ketosis, secondary ketosis, butyric acid ketosis and underfeeding ketosis (Ingvartsen, 2006). Classical or primary ketosis (also called spontaneous or production ketosis) generally occurs in cows during the first 2 to 4 weeks of lactation (Goff, 2006a). It develops when the glucose demand exceeds the gluconeogenesis capacity of the liver resulting in increased ketogenesis, and thus in high concentrations of ketone bodies in blood, milk, and urine. The disorder is mainly seen in obese cows. Secondary ketosis results from another disease that depresses feed intake and increases body fat mobilization (Ingvartsen, 2006).

Butyric acid ketosis is caused by large amounts of butyrate in the feed and probably by related depressed feed intake. Silage with high butyrate concentrations results in an increased concentration of beta-hydroxybutyrate in the blood, and such silage is consumed in lower amount then normal one. Underfeeding ketosis occurs especially in cows that are fed insufficiently. Underfed animals are deficient in glucogenic precursors and this condition then leads to increased ketogenesis (Ingvartsen, 2006).

Another form of ketosis often seen in US dairies usually occurs in cows less than 10 days in lactation. It can often be difficult to treat as is generally accompanied by some degree of fatty liver (ketosis/fatty liver complex). Ketosis, like fatty liver, also occurs during periods of elevated plasma NEFA. Clinical symptoms are similar to classical ketosis, but ketotic state was actually preceded by an increase in fat accumulation in the liver (Goff, 2006a).

Major factors directly or indirectly increasing the risk of fatty liver and ketosis are: overconditioning at calving, excessive mobilization of body fat, low nutrient intake, some nutrient or diet specific factors and management and environmental stress (Ingvartsen, 2006). Ketosis is a metabolic condition that occurs primarily when a cow is in negative energy balance immediately after calving. To support the energy demands, the body mobilizes fat reserves. Excessive release of NEFA from fat depots may overwhelm the capacity of the liver to use the fatty acids as a fuel. They are instead converted to ketone bodies (Rushen et al., 2008). Feed intake is naturally depressed by about 20% around the time of calving. Simultaneous with the feed intake depression, plasma NEFA and liver triglyceride concentrations exhibited their greatest increase (Goff, 2006a). All factors exacerbating negative energy balance and depression of feed intake around calving increase the risk of ketosis, especially in overconditioned cows.

Negative energy balance, high blood levels of NEFA or beta-hydroxybutirate and ketosis reduce the response capacity of white blood cells so that invading bacteria can out compete the immune system of the cow (Zadoks, 2006). Hypoglycemia alone is not likely to exacerbate periparturient immunosuppression, but hyperketonemia appears to have multiple negative effects on immune functions (Overton & Waldron, 2004). Impaired immune functions around calving make cow more prone to infectious diseases. Cows that have suffered ketosis have a higher risk of ketosis in the following lactation (Ingvartsen, 2006). Affected animals produce 0,7-3,3 kg/day less milk in the rest of lactation (Fourichon et al., 1999) or about 200 kg annually (Guard, 1996). In addition, prolonged and pronounced negative energy balance in early lactation adversely affects reproductive performance of cows and increases the risk of other metabolic diseases (Rushen et al., 2008).

Ewes experience ketosis typically during the last month of pregnancy and is known as pregnancy toxemia (Schlumbohm & Harmeyer, 2003). Ovine pregnancy toxemia (twin lamb disease, pregnancy disease) occurs primarily in ewes carrying more than one fetus (Andrews et al., 1996; Schlumbohm & Harmeyer, 2003). It is usually seen when the plane of nutrition has been static or falling over the last month of gestation when the requirements of the ewe are increasing to allow the growth of the fetuses (Andrews et al., 1996). In goats, two clinical form of ketosis have been described: pregnancy toxemia during the last month of pregnancy and primary ketosis during the first month of lactation (Stelletta et al., 2008).

Limitation of the availability of glucose has been found to be a crucial factor for the development of pregnancy toxemia. Hyperketonemia lowered plasma glucose concentration and depressed endogenous glucose production in sheep by approximately 30%, and this facilitates the onset of pregnancy toxemia (Schlumbohm & Harmeyer, 2003).

Mortality is high if pregnancy toxemia is not treated, with only about 20% of affected ewes recovering without treatment. Encephalopathy can results from depressed glucose metabolism in the brain (Andrews et al., 1996).

2.3. Rumen acidosis

Ruminal acidosis is a nutritional disorder of ruminants generally resulting from ingestion of large amounts of feeds rich in readily fermentable carbohydrates, particularly when animals have not been gradually acclimated to those feeds (Bramley et al., 2008). The disorder usually occurs in high-producing dairy cows. Subacute or subclinical ruminal acidosis (SARA) is considered to be one of the major threats to the welfare of lactating dairy cows and may affect up to 20% of cattle in early to mid lactation (Rushen et al., 2008). Subclinical rumen acidosis is defined as a condition where rumen fluid pH is below 6,0 while acute rumen acidosis is when rumen pH is below 5,5 associated with rumen motility that is weak or ceased (Ingvartsen, 2006).

Rumen acidosis classically occurs when an animal consumes the excess of grain (Stone, 2004). Fermentation of the high grain diet reduces rumen pH which can cause undesirable changes in microbial populations within the rumen (Goff, 2006a). Rumen pH is lowered due to large quantities of volatile fatty acids and lactic acid produced during grain fermentation (Bramley et al., 2008). Lactic acid production is not the hallmark of rumen acidosis in dairy cows, as observed in beef feedlot cattle. Instead, it is the total organic acid load induced by the high grain diet combined with the inability of the cow to buffer the acids with salivary secretions (Goff, 2006a; Stone, 2004). However, lactate accumulation may occur in cows after calving if the shift in fermentable carbohydrates between the diets fed before and after calving is too dramatic (Stone, 2004).

Low rumen pH can cause rumenitis, metabolic acidosis, lameness, hepatic abscesses formation, pneumonia, and even death (Bramley et al., 2008). Reduced ruminal efficiency, liver and lung abscesses, and laminitis all can be related to SARA (Stone, 2004).

2.4. Laminitis

Laminitis and laminitis-related hoof problems (sole ulcer, white line abscess, solar hemorrhage, etc.) are one of the leading causes of lameness in cows. Laminitis has been associated with nutrition, and specifically with ruminal acidosis either in its acute or subacute form. Exact relationship between laminitis and SARA is not known. One of the theories states that damage of the ruminal epithelium induced by acidosis allows absorption of histamine and endotoxins into the blood. These and possibly other compounds affect circulation within the hoof and cause inflammation leading to the condition known as laminitis (Stone, 2004). Cows fed the higher level of crude protein may have increased incidence and duration of lameness. It is considered that products of degradation of protein excess in the rumen may be the causative agents for lameness.

It is widely accepted that nutrition helps maintain claw health through the production of good quality horn (Baird & Muelling, 2009). For example, heifers fed on hay before calving might have better foot health than those fed silage, even if both groups are being fed the same diets postpartum.

Forage type and level of moisture in the diet could influence lameness and this effect is exerted even before first calving (Logue & Offer, 2001).

It has also been shown that nutritional supplements such as biotin and zinc (Zn) can help reduce lameness through improving claw horn quality (Baird & Muelling, 2009; Goff, 2006a). Biotin is essential for two major metabolic pathways in keratinisation, keratin protein synthesis and lipogenesis. Other vitamins also may play important roles in maintaining claw integrity, including vitamin A and vitamin E. Zinc, as a component of many enzyme systems, has a role in major functions during keratinisation including the formation of structural proteins (Baird & Muelling, 2009). This element improves claw integrity by speeding wound healing and epithelial tissue repair. Other trace minerals that impact claw condition include iodine (I), selenium (Se), copper (Cu), manganese (Mn) and Co. Calcium and phosphorus (P) are also needed for normal claw growth and integrity.

Laminitis reduces profitability of the dairy herd. It is estimated that 15% of cows culled for slaughter are culled due to laminitis. In clinically lame cows, milk yield was reduced from 4 months before and for the 5 months after treatment. The total mean estimated reduction in milk yield per lactation was approximately 360 kg (Green et al., 2002).

2.5. Displaced abomasum

Displaced abomasum is a multifactorial disease where the abomasum is dilated as a result of gas accumulation and dislocated to the left (left-displaced abomasum) or to the right (right- displaced abomasum) in the stomach in relation to the normal placing (Ingvartsen, 2006). The passage of feed to the intestines is partly or totally blocked. Approximately 80-90% of the incidences are left-displaced abomasums. The disease is most frequent in high producing cows in early lactation and 80–90% of the cases are seen in the first 4 weeks postpartum (Shaver, 1997). It occurs in approximately 3,5% of dairy cows each year (Goff, 2006a).

Nutrition has been implicated as a major risk factor in the etiology of displaced abomasums, but the precise cause is still unclear. Low feed intake and increased negative energy balance prepartum have been found to increase the risk of displaced abomasums in cows (Goff, 2006a; Ingvartsen, 2006). Cows fed high concentrate diets in early lactation or diets with inadequate particle size are also at increased risk of displaced abomasums (Goff, 2006a). Sudden changes in the diet and rapid increase in concentrate allowance in early lactation are also the risk factors. Some feeds increase the risk of the disease compared to others. Thus, a higher risk of displaced abomasum has been found for cows fed silage compared to those fed hay, probably because silage is often finely chopped. Risk can be almost eliminated if cow eats a kilogram of straw daily (Ingvartsen, 2006).

Displaced abomasum develops if three major conditions are met. The first one is reduced contractility and atony of abomasum which is gas-dilated. The next condition is that the mesentery shall stretch for the abomasum to be able to dislocate, and a third condition is the space in the abdominal cavity (Ingvartsen, 2006). The conditions leading to atony and reduced motility are still unclear, but hypocalcaemia around calving is a possible factor. Reduction in blood Ca concentration around calving results in a reduction of abomasal contractility, which can lead to its atony and dilatation (Shaver, 1997). Hypocalcemic cows after calving have 3,4 to 4,8 times increased risk of development of abomasal displacement (Massey et al., 1993).

Volatile fatty acids (VFA) in the abomasum also have been reported to reduce abomasal motility. Effects of VFA on motility may be exacerbated by low ruminal absorption of VFA during the transition period (Shaver, 1997). Hypocalcaemia also may play a role in this process. Due to reduced feed consumption, slow contractions and inadequate filling of the rumen which therefore does not reach the ventral abdominal wall, empty space appears for movement of abomasum. Then usually more fatty acids escape absorption in the rumen and reach the abomasum. Those VFA, along with hypocalcaemia and accumulated gases, contribute to reducing abomasum contractility and development of atony (Goff & Horst, 1997c). Thus, inadequate feed consumption and insufficient rumen fill with reduced motility and strength of abomasal contractions together contribute to the onset of this disorder (Goff & Horst, 1997c).

Low intake of concentrates during the prepartum period also may increase the risk of left displaced abomasum because absorptive capacity of the ruminal papillae is not increased sufficiently and microbial population of the rumen is not adapted prior to intake of high energy postpartum diets (Shaver, 1997). Too rapid increase of concentrates after calving may reduce roughage intake and potentially increase the risk of displaced abomasum (Ingvartsen, 2006). Uncomplicated ketosis, fatty liver, retained placenta, and metritis are also risk factors for left displaced abomasums (Bobe et al., 2004; Shaver, 1997).

Treatment of displaced abomasum often requires surgical treatment, and deaths are not uncommon so that potential losses due to disease are very high. In addition, cows that recover from abomasal displacement produce about 350 kg less milk during the next month (Shaver, 1997) or 0,8-2,5 kg/day in the rest of lactation (Fourichon et al., 1999). According to Guard (1996) the average losses amounted to about 380 kg of milk annually.

2.6. Milk fat depression

Nutrition influences both the quantity and composition of milk fat. In modern dairy production cows are fed diets with high level of concentrates to maximize milk production but such diets often causes drop in milk fat. Condition is known as milk fat depression (MFD) or low-fat milk syndrome. Although milk volume and yield of other milk constituents may not be affected or may be even increased, depression of fat in milk can be a serious economic problem for dairy producers. This syndrome is not a disease, but rather metabolic consequence of attempts to reach the higher milk production in animals.

Milk fat depression refers to a marked reduction in milk fat yield with no change in milk yield or yield of other milk components (Harvatine et al., 2009). It represents a level of milk fat production below the genetic potential of the cow, usually as below 3,2% fat in milk for Holstein, or below 4,2% fat in the milk for a Jersey herd. When problem with MFD exists, the ratio of milk fat/milk protein will be less than 1,0 for Holstein herds. In severe cases the ratio will be less than 0,8. Milk fat yield can be reduced up to 50% and the effect is specific for milk fat (Bauman et al., 2008). The problem has been observed in many feeding situations and dietary conditions, including high level of concentrates and low level of dietary fiber, and diets supplemented with unsaturated oils. The fat content of milk can also be affected by the physical characteristics of the roughage, including grinding and pelleting (Perfield & Bauman, 2005). In diet-induced MFD, the yield of all individual fatty acids is reduced, but the decline is greatest for short- and medium-chain fatty acids that are synthesized *de novo* in mammary gland (Bauman et al., 2008).

It is generally accepted that induction of MFD requires both an altered rumen environment and the presence of unsaturated fat in the rumen which biohydrogenation pathways are altered (Perfield & Bauman, 2005). Changes in ruminal microbial processes are an essential component for the development of MFD and are often associated with a decrease in rumen pH and a shift in the acetate/propionate ratio.

Historically, there have been several theories proposed to explain the basis for diet-induced MFD. The biohydrogenation theory was proposed by Bauman & Griinari (2001) focusing on the latest research in MFD problem. The authors sugested that "under certain dietary conditions the pathways of rumen biohydrogenation are altered to produce unique fatty acid intermediates which are potent inhibitors of milk fat synthesis". The theory was born after identification of *trans*-10,*cis*-12 conjugated linoleic acid (CLA) as a highly potent inhibitor of milk fat synthesis (Bauman & Griinari, 2001). This isomer is a rumen precursor of trans-10 18:1 in biohydrogenation pathway that may arise under certain dietary conditions, including high concentrate/low fiber diets. The biohydrogenation of linoleic acid in rumen involves the formation of *trans*-11 18:1 and *cis*-9,*trans*-11 CLA as the main intermediates. However, under certain dietary situations the other pathways of biohydrogenation can happen and some of the intermediates produced during the process may be potent inhibitors of milk fat synthesis (Bauman & Griinari, 2001). The most extensively studied is *trans*-10,*cis*-12 CLA. Minimal quantities of this isomer markedly inhibited milk fat synthesis and a curvilinear reduction in milk fat yield occurred with increasing *trans*-10,*cis*-12 CLA in the rumen content (Baumgard et al., 2001).

Isomer *trans*-10,*cis*-12 CLA alone does not completely explain the extent of the decrease in milk fat. Additional biohydrogenation intermediates produced in the rumen probably inhibit milk fat synthesis and two of them (*trans*-9,*cis*-11 and *cis*-10,*trans*-12 CLA) have already been identified (Bauman et al., 2008).

Consequences of MFD are almost exclusively related to losses in yield of milk fat and lower price of milk. As problem is normally reversible, health consequences are usually not a matter of concern. However, MFD inducing diet is usually risky for rumen homeostasis due to high level of carbohydrates and low level of effective fiber which may reduce ruminal pH and cause acidosis (Bergen, 2009). If acidosis is not compensated and drop in ruminal pH is not prevented, the condition may become an important animal health issue.

2.7. Parturient hypocalcaemia

The lack of Ca in the diet does not lead to any changes in health or production for a long time because it is compensated by mobilization of Ca reserves in the skeleton. The change of Ca content in the blood may not occur until the cows show symptoms of osteoporosis and bone fractures. A unique example, however, is an acute hypocalcaemia in dairy cows that occurs in the periparturient period as a result of sudden loss of great amounts of Ca in colostrum along with temporary dysfunction of the mechanisms of Ca mobilization from bones. From a relative inactivity in the dry period, these mechanisms often fail to achieve full activity quickly enough to maintain normocalcaemia around calving (Horst et al., 1994). More pronounced hypocalcaemia causes progressive neuromuscular dysfunction which is manifested as a specific clinical syndrome known as paresis puerperalis (milk fever, parturient paresis, calving paralysis). More often, however, and in a larger number of cows, hypocalcaemia exists in a subclinical form with very few or no symptoms (subclinical hypocalcaemia).

Hypocalcaemia was defined as the content of total Ca in blood below 2 mmol/L with or without clinical signs of paresis (Oetzel et al., 1988; Radostits et al., 2000), which is roughly equivalent to 1 mmol/L of ionized Ca (Massey et al., 1993; Oetzel, 1996).

Milk fever (MF) is non-febrile disease of adult dairy cows accompanied by general weakness, circulatory collapse and depression of sensation (Oetzel, 1988). It is one of the most common metabolic disorders in intensive dairy production (Horst, 1986) and a typical nutritional disorder (Ender et al., 1971). The disease attacks 5-10% of adult dairy cows in the U.S. and Europe annually, primarily those with high productivity. The classic syndrome of MF occurs immediately before or after parturition with about 87% of cases in the first 48 hours after calving and only about 9% cases before or during the parturition (Oetzel, 1988). The disease is characterized by pronounced hypocalcaemia which is accompanied by hypophosphataemia in most cases, and by more or less serious hypermagnesaemia (Phillippo et al., 1994).

Phenomena like paresis may rarely occur in beef cows, ewes (before or after lambing) and more frequently in dairy goats (Oetzel, 1988). In mares, the disease can occur several weeks after birth and is known as milk tetany or eclampsia (Radostits et al., 2000).

Milk fever has been recognized as a nutritional disorder with various endocrine abnormalities which occur more or less as a secondary phenomenon (Ender et al., 1971). The characteristic of the disease is an acute inability to mobilize Ca from bone (Horst, 1986) despite apparently normal endocrine dynamics in most affected cows (Radostits et al., 2000). The problem seems to lie in the reduced sensitivity of target tissues (bone, kidneys and intestines) on calcitropic hormones due to reduced number of receptors or because of their dysfunction (Horst et al., 1994).

Widely accepted theory for a long time was that MF was caused by a high content of Ca in the diet consumed before calving, as well as by unfavorable ratio of Ca/P. However, further studies emphasized importance of the high content of monovalent cations in the diet, mainly potassium (Block, 1984; Ender et al., 1971; Goff & Horst, 1997b; Horst & Goff, 1997). As a relative excess of cations in the diet affects blood acid-base balance toward alkalosis, the disruption of integrity of tissue receptors is related to metabolic alkalosis that normally exists in cows at the time of calving (Goff, 2000). Thus, the primary cause of MF is considered to be the temporary inability of cow tissues to adequately respond to stimuli generated by calcitropic hormones, mainly parathyroid hormone (PTH) (Goff et al., 1991; Goff, 2000).

In addition to monovalent ions, risk factors also include dietary magnesium (Sansom et al., 1983), dietary P (Curtis et al., 1984), and possibly vitamin D (Goff, 2000; Horst, 1986). Animal factors as the risks for hypocalcaemia include age (Curtis et al., 1984; Dishington, 1974) and breed (Goff, 2000; Oetzel, 1988). Adult cows are at increased risk for the disease. Occurrences are more frequent from the third lactation onwards, reaching the plateau of 12-15% in cows 6-10 years of age (Curtis et al., 1984; Dishington, 1974). The most predisposed breeds for milk fever are Jersey and Guernsey, followed by Holstein and Brown Swiss. In the case of Jerseys possible reasons are considered to be the higher milk production per unit of body weight (Oetzel, 1988) and a higher content of Ca in the colostrum (Goff, 2000). In addition, the number of receptors in the intestine is about 15% lower in Jerseys than in Holsteins (Goff, 2000; Horst et al., 1997).

Depending on the age, breed, feeding regimen and housing conditions, as many as 50-80% of cows can suffer MF annually in some herds (Oetzel, 1988). Concerning that well-timed treatment of MF easily solves the problem and is relatively inexpensive, the disease was not considered as a factor of economic importance in dairy production for a long time (Radostits et al., 2000). However, later findings confirmed its close relationship with other health disorders in the puerperium (Curtis et al., 1985; Massey et al., 1993). Many of these disorders are now considered as complications of hypocalcaemia. They generally occur as a complex, rarely as isolated diseases (Curtis et al., 1985) and include most of metabolic and reproductive disorders in periparturient period: dystocia, retained placenta, metritis, uterine prolapse, ketosis, mastitis and displaced abomasum (Horst et al., 1997). Risk of occurrence of each of them is much higher in hypocalcemic cows (Curtis et al., 1985). A well known physiological effect of hypocalcaemia is atony of smooth and skeletal muscles because their contractility is a function of Ca concentration in extracellular fluid (Ramberg et al., 1984). Most of the possible complications are related specifically to reduced contractility and to atony of smooth muscles of digestive and reproductive tract (Beede, 1995).

Spontaneous recovery of MF is usually not possible and approximately 75% of affected cows die if treatment fails (Oetzel, 1988). Even under normal conditions about 8% of treated cows still die due to various complications, 12% of them are culled (Guard, 1996) and about 25% require another treatment. Due to increased susceptibility to other health disorders and possible complications, the production life of cows that experienced MF was reduced by an average of 3-4 years (Horst et al., 1997).

Subclinical hypocalcaemia influences in the same manner as the clinic one, but to a lesser extent. As is more common in the herd, adverse effects of subclinical hypocalcaemia on herd economy can be equal or even greater than the effects of MF due to its broader influence on feed intake, secondary disease conditions, and milk production during early lactation (Horst et al., 1994).

2.8. Hypomagnesaemia

Magnesium is an essential mineral with many functions in the body but its homeostasis is not hormonally regulated. The concentration of Mg in the blood is dependent exclusively on its absorption from the diet. If Mg secretion in milk and its endogenous losses exceeds absorption from the forestomachs hypomagnesaemia occurs because of the lack of hormonal control. Hypomagnesaemia is common problem in ruminants and may be one of the major health problems of cattle, sheep and goats in large scale production systems especially in temperate climates (Mayland, 1988; Stelletta et al., 2008).

Clinical hypomagnesaemia is also called hypomagnesemic tetany, grass tetany, spring tetany, or lactational tetany. Clinical signs usually occur when the animal is both hypomagnesaemic and hypocalcaemic (Robinson et al., 1989). In its subclinical form, however, hypomagnesaemia may be risk factor for other diseases. Around calving it may be the important risk factor for MF (Sansom et al., 1983; Schonewille et al., 2008).

Hypomagnesaemia may develop due to Mg deficiency in the diet or due to its low utilization in the gut. It may also occur because of a need for increased amounts of Mg during parturition and early lactation (Mayland, 1988). Clinical hypomagnesaemia in cows with plasma Mg concentrations bellow 0,4 mmol/L is manifested as grass tetany (Schonewille et al., 2008). Grass tetany has been investigated extensively, but its complex etiology is not well understood. Large number of factors influences the development of the disease (Robinson et al., 1989). It appears within 2-4 weeks after cattle or sheep have been turned out to rapidly growing pasture. Ewes with twins are more susceptible to grass tetany than are ewes with single lamb (Mayland, 1988). Specifically, older lactating animals consuming lush, intensively fertilized, cool-season grasses during the early spring are the most frequently affected (Robinson et al., 1989). Older animals also have reduced ability to mobilize body reserves of Mg. Lower average environmental temperatures (<14°C) facilitate the onset of the disorder (Mayland, 1988). However, periods of rapid plant growth during any season, resulting in forage of low Mg and high moisture, high nitrogen (N) and K, present dietary conditions that increase the potential for the development of tetany (Robinson et al., 1989). Grass tetany also may develop if animals graze forage that is high in digestible protein, but low in digestible energy (Mayland, 1988).

Grass tetany generally occurs when the dietary intake of total Mg is not particularly low, but factors which increase the animal's requirement for Mg or reduce the availability of dietary Mg are present (Mayland, 1988). The Mg concentrations in forage and subsequently in the blood of cattle are influenced strongly by high amounts of fertilizer K and, to some extent, fertilizer N. Absorption of Mg by plants is reduced by high levels of K in the soil (Mayland, 1988; Robinson et al., 1989). Mg absorption by ruminants is also reduced by a high intake of K (Schonewille et al., 2008), and by high content of Ca and P in the diet (Sansom et al., 1983). High concentrations of N in the forage may additionally decrease the availability of Mg (Mayland, 1988).

Absorption of Mg is dependent on the concentration of Mg in the rumen fluid and the functionality of the Mg active transport process across the rumen wall. The active transport mechanism is critical for the animal when dietary Mg concentration is less than 0,25%. Several factors, such as dietary K, can inhibit Mg absorption by this pathway. At high concentration of Mg in the rumen it will be absorbed by passive transport, and this mechanism is not affected by high K but only by the solubility of Mg in the rumen fluid. This mechanism requires dietary Mg to be not less than 0,35% (Goff, 2006b).

Mg deficiency results in reduced appetite which decreases total nutrient intake (Robinson et al., 1989), thus chronic hypomagnesaemia results in reduced feed intake and milk production. It may progress quickly into acute hypomagnesaemia which terminates in convulsions, coma and death. One-third of animals with clinical symptoms die (Mayland, 1988). Even mild form of hypomagnesaemia (Mg <0,85 mmol/L) significantly reduces the mobilization of Ca from the skeleton around calving (Goff, 2000; Sansom et al., 1983). That is why hypomagnesaemia may be a risk factor for parturient hypocalcaemia and milk fever. According to Sansom et al. (1983) possible reasons for inhibition of Ca mobilization are considered to be:

a) influence on PTH secretion; there are indications that serious Mg deficiency may impair or completely interrupt secretion of PTH,

b) interference with the action of PTH on target tissues; the integrity of interaction between PTH and its receptors is disrupted and activity of initiated enzymes is disabled,

c) interference with the metabolism of vitamin D; the first phase of vitamin D hydroxylation in the liver requires Mg^{2+} ions, while the second phase in the kidney takes place in the presence of PTH.

2.9. Udder edema

Udder edema occurs sporadically in cows and heifers near parturition, peaking in severity during the immediate prepartum period. Unless complicated, recovery of edema is spontaneous within a few days after calving. However, if edema is sufficiently severe it can interfere with suckling by the calf, milking, and may cause other complications including udder inflammation. Edema is caused by excessive accumulation of fluid underneath the skin and at least some mammary edema is associated with pregnancy and parturition, especially in primigravid heifers (Malven et al., 1983).

Investigations of the causes of udder edema and the development of strategies to reduce its prevalence have progressed in very limited degree during last several decades (Goff, 2006a). Some authors (Schmidt & Schultz, 1959) proposed that severity of udder edema is based on an inherent physiological phenomenon, because a cow may tends to have the same amount of edema each year regardless of the feeding. Feeding program had little if any effect on udder edema (Randalu et al., 1974). "Steaming up" method of feeding dry pregnant cows and high level of concentrates in their diet was initially blamed for high incidences of udder edema. However, investigation found that the amount of edema at calving was not related to grain feeding during the dry period or after calving (Schmidt & Schultz, 1959). Edema also is not correlated with the body condition of the cows (Schmidt & Schultz, 1959). However, addition of Na or K to the diet before calving can increase the incidence and severity of udder edema in dairy cows and first calf heifers (Nestor et al., 1988). In practical terms, high level of NaCl in the feed during the dry period may be one of the major factors in the development of udder edema around calving. Conflicting results from some earlier studies on effect of grain on udder edema could be explained with salt content of the diets. However, KCl as a replacement for NaCl results in edema of about the same severity (Randalu et al., 1974).

Cows with severe edema that requires veterinary treatment are more likely to have additional health problems. Authors found higher culling rates for cows with more severe edema. Thus, there would be a substantial economic benefit to minimizing edema in dairy herds (Dentine & McDaniel, 1984).

2.10. Retained placenta

Inadequate antioxidant status or "oxidative stress" of the cow contributes to a poorly functioning immune system and increases the risk of mastitis as well as retained placenta (retained fetal membranes, placental retention). Fetal membranes are retained if not expulsed longer than 12 hours after parturition. Se and vitamin E are important dietary antioxidants and their low levels in the diet are associated with a high incidence of mastitis and retained fetal membranes (Goff, 2006a). Subsequently, addition of vitamin E or Se may improve antioxidant status and decrease the incidence of those diseases. The similar effects can be obtained by addition of beta-carotene in the diet. Blood lymphocyte proliferation was higher in cows supplemented with beta-carotene, and phagocytic activity of blood neutrophils was enhanced as well as intracellular killing by blood neutrophils. Therefore, dietary beta-carotene can elevate blood beta-carotene and enhance peripartum host defense mechanisms by enhancing lymphocyte and phagocyte function (Michal et al., 1994).

Hypocalcaemia, among others, may be an important risk factor in the development of retained placenta. Muscle weakening or absence of uterine contractions in hypocalcemic animals does not contribute to the expulsion of fetal membranes (Goff & Horst, 1997c; Oetzel, 1988). In cows with hypocalcaemia placental retention is 3,2 to 4 times more frequent than in normocalcemic cows (Curtis et al., 1985). Hypocalcaemia also delay the physiological involution of uterus and increases the incidence of metritis (Beede, 1995). Moreover, hypocalcaemia is considered as one of the main causes of uterine prolapse, also due to loss of uterine muscles tone.

Cows with retained placenta were 3 times more likely to develop mastitis than animals without retained placenta (Overton & Waldron, 2004). It has been reported that placental retention, metritis and mastitis predisposes dairy cows to foot problems. Moreover, uncomplicated ketosis, retained placenta, metritis, and hypocalcaemia at parturition are also risk factors for left displaced abomasums (Shaver, 1997). Cows with retained placenta or metritis produce 0,3-2,3 kg/day less milk during subsequent lactation (Fourichon et al., 1999). In the case of retention alone, Guard (1996) stated that losses are on average 350 kg of milk annually.

2.11. Metritis

Metritis and endometritis are inflammatory uterine diseases. They frequently occur soon after calving and may severely compromise reproductive performances. Metritis and endometritis refer to the inflammation of the uterus and of its endometrial lining - both conditions are referred to subsequently as metritis (Urton et al., 2005). Younger cows were more likely to have dystocia or assisted deliveries, while older cows were most likely to have retained placenta and metritis (Lewis, 1997).

In some herds, up to 40% of the postpartum cows may be diagnosed with, and treated for uterine infections. However, the exact causes of uterine infections are unknown but are associated with several factors. Cows with dystocia, retained placenta, or stillbirths, and other metabolic disorders are more likely to develop metritis than healthy cows. Impaired immune functions before and after calving seem to predispose cows to severe uterine infections. It has been suggested that the function of neutrophils is impaired in cows that develop uterine infections. Thus, methods for regulating immune function in periparturient cows may have potential for preventing uterine infections. However, prevention of uterine infections is still difficult because the primary causes cannot be defined clearly (Lewis,

1997). Malnutrition influences the ability of the immune system to function, which affects the incidence of diseases such as mastitis and metritis (Goff, 2006a).

Cows suffering metritis exhibit reduced milk yield and reproductive performances (Urton et al., 2005). Few cows die from uterine infections, but affected animals are more likely to be culled for poor reproductive performances. The estimated cost to producers for each cow with metritis was 106 U.S. dollars (Lewis, 1997). Many of the financial costs of metritis are indirect, such as increased days open or predisposition to other diseases, and are thus difficult to measure. As mentioned previously, metritis is one of the diseases that predispose dairy cows to foot problems and left displaced abomasums (Shaver, 1997).

In contrast, the effects of metritis on milk production can be measured immediately, and losses during first four months after calving can be almost 270 kg. Also, besides reduced milk production in sick animals, some pharmaceuticals for treatment of uterine infections contaminate milk with residues, and the milk must be discarded. This is how uterine infections may have an indirect effect on milk production (Lewis, 1997).

3. USE OF BLOOD AND MILK ANALYSES TO EVALUATE NUTRITIONAL AND DISEASE STATUS

If recognition of subclinical diseases is difficult, the condition may be confirmed by analyzing blood, milk or sometimes urine, although some of them are still difficult to diagnose in practice, including rumen acidosis and fatty liver (Ingvartsen, 2006). Blood and milk analyses as tools to evaluate nutritional and disease status of individual animal or whole herd are aimed to help making decisions for improvement of nutritional strategies and production management. However, many problems are associated with interpretation of results of laboratory analyses, including Compton metabolic profile test, as well as other tools for assessment of metabolic status of dairy animals.

3.1. Milk analyses

Producers can use changes in milk production or in milk composition to monitor the health of their animals, but these tools are not always completely reliable because milk production corresponds poorly with mild or subclinical infectious disease. Nevertheless, measuring energy balance from changes in milk composition, most likely changes in milk fat and protein contents, could provide a cheap and reliable estimator of energy balance. A study clearly showed that there was a strong relationship between energy balance and milk composition under stable feeding conditions (Friggens et al., 2007). Milk composition varies with energy status and was proposed for measuring energy balance on-farm. Ratio milk fat/protein below 1,4 indicates the optimal or positive energy balance, and above it balance is negative (Pehrson, 1996; Zadoks, 2006). During peak lactation many cows with negative energy balance have this ratio in milk even above 2, 1 (Pehrson, 1996).

The biological basis for the relationship between fat/protein ratio and energy status of the animal can be found in two physiological features: a) mechanism that maintains milk energy output by increasing milk fat content when yield is compromised due to a deficit in energy supply, and b) decreased milk protein content under negative energy balance (Friggens et al., 2007). Negative energy balance also may be reflected in elevated ketone bodies which are excreted in milk and urine. Milk has about half the ketone level of blood, and is recommended to check milk ketone levels for detection of ketosis.

Milk fat depression is an example of disorder that can be diagnosed exclusively by milk analysis. Specific blood test for MFD diagnosis does not exists, but milk analyses are usually enough to evaluate risk for or presence of MFD in a herd. Every drop in milk fat percentages with no changes in the content of other milk constituents and milk yield can be considered as MFD. In serious cases, diet-induced MFD can result in reduction of milk fat yield of up to 50% or even more (Bauman & Griinari, 2001; Bauman et al., 2008) and represents one of the greatest risks for production economy.

3.2. Blood analyses

Analyses of blood constituents have long been a matter of concern as a possible tool for assessment of diet adequacy or metabolic and health status of animals. Compton metabolic profile test (CMPT) originally involved the analysis of a set of blood variables from three groups of seven cows, one near peak lactation, and others in midlactation and in late dry period (Kronfeld et al., 1982). The samples are collected at least three times yearly: summer, autumn and winter, or when nutritional imbalance is expected, using the same procedures and timings (Radostits et al., 2000). Means of variables are calculated for each lactational group in each herd, and this set of means constitutes the metabolic profile. Each group mean is compared to reference ranges determined from corresponding means plus or minus two standard deviations that are based on group-means for all herds (Kronfeld et al., 1982). The test was based on the concept that laboratory measurement of certain components of the blood will reflect the nutritional status of an animal with or without presence of clinical abnormalities. However, the results of research indicate that CMPT may be useful only as an aid in the diagnosis of the nutritional imbalances and production diseases. It must be carefully planed and is still expensive. Laboratories with automated analytical equipment should be available and this is often a major limiting factor (Radostits et al., 2000).

Compton metabolic profile test includes following analyses: blood glucose, packed cell volume, hemoglobin, blood urea nitrogen, serum inorganic phosphate, serum Ca, Mg, K and Na, total serum protein, albumin and globulins, serum Cu and iron (Fe), and plasma NEFA. Obtained results are interpreted with other relevant information taken on the day of sampling related to individual animals or to the herd: age, milk yield, days in milk, concentrates and forage intake, etc (Radostits et al., 2000).

Workers tested the CMPT as a guide to the nutritional status of dairy cows, and found that blood variables were not reliable predictors of energy and nutrients consumption relative to the requirements. Then they supposed that, for prediction of nutritional status, selection of blood variables should be probably different. Those workers have suggested, however, that protein intake may be reflected in blood concentrations of urea, albumin, and hemoglobin. Similarly, blood glucose concentration has attracted attention as an index of energy intake, although results from study to study have been inconsistent (Kronfeld et al., 1982).

Despite limitations and unconformity of metabolic profile tests, blood analyses still can help in diagnostics of some nutritional imbalances. Negative energy balance in postpartum cows can be detected by change in concentration of some blood metabolites (Rushen et al., 2008). As condition that affect immune functions and predispose animals to many metabolic disorders, it is important that negative energy balance is reflected in a higher levels of NEFA and beta-hydroxybutyric acid (BHBA) in the blood. To screen a herd for negative energy balance, Zadoks (2006) recommends having at least 12 animals tested. Testing of NEFA is done 2-14 days before calving and BHBA testing at 2-21 days after calving. If more than 10-15% of animals have NEFA levels above 0,40 mEq/L or BHBA levels above 1,36 mmol/L, the herd is considered to be suffering from excessive negative energy balance. Ketosis and fatty liver are diseases typically associated with negative energy balance for long periods of time. Increased ketone bodies and NEFA in the blood, and decreased blood glucose level are common findings in cows with ketosis and fatty liver. Values of glucose below 2,2 mmol/L are considered subnormal. However, fatty liver in cows currently can be diagnosed only by liver biopsy (Bobe et al., 2004).

The concentration of BHBA in the blood may be taken as a guide for correction of ration supplement for ewes during the final weeks of pregnancy to prevent pregnancy toxemia. It has been recommended to check 10% of the flock and feed is increased if blood concentration of BHBA exceeds 0,8 mmol/L (Morgante, 2004). Besides hypoglycemia and hyperketonaemia, plasma cortisol levels also may be elevated in sheep with pregnancy toxemia (Andrews et al., 1996). Early in the disease, both does and ewes will show a positive test for ketone bodies in the urine (Morgante, 2004).

There are no specific blood variables for detecting metabolic diseases like laminitis, displaced abomasums, udder edema, retained placenta and metritis. However, if these diseases are herd problems, blood metabolic profile should be checked for more than one variable, and carefully used as an aid for development of preventive strategies. When laminitis is a problem, a metabolic profile on both dry and lactating cows as well as springing heifers should include both red and white blood cells count, packed cell volume, Se, Zn, Cu, Fe, blood urea nitrogen, vitamins A and E, and beta-carotene. When retained placenta is a herd problem, blood profile should include serum minerals, Se, vitamin E, and beta-carotene. In individual cases, blood urea nitrogen and packed cell volume should be included. Metritis requires testing metabolic profile on dry and fresh cows which should include white blood cells count, Se, Zn, Cu, Fe, Mg, blood urea nitrogen, vitamins A and E, and beta-carotene.

Use of blood analyses to evaluate degree of hypocalcaemia is sufficient diagnostic test for this disorder. Concentration of Ca may be determined in whole blood, but more often determination is performed in plasma or serum. When there are no disturbances in acid-base status and protein metabolism, ionized and total Ca are strongly correlated and therefore total Ca concentrations in plasma may serve as an acceptable diagnostic value (Oetzel, 1988).

Hypocalcaemia means that the content of total Ca in the blood is below 2 mmol/L, or below 1 mmol/L of ionized Ca (Massey et al., 1993; Oetzel, 1996). Signs of paresis can occur at total Ca level of 1,9 mmol/L, but most of the cows remain on their feet down to 1,0-1,25 mmol/L. A reduction in the plasma Ca concentration at parturition is usually accompanied by hypophosphataemia and hypermagnesaemia (Phillippo et al., 1994). However, this is not always the case, especially if hypocalcaemia is not severe and is not clinically manifested (Crnkic et al., 2010; Joyce et al., 1997; Oetzel et al., 1988).

Blood analysis is also sufficient diagnostic test for hypomagnesaemia. Based on blood serum or plasma Mg levels, animals may be normally magnesemic (0,74-1,23 mmol/L), chronically hypomagnesemic (0,20-0,74 mmol/L) or acutely hypomagnesemic (<0,20 mmol/L) (Mayland, 1988). Sampling and analysis of the blood of several cows within 12 hours after calving is a good indicator of Mg status of the periparturient period. If serum Mg concentrations are less than 0,82 mmol/L in 90% of cows it suggests insufficient dietary Mg absorption (Goff, 2006a).

4. NUTRITIONAL STRATEGIES TO REDUCE METABOLIC DISEASES INCIDENCES

Although metabolic disorders are not easily categorized as to their cause, nutritional strategies have been developed to help prevent many of these disorders (Goff, 2006a). Scientific justification of those measures can be found in many research articles published in scientific journals during the last decades. However, problems in application of the strategies in the field still exist, as well as their limitations and possible negative consequences on milk production economy in certain circumstances. Strategies are based on major factors directly or indirectly increasing the risk of diseases such as overconditioning at calving, excessive mobilization of body fat, low nutrient intake, nutrient or diet specific factors and management and environmental stress (Ingvartsen, 2006). Strategies are expected to be able to reduce morbidity and at the same time improve reproduction and production.

To prevent metabolic disorders in the periparturient period nutritional strategies must start prior to calving. In a survey conducted by Curtis et al. (1985) consumption of nutrients before calving was directly related to the occurrence of metabolic disorders, and directly or indirectly to the occurrence of reproductive disorders after parturition. In most cases, these disorders occur as a complex and many of them are interrelated in their occurrence (Beede, 1995; Curtis et al., 1985). Consequently, strategies to reduce one disease can help preventing others. For example, strategies to reduce liver triglyceride accumulation at calving may decrease incidence of ketosis, etc.

Goff & Horst (1997c) recommended that three basic physiological functions must be maintained during the periparturient period if disease is to be avoided: adaptation of the rumen to lactation diets that are high in energy density, maintenance of a strong immune system, and maintenance of normocalcaemia. Whenever one or more of these functions are impaired, the incidence of both metabolic and infectious diseases is increased. However, measures to maintain these physiological functions must be conducted carefully avoiding other risk factors such as overconditionig and excessive mobilization of body fat around calving.

Metabolic disorders that may appear in the rest of lactation also require adequate prevention strategy to be controlled, including rumen acidosis, laminitis, hypomagnesaemia and milk fat depression.

4.1. Adaptation of the rumen

Fully adapting the rumen flora to a high starch diet that will be fed after parturition requires about 3 to 4 weeks, and full development of rumen papillae requires about 5 weeks of concentrate feeding (Goff & Horst, 1997c). It is, therefore, important to start increasing concentrates in the diet 3-4 weeks before calving and continue during the first 1-2 weeks after parturition to fully adapt rumen to lactation diet. If fresh cow is abruptly switched to a high starch lactation diet, the risk of developing rumen acidosis exists because the lactate production and accumulation. During the rest of lactation, the most reliable means of preventing rumen acidosis is to apply feeding methods that ensure a more even distribution of feed intake over the day. Feeding TMR only once or twice a day may result in that cows eat an excessive amount of feed during a short period of time, because cows are strongly attracted by the arrival of fresh food (Rushen et al., 2008). Buffering agents such as Na-bicarbonate or alkalinizing agents such as Mg-oxide are added to high concentrate ration to reduce the risk of acidosis (Goff, 2006a). Na-bicarbonate should be supplemented particularly to corn silage-based diets at the rate of approximately 0,8 to 1% of DM (Stone, 2004). Too rapid increase in concentrate allowance during early lactation may reduce roughage intake and increase the risk of not only rumen acidosis, but also displaced abomasums (Ingvartsen, 2006).

A highly significant relationship between forage neutral detergent fiber (NDF) content in the diet and ruminal pH has been found. NRC (2001) recommended 19% of forage NDF as absolute minimum when formulating rations in the field. A system of "physically effective" NDF (peNDF) relates the ability of a feedstuff to stimulate chewing relative to a hypothetical long grass hay containing 100% NDF. The peNDF of a feed is the product of its physically effective fiber (pef) and NDF content. The diet should contain about 22% peNDF to maintain ruminal pH of 6,0 (Stone, 2004).

Gradual adjustment of the rumen to a high concentrate diet can help avoiding serious drop in feed intake around calving and minimize body fat mobilization in early lactation. The plasma NEFA concentration is negatively correlated with DMI and depression in feed intake around the time of calving was largely responsible for fatty liver development (Goff, 2006a). Therefore, all nutritional measures that prevent drop in DMI before parturition may be useful in prevention of fatty liver. However, it has been assumed that increased energy and nutrient density of the diet may assure maintenance of the same intake of nutrients and energy despite lower DMI around calving, and decrease rate of lipid mobilization. Increase in nutrient density during the last 2–3 weeks prepartum by increasing concentrates in the ration has been referred to as "steaming-up" or "close-up" diet (Ingvartsen, 2006). Use these diets should not last for too long because of risk of overfeeding energy and development of obesity. NRC (2001) recommends increasing the energy content of the precalving diet from 5,2 MJ/kg of DM during the "far-off" dry period, to 6,8 MJ/kg of DM for the 3 weeks before calving. This strategy is thought to prepare the cow for the metabolic demands of early lactation and thereby minimize the need for body tissue mobilization (Urton et al., 2005).

As fatty liver and ketosis in cows often occur as a complex, all measures to reduce fatty liver incidence may decrease incidence of ketosis (Grummer, 1993). Moreover, positive effects on ketosis incidence and lipid transport also have been seen when niacin or rumen-protected choline is fed to dry and fresh cows (Goff, 2006a). The treatment of pregnancy toxemia in ewes is usually unsuccessful, therefore the prevention is of key importance to reduce occurrence of the disease. Grain is a high source of available energy. Feeding 0,5-1 kg of grain daily along with high quality hay during the last four to six weeks of pregnancy will help prevent pregnancy toxemia.

Manipulation of the nutritional program of dairy cows affects rumen health, which influences hoof health (Stone, 2004). It is very evident that feeding diets that cause drop in rumen pH will result in increase in laminitis cases. To minimize the drop in rumen pH it is necessary to limit amount of concentrate fed per meal to no more than 3,6 kg and provide fresh feed at the bunk throughout the day. Diet should contain at least 25% total NDF or 19% acid detergent fiber (ADF), with non-fiber carbohydrate (NFC) levels between 35 and 40% or non-structural carbohydrate (NSC) levels between 30 to 35%, and minimize abrupt changes of ration. Adding buffers to diets may help maintain claw integrity as buffers, such as Na-bicarbonate, minimize the drop in rumen pH. Supplements that contain combinations of complexed trace minerals Zn, Mn, Cu and Co positivelly influence claw health, and it has been advised to feed them when lameness is problem in a herd.

Increased energy content of the diet fed during the prepartum period was also associated with decreased incidence of displaced abomasum (Curtis et al., 1985). Risk of displaced abomasum for cows fed silage can be almost eliminated if every cow eats a kilogram of straw daily (Ingvartsen, 2006).

4.2. Maintenance of immunity

Maintaining of a strong immune system may help to reduce incidence of infectious diseases after parturition including those related to nutritional factors: retained placenta, metritis, mastitis, laminitis, etc. Inadequate nutrition may contribute depression of the immune system that occurs around calving time (Rushen et al., 2008) when stress of parturition and metabolic challenges experienced by the dairy cow at the onset of milk production impair immune cell function (Goff, 2006a).

Dietary factor that may influence immune functions include vitamin E and Se as the most important vitamin and mineral related to immunity. To a lesser extent vitamin A and beta-carotene, Cu and Zn may play a role (Zadoks, 2006). Metabolic conditions related to parturition also may exacerbate immunosuppression, including negative energy balance (Zadoks, 2006), hypocalcaemia (Ganjkhanlou et al., 2010), and hyperketonemia (Overton & Waldron, 2004). Therefore, strategies to maintain strong immune system must include feeding adequate or increased levels of all dietary components that influence immunity, as well as prevention of metabolic disorders that may exacerbate immunosuppression.

Most measures explained in the preceding section related to adaptation of the rumen to lactation diets may help maintain immune functions throughout reduction in negative energy balance and hyperketonemia. Measures to reduce hypocalcaemia around calving are presented in the following section. Feeding vitamin E and Se have long been associated with reducing incidences of retained placentas and uterine infections. Increased protein content of the diet fed during the prepartum period was also associated with decreased incidences of retained placenta (Curtis et al., 1985).

Although feeding is generally not related to the incidence and severity of udder edema, some dietary measures can help prevention of the problem. Restriction of the salt NaCl in the diet reduces the severity of udder edema (Randalu et al., 1974). Restriction or exclusion of other Na or K sources from the diet is also recommended until the levels of dietary Na and K are not in line with dietary recommendations. NRC (2001) recommends at least 0,1% Na in dietary DM for close-up dry cow diet and at least 1% K. As membrane permeability may be a primary cause of udder edema, vitamin E supplementation precalving may be a supportive preventive measure (Thomas et al., 1990). More recently, feeding anionic diets or diets with additional antioxidants have shown some promise in reducing udder edema (Goff, 2006a). Prevention of udder edema is important as condition has temporary effects of pain and stress for the cow, and may increase risk of udder inflammation and additional health problems.

4.3. Prevention of parturient hypocalcaemia

Concerning effects of hypocalcaemia on many physiological functions it is obvious that health, production and reproduction can be compromised in hypocalcemic cows even in the absence of clinical signs of paresis (Oetzel, 1996). Therefore, measures to prevent decline of blood Ca in cows after calving can improve milk production in herds that apparently have no problem with this disease. Prevention aims to increase the mobilization of Ca from the skeleton, or its absorption in the gastrointestinal tract, or both. The most important nutritional measures for prevention of hypocalcaemia include manipulation of minerals in the diet which is nutritionally balanced in accordance with requirements of the cows several weeks before calving. This method of prevention is based on limiting the consumption of Ca or increasing the content of anions in the diet.

4.3.1. Anionic diet

An anionic, or acidic, diet is one supplemented with anionic salts to provide more anions (Cl- and S^{2-}) relative to the cations (Na^+ and K^+). It has a negative dietary cation-anion difference (DCAD), calculated as mEq/kg of DM = (Na + K) – (Cl + S). It is well known that anionic diets help maintain blood Ca at parturition and prevent milk fever when fed to cows during the last several weeks of pregnancy (Block, 1984). Feeding an anionic diet before parturition has been advised if the incidence of milk fever in a herd exceeds 10%, and also when it is desired to improve the health status and production in herds in which MF is not a serious problem (Horst et al., 1994; Oetzel, 1993).

Diet with excess of anions, mainly Cl- and S^{2-}, relative to the cations Na^+ and K^+ acidifies the body and is therefore considered as acidogenic (Goff & Horst, 1997b). The two PTH-dependent functions, including bone resorption and production of 1,25-$(OH)_2$D, were enhanced after feeding acidogenic diet which contributes to lower incidence of hypocalcaemia after calving (Goff, 2000). However, the diets of dry cows in the field are always more or less alkalogenic and have a positive DCAD. When feeding diets with positive DCAD the acid-base balance tends to metabolic alkalosis, and vice versa. Formulation of acidogenic diet is only possible by supplementation with chloride and sulfate in a quantity that provides a relative excess of Cl- and S^{2-} in the diet. Manipulation of dietary cations and anions, however, is limited by metabolic requirements and by tolerance levels of minerals in the diet.

The literature offered a dozen different formulas to calculate DCAD, which include some or all dietary minerals with or without the use of coefficients related to their utilization from the diet or the degree of influence on acid-base status:

A. (Na+K) – (Cl+S) Ender et al. (1971)

B. (Na+K) – (Cl+S+P) Lomba et al. (1978)

C. (Na+K) – Cl Gaynor et al. (1989)

D. (Na+K+Ca+0,3P) – (Cl+S) Goff et al. (1991)

E. (Na+K+0,38Ca+0,3Mg) – (Cl+0,6S+0,5P) Goff (1992)

F. (Na+K+0,38Ca+0,25P+0,3Mg) – (Cl-0,6S) Goff (1992)

G. (Na+K+0,38Ca+0,3Mg) – (Cl-0,6S) Horst et al. (1997)

H. (Na+K+0,15Ca+0,15Mg) – (Cl+0,2S+0,3P) Oetzel (2000)

I. (Na+K+0,15Ca+0,15Mg) – (Cl+0,25S+0,5P) Goff (2000)

J. (Na+K) – (Cl+0,6S) Goff et al. (2004)

Equations A, B, and C imply the complete absorption of each dietary element that can be considered accurate only in the case of equation C. Equations E and G take into account the coefficients of utilization of those elements that are not completely utilized. Equations H and I could be considered biologically and functionally most accurate because they include the degree of influence of individual ions on acid-base status, and are physiologically the most relevant (Horst & Goff, 1997; Goff, 2000).

Equation F can be used to assess the risk of developing MF. All the elements that increase the risk of MF are on the left side, while those who decrease the risk are on the right side of equation. Equation D can serve the same purpose (Goff et al., 1991). The last equation J, (Na + K) – (Cl + 0,6S), was the most recently proposed by Goff et al. (2004).

Oetzel (1991) collected data from 75 published experiments and determined the risk factors of diet in the development of MF using meta-analysis technique. Comparing the three different equations to calculate DCAD he found that the equation (Na + K) - (Cl + S) was strongly correlated with the appearance of MF. The author in the same study well justified the inclusion of S in the calculation of DCAD. Charbonneau et al. (2006) used the same technique but found that the equation (Na + K) – (Cl + 0,6S) was the most highly associated with clinical milk fever (R^2 = 0,44) and urinary pH (R^2 = 0,85).

Mineral content in feed is usually expressed as a percentage so it is necessary to convert it in mEq/kg for calculation of DCAD. Equivalent weight (g/Eq) of each element is calculated by dividing the relative atomic weight (Ar) with its valence. For example, Ar of S is 32,06. Dividing 32,06 by 2 we get the equivalent weight of S which is 16,03 g/Eq. Factor for conversion of percentage in mEq/kg is calculated by dividing the 10.000 with equivalent weight of the element. For S it is 10.000/16,03 = 623,83 (Table 1). Multiplying this factor with percentage of the element in dietary DM its content is expressed in mEq/kg DM. When this is calculated for each element in the equation, then the sum of cations is subtracted from the sum of anions and DCAD value is obtained. The example of calculating DCAD of alfalfa hay and rapeseed meal using equation (Na + K) - (Cl + S) is given in Table 2.

Table 1. Mineral elements in the diet used for calculation of DCAD

Element	A_r	Valence	Equivalent weight (g/Eq)	Conversion factor (from % in mEq/kg)
Na	22,99	1	22,99	434,98
K	39,09	1	39,09	255,77
Ca	40,08	2	20,04	499,00
Mg	24,31	2	12,16	822,88
Cl	35,45	1	35,45	282,06
S	32,06	2	16,03	623,83
P	30,97	3	10,32	968,56

Table 2. Example of calculating DCAD in alfalfa hay and rapeseed meal

Alfalfa hay	K	Na	Cl	S
Content in DM, %	1,71	0,12	0,38	0,28
Factor for multiplication (F)	255,77	434,98	282,06	623,83
Content in mEq/kg DM (% × F)	437,37	52,20	107,18	174,67
DCAD mEq/kg DM [(Na+K) – (Cl+S)]	489,57 – 281,85 = + 207,72			
Rapeseed meal				
Content in DM, %	1,36	0,10	0,11	1,25
Factor for multiplication (F)	255,77	434,98	282,06	623,83
Content in mEq/kg DM (% × F)	347,85	43,50	31,03	779,79
DCAD mEq/kg DM [(Na+K) – (Cl+S)]	391,35 – 810,82 = - 419,47			

The example in Table 2 clearly shows the difference in the alkali-acid production potential of the two feeds depending on their mineral content. A positive DCAD of alfalfa hay is result of the relatively high K content. Extremely negative DCAD of rapeseed meal is result of high S content.

Most diets commonly used for dry cows have DCAD value of +100 to +250 mEq/kg DM (Oetzel, 2000) and even up to +500 mEq/kg DM (Pehrson et al., 1999). By choosing appropriate feeds DCAD can be reduced, but not sufficiently to influence Ca metabolism. It is therefore necessary to enrich the diet with Cl and S by adding anionic salts in a quantity that provides a negative DCAD. That is, indeed, a standard diet that contains all essential nutrients required by dry cow and was further enriched with sources of Cl and S in amount that causes a mild metabolic acidosis.

Acidogenic, or anionic mineral salts are chlorides and sulfates with a high content of Cl and S, not containing Na and K. Alkalogenic salts are, on the other hand, Na and K salts that contain organic anion which is metabolized and are also called cationic salts (Na_2CO_3, N_aHCO_3, K_2CO_3, $KHCO_3$, etc). They have the effect opposite to anionic salts and are undesirable in the diet of dry cows.

To formulate an anionic diet three chlorides and three sulfates are commonly used (Table 3). Two or more of them are usually added depending on mineral content of the diet. Ca and Mg salts also serve to meet requirements in these two minerals whose content in anionic diet should be higher than in standard diet (NRC, 2001). Ammonium salts, besides Cl and S, are also a source of non-protein nitrogen.

Table 3. Characteristics and chemical composition of anionic salts

Salt	M_r	Val	g/Eq	DM, %	% of DM					DCAD mEq/kg DM
					Ca	Mg	N	Cl	S	
$MgSO_4x7H_2O$ [1]	246,51	2	123,26	48,83		20,20			26,63	-16.613
$MgCl_2x6H_2O$ [2]	203,33	2	101,67	46,83		25,53		74,46		-21.002
$CaSO_4x2H_2O$ [3]	172,18	2	86,09	79,07	29,44				23,55	-14.691
$CaCl_2x2H_2O$ [4]	147,02	2	73,51	75,49	36,11			63,88		-18.018
$(NH_4)_2SO_4$ [5]	132,14	2	66,07	100,0			21,20		24,26	-15.134
NH_4Cl [6]	53,49	1	53,49	100,0			26,19	66,27		-18.692

[1]magnesium sulfate heptahydrate (Epsom salt), [2]magnesium chloride hexahydrate, [3]calcium sulfate dihydrate (gypsum), [4]calcium chloride dihydrate, [5]ammonium sulfate, [6]amonium chloride, Mr – relative molecular weight of a substance (sum of atomic weights of all atoms in the molecule), Val – valence, g/Eq – equivalent weight (M_r/valence of cation), DCAD = (Na+K) – (Cl+S).

Acid production potential or acidifying activity of certain salt depends on preferential absorption of anions in relation to cations which the salt is consisted of (Goff et al., 1991). Phosphates, for example, have a weak acidifying activity as they are absorbed only slightly better than the corresponding cations (Horst et al., 1997). NaCl is neutral salt as both elements are absorbed completely and none is metabolized, so it does not affect neither DCAD nor acid-base status.

Anionic salts are not harmless substances and require caution for the use and handling them. The main limiting factors are bitter and salty taste and potential toxicity of higher doses. NH4Cl is considered the most toxic, followed by CaCl2. The combination of salts is the best solution because it reduces the possibility of harmful effects (Oetzel, 2000). The most acceptable method of administration is mixing salts in the feed.

An alternative to the anionic salts is adding mineral acids directly into a diet. Very potent in causing systemic acidosis is hydrochloric acid (HCl), while sulfuric acid (H2SO4) is less efficient (Goff & Horst, 1997a). An additional advantage of mineral acids is that cows prefer diets with acidic taste rather than bitter-salty taste of anionic salts. Goff & Horst (1998) found that efficient prevention of MF can be achieved with only 1,5 Eq of HCl in the diet, and that with no adverse consequences up to 2,5 Eq of the acid can be fed daily. Moreover, the addition of HCl in the diet had, unlike anionic salts, a beneficial effect on the consumption of feed. For safety, it is not advisable to use pure acid or keep it on farm, but rather as commercial products in the form of acidified feeds where the acid is mixed with some feed as a carrier. Byproducts of fermentation are usually used for this purpose, and also soybean meal or sugar beet pulp (Goff & Horst, 1997a, 1998).

Feeding anionic diet few weeks before calving significantly reduces the incidences of MF and subclinical hypocalcaemia in the herd, and improves milk production and reproductive performances of cows in the subsequent lactation (Beede et al., 1991; Block, 1984; DeGroot et al., 2010). Metabolic acidosis caused by anionic diet is subclinical, mild and compensated, and with no significant impact on animal health, but may influence some physiological functions if lasts long enough. Increased sensitivity of PTH receptors in renal tissue contributes to increased production of 1,25-(OH)2D before parturition (Gaynor et al., 1989; Goff et al., 1991). Cows fed anionic diet have a higher content of 1,25-$(OH)_2D$ in plasma, although changes in secretion of PTH does not occur (Phillippo et al., 1994). The active form of vitamin D takes part in osteolysis with PTH (Horst et al., 1994; Horst, 1986) and stimulates Ca reabsorption in renal tubules (Goff, 1992). Since 1,25-$(OH)_2D$ is important for the functioning of both Ca compensation mechanisms, change in its secretion is considered as an important effect of anionic diet.

Oetzel (1993) recommended the application of anionic salts in the herds with high incidence of MF (>10%) if not possible to formulate low Ca diet, or to improve production and reproduction in the herd which apparently does not have high incidence of MF. Adding anions in the diet is, therefore, to be considered not only for preventing MF, but also for prevention of subclinical hypocalcaemia which is responsible for the frequent occurrence of other metabolic and reproductive disorders in puerperium (Horst et al., 1994). Beede (1992) states that even in well-kept herds, where cows are in proper condition with no major problems with metabolic disorders, an additional 250-500 kg of milk can be achieved by using anionic salts in diets before parturition probably because lower incidence of subclinical hypocalcaemia.

Based on results of earlier studies, Horst et al. (1997) found that the most effective diets for prevention of MF are those with DCAD from -50 to -100 mEq/kg DM, while others state that it is between -50 and -150 (Beede, 1995), or -200 (Horst et al., 1994). However, Beede et al. (1991) achieved a good preventive effect with DCAD -250 mEq/kg, and Goff et al. (1991) with -230 mEq/kg. Too low DCAD, irrespective of the amount of added salts, can reduce feed consumption because too strong metabolic acidosis which depresses the appetite in animals (Goff & Horst, 1997b). Preventive effect of anionic diet on MF incidence occurs only if DCAD is below -40 mEq/kg DM in most cases, regardless of the type and quantity of added salts.

A risk for many health and reproductive disorders in puerperium is increased in cows with hypocalcaemia (Curtis et al., 1985; Massey et al., 1993). The prevention of hypocalcaemia can, in addition to better health status of cows in puerperium, improve milk production and increase reproduction performances in the herd (Beede et al., 1991; Block, 1984; DeGroot et al., 2010). Comparing the milk production in healthy cows with those who suffered MF, Block (1984) found greater milk production by about 14% in healthy animals. The same author found that cows fed anionic diet in the dry period produced on average 486 kg (7,3%) more milk than those fed cationic diet. Beede et al. (1991) conducted a field experiment with 510 cows and showed three positive effects of anionic diet: enhanced Ca metabolism in periparturient period, increased milk production in subsequent lactation, and better reproductive performances. Cows fed anionic diet before parturition produced 327 kg more milk (about 3,6%) than control cows.

Because of unpleasant taste, anionic salts can have depressing effect on feed intake (Joyce et al., 1997; Vagnoni & Oetzel, 1998), but this is minimal if salts are properly used (Block, 1984; Moore et al., 2000; Oetzel et al., 1988). Up to 300 mEq of anions/kg DM (3-3,5 Eq of salts per day per cow) can be added to the diet without depressing effect on feed intake (Horst et al.,

1994; Oetzel, 2000). Intake also could be influenced by the type of salt because more acidic salts have less favorable taste. Feed intake depression was at least expressed with $MgSO_4$, and much more with other salts, mainly $CaCl_2$ and NH_4Cl (Oetzel & Barmore, 1993). Last two also have much higher acidifying activity in relation to MgSO4.

The best and easiest method of application of anionic salts is mixing in TMR which effectively masks their taste. In the conventional feeding system application of anionic salts is difficult, but still possible. The salts should be manually mixed in silage or haylage, or in concentrates (Oetzel & Barmore, 1993). In the latter case, salts should be mixed in sufficient amount of concentrates containing palatable feeds, at least 2,3-2,6 kg (Oetzel & Barmore, 1993; Pehrson et al., 1999).

Oetzel (2000) finds that preventive effect of anionic salts can be achieved if cows consume anionic diet at least 5 days before calving, while others state that at least 10 days is necessary. Since the expected and actual date of calving are matter of discrepancies in practice, it is advisable to start feeding salts 3-4 weeks before expected parturition, at least 2-3 weeks, to ensure that most of the cows consume the diet at least 10 days (Beede 1992, 1995).

NRC (2001) gives fairly broad recommendations for the content of Ca in anionic diet for dry cows, from 0,6 to 1,5% DM, which is not difficult to formulate when anionic Ca salts are used. Anionic salts should not be fed if Ca intake is below 50 g/day (Oetzel, 1993). As hypomagnesaemia negatively affects Ca metabolism at the time of calving, many authors emphasize the importance of adequate content of Mg in dry cow diet (Sansom et al., 1983; Thilsing-Hansen et al., 2002). Recommendation has increased from earlier 0,20-0,25% to the current 0,35-0,40% DM either in standard or in anionic diet (NRC, 2001). Increasing dietary Mg up to 0,40% DM has no negative consequences so, even if not prove useful, poses no risk or any practical problem in the formulation of a diet when anionic Mg salts are used (Wang & Beede, 1992). On the other hand, the Mg level of 0,4% DM has been also set as the maximum tolerable (NRC, 2001).

If we assume that DCAD in efficient anionic diet should be around -100 mEq/kg DM, it means that the initial DCAD should not be greater than +200 to +250 (Horst et al., 1997). Therefore, the diet must be formulated using feeds with lower content of K, such as corn silage, grass hay, etc. If necessary, part of the forage with high content of K can be replaced with concentrates rich in fiber, such as brewers' grains, sugar beet pulp, malt sprouts, etc. The most common problem is grass silage and alfalfa silage because they can contain more than 3% K. Most cereal grains have almost neutral DCAD, while it is more negative only in rapeseed meal and brewers grains primarily due to high content of S (NRC, 2001).

In field conditions, a marked decrease in feed intake may occur when anionic salts are included in the diet for the first time, therefore, adjusting cows should last at least three days (Oetzel, 2000)

The most accurate biological indicator of the degree of acidification of the body is pH of urine (Vagnoni & Oetzel, 1998). Urinary pH in cows was decreased linearly with decreasing DCAD (Charbonneau et al., 2006; Ganjkhanlou et al., 2010). The optimum pH of urine to prevent puerperal hypocalcaemia in Holsteins fed anionic diet is 6 - 6,5 (Horst & Goff, 1997) or wider, 6 - 7 (Moore et al., 2000). In Jersey cows, however, the pH of urine is necessary to reduce to 5,8 - 6,2. If the pH is lower than the specified, amount of anionic salts should be reduced and vice versa.

4.3.2. Low calcium diet

Maintenance requirements of the cow and requirements of fetus during high pregnancy are usually satisfied with 35-45 g Ca/day (NRC, 2001). The content of Ca in the diet is often beyond that, and the requirements in this case are met mainly by passive absorption, while mechanisms of mobilization of body reserves and active absorption of Ca in the intestine are suppressed. Inactivity of these mechanisms in the dry period makes cows difficult to adapt to the sudden loss of Ca in colostrum at the moment of mammary gland activation. A few days required for starting these mechanisms causes temporary crisis in Ca homeostasis that results in decline of Ca in the blood (Goff, 1992).

When Ca content in the diet is below the minimum requirement, the animal is dependent on Ca mobilization from bone with simultaneous increase of active Ca absorption in the gut (Horst, 1986). These mechanisms can be activated before parturition and maintained active by constant stimulation up to the critical moment of parturition. Restriction of Ca intake to below 20 g/day caused negative balance of Ca and stimulates the secretion of PTH which increases tubular Ca reabsorption, bone resorption and production of 1,25-(OH)2D (Goff, 1992; Horst et al., 1997). This allows the cow to use Ca more efficiently from the diet and from body reserves immediately after calving (Goff, 1992). Introduction of low Ca diet leads to a slight decrease of Ca and P concentrations in plasma which returns to the base level in 3-4 days due to simultaneous increase in secretion of PTH that remained increased up to calving time (Goings et al., 1974). Prolonged exposure of the tissues to elevated PTH levels after feeding low Ca diet can overcome tissue resistance to PTH that might be induced by high dietary K (Goff, 2006a). This is the way to reduce or avoid period of adaptation to sudden losses of Ca which usually takes several days (Goff, 1992). Rations formulated to contain less than 20 grams of Ca proved to be very effective in preventing MF and significantly reduce the occurrence of hypocalcaemia too (Goings et al., 1971, 1974; Wiggers et al., 1975). For the best effect, restrictions of Ca consumption in the dry period should last at last 7-10 days (Goings et al., 1974; Wiggers et al., 1975) and be folowed by a high Ca diet immediately after calving (Horst et al., 1994; Oetzel, 2000). Cows fed in this way at the moment of calving mobilize about 10 g of Ca from the skeleton daily that is enough to prevent the occurrence of MF (Wiggers et al., 1975). Increasing Ca in the diet above daily needs and manipulation of its content within these limits does not affect the blood Ca status if other dietary minerals are maintained unchanged (Goff & Horst, 1997b).

Feeding low Ca diet to dry cows is considered the traditional way of preventing MF. For this reason, the use of feeds rich in Ca is usually avoided and the diet is formulated with corn silage and grass hay. Consumption of Ca can be reduced to 50-60 g/day in this way what often gives good results in the field (Goff, 1992; Horst et al., 1997). However, significant stimulation of parathyroid gland and complete preventive effect can be achieved only if Ca consumption is restricted to below 20 g/day. The formulation of such diet is, unfortunately, unrealistic in practice and that is the reason for its limited application (Goff, 1992). This type of diet has no adverse effects on production in the subsequent lactation (Goings et al., 1974), but the problem is, in addition to impracticality, that it may not last longer than 2-3 weeks because negative balance of Ca can be too exhausting for body reserves (Van Saun & Sniffen, 1996).

In the strategy for prevention of MF, Horst et al. (1997) recommended measures depending on the baseline cation-anion difference of the diet (Fig. 1). If it is less than 250 mEq/kg DM, the use of chloride and sulfate is justified without danger of low feed intake. If this value is above 250 mEq/kg DM it is necessary to consider other preventive measures, such as low Ca diet, short-term administration of oral Ca salts or some pharmacologic therapy (vitamin D analogs and active metabolites, PTH injections, etc.).

Thilsing-Hansen et al. (2002) rewieved research conducted over the past 50 years and concluded that any measures for preventing MF, even if used under ideal conditions, rarely reach preventive effect of 100%. The same authors calculated that efficiency of low Ca diets is best, reaching 80-100%, and anionic diets efficiency was 65-80%.

4.4. Prevention of hypomagnesaemia

Commonly used parameter to characterize the grass tetany potential of forage is the ratio [K/(Ca+Mg)] (Mayland, 1988). Forages containing less than 0,2% Mg and a "tetany ratio" [K/(Ca+Mg)] greater than 2,2 have increased risk of inducing grass tetany (Crawford et al., 1998). The fertilizers containing N and K are the most important factors increasing

[K/(Ca+Mg)] ratio in forages. Key step in nutritional prevention of hypomagnesaemia is finding a forage with lower ratio [K/(Ca+Mg)], or modify this ratio in the diet by adding more Ca and Mg supplements. Commercial grade MgO, $MgCl_2$, $MgCO_3$ and $MgSO_4$ are good sources. As hypomagnesaemia may induce hypocalcaemia, including Ca in supplements to prevent Mg tetany may have a beneficial effect (Robinson et al., 1989).

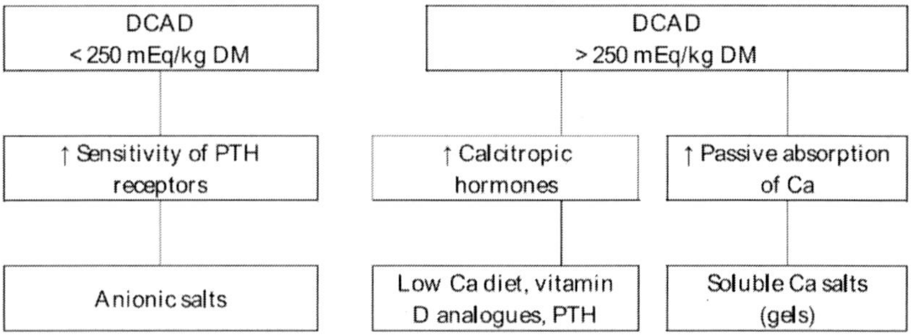

Figure 1. Strategy for preventing milk fever (adapted from Horst et al., 1997)

NRC (2001) recommends Mg level of 0,35-0,40% DM in the diet for dry cows, although studies showed no benefit from increasing dietary Mg above 0.2% DM when diets contain less than 3% K (Wang & Beede, 1992). Van Saun & Sniffen (1996) recommended increasing dietary Mg above 0,20% when the content of K exceeds 1,2 to 1,5% DM. However, Goff (2006a) recommends that Mg content of the close-up dry cow ration and the early lactation ration should be between 0,35 and 0,40% as insurance against the possibility that the active transport processes for Mg absorption are impaired. In pregnant cows high Ca intake decreased utilization of dietary Mg and increases its excretion in urine (Sansom et al., 1983) that can be even worsened in metabolic acidosis (Wang & Beede, 1992). So it seems logical to increase Mg content in anionic close-up diet because it usually contains more Ca.

4.5. Prevention of milk fat depression

As milk fat depression has been observed over a wide range of feeding situations, this problem on dairy farms remains one of the more challenging tasks within overall nutritional management of dairy cows. Diets high in concentrates and low in fiber, and diets supplemented with plant or fish oil are the most often associated with MFD, but many other dietary factors also can affect milk fat synthesis including those which are able to alter rumen environment and those related to supply of polyunsaturated fatty acids (PUFA). Factors that can alter rumen environment include low level of physically effective NDF, feed particle size, total fiber in the diet, starch and non-structural carbohydrates, feeding pattern etc. On the other side, factors related to supply of PUFA are variation in fat content and fatty acid composition of feed ingredients, amount and availability of PUFA, and also feeding pattern.

Factor that alter rumen environment are first to consider in nutritional strategy to prevent or solve the problem of MFD, and those who affect rumen pH are the most important. Lower then normal ruminal pH, even without signs of acidosis, causes the change in bacterial population favoring those that have alternative pathways of biohydrogenation of dietary fatty acids. Inclusion of adequate level of so-called "effective fiber" and appropriate buffers in lactation diets can prevent drop in ruminal pH and markedly decrease MFD (Bergen, 2009). Forages, including long stemmed hay, are the main sources of effective fiber, thus the best method to maintain an adequate fat percentage in milk is to feed a balanced ration with adequate forage. The low levels of effective fiber may result from overfeeding of concentrates or the lack of forage, from consumption of large amounts of lush pasture and from silage or haylage that is too finely chopped (Perfield & Bauman, 2005).

Additional management practices to maintain a stable milk fat percentage in dairy herds include regular feeding of diet without abrupt changes, as well as feeding buffers such as Na-bicarbonate and/or Mg-oxide. Buffers are particularly useful when more than 5,5 kg of concentrate is fed per feeding and when frequent changes in diet are made.

Adding different sources of fat in dairy cow rations is a practice which has been favored when it is necessary to increase the energy consumption. However, higher amounts of fat can inhibit the activity of rumen bacteria and reduce the efficiency of fiber digestion, thus leading to a reduction in milk fat content (Schroeder et al., 2004). Although unsaturated vegetable oils can have a number of positive effects on fatty acid composition of milk fat, a negative effect on fat content must not be neglected due to the fact that the fat in milk is one of the main determinants of the price of milk and profits that farmers make. High levels of unsaturated oil in the diet can reduce fat content in milk along with possibly other adverse effects on production performances, such as drop in milk yield, decreased protein content in milk, low feed intake, etc. (Shingfield et al., 2006). Loor et al. (2002) found that the upper limit for dietary supplementation of unsaturated vegetable oils is 3,5% DM without serious nefative consequences on production. If fish oil is the supplement, it is at most 1% DM (Donovan et al., 2000).

REFERENCE

1. Andrews, A.H., Holland-Howes, V.E. & Wilkinson, J.I.D. (1996). Naturally Occurring Pregnancy Toxaemia in the Ewe and Treatment with Recombinant Bovine Somatotropin. *Small Ruminant Research*, Vol.23, No.2-3, (January 1997), pp. 191-197, ISSN 0921-4488
2. Baird, L.G. & Muelling, C.K.W. (2009). Risk Factors, Pathogenesis and Prevention of Subclinical Laminitis in Dairy Cows. *Proceedings of the CanWest Veterinary Conference*, pp. 1-10, Alberta, Canada, October 17-20, 2009
3. Bauman, D.E. & Griinari, J.M. (2001). Regulation and Nutritional Manipulation of Milk Fat: Low-Fat Milk Syndrome. *Livestock Production Science*, Vol.70, No.1-2, (July 2001), pp. 15-29, ISSN 1871-1413
4. Bauman, D.E., Perfield, II J.W., Harvatine, K.J. & Baumgard, L.H. (2008). Regulation of Fat Synthesis by Conjugated Linoleic Acid: Lactation and the Ruminant Model. *Journal of Nutrition*, Vol.138, No.2, (February 2008), pp. 403-409, ISSN 0022-3166
5. Baumgard, L.H., Sangster, J.K. & Bauman, D.E. (2001). Milk Fat Synthesis in Dairy Cows is Progressively Reduced by Increasing Supplemental Amounts of trans-10,cis-12 Conjugated Linoleic Acid (CLA). *Journal of Nutrition*, Vol.131, No.6, (June 2001), pp. 1764-1769, ISSN 0022-3166
6. Beede, D.K., Wang, C., Donovan, G.A., Archibald, L.F. & Sanchez, W.K. (1991). Dietary Cation-Anion Difference (Electrolite Balance) in Late Pregnancy. *Proceedings of the 28th Florida Dairy Production Conference*, pp. 98-104, Gainesville, FL, April 9-10, 1991
7. Beede, D.K. (1995). Macromineral Element Nutrition for the Transition Cow: Practical Implications and Strategies. *Proceedings of The Tri-State Dairy Nutrition Conference*, pp. 175-188, Fort Wayne, IN, USA, May 11-13, 1995
8. Beede, D.K. (1992). The DCAD Concept: Transition Rations for Dry Pregnant Cows. *Feedstuffs*, Vol.64, No.53, (December 1992), pp. 12-19, ISSN 0014-9624 Bergen, W.G. (2009). Milk-Fat Depression and Lipid Repartitioning in Lactating Dairy Cows. *Journal of Nutrition*, Vol.139, No.5, (May 2009), pp. 826-827, ISSN 0022-3166
9. Bertics, S.J., Grummer, R.R., Cadorniga-Valino, C. & Stoddard, E.E. (1992). Effect of Prepartum Dry Matter Intake on Liver Triglyceride Concentration and Early Lactation. *Journal of Dairy Science*, Vol.75, No.7, (July 1992), pp. 1914-1922, ISSN 0022-0302
10. Block, E. (1984). Manipulation Dietary Anions and Cations for Prepartum Dairy Cows to Reduce Incidence of Milk Fever. *Journal of Dairy Science*, Vol.67, No.12, (December 1984), pp. 2939-2948, ISSN 0022-0302
11. Bobe, G., Young, J.W. & Beitz, D.C. (2004). Invited Review: Pathology, Etiology, Prevention, and Treatment of Fatty Liver in Dairy Cows. *Journal of Dairy Science*, Vol.87, No.10, (October 2004), pp. 3105-3124, ISSN 0022-0302
12. Bramley, E., Lean, I.J., Fulkerson, W.J., Stevenson, M.A., Rabie, A.R. & Costa, N.D. (2008). The Definition of Acidosis in Dairy Herds Predominantly Fed on Pasture and Concentrates. *Journal of Dairy Science*, Vol.91, No.1, (January 2008), pp. 308-321, ISSN 0022-0302
13. Charbonneau, E., Pellerin, D. & Oetzel, G.R. (2006). Impact of Lowering Dietary Cation-
14. Anion Difference in Nonlactating Dairy Cows: A Meta-Analysis. *Journal of Dairy Science*, Vol.89, No.2, (February 2006), pp. 537-548, ISSN 0022-0302
15. Crawford, Jr.R.J., Massie, M.D., Sleper, D.A. & Mayland, H.E. (1998). Use of an ExperimentalHigh-Magnesium Tall Fescue to Reduce Grass Tetany in Cattle. *Journal of Production Agriculture*, Vol.11, No.4, (October-December 1998), pp. 491-496, ISSN 0890-8524
16. Crnkic, C., Muratovic, S., Piplica, S., Kavazovic, A. & Kutlaca, S. (2010). Blood Plasma Mineral Profile and Health Status in Postpartum Cows Fed an Anionic Diet Before Parturition. *Turkish Journal of Veterinary and Animal Science*, Vol.34, No.3, (June 2010), pp. 255-260, ISSN 1300-0128
17. Curtis, C.R., Erb, H.N., Sniffen, C.J., Smith, R.D. & Kronfeld, D.S. (1985). Path Analysis of Dry Period Nutrition, Postpartum Metabolic and Reproductive Disorders, and Mastitis in

Holstein Cows. *Journal of Dairy Science*, Vol.68, No.9, (September 1985), pp. 2347-2360, ISSN 0022-0302
18. Curtis, C.R., Erb, H.N., Sniffen, C.J. & Smith, R.D. (1984). Epidemiology of Parturient Paresis: Predisposing Factors with Emphasis on Dry Cow Feeding and Management. *Journal of Dairy Science*, Vol.67, No.4, (April 1984), pp. 817-825, ISSN 0022-0302
19. DeGroot, M.A., Block, E. & French, P.D. (2010). Effect of Prepartum Anionic Supplementation on Periparturient Feed Intake, Health, and Milk Production. *Journal of Dairy Science*, Vol.93, No.11, (November 2010), pp. 5269-5279, ISSN 0022-0302 (abstract)
20. Dentine, M. R. & McDaniel, B. T. (1984). Associations of Subjective Udder Edema Scores and Descriptive Trait Codes for Udder Types. *Journal of Dairy Science*, Vol.67, No.1, (January 1984), pp. 208-215, ISSN 0022-0302
21. Dishington, I.W. (1974). The Role of Age on the Induction of Hypocalcaemic Paresis Puerperalis in Dairy Cows. *Nordisk Veterinaermedicin*, Vol.26, No.3, (March 1974), pp. 205-210, ISSN 0029-1579
22. Donovan, D.C., Schingoethe, D.J., Baer, R.J., Ryali, J., Hippen, A.R. & Franklin, S.T. (2000). Influence of Dietary Fish Oil on Conjugated Linoleic Acid and Other Fatty Acids in Milk Fat from Lactating Dairy Cows. *Journal of Dairy Science*, Vol.83, No.11, (November 2000), pp. 2620-2628, ISSN 0022-0302
23. Ender, F., Dishington, I.W. & Helgebostad, A. (1971). Calcium Balance Studies in Dairy Cows Under Experimental Induction and Prevention of Hypocalcaemic Paresis Puerperalis. *Zeitschrift fur Tierphysiologie, Tierernahrung und Futtermittelkunde*, Vol.28, No.1-5, (January-December 1971), pp. 233-256, ISSN 0044-3565
24. Fourichon, C., Seegers, H., Bareille, N. & Beaudeau, F. (1999). Effects of Disease on Milk Production in the Dairy Cow: A Review. *Preventive Veterinary Medicine*, Vol.41, No.1, (June 1999), pp. 1-35, ISSN 0167-5877
25. Friggens, N.C., Ridder, C. & Løvendahl, P. (2007). On the Use of Milk Composition Measures to Predict the Energy Balance of Dairy Cows. *Journal of Dairy Science*, Vol.90, No.12, (December 2007), pp. 5453-5467, ISSN 0022-0302
26. Ganjkhanlou, M., Nikkhah, A. & Zali, A. (2010). Effect of Dietary Cation-Anion Balance on
27. Milk Production and Blood Mineral of Holstein Cows During the Last Two Months of Pregnancy. *African Journal of Biotechnology*, Vol.9, No.36, (September 2010), pp. 5983-5988, ISSN 1684–5315
28. Gaynor, P.J., Mueller, F.J., Miller, J.K., Ramsey, N., Goff, J.P. & Horst, R.L. (1989). Parturient Hypocalcaemia in Jersey Cows Fed Alfalfa Haylage Based Diets With Different Cation
29. to Anion Ratios. *Journal of Dairy Science*, Vol.72, No.10, (October 1989), pp. 2525-2531, ISSN 0022-0302
30. Goff, J.P., Horst, R.L., Mueller, F.J., Miller, J.K., Kiess, G.A. & Dowlen, H.H. (1991). Addition of Chloride to a Prepartal Diet High in Cations Increases 1,25-dihydroxyvitamin D Response to Hypocalcaemia Preventing Milk Fever. *Journal of Dairy Science*, Vol.74, No.11, (November 1991), pp. 3863-3871, ISSN 0022-0302
31. Goff, J.P., Ruiz, R. & Horst, R.L. (2004). Relative Acidifying Activity of Anionic Salts Commonly Used to Prevent Milk Fever. *Journal of Dairy Science*, Vol.87, No.5, (May 2004), pp. 1245-1255, ISSN 0022-0302
32. Goff, J.P. (1992). Cation-Anion Difference of Diets and Its Influence on Milk Fever and Subsequent Lactation: the Good and the Bad News. *Proceedings of Cornell Nutrition Conference For Feed Manufacturers*, pp. 148-159, ISSN 0885-7687, Rochester, N.Y, USA, October 13-15, 1992
33. Goff, J.P. (2000). Pathophysiology of Calcium and Phosphorus Disorders. *Veterinary Clinics of North America: Food Animal Practice*, Vol.16, No.2, (July 2000), pp. 319-337, ISSN 0749-0720
34. Goff, J.P. (2006a). Major Advances in Our Understanding of Nutritional Influences on Bovine Health. *Journal of Dairy Science*, Vol.89, No.4, (April 2006), pp. 1292-1301, ISSN 0022-0302

35. Goff, J.P. (2006b). Mineral Disorders of the Transition Period: Origin And Control. *Proceedings of XXIV World Buiatrics Congress*, pp. 1-9, ISBN 2-903 62 3407, Nice, France, October 15-19, 2006
36. Goff, J.P. & Horst, R.L. (1997a). Comparison of Sulfuric Acid with Hydrochloric Acid as a Source of Acidifying Anions. *Journal of Animal Science*, Vol.75, Suppl.1, (January 1997), pp. 98, ISSN 0021-8812
37. Goff, J.P. & Horst, R.L. (1997b). Effects of the Addition of Potassium or Sodium, but not Calcium, to Prepartum Rations on Milk Fever in Dairy Cows. *Journal of Dairy Science*, Vol.80, No.1, (January 1997), pp. 176-186, ISSN 0022-0302
38. Goff, J.P. & Horst, R.L. (1997c). Physiological Changes at Parturition and their Relationship to Metabolic Disorders. *Journal of Dairy Science*, Vol.80, No.7, (July 1997), pp. 1260-1268, ISSN 0022-0302
39. Goff, J.P. & Horst, R.L. (1998). Use of Hydrochloric Acid as a Source of Anions for Prevention of Milk Fever. *Journal of Dairy Science*, Vol.81, No.11, (November 1998), pp. 2874-2880, ISSN 0022-0302
40. Goings, R.L., Jacobson, N.L. & Littledike, E.T. (1971). Responses of Cows to a Very Low Calcium Intake at Parturition. *Journal of Dairy Science*, Vol.54, No.5, (May 1971), pp. 791-792, ISSN 0022-0302
41. Ganjkhanlou, M., Nikkhah, A. & Zali, A. (2010). Effect of Dietary Cation-Anion Balance on
42. Milk Production and Blood Mineral of Holstein Cows During the Last Two Months of Pregnancy. *African Journal of Biotechnology*, Vol.9, No.36, (September 2010), pp. 5983-5988, ISSN 1684–5315
43. Gaynor, P.J., Mueller, F.J., Miller, J.K., Ramsey, N., Goff, J.P. & Horst, R.L. (1989). Parturient Hypocalcaemia in Jersey Cows Fed Alfalfa Haylage Based Diets With Different Cation to Anion Ratios. *Journal of Dairy Science*, Vol.72, No.10, (October 1989), pp. 2525-2531, ISSN 0022-0302
44. Goff, J.P., Horst, R.L., Mueller, F.J., Miller, J.K., Kiess, G.A. & Dowlen, H.H. (1991). Addition of Chloride to a Prepartal Diet High in Cations Increases 1,25-dihydroxyvitamin D Response to Hypocalcaemia Preventing Milk Fever. *Journal of Dairy Science*, Vol.74, No.11, (November 1991), pp. 3863-3871, ISSN 0022-0302
45. Goff, J.P., Ruiz, R. & Horst, R.L. (2004). Relative Acidifying Activity of Anionic Salts Commonly Used to Prevent Milk Fever. *Journal of Dairy Science*, Vol.87, No.5, (May 2004), pp. 1245-1255, ISSN 0022-0302
46. Goff, J.P. (1992). Cation-Anion Difference of Diets and Its Influence on Milk Fever and Subsequent Lactation: the Good and the Bad News. *Proceedings of Cornell Nutrition Conference For Feed Manufacturers*, pp. 148-159, ISSN 0885-7687, Rochester, N.Y, USA, October 13-15, 1992
47. Goff, J.P. (2000). Pathophysiology of Calcium and Phosphorus Disorders. *Veterinary Clinics of North America: Food Animal Practice*, Vol.16, No.2, (July 2000), pp. 319-337, ISSN 0749-0720
48. Goff, J.P. (2006a). Major Advances in Our Understanding of Nutritional Influences on Bovine Health. *Journal of Dairy Science*, Vol.89, No.4, (April 2006), pp. 1292-1301, ISSN 0022-0302
49. Goff, J.P. (2006b). Mineral Disorders of the Transition Period: Origin And Control. *Proceedings of XXIV World Buiatrics Congress*, pp. 1-9, ISBN 2-903 62 3407, Nice, France, October 15-19, 2006
50. Goff, J.P. & Horst, R.L. (1997a). Comparison of Sulfuric Acid with Hydrochloric Acid as a Source of Acidifying Anions. *Journal of Animal Science*, Vol.75, Suppl.1, (January 1997), pp. 98, ISSN 0021-8812
51. Goff, J.P. & Horst, R.L. (1997b). Effects of the Addition of Potassium or Sodium, but not Calcium, to Prepartum Rations on Milk Fever in Dairy Cows. *Journal of Dairy Science*, Vol.80, No.1, (January 1997), pp. 176-186, ISSN 0022-0302
52. Goff, J.P. & Horst, R.L. (1997c). Physiological Changes at Parturition and their Relationshipto Metabolic Disorders. *Journal of Dairy Science*, Vol.80, No.7, (July 1997), pp. 1260-1268,ISSN 0022-0302

53. Goff, J.P. & Horst, R.L. (1998). Use of Hydrochloric Acid as a Source of Anions forPrevention of Milk Fever. *Journal of Dairy Science*, Vol.81, No.11, (November 1998), pp.2874-2880, ISSN 0022-0302
54. Goings, R.L., Jacobson, N.L. & Littledike, E.T. (1971). Responses of Cows to a Very LowCalcium Intake at Parturition. *Journal of Dairy Science*, Vol.54, No.5, (May 1971), pp. 791-792, ISSN 0022-0302
55. Green, L.E., Hedges, V.J., Schukken, Y.H., Blowey, R.W. & Packington, A.J. (2002). TheImpact of Clinical Lameness on the Milk Yield of Dairy Cows. *Journal of Dairy Science*,Vol.85, No.9, (September 2002), pp. 2250-2256, ISSN 0022-0302
56. Grummer, R.R. (1993). Etiology of Lipid-Related Metabolic Disorders in PeriparturientDairy Cows. *Journal of Dairy Science*, Vol.76, No.12, (December 1993), pp. 3882-3896,ISSN 0022-0302
57. Guard, C.L. (1996). Fresh Cow Problems are Costy; Culling Hurts the Most. *Hoard'sDairyman*, Vol.141, pp. 8, ISSN 0018-2885
58. Harvatine, K.J., Boisclair, Y.R. & Bauman, D.E. (2009). Recent Advances in the Regulation ofMilk Fat Synthesis. *Animal*, Vol.3, No.1, (January 2009), pp. 40-54, ISSN 1751-7311
59. Horst, R.L. (1986). Regulation of Calcium and Phosphorus Homeostasis in Dairy Cow.*Journal Of Dairy Science*, Vol.69, No.2, (February 1986), pp. 604-616, ISSN 0022-0302
60. Horst, R.L. & Goff, J.P. (1997). Milk Fever and Dietary Potassium. *Proceedings of CornellNutrition Conference for Feed Manufacturers*, pp.181-189, ISSN 0885-7687, Rochester, NY,USA, October 21-23, 1997
61. Horst, R.L., Goff, J.P., Reinhardt, T.A. & Buxton, D.R. (1997). Strategies for Preventing MilkFever in Dairy Cattle. *Journal of Dairy Science*, Vol.80, No.7, (July 1997), pp. 1269-1280,ISSN 0022-0302
62. Ingvartsen, K.L. (2006). Feeding- and Management-Related Diseases in the Transition Cow:Physiological Adaptations Around Calving and Strategies to Reduce Feeding-RelatedDiseases. *Animal Feed Science and Technology*, Vol.126, No.3-4, (March 2006), pp. 175-213,ISSN 0377-8401
63. Johnson, E.H., Al-Habsi, K., Kaplan, E., Srikandakumar, A., Kadim, I.T., Annamalai, K., Al-Busaidy, R. & Mahgoub, O. (2004). Caprine Hepatic Lipidosis Induced Through theIntake of Low Levels of Dietary Cobalt. *The Veterinary Journal*, Vol.168, No.2, (September2004), pp. 174-179, ISSN 1090-0233
64. Joyce, P.W., Sanchez, W.K. & Goff, J.P. (1997). Effect of Anionic Salts in Prepartum DietsBased on Alfalfa. *Journal of Dairy Science*, Vol.80, No.11, (November 1997), pp. 2866-2875, ISSN 0022-0302
65. Kronfeld, D.S., Donoghue, S., Copp, R.L., Stearns, F.M. & Engle, R.H. (1982). NutritionalStatus of Dairy Cows Indicated by Analysis of Blood. *Journal of Dairy Science*, Vol.65,No.10, (October 1982), pp. 1925-1933, ISSN 0022-0302
66. Lewis, G.S. (1997). Uterine Health and Disorders. *Journal of Dairy Science*, Vol.80, No.5, (May1997), pp. 984-994, ISSN 0022-0302
67. Logue, D.N. & Offer, J.E. (2001). Guest Editorial: The Effect of Forage Type on Foot Health inDairy Heifers. *The Veterinary Journal*, Vol.162, No.1, (July 2001), pp. 7-8, ISSN 1090-0233
68. Lomba, F., Chauvaux, E.T., Lengele, L. & Bienfet, V. (1978). Calcium Digestibility in Cows asInfluenced by the Excess of Alkaline Ions Over Stable Acid Ions in Their Diets. *The British Journal of Nutrition*, Vol.39, No.3, (May 1978), pp. 425-429, ISSN 0007-1145
69. Loor, J. J., Quinlan, L. E., Bandara, A. B. P. A. & Herbein, J. H. (2002). Distribution of transvaccenic acid and cis-9,trans-11-conjugated linoleic acid (rumenic acid) in blood plasma lipid fractions and secretion in milk fat of Jersey cows fed canola or soybean oil. *AnimalResearch*, Vol.51, No.2, (March-April 2002), pp. 119-134, ISSN 1627-3583
70. Malven, P.V., Erb, R.E., D'amico, MF, Stewart, T.S. & Chew, B.P. (1983). Factors Associatedwith Edema of the Mammary Gland in Primigravid Dairy Heifers. *Journal of DairyScience*, Vol.66, No.2, (February 1983), pp. 246-252, ISSN 0022-0302
71. Massey, C.D., Wang, C., Donovan, G.A. & Beede, D.K. (1993). Hypocalcaemia at Parturitionas a Risk Factor for Left Displacement of the Abomasum in Dairy Cows. *Journal*

of theAmerican Veterinary Medical Association, Vol.203, No.6, (September 1993), pp. 852-853,ISSN 0003-1488

72. Mayland, H.F. (1988). Grass Tetany. In: *The Ruminant Animal; Digestive Physiology andNutrition*, D.C. Church, (Ed.), 511-531, Prentice Hall, ISBN 0881337404, EnglewoodCliffs, N.J., USA
73. Michal, J.J., Heirman, L.R., Wong, T.S., Chew, B.P., Frlgg, M. & Volker, L. (1994).Modulatory Effects of Dietary Beta Carotene on Blood and Mammary LeukocyteFunction in Periparturient Dairy Cows. *Journal of Dairy Science*, Vol.77, No.5, (May1994), pp. 1408-1421, ISSN 0022-0302
74. Moore, S.J, Van de Haar, M.J., Sharma, B.K., Pilbeam, T.E., Beede, D.K., Bucholtz, H.F.,Liesman, J.S., Horst, R.L. & Goff, J.P. (2000). Effects of Altering Dietary Cation-AnionDifference on Calcium and Energy Metabolism in Peripartum Cows. *Journal of DairyScience*, Vol.83, No.9, (September 2000), pp. 2095-2104, ISSN 0022-0302
75. Morgante, M. (2004). Digestive Disturbances and Metabolic-Nutritional Disorders. In: *Dairysheep nutrition*. G. Pulina (Ed.), 165-191, CAB International, ISBN 0851996817, Wallingford, UK.
76. Nestor, Jr.K.E., Hemken, R.W. & Harmon, R.J. (1988). Influence of Sodium Chloride andPotassium Bicarbonate on Udder Edema and Selected Blood Parameters. *Journal of DairyScience*, Vol.71, No.2, (February 1988), pp. 366-372, ISSN 0022-0302
77. NRC, (2001). *Nutrient Requirement Of Dairy Cattle*. 7th rev.ed., National Academy Press,ISBN 0309069971, Washington, DC, USA
78. Oetzel, G.R. (1988). Parturient Paresis and Hypocalcaemia in Ruminant Livestock. *VeterinaryClinics of North America: Food Animal Practice*, Vol.4, No.2, (July 1988), pp. 351564, ISSN0749-0720
79. Oetzel, G.R. (1991). Meta-Analisis of Nutritional Risk Factor for Milk Fever in Dairy Cattle.*Journal of Dairy Science*, Vol.74, No.11, (November 1991), pp. 3900-3912, ISSN 0022-0302
80. Oetzel, G.R. (1993). Use of Anionic Salts for Prevention Milk Fever in Dairy Cattle.*Compendium On Continuing Education For The Practicing Veterinarian*, Vol.15, No.8,(August 1993), pp. 1138-1146, ISSN 0193-1903
81. Oetzel, G.R. (1996). Effect of Calcium Chloride Gel Treatment in Dairy Cows On Incidenceof Periparturient Diseases. *Journal of the American Veterinary Medical Association*, Vol.209,No.5, (September 1996), pp. 958-961, ISSN 0003-1488
82. Oetzel, G.R. (2000). Management of Dry Cows for the Prevention of Milk Fever and OtherMineral Disorders. *Veterinary Clinics of North America: Food Animal Practice*, Vol.16, No.2,(July 2000), pp. 369-386, ISSN 0749-0720
83. Oetzel, G.R. & Barmore, J.A. (1993). Intake of a Concentrate Mixture Containing VariousAnionic Salts Fed to Pregnant, Nonlactating Dairy Cows. *Journal of Dairy Science*, Vol.76,No.6, (Juny 1993), pp. 1617-1623, ISSN 0022-0302
84. Oetzel, G.R., Olson, J.D., Curtis, C.R. & Fettman, M.J. (1988). Ammonium Chloride andAmmonium Sulfate for Prevention of Parturient Paresis in Dairy Cows. *Journal of DairyScience*, Vol.71, No.12, (December 1988), pp. 3302-3309, ISSN 0022-0302
85. Overton, T.R. & Waldron, M.R. (2004). Nutritional Management of Transition Dairy Cows:Strategies to Optimize Metabolic Health. *Journal of Dairy Science*, Vol.87, (E. Suppl.),(July 2004), pp. E105-E119, ISSN 0022-0302
86. Pehrson, B. (1996). Milk Analysis as an Indicator of the Nutritional and Disease Status ofDairy Cow. In: *Recent Advances in Animal Nutrition*, P.C. Garnsworthy, J. Wiseman, W.Haresign, (Eds.), 113-129, Nottingham University Press, ISBN 1897676034
87. Pehrson, B., Svensson, C., Gruvaeus, I. & Virkki, M. (1999). The Influence of Acidic Diets on the Acid-Base Balance of Dry Cows and the Effect of Fertilization on the Mineral Content of Grass. *Journal of Dairy Science*, Vol.82, No.6, (June 1999), pp. 1310-1316, ISSN 0022-0302
88. Perfield, II J.W. & Bauman, D.E. (2005). Current Theories and Recent Advances in theBiology of Milk Fat Depression. *Proceedings of Cornell Nutrition Conference for FeedManufacturers*, pp. 95-106, ISBN 9998258294

89. Phillippo, M., Reid, G.W. & Nevison, I.M. (1994). Parturient Hypocalcaemia in Dairy Cows:Effects of Dietary Acidity on Plasma Minerals and Calciotrophic Hormones. *Research inVeterinary Science*, Vol.56, No.3, (May 1994), pp. 303-309, ISSN 0034-5288
90. Radostits, O.M., Gay, C.C., Blood, D.C. & Hinchcliff, K.W. (2000). *Veterinary Medicine: ATextbook of the Diseases of Cattle, Sheep, Pigs, Goats and Horses*. 9th Edition. 1417-1476, WBSaunders, ISBN 0702026042, London, UK
91. Ramberg, Jr.C.J., Johnson, E.K., Fargo, R.D. & Kronfeld, D.S. (1984). Calcium Homeostasis inCows with Special Reference to Parturient Hypocalcaemia. *American Journal ofPhysiology*, Vol.246, No.5, (May 1984), pp. R698-R704, ISSN 0363-6119
92. Randalu, W.E., Hemken, R.W., Bull, L.S. & Douglas, L.W. (1974). Effect of Dietary Sodiumand Potassium on Udder Edema in Holstein Heifers. *Journal Of Dairy Science*, Vol.57,No.4, (April 1974), pp. 472-475, ISSN 0022-0302
93. Robinson, D.L., Kappel, L.C. & Boling, J.A. (1989). Management Practices to Overcome theIncidence of Grass Tetany. *Journal of Animal Science*, Vol.67, No.12, (December 1989), pp.3470-3484, ISSN 0021-8812
94. Rushen, J., de Passillé, A.M., von Keyserlingk, M.A.G. & Weary, D.M. (2008). *The Welfare ofCattle*. Animal Welfare Series, Vol.5., C.Phillips (Ed.), Springer, ISBN 978-1-4020-6558-3,Dordrecht, The Netherlands
95. Sansom, B.F., Manston, R. & Vagg, M.J. (1983). Magnesium and Milk Fever. *VeterinaryRecord*, Vol.112, No.19, (May 1983), pp. 447-449, ISSN 2042-7670 Production of Dairy Cows. *JournalOf Dairy Science*, Vol.42, No.1, (January 1959), pp. 170-179, ISSN 0022-0302
96. Schonewille, J.T. , Everts, H. , Jittakhot, S. & Beynen, A.C. (2008). Quantitative Prediction ofMagnesium Absorption in Dairy Cows. *Journal of Dairy Science*, Vol.91, No.1, (January2008), pp. 271-278, ISSN 0022-0302
97. Schroeder, G.F., Gagliostro, G.A., Bargo, F., Delahoy, J.E. & Muller, L.D. (2004). Effects of FatSupplementation on Milk Production and Composition by Dairy Cows on Pasture: AReview. *Livestock Production Science*, Vol.86, No.1-3, (March 2004), pp. 1-18, ISSN 1871-1413
98. Shaver, R.D. (1997). Nutritional Risk Factors in the Etiology of Left Displaced Abomasum inDairy Cows: A Review. *Journal of Dairy Science*, Vol.80, No.10, (October 1997), pp. 2449-2453, ISSN 0022-0302
99. Stelletta, C., Gianesella, M. & Morgante, M. (2008). Metabolic and Nutritional Diseases. In:*Dairy Goats Feeding And Nutrition*. A.Cannas, G.Pulina (Eds.), 263-288, CABInternational, ISBN 1845933486, Wallingford, UK
100. Stone, W.C. (2004). Nutritional Approaches to Minimize Subacute Ruminal Acidosis andLaminitis in Dairy Cattle. *Journal of Dairy Science*, Vol.87, E. Suppl., (July 2004), pp. E13-E26, ISSN 0022-0302
101. Thilsing-Hansen, T., Jorgensen, R.J. & Ostergaard, S. (2002). Milk Fever Control Principles: AReview. *Acta Veterinaria Scandinavica*, Vol.43, No.1, (March 2002), pp. 1-19, ISSN 1751-0147
102. Thomas, D.G.; Miller, J.K. Mueller, F.J. & Madsen, F.C. (1990). Udder Edema Reduced byPrepartum Vitamin E Supplementation. *Journal of Dairy Science*, Vol.73, Suppl.1, pp. 271,ISSN 0022-0302
103. Urton, G., von Keyserlingk, M.A.G. & Weary, D.M. (2005). Feeding Behavior Identifies DairyCows at Risk for Metritis. *Journal of Dairy Science*, Vol.88, No.8, (August 2005), pp. 2843-2849, ISSN 0022-0302
104. Vagnoni, D.B. & Oetzel, G.R. (1998). Effect of Dietary Cation-Anion Difference on the Acid-Base Status of Dry Cows. *Journal of Dairy Science*, Vol.81, No.6, (June 1998), pp. 1643-1652, ISSN 0022-0302
105. Van Saun, R.J. & Sniffen, C.J. (1996). Nutritional Management of the Pregnant Dairy Cow toOptimize Health, Lactation and Reproductive Performance. *Animal Feed Science and Technology*, Vol.59, No.1-3, (June 1996), pp. 13-26, ISSN 0377-8401
106. Wang, C. & Beede, D.K. (1992). Effects of Diet Magnesium on Acid-Base Status and CalciumMetabolism of Dry Cows Fed Acidogenic Salts. *Journal of Dairy Science*, Vol.75, No.3,(March 1992), pp. 829-836, ISSN 0022-0302

107. Wiggers, K.D., Nelson, D.K. & Jacobson, N.L. (1975). Prevetion of Parturient Paresis by aLow-Calcium Diet Prepartum: A Field Study. *Journal of Dairy Science*, Vol.58, No.3,(March 1975), pp. 430-431, ISSN 0022-0302
108. Zadoks, R.N. (2006). The Role of Nutrition in Udder Health. Proceedings of High Plains DairyConference, pp. 115-128, Albuquerque, NM, March 17, 2006

CHAPTER 2

Protein Sources in Ruminant Nutrition

Monica I. Cutrignelli, Serena Calabrò, Raffaella Tudisco, Federico Infascelli and Vincenzo Piccolo

Department of Animal Science and Food Control, University of Naples Federico II, Naples, Italy

1. INTRODUCTION

Since 2001 the European Commission banned the use of meat and bone meal and its by-products in diets for livestock animals (EC directive 999/2001) in order to assure consumer safety on animal products. Consequently, soybean meal became the most utilised protein source in the intensive livestock systems.

Moreover, the proteins of this source are low degradable in the rumen and well proportioned to the non structural carbohydrates (NSC).

Soybean meal solvent extract (s.e.) is a by-product of oil industry, where soybean seeds are treated with organic solvents (e.g. hexane) and subsequently with high temperature. For this reason soybean meal has been banned in the organic livestock (EC directive 2092/1991; EC directive 834/2007).

Even if in Europe the high part of soybean is imported, soybean solvent extract represents the less expensive protein source for its high crude protein content (44-50 % as fed). However soybean meal costs and availability are strongly related with the price development of agricultural commodities on the world market (Jezierny et al., 2010). Factors which may influence world market prices include variations in population and economic growth, changes in consumer's product preferences, but world market prices are also dependent on weather conditions (Gill, 1997; Trostle, 2008).

Finally, another factor has to be evaluated that is the genetically modification (GM) of soybean. Indeed, public concerns are increasing in GM food consumption due to the fact, even if for several years no direct evidence that it may represent a possible danger for health has been reported, recently, a number of papers have been published with controversial results.

Thus, the search for alternative protein sources has led to an increasing interest in the use of grain legumes, as they supply the important source of plant protein. The botanical family of grain legumes is known as Fabaceae, also referred to as Leguminosae. Grain legumes are cultivated primarily for their seeds which are harvested at maturity, and which are rich in protein and energy. The mature dry seeds of grain legumes are used either as animal feed ingredient or for human consumption (Singh et al., 2007). Beans, lentils and chickpeas are utilised exclusively for human nutrition, while the other grains are used in animal feeding too.

In Italy grain legumes cultivation is progressively increased due the presence of new cultivars more hardy and productive. These new cultivars were selected principally in France and are characterised by lower water requirements, higher production and higher resistance to the parasitic infestations and to the adverse environmental conditions.

Generally, legumes are characterised by their ability to use atmospheric nitrogen as a nutrient due to the symbiosis with nitrogen-fixing bacteria from the Rhizobium species (Sprent and Thomas, 1984; Zahran, 1999). Therefore, unlike other cultivated plants, legume crops need less nitrogen fertiliser for optimal growth, and the use of legumes in crop rotation systems reduces the need of nitrogen fertiliser in subsequent crops (López-Bellido et al., 2005). Nitrogen benefits in legume-cereal rotation systems have been attributed not only to the transfer of biologically fixed nitrogen (Díaz-Ambrona and Mínguez, 2001; Evans et al., 2001), but also to lower immobilisation of nitrate in the soil during the decomposition of legumes compared to cereal residues (Green and Blackmer, 1995), also termed as the nitrogen-sparing effect. Thus, nitrogen benefits may result from a combination of legume nitrogen sparing effects and the bacterial nitrogen fixation (Chalk et al., 1993; Herridge et al., 1995). In addition, crop rotation and intercropping with legumes may provide successful strategies for weed suppression (Liebman and Dyck, 1993; Bulson et al., 1997). Weed growth and development may be disrupted due to varying cultivation conditions prevailing for the different crops used (e.g. fertiliser requirements, planting or maturation dates), thereby preventing domination of only a few weed species (Froud-Williams, 1988; Liebman and Janke, 1990). Due to these crop effects, cultivation of grain legumes is an important part of crop rotation, particularly in organic farming (Badgley et al., 2007).

In animal nutrition, grain legumes are mainly used as protein supplements, but also as a valuable energy source, due to their partly high contents of starch (faba bean, peas) andlipids (lupins) (Gatel, 1994; Bach Knudsen, 1997; Salgado et al., 2002a).

However, the use of grain legumes in animal nutrition has been hampered due to partially high concentrations of secondary plant metabolites, also referred as antinutritional factors (ANFs), including condensed tannins, protease inhibitors, alkaloids, lectins, pyrimidine glycosides and saponins. Possible negative effects of these secondary plant metabolites include, for example, feed refusals (tannins, alkaloids), reduced nutrient digestibility (tannins, protease inhibitors, lectins) or even toxic effects (alkaloids) (Rubio and Brenes, 1995; Lallès and Jansman, 1998; Huisman and Tolman, 2001).

The objectives of the following chapters are the comparison of chemical composition as well as the nutritive value of soybean, soybean meal solvent extract and several legume grain (e.g.: peas, lupine, faba bean). In addition, in order to evaluate the opportunity of soybean replacement with grain, the results of *in vitro* studies are described and the influence of protein sources on meat quality are discussed.

2. NUTRITIONAL CHARACTERISTICS OF SOYBEAN MEAL SOLVENT EXTRACT AND LEGUME GRAINS

The chemical composition of some grain legumes in comparison to soybean and soybean meals s.e are pictured in Table 1.

Soybeans (*Soja hispida*) are characterized by high protein (380 g/kg dry matter) and lipid (200 g/kg dry matter) concentrations, which provide high energy density (1.1-1.2 UFL/kg dry matter). In ruminant feeding, soybean integral seeds could represent the only source of protein supplementation for cattle fattening. However, in dairy cows it is preferable not to exceed the dry matter administration of soybean seeds as it may modify fatty acid profile of milk fat and worse butter consistency and conservation.

Soybean meal s.e. proteins, due to the heat treatment, are medium to low degradable in the rumen (Chaubility et al., 1991; Infascelli et al., 1995). In addition, soybean meal shows an elevated ratio of protein/non structural carbohydrates.

Legume grains are characterised by high energy density allowed to the high protein, starch and/or fat concentrations, as more than sufficient is their calcium concentration.

The proteins of legume grain are highly degradable in the rumen and digestible in the intestine. Notwithstanding, large part of legume grain shows anti-nutritional factors (i.e. lecithin, trypsin inhibitors, tannins, saponin, phytase), that are inactivated by the enzymes produced by the bacteria present in the rumen. Within the grain legumes, lupins have higher amounts of crude protein (324–381 g/kg dry matter), compared to faba beans (301 g/kg dry matter) and peas (246 g/kg dry matter) (Degussa, 2006). Jezierny et al. (2007) reported similar contents of crude protein in different batches of lupins, faba beans and peasaveraging 387, 308 and 249 g/kg dry matter, respectively. In comparison to soybean meal, faba beans and peas contain between 45 to 55% and lupins (L. albus) even up to 70% of its crude protein content (Degussa, 2006).

The ether extract content in peas and faba beans is generally rather low compared to lupins; crude fat contents of faba beans and peas range from 15 to 20 g/kg dry matter, thus being in a similar range as values for soybean meal (15–28 g/kg dry matter) (DLG, 1999; Jezierny et al., 2007). In lupins, the crude fat content varies between cultivars, with values of about 57 g/kg dry matter (*L. luteus, L. angustifolius*) to 88 g/kg dry matter (*L. albus*) (DLG, 1999).

The carbohydrate fraction includes the low molecular-weight sugars, starch and various non-starch-polysaccharides (NSP) (Bach Knudsen, 1997). The NSP and lignin are the principal components of cell walls and are commonly referred to as dietary fibre (Theander et al., 1989; Canibe and Bach Knudsen, 2002). Generally, faba beans and peas are rich in starch (422–451 and 478–534 g/kg dry matter, respectively) (DLG, 1999; Jezierny, 2009), whereas lupins have comparatively low levels of starch (42–101 g/kg dry matter) (DLG, 1999; Jezierny, 2010). However, it needs to be emphasized that the determination of starch in grain legumes may be confounded by the analytical method used (Hall et al., 2000).

Faba beans and peas contain rather low amounts of fibre fractions in comparison to lupins (Bach Knudsen, 1997; Jezierny, 2009), and, with regard to lignin content, faba beans and L. angustifolius have similar amounts of lignin (1 to 7 and 6 to 9 g/kg dry matter, respectively), whereas the lignin content in peas is of minor importance (0.4–3 g/kg dry matter) (Salgado et al., 2002a; Jezierny, 2010).

The NSP fraction of faba beans consists mainly of cellulose (89–115 g/kg dry matter), with lower levels of hemicellulose (21–57 g/kg dry matter) (Salgado et al., 2002a,b; Jezierny, 2009).

Hemicellulose contents in peas range from 23 to 95 g/kg dry matter and cellulose contents range from 52 to 77 g/kg dry matter (Salgado et al., 2002a,b; Jezierny, 2009).

Lupins contain high levels of NSP, with contents of cellulose generally being higher than hemicellulose (131 to 199 vs. 40 to 66 g/kg dry matter) (Bach Knudsen, 1997; Salgado et al., 2002a,b; Jezierny, 2009), and they also have considerable amounts of oligosaccharides (Bach Knudsen, 1997; Salgado et al., 2002a).

Comparing the different batches in each species, lupin show lower variability than faba bean and peas, probably because the genetic selection in this species was addressed principally on the reduction of secondary plant metabolites (Colombini, 2004) than to the improving of chemical characteristics.

The protein of faba beans and peas contains similar or even higher proportions of lysine (70 and 80 g/kg crude protein, respectively), when compared to protein from soybean meal s.e. (69 g/kg crude protein) or lupins (51 to 54 g/kg crude protein) (Degussa, 2006).

Table 1. Mean values of chemical composition (%dry matter) of different protein sources (Calabrò et al., 2001; Calabrò et al., 2009; Calabrò et al., 2010).

Protein sources	DM	CP	EE	Ash	NDF
Soybean	93.30	33.2	25.0	5.57	14.5
Soybean meal s.e.	89.80	41.3	1.36	6.00	22.2
Lupin Lublanc	93.26	36.92	6.78	4.33	4.33
Lupin Lutteur	94.31	35.30	5.73	4.19	4.19
Lupin Multitalia	93.97	36.67	9.54	3.92	3.92
Faba bean Irena	89.22	25.62	1.05	3.99	20.94
Faba bean Lady	88.48	25.17	1.02	4.21	25.82
Faba bean Scuro di Torre Lama	90.55	26.91	0.90	4.32	21.75
Faba bean Chiaro di Torre Lama	90.80	24.69	1.01	4.19	21.23
Faba bean ProtHABAT69	90.60	28.69	1.10	4.51	18.21
Faba bean Sicania	90.29	26.52	0.95	3.89	21.43
Peas Alembo	88.49	31.27	0.73	3.94	21.60
Peas Alliance	89.20	28.47	0.56	3.82	20.40
Peas Attika	89.68	25.04	0.81	4.01	18.09
Peas Corallo	88.80	28.50	0.80	3.69	21.82
Peas Iceberg	90.08	27.30	0.78	4.12	22.75
Peas Ideal	89.94	28.28	0.88	4.17	18.37
Peas Spirale	93.00	28.74	0.55	4.30	19.05

Comparing the different batches in each species, lupin show lower variability than faba bean and peas, probably because the genetic selection in this species was addressed principally on the reduction of secondary plant metabolites (Colombini, 2004) than to the improving of chemical characteristics.

Comparing the different batches in each species, lupin show lower variability than faba bean and peas, probably because the genetic selection in this species was addressed principally on the reduction of secondary plant metabolites (Colombini, 2004) than to the improving of chemical characteristics.

The protein of faba beans and peas contains similar or even higher proportions of lysine (70 and 80 g/kg crude protein, respectively), when compared to protein from soybean meal (69 g/kg crude protein) or lupins (51 to 54 g/kg crude protein) (Degussa, 2006).

The proportion of threonine in grain legume protein (38 to 42 g/kg crude protein) is similar to that in soybean meal (45 g/kg soybean meal) (Degussa, 2006), however, there is a severe deficiency in the sulphur containing AA methionine + cystine, while tryptophan is marginally deficient to fulfill nutrient requirements for pigs (20 to 50 kg body weight) (NRC, 1998; Degussa, 2006). In fact, apart from *L. albus*, the seeds of faba beans, peas and lupins contain less than 50% of these AA in comparison to soybean meal (Table 2), thus constraining the use of grain legumes as sole protein source in pig diets.

As concerns the fatty acid profile, the rather high proportion of essential unsaturated fatty acids of some grain legumes, e.g. some Vicia species (Akpinar et al., 2001) or *L. albus* (Erbaş et al., 2005) may be attractive both from the human and animal nutrition perspective (Bézard et al., 1994), while adverse effects of unsaturated fatty acids on meat quality should be taken into account (Wood et al., 2003). For example, in faba beans a ratio of saturated to unsaturated fatty acids of 40–60 has been reported (Akpinar et al., 2001), whereas in *L. albus*, a ratio of saturated, monounsaturated and polyunsaturated fatty acids of 13.5 to 55.4 to 31.1 has been established (Erbaş et al., 2005).

As concerns the mineral composition of soybean and grain legumes, only few data are reported in literature.

The calcium concentration ranges between 1.0 g/kg (faba beans) and 1.9 g/kg (*L. angustifolius*). Phosphorus concentration varies between 4.2 g/kg (*L. angustifolius*) and 7.6 g/kg (*L. luteus*). No extreme differences in trace mineral concentrations occurred except for the manganese concentration of *L. albus*, which containes approximately 10 times more manganese than the other legume grains (Brand et al., 2004).

Table 2. Amino acid contents of grain legumes compared to soybean meal (g/kg dry matter) (Jezierny et al., 2010) SBM= soybean meal; CP= crude protein; AA= amino acids.

	Vicia faba	*Pisum sativum*	*Lupinus albus*	*Lupinus angustifolius*	*Lupinus luteus*	SBM
CP	301	246	381	324	361	541
Indispensable AA						
Arginine	26.4	21.0	39.3	33.5	38.0	39.7
Histidine	7.8	6.1	9.3	8.8	9.7	14.4
Isoleucine	11.8	10	15.3	12.7	14.2	24.3
Leucine	21.4	17.4	27.5	21.5	24.1	40.9
Lysine	18.4	17.3	18.2	15	16.3	33.1
Methionine	2.2	2.2	2.5	2.0	2.0	7.3
Phenyl-alanine	12.6	11.7	14.9	12.5	13.6	27.2
Threonine	10.5	9.1	13.3	10.9	11.9	21.3
Tryptophan	2.6	2.2	3.0	2.6	3.0	7.4
Valine	13.3	11.4	14.5	12.5	13.6	25.5
Dispensable AA						
Alanine	11.9	10.5	12.5	10.9	11.8	23.3
Aspartic acid	31.6	28.2	38.5	31.5	35.1	62.0
Cystine	3.5	3.5	6.7	4.3	4.8	8.0
Glutamic acid	46.9	40.0	79.3	65.6	72.5	97.6
Glycine	12.2	10.6	15.0	13.4	14.3	23.0
Proline	11.8	10.2	15.3	13.5	14.3	27.5
Serine	14.1	11.5	19.0	15.3	17.0	27.3

Concerning the trace elements, Cabrera et al (2003) reported that in legumes their levels ranged from 1.5–5.0 µg Cu/g, 0.05–0.60 µg Cr/g, 18.8–82.4 µg Fe/g, 32.6–70.2 µg Zn/g, 2.7– 45.8 µg Al/g, 0.02–0.35 µg Ni/g, 0.32–0.70 µg Pb/g and not detectable—0.018 µg Cd/g. In nuts, the levels ranged from 4.0–25.6 µg Cu/g, 0.25–1.05 µg Cr/g, 7.3–75.6 µg Fe/g, 25.6– 69.0 µg Zn/g, 1.2–20.1 µg Al/g, 0.10–0.64 µg Ni/g, 0.14–0.39 µg Pb/g, and not detectable— 0.018 µg Cd/g. The authors found a direct statistical correlation between Cu–Cr, Zn–Al and Cr–Ni ($P<0.05$), and Al–Pb ($P<0.001$).

Sankara Rao and Deosthale (2006) comparing for their total ash, calcium, phosphorus, iron, magnesium, zinc, manganese, copper, and chromium contents five grain legumes (Indian legumes, chick pea, pigeon pea, green gram, and black gram), found significant varietal differences only for chromium content in black gram. The cotyledons of these legumes were significantly lower in calcium content as compared to the whole grains. The authors concluded that for human nutrition, differences in mineral composition of whole grain and cotyledons were marginal except for calcium. These legumes as whole grain and cotyledons, appeared to be significant contributors to the daily requirements of magnesium, manganese, and copper in the diet.

As concerns the mineral composition of soybean and grain legumes, only few data are reported in literature.

The calcium concentration ranges between 1.0 g/kg (faba beans) and 1.9 g/kg (L. angustifolius). Phosphorus concentration varies between 4.2 g/kg (L. angustifolius) and 7.6 g/kg (L. luteus). No extreme differences in trace mineral concentrations occurred except for the manganese concentration of *L. albus*, which containes approximately 10 times mor manganese than the other legume grains (Brand et al., 2004).

Concerning the trace elements, Cabrera et al (2003) reported that in legumes their levels ranged from 1.5–5.0 μg Cu/g, 0.05–0.60 μg Cr/g, 18.8–82.4 μg Fe/g, 32.6–70.2 μg Zn/g, 2.7– 45.8 μg Al/g, 0.02–0.35 μg Ni/g, 0.32–0.70 μg Pb/g and not detectable—0.018 μg Cd/g. I nuts, the levels ranged from 4.0–25.6 μg Cu/g, 0.25–1.05 μg Cr/g, 7.3–75.6 μg Fe/g, 25.6– 69.0 μg Zn/g, 1.2–20.1 μg Al/g, 0.10–0.64 μg Ni/g, 0.14–0.39 μg Pb/g, and not detectable—0.018 μg Cd/g. The authors found a direct statistical correlation between Cu–Cr, Zn–Al and Cr–Ni ($P<0.05$), and Al–Pb ($P<0.001$).

Sankara Rao and Deosthale (2006) comparing for their total ash, calcium, phosphorus, iron, magnesium, zinc, manganese, copper, and chromium contents five grain legumes (Indian legumes, chick pea, pigeon pea, green gram, and black gram), found significant varietal differences only for chromium content in black gram. The cotyledons of these legumes weresignificantly lower in calcium content as compared to the whole grains. The authors concluded that for human nutrition, differences in mineral composition of whole grain and cotyledons were marginal except for calcium. These legumes as whole grain and cotyledons, appeared to be significant contributors to the daily requirements of magnesium, manganese, and copper in the diet.

3. *IN VITRO* EVALUATION OF SOYBEAN AND LEGUME GRAINS

The nutritive value of a feed is assessed by its chemical composition, digestibility and level of voluntary intake. Feed evaluation methods are use to express nutritive value of feed. It is basically description of feeds interns that allow for a prediction of the performance ofanimals offered the feeds (Medsen et al., 1997).

Several methods are used in feed evaluation such as chemical analysis, rumen degradability measurement using the nylon bag technique, digestibility measurement and feed intake prediction. However, the tables supply mean values, which cannot be used for individual lots, and all the *in vivo* techniques are very expensive and time consuming while accuracy may be low.

The *in vitro* gas production technique (IVGPT, Theodorou et al., 1994) has proved to be a potentially useful technique for ruminant feed evaluation (Herrero et al., 1996; Getachew et al., 2004), as it is capable of measuring rate and extent of nutrient degradation (Groot et al. 1996; Cone et al., 1996).

To evaluate a feedstuff by IVGPT, it is incubated at 39°C and under anaerobiosis condition with buffered rumen fluid and gas produced is measured as an indirect indicator of fermentation kinetics. During the incubation the feedstuff is first degraded and the degraded fraction may either be fermented to produce gas (CO^2 and methane) and fermentation acids, or incorporated into microbial biomass.

The IVGPT is considered the most complete *in vitro* technique, because it allows to estimate the fermentation kinetics and contemporary gives information on the fermentation products (degradability of dry matter, volatile fatty acids).

The IVGPT has been used by our research group for many years in order to investigate feed fermentation kinetics.

In particular, in this chapter we report some results obtained incubating different protein sources (i.e. soybean and legumes grains) in order to compare their fermentation kinetics (Calabrò et al., 2001a), to assess the effect of some technological treatments (i.e. crushing and flaking) on the carbohydrates fermentation kinetics (Calabrò et al., 2001b) and to test different legumes grain cultivars (Calabrò et al., 2009).

Calabrò et al. (2001a), in order to study the *in vitro* fermentation characteristics and kinetics proposed the following protocol: the samples (about 1.00 g), ground to pass a 1 mm screen, were incubated in triplicate at 39°C in 120 ml serum bottles under anaerobic conditions. Rumen liquor for the inoculum was collected from four buffaloes fed a standard diet, and immediately transported to the laboratory where it was homogenised and filtered.The gas measurements was made at 2-24 time intervals using a manual a pressure transducer (figure 1).

The cumulative gas produced at each time was fitted to the Groot et al. (1996) model which estimates the asymptotic value (A, ml/g), the time after incubation at which A/2 is formed (B, h), the time to reach the maximum rate (tmax, h) and the maximum rate (Rmax, ml/h). At the end of incubation, the degraded organic matter (dOM, %) was calculated as a difference between incubated and residual OM (filtering the bottle content through pre- weighed glass crucibles and burning at 550°C for 3 hours), and pH and volatile fatty acid concentration (VFA, mM/g) were determined, using a pH-meter and a gas chromatography, respectively.

Several concentrate ingredients such as cereals and grain legumes, used in ruminant diets in order to increase production levels, were evaluated. In particular barley, maize, hard wheat, soft wheat, oats, faba bean and pea were used as test substrates. Carbohydrates fractionation was carried out according to Cornell Net Carbohydrate and Protein System (CNCPS, Sniffen *et al.*, 1992).

The Cornell Net Carbohydrate and Protein System was developed to predict requirements, feed utilization, animal performance and nutrient excretion for dairy and beef cattle and sheep, using accumulated knowledge about feed composition, digestion, and metabolism in supplying nutrients to meet requirements.

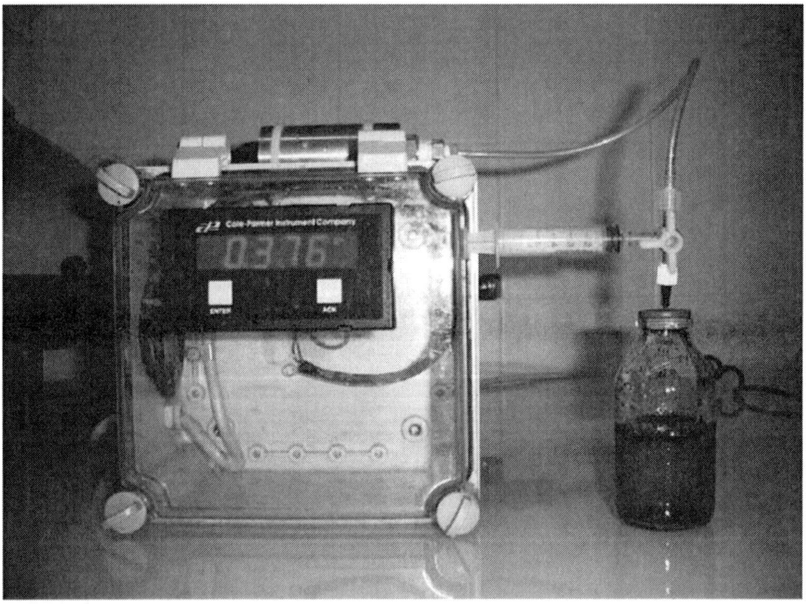

Figure. 1. Pressure transducer.

The CNCPS partitions crude protein into fractions A, B, and C, depending on their rate and extent of degradability in the rumen (NRC, 2001). Fraction A represents the non-protein N (NPN) (ammonia, peptides, amino acids) and is considered to be completely soluble; fraction B, subdivided into B1, B2, and B3, consists of true protein with progressively declining ruminal degradability. Fraction C is unavailable true protein. Broadly, these crude protein fractions are categorized into rumen degradable protein (RDP) and rume undegradable protein (RUP). The rumen degradable protein meets protein requirements for ruminal microbial growth and protein synthesis.

Once reaching the rumen, feed and protein degradation is a function of microbial activity. Rumen microbial activity, growth and protein synthesis is primarily limited by the rate and extent of carbohydrate fermentation in the rumen. Consequently, dietary fiber fractions in the forage determine the animal response to feed.

Microbial protein and rumen undegradable protein reaching the small intestine are absorbed to meet the ruminant's protein requirement. When rumen degradable protein exceeds the capacity of the rumen microbes to assimilate it, ammonia builds up in the rumen. This is followed by absorption of ammonia into the blood, conversion into urea by the liver, and excretion in the urine. The conversion of ammonia to urea costs the dairy cow energy that could otherwise be used for milk production. This loss of dietary crude protein and energy reduces the utilization efficiency of rumen degradable protein and therefore, reduced ruminant production (NRC, 2001). It also causes a negative energy balance that leads to a reduced fertility.

In the study of Calabrò et al. (2001a), the CNCPS carbohydrate fractions (table 3) was consistent with the values reported by Sniffen *et al.* (1992).

Oats (table 4) had the lowest potential gas production (A: 251 *ml/g*, $P<0.0$) and a very fast fermentation process (evidenced by low B and high RM). This result was probably due to its high soluble sugars contents (A fraction). Interestingly, oat values were very high compared to *in situ* observations. As reported by Van Soest *et al.*, (1992), maize proved to be the slowest because of its high starch content (B1 fraction), which degraded slowly. Hard and soft wheat showed very similar fermentation characteristics according to their chemical composition. Barley has fermentation kinetics between those of wheat and maize. Results related to these last three grains were similar to the results obtained *in vitro* using the IGPT (Mould *et al.*, 2005).

Legumes were characterised by high gas production alongside a rather slow fermentative process. However, interpretation of faba bean and pea data was complicated by their high protein content, whose degradation can interact with gas production (Schofield., 2000).

Organic matter degradability for all tested grains was always higher than 95%.

Table 3. Chemical composition and CNCPS fraction of the tested grains.

	Barley	Maize	Oats	Hard wheat	Soft wheat	Faba bean	Pea
CP (% DM)	11.0	8.79	12.2	14.4	13.5	29.8	23.9
A	3.80	0.49	12.6	2.40	5.00	6.00	18.0
B1	66.4	75.4	45.6	67.6	65.7	43.6	40.8
B2	11.4	6.14	16.0	9.86	9.64	9.99	7.92
C	3.60	2.57	6.77	2.64	2.16	6.41	4.18

At 72 h the pH always remains good to guarantee microbial activity.

Oats reached the maximum rate before all the other grains (figure 2) and remained the highest until 12 incubation hours, subsequently decreasing below the soft wheat level. Until 6 hours of incubation barley was faster than the two wheat, later becoming slower until the end; up to this time the three sharp curves were practically overlapping, suggesting that the fermentation trend was similar for the three grains. Maize, faba bean and pea always showed the slowest fermentation rate with overlapping trends: fermentation slowly reaches its maximum rate and slowly decreases. The results of this investigation evidenced the validity of the IVGPT in describing the kinetics fermentation of the examined grains.

The range of rates obtained with the GPT (oats>wheat>barley> maize> legumes) approximately reflects that reported by other authors cited by Sniffen *et al.* (1992): wheat>oats>barley>maize.

In the second investigation Calabrò et al. (2001b) evaluated the effects of flaking and crushing on some cereal and legume grains, using the *in vitro* cumulative gas production technique (Theodorou *et al.*, 1994).

The grains used in animal nutrition are commonly subjected to different technological treatments (i.e.: grounding, crushing, flaking, extrusion, cooking, micronization). Flaking is a hydrothermal-mechanical process which causes starch gelatinization and increases the enzymatic hydrolysis rate of the polysaccharide, which in turn favours the rumen microbial attack. Crushing differs from flaking in the shorter exposure to vapour. Also in this case, the obtained product presents a larger surface area compared to the primary grain, and a modified starch, which promotes faster fermentation kinetics.

Several investigations to evaluate the effect of flaking on organic matter degradability in the rumen, indicated the considerable influence of the primary grain (Arieli *et al.*, 1995; Bittante *et al.*, 1989); by contrast, little was known about the effects of crushing.

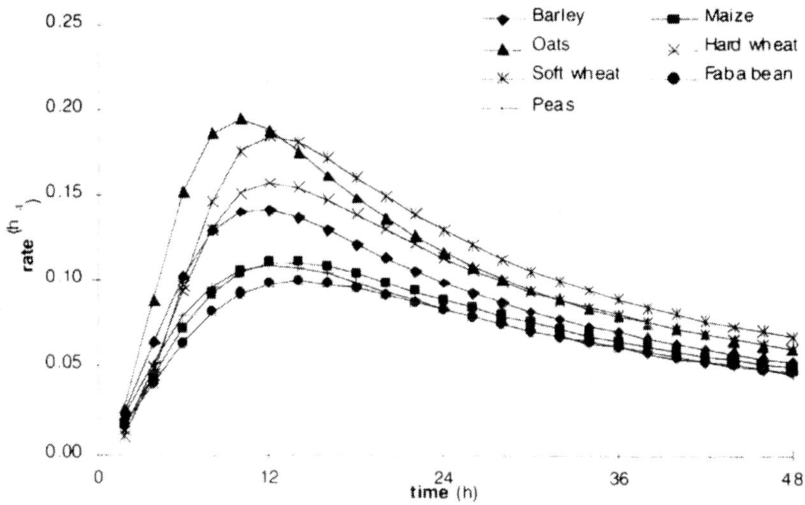

Figure. 2. Gas production fractional rate plots of different grains.

Table 4. In vitro fermentation characteristics of whole and flaked grains.

		VC_{501}	A	B	RM	dOM	Y	VFA
		ml/g	*ml/g*	*h*	*h⁻¹*	%	*ml/g*	*mM/g*
Maize	wholegrain	340	362	14.9	0.078	92.9	363	99.1
	flaked grain	338	347	13.2	0.096	96.6	343	83.3
	whole + flaked	339^{A}	355^{A}	14.1^{B}	0.087^{B}	94.8	353^{A}	91.2^{A}
Barley	wholegrain	304	323	10.2	0.125	90.6	341	92.4
	flaked grain	346	369	11.7	0.104	97.9	359	94.8
	whole + flaked	325^{B}	346^{A}	11.0^{C}	0.115^{A}	92.6	350^{A}	93.6^{A}
Soybean	wholegrain	204	236	15.5	0.051	96.7	209	71.6
	flaked grain	225	280	19.8	0.039	86.8	257	76.0
	whole + flaked	215^{C}	258^{B}	17.6^{A}	0.045^{C}	91.8	233^{B}	73.8^{B}
Main effect								
	grain	***	***	***	***	n.s.	***	**
	treatment	***	**	**	n.s.	n.s.	**	n.s.
Interaction								
	grain. x treat.	***	***	***	***	***	***	n.s.
Var.er.		87.9	135	1.21	0.00009	10.5	206	103

OMCV, cumulative gas production related to incubated OM at 120 h; A, potential gas production; B, time at which A/2 is produced; Rmax, maximum fermentation rate; Y, yield related to the degraded OM; VFA, volatile fatty acids related to degraded OM.
*a,b,c: P<0.05; A,B: P<0.001ns, not significant;***: p < 0.001.*

In this experiment three whole and flaked grains (barley, maize, soybean) and four whole and crushed grains (barley, oats, maize, faba bean) were used as test substrates.

Fermentation characteristics (Table 4 and 5) were affected by the grain type and treatment; in addition, the two factors interacted with each other.

Table 5. In vitro fermentation characteristics of whole and crushed.

		OMC V	A	B	Rmax	dOM	Y	VFA
		ml/g		*H*	*h^{-1}*	*%*	*ml/g*	*mM/g*
Oats	whole grain	321	334	10.6	0.107	87.8	362	90.6
	crushed grain	272	311	10.3	0.106	86.9	332	90.5
	whole +crushed	296^{B}	322^{B}	10.4^{C}	0.0106^{Aa}	87.3^{aCD}	347^{B}	90.5^{AB}
Faba bean	whole grain	335	360	14.4	0.071	96.7	344	94.3
	crushed grain	325	367	14.8	0.062	89.9	359	79.3
	whole +crushed	330^{A}	364^{A}	14.6^{A}	0.066^{B}	93.3^{B}	352^{B}	86.8^{AbB}
Maize	whole grain	324	348	12.2	0.099	97.6	331	90.6
	crushed grain	327	350	13.4	0.092	96.7	338	100
	whole +crushed	326^{Ab}	349^{A}	12.8^{B}	0.096^{AbC}	97.2^{A}	335^{B}	95.5^{aA}
Barley	whole grain	310	327	8.68	0.138	81.5	398	77.3
	crushed grain	370	386	9.59	0.128	90.1	395	91.7
	whole +crushed	340^{Aa}	357^{A}	9.13^{D}	0.133^{D}	85.8^{bD}	397^{A}	84.5^{B}
Main effect								
	Grain	***	***	***	***	***	***	***
	Treatment	n.s.	n.s.	*	n.s.	n.s.	n.s.	n.s.
Interaction								
	gran x treat	***	***	n.s.	n.s.	***	***	n.s.
	Var. er	204	398	0.73	$1 \cdot 10^{-1}$	1.85	437	48.19

OMCV, cumulative gas production related to incubated OM at 120 h; A, potential gas production; B, time at which A/2 is produced; Rmax, maximum fermentation rate; Y, yield related to the degraded OM; VFA, volatile fatty acids related to degraded OM.
*a,b,c: P<0.05; A,B: P<0.001ns, not significant;***: p < 0.001.*

Flaking increased the potential gas production (A) for soybean and barley (P< 0.01), and slowed their fermentation kinetics (increased t/1/2 and decreased RM values). By contrast, maize flakes showed lower gas production (P<0.05) and the fermentation process was faster (P<0.01), compared to whole grain.

Organic matter degradability decreased after the treatment for soybean (P<0.01), while it increased for barley (P<0.01) and, not significantly, for maize. The slower fermentative process for barley flakes (increased t/1/2and decreased RM values) agrees with the results of other authors on treated cereals for both *in vitro* (Bittante et al., 1989) and *in situ* experiments (Arieli et al., 1995).

Comparing each grain with the respective flakes, OM degradability did not strictly follow the trend of potential gas production, confirming that not all the degraded OM is fermented. This result holds mainly for soybean flakes, which had a higher A (P<0.01) and a lower dOM (P<0.01) compared to the whole grain.

Such decreased OM degradability was probably due to the thermal treatment causing a decrease in the rapidly degradable fraction (Sarubbi, 1999).

Overall, the two cereals showed higher gas and VFA production, as well as a faster fermentative process.

Crushing showed slighter effects compared to flaking, and also the results of the grain x treatment interaction were less important.

The different technological condition of the two treatments (less drastic for crushing) may well have contributed to this trend. In particular, crushed barley presents, with respect to the whole grain, the same behaviour as the flakes (higher A and dOM, P<0.001; higher *t*/1/2 and lower RM, not significant).

Interestingly, there was a decrease in faba bean degradability due to the treatment (P<0.01), which agrees with the results of a contemporaneous *in situ* trial (Sarubbi, 1999). Besides, faba bean also showed a lower VFA concentration, which was probably also caused by the considerable decrease in their protein degradability (P<0.001) as found *in situ* (Sarubbi, 1999). As usually observed *in vivo* in young bulls fed a cereal-rich diet, pH values were always quite low.

In the third study Calabrò *et al.* (2009) evaluated the fermentation characteristics of different cultivars of grain legumes using IVGPT.

Three grain legumes were tested: lupine (Lupinus spp.) (Lublanc, Luteur, Multitalia), faba bean (*Vicia faba* L.) (Chiaro di Torre Lama, Irena, Lady, ProtHABAT69, Scuro di Torre Lama, Sicania) and peas (Pisum sativum) (Alembo, Alliance, Attika, Corallo, Iceberg, Ideal, Spirale).

The fermentation characteristics are reported in table 6.

The values of pH ranged between 6.35 and 6.72, indicating a normal pattern of fermentation, and were consistent to the crude protein content.

As regards faba bean, the cultivar "Scuro di Torre Lama" showed significantly (P<0.01) lower values of dOM and OMCV than the other 5 cultivars.

In the case of lupine the cultivar "Lublanc" had lower (P<0.01) OMCV than the other 2 cultivars and for peas the cultivar "Spirale" produced less gas and showed a faster kinetics than the other 6 cultivars.

As expected, the OM degradability resulted very high in any case. However, comparing the pools of the grain legumes, dOM was in each case lower than that of soybean meal. OMCV was significantly (P<0.01) higher for pea than faba bean (330 vs. 316 ml/g, P <0.05) and lupine (330 vs. 258 ml/g, P <0.01). Gas production of peas was always higher than that of lupine, faba bean and also soybean meal according to the results of Buccioni et al. (2007) who studied the *in vitro* fermentation of soybean meal, faba bean and pea, and found in the latter the best balance between energy and nitrogen inputs.

Table 6. Fermentation characteristics of the different grain legume cultivars and soybean meal.

Cultivar	pH	dOM	OMCV	Yield	A	B	tmax	Rmax
		%	*ml/g*	*ml/g*	*ml/g*	*h*	*h*	*ml/h*
				Faba bean				
Irene	6.46	92.9	370	397	328	22.9	12.42	9.14
Lady	6.35	93.3	354	363	333	24.5	15.67	9.26
Scuro di Torre Lama	6.49	87.8	308	351	269	22.0	15.39	8.84
Chiaro di Torre Lama	6.41	91.8	348	379	310	23.1	13.20	8.69
ProtHABAT69	6.47	93.8	359	383	303	20.2	12.49	10.02
Sicania	6.40	92.9	324	349	299	21.0	12.83	9.71
MSD	0.135	3.19	60.1	49.3	60.1	3.90	3.65	1.89
				Lupine				
Lublanc	6.63	93.4	256	279	283	26.1	10.25	6.73
Luteur	6.69	92.4	275	298	309	25.6	5.58	7.83
Multitalia	6.72	91.2	273	297	303	27.0	8.48	7.08
MSD	0.219	5.40	26.6	91.5	13.9	45.8	15.6	4.74
				Peas				
Alembo	6.57	99.0	406	410	361	20.6	12.52	11.73
Alliance	6.49	99.3	397	396	358	20.1	11.99	11.72
Attika	6.57	98.4	397	404	360	20.5	11.82	11.46
Corallo	6.53	98.9	393	394	365	22.3	11.42	10.38
Iceberg	6.55	98.8	381	385	347	21.0	12.45	10.86
Ideal	6.58	97.0	371	383	336	20.7	13.24	11.06
Spirale	6.58	98.8	344	343	310	17.1	10.66	12.14
MSD	0.188	2.81	52.5	53.5	3.68	74.1	5.22	2.47
Faba bean[1]	6.52Ab	90.9b	368B	405A	321a	21.1ab	12.4	10.0Ab
Lupine[1]	6.64B	91.8ab	284C	309B	293b	24.4a	9.03	7.42B
Peas[1]	6.60a	95.1a	394A	413A	336a	18.2b	11.4	12.6Aa
Soybean meal[2]	6.73	96.5	295	306	323	18.7	6.01	10.67
MSE	0.001	2.31	56.2	30.5	101	4.40	2.99	0.56

MSD: Minimum Significant Differences for P<0.01. MSE: Mean Square Error.,In the column A,B,C: P<0.01; a,b,c: P<0.05. [1]Data obtained from the grain legumes incubated in vitro as a pool. [2]Data not statistically assessed.

The slower fermentation kinetics of faba bean may be due to the content in polyphenols while that of lupine may be caused by the very low starch content (INRA, 1988).

From the data obtained, the authors concluded that the tested grain legumes show only few differences compared to soybean meal (higher dOM and lower OMCV), consequently they may be considered in replacing, totally or partially, soybean.

The reported results are of particular importance as well as highlight the differences in dietary and nutritional characteristics of soybean and legume grains, provide data on the effects of the treatments of these feedstuffs and the differences among the cultivars present on the European market.

With this information, the nutritionist may from time to time choose the most suitable protein source in order to satisfy animal requirements ensuring the simultaneous availability of nitrogen and energy for the bacteria present in the rumen with beneficial effects on livestock production and environmental impact.

4. EFFECT OF DIFFERENT PROTEIN SOURCES ON ANIMAL PERFORMANCE

Several researches have been carried out in order to compare the nutritional characteristics of soybean solvent extract and legume grains, as faba bean, peas and lupine, for ruminant feeding.

Di Francia et al. (2007) evaluated the effect of partial replacement of soybean cake with extruded peas in the diet of lactating buffalo cows on milk yield and quality over the first 100 days of lactation. Their results showed that peas could represent an attractive GMO free protein feed when approaching the problem of the choice of a protein source alternative to soybean in diet formulation for buffalo cows raised in organic farms.

On the other hand Morbidini et al (2005) investigating on the effect of two different fattening diets with different protein sources - soybean meal and flaked faba bean - on slaughtering performance and carcass quality in light Apennine and Italian Merino lambs, observed that the use of faba beans lightly depressed growth performance (Morbidini et al., 2004) and slaughtering weight, even if did not affect the carcass quality.

In order to make a contribution on this issue, our research group conducted a trial on the effect of protein source on growth performance and meat quality of Marchigiana young bulls (Cutrignelli et al 2008 a, b).

The trial was carried out on a farm situated at 700 m a.s.l. in Campania Region (Southern Italy), where 12 weaned young bulls (129 d of age) were equally divided into two groups. Each animal was placed in individual box up to the slaughtering weight (620 kg).

The groups were fed diets with the same protein and energy concentrations and the same forage/concentrate ratios (F/C), but differing in protein source: faba bean (*Vicia faba minor* L.) vs soybean meal (*Soja hispida*).

All the animals were regularly weighed until the body weight (BW) of 620 kg fixed in advance as slaughter weight, was reached. All animals were slaughtered in an authorized slaughterhouse according to EU legislation (EU Regulation EC No 882/2004).

Live animals and carcasses were weighed and measured according to ASPA (1991). After 9 days of refrigeration at 4 ± 1 °C, dissection of the carcasses was carried out.

Samples of Longissimus thoracis (LT), Semitendinosus (ST), Iliopsoas plus Psoas minor (IP) muscles, perirenal (PF) and subcutaneous (SF) adipose tissues were collected and rapidly transported, upon refrigeration temperature, to the laboratories for the chemical analysis in order to evaluate the rheological (water holding capacity) and nutritional (chemical composition fatty acid profile, cholesterol and hydrossiproline contents) characteristics of meat.

The protein source did not affect any infra vitam parameters except the body weight (BW) at 180 d of age (173 vs. 186 kg, for group FB and SB, respectively; $P<0.01$).

The difference was probably due to the higher non-protein nitrogen (NPN) concentration of the faba bean than the soybean meal (about 12 vs 1.3% of crude protein, respectively). Indeed, in the months immediately following the weaning it should be preferable to administer diets with higher rumen undegradable protein content because at this age the rumen is not yet perfectly functional and microbial protein synthesis is less efficient. Furthermore, for the same reason, the animals in this period are probably unable to neutralize possible anti-nutritional factors of faba beans.

As respects the influence of the replacement of soybean meal solvent extract with legumes seeds, the literature results are contrasting. Moss et al. (1997) found no significant effects on weight gain and feed intake when soybean meal was replaced on an iso-protein basis by lupin seeds in diets for growing bulls (BW from 182 to 243 kg); similar results are reported by Kwak and Kim (2001) on Korean native bulls (BW from 247 to 427 kg) utilising two different concentrations (15 and 30%) of flaked lupin. Instead, according to our results, Murphy and McNiven (1994) found significantly higher weight gain in growing steers (BW from 235.2 to 343.7 kg) fed soybean meal vs raw or roasted lupin although the differences were not significant in the finishing phase (final BW 503.4 kg).

No significant differences between the groups were found for carcass measurements. In each case the carcass measurements of this trial ranged into the interval indicated by Keane (2003) for European/North American breeds.

No differences were found in dressing out, organs and tare incidence on net weight. Protein source influenced neither body and carcasses conformations nor dressing out with the exception of the incidence of long bones showed a significant difference (6.2 vs 6.7 for faba and soybean, respectively; $P<0.05$).

The first quality meat cuts were acceptable in both groups (58.1 and 57.8%, for faba and soybean, respectively). Concerning the comparison between the two protein sources, only the incidence of long bones showed a significant difference (6.2 vs 6.7 for faba and soybean, respectively; $P<0.05$). It is important to underline that the sample cut measurements were contradictory and conflicted with the data obtained from total carcass dissection. The results of the sample cut dissection indicated a significant ($P<0.05$) difference between groups faba and soybean in meat incidence (69.1 vs 66.9, for faba and soybean, respectively); while no differences were found between groups at carcass dissection. Moreover, sample cut of soybean group showed in the meantime the smallest meat incidence and the highest LT area (88.4 vs. 84 for soybean and faba group, respectively).

The animals fed faba bean showed significantly higher water losses, measured with the compression method (WHC 7.6 vs. 5.7% for faba and soybean group, respectively; $P<0.01$). Our grilling loss data (31.7 vs. 28.3 % for faba and soybean group, respectively; $P>0.01$) were higher than those reported by Sami *et al.* (2004) in Simmental young bulls but they were in agreement with those reported by Pen *et al.* (2005) cooking ST samples of Holstein steers in an oven, on the contrary our drip loss data resulted higher than that reported by these authors.

The chemical composition of LT was not statistically different between groups. Meat from both groups showed a very low fat content (<3%) and higher protein concentrations than the Holstein steers (Pen *et al.*, 2005), confirming the high quality of the Marchigiana meat.

Hydroxyproline (60.0 vs. 62.6 mg $100g^{-1}$ of meat, in group faba and soybean, respectively) contents were not influenced by protein sources. Regarding the differences registered for this parameter among the texted muscles (LT: Longissimus thoracis; ST: Semitendinosus; IP: Iliopsoas plus Psoas minor) the IP samples, which correspond to the tenderloin, showed in all the groups significantly ($P<0.01$) lower hydroxyproline concentrations than the other two muscles; also between the LT and ST muscles the differences were statistically significant ($P<0.05$) being lower for the former.

Cholesterol content (56.3 vs.55.1 mg $100g^{-1}$ of meat, in group faba and soybean, respectively) was not influenced by protein sources according to the observations of Cutrignelli (2000) on Podolian young bulls and by Poli *et al.* (1996) on Chianina young bulls.

Considering the differences among the muscles, cholesterol values were significantly ($P<0.05$) lower for IP. Cifuni *et al.* (2004) found no differences in cholesterol contents among muscles, while Rusman *et al.* (2003) found significant differences. This contradiction is probably due to the different muscles analysed in each experiment. As theorised by Wheeler *et al.* (1987) the cholesterol content may be affected by the different physiological function of the muscles. In both groups, and especially for LT and ST muscles, the cholesterol contents were slightly higher than the value (less than 50 $mg100g^{-1}$ of muscle) indicated by the Protected Geographical Indication (PGI) of the "Vitellone Bianco dell'Appennino Centrale" (Council Regulation EEC No 2081/92; Floroni, 2002). Nevertheless, our results are very close to those reported for Italian meat breeds (Poli et al. 1996; Cifuni *et al.*, 2004) and lower than those from other breeds (Migdal *et al.*, 2004).

Regarding the fatty acids profile in both groups and in each analysed tissues (intramuscular, perirenal and subcutaneous adipose tissues) palmitic (C16:0), stearic (C18:0) and oleic (C18:1) acids were the most widely represented fatty acids. In particular, in intramuscular fat, the sum of oleic, stearic and palmitic acids represents over 50% of total fatty acids, according to the observations of Cifuni *et al.* (2004) and Migdal et al. (2004).

Comparing the data of groups faba and soybean, the only significant ($P<0.01$) difference was for stearic acid being higher for bulls fed faba bean than for those fed soybean meal solvent extract. However this result did not significantly affect the SFA concentration, or AI and TI indexes (table 7).

Protein source did not affected the fatty acids composition of the analysed adipose tissues. This is probably due to the very low concentration of phospholipids in subcutaneous and perirenal fat tissue.

Comparing the fatty acid composition of the three adipose tissues the following differences were noted:

- the erucic (C22:1), docosahexaenoic (DHA, C22:6, ω-3) eicosapentaenoic (C 20:5, ω-3) and docosapentaenoic (C 22:5, ω-3) acids, were present only in intramuscular fat, due to the lower level of intramuscular fat compared to the two adipose tissues. Lower fat content is associated with fewer and smaller adipocytes containing fewer triglycerides, accompanied by a relative increase in the proportion of phospholipids in total lipids and an increased PUFA content (Scollan *et al.*, 2006);

- the proportion of total SFA was minimal in LT and maximal in SF (mean values 46.56 *vs* 54.50 and 59.35% of total fatty acids; in LT, PF and SF, respectively, Figure 1); the single saturated fatty acids amounts also showed a similar trend. These data are partly in contrast with those of Kim *et al.* (2004) who in Hanwoo cattle found a higher SFA proportion in perirenal than in subcutaneous fat. Even if the differences in the fatty acids profile between internal and external fat deposits have not been fully elucidated, Eguinoa *et al.* (2003) suggested that the lipogenic enzyme activities per cell could be influenced by several factors such as adipocite size, nutrient supply, etc.;
- the proportion of total polyunsaturated fatty acids (PUFA) was higher in LT and lower in the perirenal adipose tissue (mean values 22.23 *vs* 11.36 *vs* 11.02% of total fatty acids, respectively in LT, SF and PF, Figure 1). This difference may be primarily ascribed to the lower concentration of ω-6 PUFA and to the absence DHA in SF and PF adipose tissues;
- the atherogenic and thrombogenic indexes were considerably higher in SF and PF tissues than in LT. This is due to the high SFA concentration of SF and PF adipose tissues as well as to the high PUFA concentration of LT.

The LT fatty acid profile was similar to those reported by Raes *et al.* (2003) for Longissimus lumborum of Belgian Blue and Limousin beef (SFA: 338 and 506 mg $100g^{-1}$ edible portion; MUFA: 323 and 554 mg $100g^{-1}$ edible portion; PUFA: 195 and 195 mg $100g^{-1}$ edible portion; in Belgian Blue and Limousin bulls, respectively) and different from values found in Irish and Argentine beef. The latter showed significantly higher total intramuscular fatty acid content compared to the former, probably due to genetic selection.

This observation confirms that of Carnovale and Nicoli (2000) who concluded that Italian meat showed a favourable intramuscular fatty acid composition with high PUFA content. Muscle with a high percentage of unsaturated fatty acids (UFA) generally scored higher in taste panel evaluation (Westerling and Hedrick, 1979) and food with high UFA, especially PUFA, is good for human health (Rusman *et al.*, 2003). The ω-6/ω-3 ratio was higher than the value (less than 3) reported by Scollan *et al.* (2006) but lower than that registered by Warren et al. (2003) for steers fed corn silage and concentrates (8.9).

It has to be underlined that the fatty acid profile o the food arouse high interest in human medicine due to their influence on the functionality of the cardio-circulatory apparatus. A number of epidemiological researches put in evidence that diets with high content of saturated fatty acids (SFA) were associated with high levels of serum cholesterol (especially of low density lipoprotein, LDL) which appear important in atheroma. Successively, Ulbricht and Southgate (1991) reported that:

1) diets high in C18:0, stearic acid, do not raise serum cholesterol;
2) short-chain SFA (C 10 and below) likewise do not raise blood cholesterol, so the putative atherogenic SFA are C12:0 (lauric), C14:0 (myristic) and C16:0 (palmitic). Myristic acid is the most atherogenic, with about four times the cholestrol-raising potential of palmitic acid.

PUFA are considered protective factors: ω-6 fatty acids show mainly anti-atherogenic activity while ω-3 fatty acids have anti-thrombogenic activity. More recently, high prominence is attributed to the role developed by the MUFA, and particularly by the oleic acid that, reducing the oxidation of the cholesterol LDL, may slow the progression of atherosclerosis.

As was pointed out above, the P/S (poliunsaturated/saturated) ratio is not suitable measure of the atherogenicity or thrombogenicity of a diet or foods. Currently they are expressed as follows:

Index of atherogenicity: C12:0 + (4 x C14:0) + C16:0/ ω-3 + ω-6 + MUFA

Index of thrombogenicity: C14:0 + C16:0 +C18:0/(0,5 x C18:1) + (0,5 other MUFA) + 0,5(ω-6) + 3(ω-3) + (ω-3/ ω-6).

The AI of the meat in this trial was particularly interesting, rather lower than the data reported by Ulbricht and Southgate (1991) for raw minced beef and than those reported by Badiani et al. (2002) for cooked beef (0.72 and 0.77, respectively); our data were similar to those of Poli et al. (1996) on Chianina young bulls (AI: 0.58).

Table 7. Fatty acid profile of Longissimus thoracis muscle (mg $100g^{-1}$ of edible part).

	FB	SB	Significance
SFA	365.1 ± 39.1	330.6 ± 26.9	Ns
MUFA	244.3 ± 39.5	271.7 ± 29.4	Ns
PUFA	186.6 ± 20.5	201.9 ± 15.7	Ns
ω-6	174.2 ± 19.9	188.3 ± 12.9	Ns
ω-3	12.39 ± 1.5	13.57 ± 0.8	Ns
AI	0.54 ± 0.04	0.49 ± 0.07	Ns
TI	1.46 ± 0.12	1.20 ± 0.09	Ns

FB: faba bean; SB: soybean meal solvent extract;AI: atherogenic index; TI thrombogenic index. ns: not significant.

However, the index of thrombogenicity in this trial was higher than the findings of the above-cited authors (1.27 and 1.30 for Ulbricht and Southgate, 1991 and Poli *et al.*, 1996, respectively); only the TI (1.77) reported by Badiani *et al.* (2002) was similar to our data.

These results show that the faba bean could be used as an alternative protein source to soybean meal solvent extract as it did not affect the growth rate (body weight, daily weight gain and biological efficiency of growth) or the feed conversion indexes during the whole experimental period, and offers decided agronomical, economical and healthy advantages. Nevertheless, in the first period after weaning the faba bean reduced the growth rate, probably due to the higher concentrations of NPN and anti-nutritional factors. It might be useful in this period to use this protein source associated with other richer in rume undegradable.

Our results contribute to show that both protein sources (soybean meal and faba bean) could be utilised in the diet for young Marchigiana bulls ensuring the high quality of the meat obtained by this breed.

Although the meat of group fed faba bean had significantly higher concentrations of stearic acid compared to the level found in soybean group, neither the atherogenic and thrombogenic indexes, nor the cholesterol content were influenced.

As regards the effect of protein source on organolepitc characteristics, our results on hydroxyproline content and water holding capacity were conflicting. While the meat of the group fed soybean meal solvent extract showed a potential low tenderness (higher level of hydroxiproline) the water holding capacity measured by compression was lower for the group receiving faba bean.

From our results it is also possible to formulate a favourable assessment of the nutritional characteristics of the of meat Marchigiana young bulls. Indeed, the cholesterol values were very close to those indicated by the PGI of the "Vitellone Bianco dell'Appennino Centrale" and lower than those found in other breeds. Moreover, the fatty acids profile of LT confirms that the meat of the Italian breed specialised in meat production has higher unsaturated fatty acids concentration and lower saturated fatty acids levels, which in turn ensures medium-low atherogenic and thrombogenic indexes.

5. CONCLUSIONS

The protein source largely used in ruminant nutrition (soybean, soybean meal solvent extract and grain legumes) are characterised by several differences in chemical composition, in particular as concerns crude protein, ether extract and carbohydrates concentrations. Also the technological treatments as flaking and crushing could affect these parameters. In addition, mainly concerning the grain legumes, the contents of different nutrients is affected by the cultivar.

In order to better estimate the nutritive value of protein source, the *in vitro* gas production technique seems to be the most useful methods, as it allow to study also the kinetics of degradation in the rumen. The knowledge of the latter phenomenon is particularly useful, since the nutritionist may choose to treat a particular protein source in order to synchronize the availability of nitrogen and energy for rumen bacteria.

From our *in vitro* data, faba bean, lupin and peas, even if from different cultivars, may be considered in replacing, totally or partially, soybean as only few differences were found among these feeds.

In addition, our in vivo studies concerning the infra vitam and post mortem performances of young bulls fed either faba bean or soybean showed:

- no influence of protein source on growth rate (body weight, daily weight gain an biological efficiency of growth) and feed conversion indexes;
- no influence of protein source on the nutritional characteristics of meat. Indeed, even if the meat of group fed faba bean had significantly higher concentrations of stearic acid compared to the level found in soybean group, neither the atherogenic and thrombogenic indexes, nor the cholesterol content were influenced;
- conflicting results on hydroxyproline content and on water holding capacity. Indeed, while the meat of the group fed soybean showed a potential low tenderness (higher level of hydroxiproline), the water holding capacity measured by compression was lower for the group receiving faba bean.

From these results, it can be concluded that the use of grain legumes as a protein source in ruminant diets could be used as alternative to soybean meal. However, mainly in the intensive livestock system the soybean meal solvent extract represents the protein source par excellence for the high protein content (in particular undegradable fraction), the absence of anti-nutritional factors and the extremely favourable biological value of its proteins. Not insignificant, then is the economic assessment, soybean is, especially in the areas where the ruminants breeding has been particularly developed, the protein source with the quality/price ratio more favourable.

By contrast, it has to be underlined that biotech herbicide tolerant soybean continued to be the principal biotech crop in 2010, occupying 73.3 million hectares or 50% of global biotech area. Farm animals are currently fed soybean and soybean meal developed from genetic transformation. Europe is strongly dependent upon the American continent for its protein requirements amounting up to 90 to 95%.

Thus, the possible risk connected to genetically modified organisms use in animal breeding has led to the reconsideration of animal production processes with special reference to the use of alternative protein sources (e.g. faba beans, dried peas, lupine seeds, chickpeas) able to replace soybean. These legumes have agronomic importance because they improve soil fertility and reduce nitrogenous dressing, with positive effects on environmental pollution. Moreover, they need a limited initial investment for their modest requirements of chemical and energetic inputs and their short culture cycle.

REFERENCES

1. A.S.P.A. (1991). Metodologie relative alla macellazione degli animali di interesse zootecnico ed alla valutazione e dissezione della loro carcasse. Ed. ISMEA, Roma.
2. Akpinar, N., Akpinar, M.A. & Türkoglu, S. (2001). Total lipid content and fatty acid composition of the seeds of some Vicia L. species. Food Chem., 74, pp. 449–453.
3. Arieli, A., Bruckental, I., Kedar, & I., Sklan, D. (1995). In sacco disappearance Of starch nitrogen and fat in processed grains. Anim. Feed sci. Techn., 51, pp. 287-295.
4. Bach Knudsen, K.E. (1997). Carbohydrate and lignin contents of plant materials used in animal feeding. Anim. Feed Sci. Technol., 67, pp. 319–338.
5. Badgley, C., Moghtader, J., Quintero, E., Zakem, E., Chappell, M.J., Avilés-Vázquez, K., Samulon, A. & Perfecto, I., (2007). Organic agriculture and the global food supply. Renew. Agric. Food Sys., 22, pp. 86–108.
6. Badiani, A., Stipa, S., Bitossi, F., Gatta, P.P., Vignola, G. & Chizzolini, R., (2002). Lipid composition, retention and oxidation in fresh and completely trimmed beef muscles as affected by common culinary practices. Meat Sci., 60, pp. 169-186.
7. Bézard, J., Blond, J.P., Bernard, A. & Clouet, P., (1994). The metabolism and availability of essential fatty acids in animal and human tissues. Reprod. Nutr. Dev., 34, pp. 539– 568
8. Bittante, G., Chies, L., Fasone, V., Sinatra, M.C. & Gallo, L. (1989). Degradabilità e digeribilita in vitro della sostanza organica di cereali fioccati e/o macinati. Zoot. Nutr. Anim., 15, pp. 613-626.
9. Brand, T.S., Brandt, D.A. & Cruywagen, C.W. (2004). Chemical composition, true metabolisable energy content and amino acid availability of grain legumes for poultry. South African Journal of Animal Science, 34, pp. 116-122.
10. Boccioni, A., Minieri, S., Petacchi, F. & Antongiovanni, M. (2007). Il favino (Vicia faba minor) e il pisello proteico (Pisum sativum) in sostituzione della farina di estrazione di soia nell'alimentazione zootecnica. Valutazione comparativa delle caratteristiche nutrizionali con metodi di laboratorio. Arsia, Toscana.
11. Bulson, H.A.J., Snaydon, R.W. & Stopes, C.E., (1997). Effects of plant density on intercropped wheat and field beans in an organic farming system. J. Agric. Sci., 128, pp. 59–71.
12. Cabrera, C., Lloris, F., Giménez, R., Olalla, M. & López, M.C. (2003). Mineral content in legumes and nuts: contribution to the Spanish dietary intake. The Science of total environment, 308, pp. 1-14.
13. Calabrò et al. (2010) www.progettobufale.it/pubblicazionenew.html

14. Calabrò, S., Marchiello, M. & Piccolo, V. (2001a). In vitro fermentation of some cereals and legumes grains. Proc. XIII Congr. Naz. ASPA , Firenze (Italy), pp. 153-155.
15. Calabró, S., Piccolo, V., Bovera, F. & Sarubbi, F. (2001b). In vitro fermentation kinetics of crushed and flaked grains. Proc. VI World Buffalo Congr., Maracaibo (Venezuela), pp. 478-484.
16. Calabrò, S., Tudisco, R., Balestrieri, A., Piccolo, G., Infascelli, F. & Cutrignelli, M.I. (2009). Fermentation characteristics of different grain cultivars with the in vitro gas production technique. Ita, J. of Anim. Sci., 8, pp. 280-282.
17. Canibe, N., Bach Knudsen, K.E., (2002). Degradation and physicochemical changes of barley and pea fibre along the gastrointestinal tract of pigs. J. Sci. Food Agric., 82, pp. 27–39.
18. Carnovale, E. & Nicoli, S., (2000). Changes in fatty acid composition in beef in Italy. J. Food Compos. Anal., 13, pp. 505-510.
19. Chalk, P.M., Smith, C.J., Hamilton, S.D. & Hopmans, P., (1993). Characterization of the N benefit of a grain legume (Lupinus angustifolius L.) to a cereal (Hordeumvulgare L.) by an in situ 15N isotope dilution technique. Biol. Fertil. Soils, 15, pp. 39–44.
20. Chaubility, U.B., Gupta, R., Bird, S.H. & Mudgal, V.D., (1991). In sacco dry matter disappearance of different protein cakes in the rumen of male buffalo and their solubility in water. Proceedings of the 3rd World Buffalo Congress, Paestum (Italy), p. 165, Abstract.
21. Chumpawadee, S., Chantiratikul, A. & Chantiratikul, P. (2007). Chemical composition and nutritional evaluation of protein feeds for ruminants using an in vitro gas production technique. Journal of Agricultural Technology pp. 191-202
22. Cifuni, G.F., Napolitano F., Riviezzi, A.M., Braghieri, A. & Girolami, A., (2004). Fatty acid profile, cholesterol content and tenderness of meat from Podolian young bulls. Meat Sci., 67, pp. 289-297.
23. Colombini, S.(2004). Resa proteica di alcune varietà di pisello, fava e lupino nell'Italia settentrionale. Informatore agrario, 60, pp. 73-75
24. Cone, J.W., van Gelder, A.H., Visscher, G.J.W. & Oudshoorn, L. (1996). Influence of rumen fluid and substrate concentration on fermentation kinetics measured with a fully automated time related gas production apparatus. Animal Feed Science Technology, 61, pp. 113-128.
25. Council Regulation (EC) No 834/2007 on organic production and labelling of organic products and repealing Regulation (EEC) No 2092/91 L 189/1- L 189/23.
26. Council Regulation (EEC) No 2092/91 of 24 June 1991 on organic production of agricultural products and indications referring thereto on agricultural products and foodstuffs L 198 , 22/07/1991 P. 0001 – 0015
27. Cutrignelli, M.I., (2000). Caratteristiche del taglio campione di mezzene di vitelloni Podolici sottoposti a differenti sistemi di allevamento. Proc. Nat. Congr: Parliamo di. on Breeding in the third Millennium, Fossano (CN), Italy, pp. 127-134
28. Cutrignelli, M.I., Piccolo, G., Bovera, F., Calabrò, S., D'Urso, S., Tudisco, R. & Infascelli, F. (2008a) Effects of two protein sources and energy level of diet on the performance of young Marchigiana bulls. 1. Infra vitam performance and carcass quality. Ital.J.Anim.Sci., 7, pp. 259-270
29. Cutrignelli, M.I., Calabrò, S., Bovera, F., Tudisco, R., D'Urso, S., Marchiello, M., Piccolo, V. & Infascelli, F. (2008b) Effects of two protein sources and energy level of diet on theperformance of young Marchigiana bulls. 2. Meat qualità. Ital.J.Anim.Sci., 7, pp. 271-285.
30. Degussa, (2006). The amino acid composition of feedstuffs. 5th rev. ed., Degussa AG, Feed Additives, Hanau, Germany.
31. Di Francia, A., De Rosa, G., Masucci, F., Romano, R., Borriello, I. & Grassi, C. (2007). Effect of Pisum sativum as protein supplement on buffalo milk production. Ital.J.Anim.Sci., 6, pp. 472-475
32. Díaz –Ambrona, C.H. & Mínguez, M.I. (2001). Cereal-legume rotations in a Mediterranean environment: biomass and yield production. Field Crops Research, 70, pp. 139- 51

33. DLG, (1999). Kleiner Helfer für die Berechnung von Futterrationen. Wiederkäuer und Schweine. DLG-Verlag, 10. Auflage, Frankfurt am Main, Germany.
34. Erbas, M., Certel, M. & Uslu, M.K., (2005). Some chemical properties of white lupin seeds (Lupinus albus L.). Food Chem., 89, pp. 341–345.
35. European Commission directive, 999/2001. Regulation (EC) No. 999/2001 of the European Parliament and of the Council of 22 May 2001 laying down rules for the prevention, control and eradication of certain transmissible spongiform encephalopathies. Official Journal of the European Communities L 147/1–L 147/40.
36. European Commission, 1992. Council Regulation of 14 July 1992 concerning the protection nof geographical indications and designations of origin for agricultural products and foodstuffs, 92/2081/EC. Official Journal, L 208, 24/07/1992, pp. 1-8.
37. European Commission, (2004). European Parliament and Council Regulation of the of 29 April 2004 concerning the official controls performed to ensure the verification of compliance with feed and food law, animal health and animal welfare rules, 882/2004/EC. Official Journal, L 165, 30/04/04, pp 1-141.
38. Evans, J., McNeil', A.M., Unkovich, M.J., Fettell, N.A. & Heenan, D.P., (2001). Net nitrogen balances for cool-season grain legume crops and contributions to wheat nitrogen uptake: a review. Aust. J. Exp. Agric., 41, pp. 347–359.
39. Floroni, A. (2002). I.G.P. "Vitellone Bianco dell'Appennino Centrale". L'Allevatore, 58, pp. 12- 13.
40. Froud-Williams, R.J., (1988). Changes in weed flora with different tillage and agronomic management systems. In: Altieri, M.A. & Liebman, M. (Eds.), Weed Management in Agroecosystems: Ecological Approaches. CRC, Boca Raton, Florida, pp. 213–236.
41. Gatel, F., (1994). Protein quality of legume seeds for non-ruminant animals: a literature review. Anim. Feed Sci. Technol., 45, pp. 317–348.
42. Getachew, G., Robinson, P.H., DePeters, E.J. & Taylor, S.J. (2004). Relationships between chemical composition, dry matter degradation and in vitro gas production of several ruminant feeds. Animal Feed Science Technology, 111, pp. 57-71.
43. Gill, C., (1997). World feed panorama. High cost of feedstuffs: global impact, response. Feed International, 18, pp. 6–16.
44. Green, C.J. & Blackmer, A.M., (1995). Residue decomposition effects on nitrogen availability to corn following corn or soybean. Soil Sci. Soc. Am. J., 59, pp. 1065–1070.
45. Groot, J.C.J., Cone, J.W., Williams, B.A. & Debersaques, F.M.A., (1996). Multiphasic analysis of gas production kinetics for in vitro fermentation of ruminant feedstuff. Anim. Feed Sci. Tech., 64, pp. 77-89.
46. Herrero, M., I. Murray, R.H. Fawcett & Dent, J.B. (1996). Prediction of in vitro gas production and chemical composition of kikuyu grass by near-infrared reflectance spectroscopy. Animal Feed Science Technology, 60, pp. 51-67.
47. Herridge, D.F., Macellos, H., Felton, W.L., Turner, G.L. & Peoples, M.B., (1995). Chickpea increases soil-N fertility in cereal systems through nitrate sparing and N2 fixation. Soil Biol. Biochem., 27, pp. 545–551.
48. Huisman, J. & Tolman, G.H., (2001). Antinutritional factors in the plant proteins of diets for non-ruminants. In: Garnsworthy, P.C. & Wiseman, J. (Eds.), Recent Developments in Pig Nutrition, vol. 3. Nottingham University Press, Nottingham, pp. 261–322.
49. Infascelli, F., Di Lella, T. & Piccolo, V., (1995). Dry matter, organic matter and crude protein degradability of high protein feeds in buffaloes and sheep. Zoot. Nutr. Anim., 21, pp. 89–94.
50. INRA (1988). Alimentation des bovins, ovins & caprins. Paris.
51. Jezierny, D., Mosenthin, R. & Bauer, E. (2010). The use of grain legumes as a protein source in pig nutrition: A review. Animal Feed Science and Technology, 157, pp. 111–128.
52. Jezierny, D., (2009). In vivo and in vitro studies with growing pigs on standardised ileal namino acid digestibilities in grain legumes. Ph.D. Thesis. University of Hohenheim, Stuttgart, Cuvillier Verlag Göttingen, Germany.
53. Jezierny, D., Mosenthin, R., Eklund, M. & Rademacher, M., (2007). Determination of standardized ileal digestibilities of crude protein and amino acids in legume seeds for

growing pigs. Proceedings of the 16th International Science Symposium on Nutrition of Domestic Animals, Radenci, pp. 198–203.
54. Keane, M.G., (2003). Beef production from Holstein-Friesian bulls and steers of New Zeland and European/American descent, and Belgian Blue x Holstein-Fresians slaughtered at two weights. Livest. Prod. Sci., 84, pp. 207-218.
55. Kim, C.M., Kim, J.H., Chung, T.Y. & Park, K.K., (2004). Effect of flaxseed diets an fattening response of Hanwoo cattle: 2 fatty acid composition of serum and adipose tissues. Asian-Aust. J. Anim. Sci., 9, pp. 1246-1254.
56. Lallès, J.P. & Jansman, A.J.M., (1998). Recent progress in the understanding of the mode of action and effects of antinutritional factors from legume seeds in non-ruminant farm animals. In: Jansman, A.J.M., Hill, G.D., Liebman, M. & Dyck, E.,. Crop rotation and intercropping strategies for weed management. Ecol. Appl. 3, pp. 92– 122.
57. Liebman, M. & Janke, R., (1990). Sustainable weed management practices. In: Francis, C.A., Flora, C.B. & King, L.D. (Eds.), Sustainable Agriculture in Temperate Zones. John Wiley&Sons, New York, pp. 111–143.
58. López-Bellido, F.J., López-Bellido, L. & López-Bellido, R.J., (2005). Competition, growth and yield of faba bean (Vicia faba L.). Eur. J. Agron., 23, pp. 359–378.
59. Medsen, J., T. Hvelplund & Weisbjerg, M.R. (1997). Appropriate method for the evaluation of tropical feeds for ruminants. Animal Feed Science Technology, 69, pp. 53-66.
60. Morbidini, L., Rossetti, E., Cozza, F. & Pauselli, M. (2005). Different protein source (soybean or faba bean) in postweaning diets for Apennine and Sopravissana (Italian Merino) light lamb: slaughtering performances. Ital.J.Anim.Sci., 4, pp. 360-362.
61. Morbidini, L., Pauselli, M., Rossetti, E. & Cozza, F., (2004). Diete post-svezzamento con differente fonte proteica (favino o soja) per la produzione di agnelli leggeri di razza Appenninica e Sopravissana: rilievi in vivo. Proc. Congr. "Parliamo di ... nuove normative in campo zootecnico", Cuneo, Italy, pp. 113-122.
62. Moss, A.R., Givens, D.I., Grundy, H.F. & Wheeler, K.P.A., (1997). The nutritive value for ruminants of lupin seeds from determinate plants and their replacement of soybean nmeal in diets for young growing cattle. Anim. Feed Sci. Tech., 68, pp. 11-23.
63. Mould, F.L., Kliem, K.E. Morgan, R. & Mauricio, R.M, (2005). In vitro microbial inoculum: A review of its function and properties Anim. Feed Sci. Technol., 123, pp. 31-50
64. Murphy, S.R. & McNiven, M.A., (1994). Raw or roasted lupin supplementation of grass nsilage diets for beef steers Anim. Feed Sci. Tech., 46, pp. 23-35.
65. NRC (2001). Nutrient Requirements of Dairy Cattle: Seventh Revised Edition, 2001. National Academy Press, Washington, D.C.
66. NRC, (1998). Nutrient requirements of swine. 10th ed., National Academy Press, Washington, D.C.
67. Pen, B., Oyabu, T., Hidaka, S. & Hidari, H., (2005). Effect of potato by-products based silage on growth performance, carcass characteristics and fatty acid composition of carcass fats in Holstein steers. Asian-Aust. J. Anim. Sci., 4, pp. 490-496.
68. Poli, B.M, Giorgetti, A., Bozzi, R., Funghi, R., Balò, F.& Lucifero, M., (1996). Quantity and quality of lipid fractions for human nutrition in Chianina muscles as influenced by age and nutritive level. In: G. Enne and G.F. Greppi (eds.) Food & Health: role of animal products, Proc. 31st Int. Symp. of Società Italiana per il Progresso della Zootecnia held in Milano, Italy, Elsevier publ., Paris, France, pp. 199-204.
69. Raes, K., Balcaen, A, Dirinck, P., De Winne, A., Claeys, E., Demeyer, D. & De Smet, S., (2003). Meat quality, fatty acid composition and flavour analysis in Belgian retail beef. Meat Sci., 65, pp. 1237-1246.
70. Rubio, L.A. & Brenes, A., (1995). Utilization de leguminosas-grano en nutricion animal: nproblemas y perspectives. Proc. XI Curso de Especializacion FEDNA, Barcelona, Spain. www.cirval.asso.fr.
71. Rusman, S. & Setiyono, A.S., (2003). Characteristics of Biceps femoris and Longissimus nthoracis muscles of five cattle breeds grown in a feedlot system. J. Anim. Sci., 74, pp. 59-65.
72. Salgado, P., Freire, J.P.B., Mourato, M., Cabral, F., Toullec, R. & Lallès, J.P., (2002a). Comparative effects of different legume protein sources in weaned piglets: nutrient

digestibility, intestinal morphology and digestive enzymes. Livest. Prod. Sci., 74, pp. 191–202.

73. Salgado, P., Martins, J.M., Carvalho, F., Abreu, M., Freire, J.P.B., Toullec, R., Lallès, J.P. & Bento, O., (2002b). Component digestibility of lupin (Lupinus angustifolius) and pea (Pisum sativum) seeds and effects on the small intestine and body organs in anastomosed and intact growing pigs. Anim. Feed Sci. Technol., 98, pp. 187–201
74. Sami, A.S., Augustini, C. & Schwarz, F.J., (2004). Effect of feeding intensity and time on feed on performance, carcass characteristics and meat quality of Simmental bulls. Meat Sci., 67, pp. 195-201.
75. Sankara Rao, D.S. &, Deosthale & Y.G. (2006). Mineral Composition of Four Indian Food Legumes. Journal of Food Science, 46, pp. 1962–1963
76. Sarubbi, F. (1999). Studi sull'impiego di proteasi tioliche per la previsione della degradabilita ruminale delle proteine e sule modificazioni dell' attivita ruminale indotte nei bufali da razioni arricchite con Aspergillus oryzae. PhD Thesis, Univ. Napoli, 1-23. 1999.
77. Schofield, P. (2000). Gas production methods. In "Farm Animal Metabolism and Nutriion", J.P.F. D'Mello, ed. CAB International, Wallingford, Oxon, U.K, pp. 209-232.
78. Scollan, N., Hocquette, JF., Neuerbetg, K., Dannenberger, D, Richardson, I. & Moloney, A., (2006). Innovations in beef production system that enhance the nutritional and health value of beef lipids and their relationship with meat quality. Meat Sci., 74, npp. 179-33.
79. Singh, R.J., Chung, G.H. & Nelson, R.L., (2007). Landmark research in legumes. Genome, 50, pp. 525–537.
80. Sniffen, C. J., J. D. O'Connor, P. J. Van Soest, D. G. Fox & J. B. Russell. (1992). A net carbohydrate and protein system for evaluating cattle diets: 11. Carbohydrate and protein availability. J. Anim. Sci., 70, pp. 3562.
81. Sprent, J.I. & Thomas, R.J., (1984). Nitrogen nutrition of seedling grain legumes: some taxonomic, morphological and physiological constraints. Plant Cell Environ., 7, pp. 637–645.
82. Theander, O., Westerlund, E., Åman, P. & Graham, H., (1989). Plant cell walls and nmonogastric diets. Anim. Feed Sci. Technol., 23, pp. 205–225
83. Theodorou, M.K. (1994.) A new laboratory procedure for determining the fermentation nKinetics of ruminants feeds. Ciencia e Investigcion Agraria, 20, pp. 332-334
84. Trostle, R., (2008). Global agricultural supply and demand: Factors contributing to the recent increase in food commodity prices. A Report from the Economic Research Service. United States Department of Agriculture Economic Research Service, Washington, DC
85. Ulbricht, T.L.V. & Southgate, D.A.T., (1991). Coronary heart disease: seven dietary factors. Lancet, 338, pp. 985-992.
86. Warren, HE., Enser, M., Richardson, I., Wood, JD.& Scollan, N.D., (2003). Effect of breed and diet on total lipid and selected shelf-life parameters in beef muscle. Proc Nat. Congr. BSAS, York, UK. pp. 43.
87. Westerling, D.B. & Hedrick, H.B., (1979). Fatty acid composition of bovine lipids as influenced by diet, sex and anatomical location and relationship to sensory characteristics. J. Anim. Sci., 48, pp. 1343-1348.
88. Wheeler, T.L., Davis, G.W., Stoeker, B.J.& Hatmon, C.J., (1987). Cholesterol concentration of Longissimus muscle, subcutaneous fat and serum of two beef cattle breed type. J. Anim. Sci., 65, pp.1531-1537.
89. Wood, J.D., Richardson, R.I., Nute, G.R., Fisher, A.V., Campo, M.M., Kasapidou, E., Sheard, P.R. & Enser, M., (2003). Effects of fatty acids on meat quality: a review. Meat Sci., 66, pp. 21–32.
90. Zahran, H.H., (1999). Rhizobium-legume symbiosis and nitrogen fixation under severe conditions and in an arid climate. Microbiol. Mol. Biol. Rev., 63, pp. 968–989.

CHAPTER 3

Mannan Oligosaccharides in Nursery Pig Nutrition and Their Potential Mode of Action

Veronika Halas [1,*] and Imre Nochta [2]

[1]*Department of Animal Nutrition, Faculty of Animal Science, Kaposvár University, P.O. Box 16, H-7400 Kaposvár, Hungary*

[2]*Provimi, P.O. Box 1, H-8112 Zichyújfalu, Hungary*

SIMPLE SUMMARY

The aim of the paper is to provide a review of mannan oligosaccharide products in relation to their growth promoting effect and mode of action. Mannan oligosaccharide products maintain intestinal integrity and the digestive and absorptive function of the gut in the post-weaning period in pigs and enhance disease resistance by promoting antigen presentation. We find that dietary supplementation has growth promoting effects in pigs kept in a poor hygienic environment, while the positive effect of MOS is not observed in healthy pig herds with high hygienic standards.

ABSTRACT

Mannan oligosaccharides (MOSs) are often referred to as one of the potential alternatives for antimicrobial growth promoters. The aim of the paper is to provide a review of mannan oligosaccharide products in relation to their growth promoting effect and mode of action based on the latest publications. We discuss the dietary impact of MOSs on (1) microbial changes, (2) morphological changes of gut tissue and digestibility of nutrients, and (3) immune response of pigs after weaning. Dietary MOSs maintain the intestinal integrity and the digestive and absorptive function of the gut in the post-weaning period. Recent results suggest that MOS enhances the disease resistance in swine by promoting antigen presentation facilitating thereby the shift from an innate to an adaptive immune response. Accordingly, dietary MOS supplementation has a potential growth promoting effect in pigs kept in a poor hygienic environment, while the positive effect of MOS is not observed in healthy pig herds with high hygienic standards that are able to maintain a high growth rate after weaning.

Keywords: nursery pigs; mannan oligosaccharides; gut integrity; immune response; growth performance

1. INTRODUCTION

Mannan oligosaccharides (MOSs) are often referred to as a potential alternative for antibiotic growth promoters. Although they are non-digestible oligosaccharides, the mode of action of MOSs differs from other prebiotics. By definition, prebiotics are non-digestible components of feed that stimulate the growth and/or activity of beneficial bacteria in the digestive system and promote the gut and general health of the host [1]. MOSs have recently been assigned to nutricines based on the reasoning that they are not direct nutrients either for intestinal microbiota or for the host, but potentially have a positive effect on the health and performance of farm animals [2].

The first experience with mannan products was investigating their potential to adhere to the mannose specific lectin on the surface of *E. coli* [3]. Results of *in vitro* studies suggested that dietary MOS could reduce the colonization of pathogenic bacteria in the gut, which was indeed confirmed later in animal, particularly poultry trials [4,5]. Dietary MOS supplementation became prevalent in the 90s as a growth promoter in broiler and turkey feeding [6,7] and to a lesser extent in pig feeding [8]. Relevant data show that mannan products can be efficiently used in two critical periods of swine production, *i.e.*, in piglets during the nursery period and in sows during late gestation and lactation. Numerous publications reported that supplementation of the sow diet with dietary MOS (2 g/kg or 5 g/day/sow) in the last 2–3 weeks of gestation and during lactation improved the growth rate of piglets [9,10,11]. The data on weaned pigs are less consistent in this respect. The mode of action of mannan-containing products has been investigated for approximately 20 years and the following underlying mechanisms have been proved: mannans potentially affect (i) the intestinal micro biota, (ii) the morphology of gut tissue and thus the digestibility of nutrients, (iii) the immune response of farm animals; and (iv) the supposed toxin-binding ability of mannan containing yeast cell derivates. This latter, however, is attributed principally to the β-glucan content [12], therefore in the present paper this property is not discussed. Based on the effects listed above, it is suggested that dietary MOS is able to support gut recovery of piglets after weaning.

Excellent reviews have been published on the dietary effect of MOSs [12,13,14]; however, a few details of the underlying mechanisms remained unexplained. Since some recent results have filled gaps in our knowledge, the aim of the present paper is to provide an overview on MOS products in relation to their growth promoting effect and mode of action based on the latest publications.

2. POST-WEANING CHANGES WITHIN THE GASTRO INTESTINAL TRACT OF PIGS

Weaning is a stress for the piglets that is usually associated with a dramatic feed refusal. The immediate post-weaning anorexia results in the alteration of gut integrity leading to different physiological changes in the gastro intestinal tract (GIT), such as morphological and functional changes, shift in microbiota population, and increased production of inflammatory cytokines (e.g., reviewed by Dong and Pluske [15]). Even a short period of starving or malnutrition (2–3 days) results in villus atrophy [16] that weakens the absorption capacity of the intestine by reducing the surface area of the gut wall. It is also well known that the brush border enzyme (amino-peptidase N, dipeptyl-peptidase-4 [16]; sucrase [17]) and pancreatic enzyme activity (trypsin, chymotrypsin, amylase, lipase [18]) decline drastically right after weaning. The temporal changes induced in the gut after weaning can be divided into an acute phase of 5–7 days and an adaptive phase of 9–10 days [19]. The changes in the acute phase definitely result in a poorer digestibility of nutrients. The higher rate of undigested nutrients, and particularly of protein, may lead to undesirable processes in the hindgut. The fermentation of N-containing compounds in the digesta yields ammonia that increases the gut pH and supports pathogenic bacteria. Since the microbiota in young piglets is unstable, the ability of the gut flora to block the colonization of harmful species is inadequate. Moreover, neither the innate nor the acquired immune defense of a 4-week old pig can adequately respond to a pathogen challenge. Due to the low antibody concentration of sow milk at late lactation, the level of maternal antibodies is also low in the gut and blood of piglets. This is concurrent with insufficient own antibody production at 4 weeks of age, which only begins to increase at 6–7 weeks of age.

These changes often result in post-weaning diarrhea and a drastic depression in growth rate. In the past, antibiotic growth promoters efficiently prevented the complex post-weaning symptoms. Nowadays all feed additives that promote gut health, sustain eubiosis in the intestine or boost the immune defense of the pigs can be effectively used in nursery feeds. MOSs are suggested to be one of the promising alternatives for antibiotic growth promoters.

3. SOURCE AND CHEMICAL TRAITS OF MANNAN PRODUCTS

MOSs are non-digestible carbohydrates that are composed of mannose blocks and can be found in the yeast cell wall in complex formation. The composition of the yeast cell is determined by the species, the growth phase and the environmental factors of fermentation. The cell wall is approximately 25–30% of the dry weight of the cell. *Saccharomyces cerevisiae* is a well-known yeast in the bakery and brewery industries and its derivate is used exclusively as MOS product in animal nutrition. The cell wall of *S. cerevisiae* contains both mannan-proteins and β-glucans. The main constituents of the outer cell wall are mannan polymers with α(1-6) and α(1-2) bonds or to a lesser extent α(1-3) bounded side chains (see more details in [12]). The enzymes of the host or of the intestinal bacteria are unable to cleave these bonds and thus MOS has no direct nutritive value, but it has been shown to be able to maintain the gut health. It can be concluded from the relevant literature that although mannan-containing feed additives are almost exclusively derivates of *S. cerevisiae*, due to the processing and production technology, their chemical composition and therefore biological efficiency might be different [13].

4. MODE OF ACTION OF DIETARY MOS

4.1. The Effect of Dietary MOS on the Intestinal Microbiota

The development of the beneficial microbiota and sustainment of eubiosis play a crucial role in the defense mechanisms and gut health. There is increasing evidence suggesting that, unlike in young pigs, the composition of the GIT microflora in a healthy adult host remains remarkably stable [20]. The steady state can, however, promptly change at immune suppression or during infection. In general, supporting the growth of *Bifidobacteria* and *Lactobacilli* in the hindgut is supposed to have a positive impact on the host; however, in stress situations, such as weaning, the number of beneficial bacteria often declines, while the number of harmful species, like *E. coli* and *Salmonella,* increases. *In vitro* studies show that in the presence of mannan products the enteric pathogens attach to the mannan compounds in the gut lumen instead of the epithelia, which reduces their colonization. Results of relevant studies suggest that dietary MOS supplementation can reduce the number of harmful bacteria in the hind gut if the pathogen exposure was high, such as in post-infection [21,22,23,24]. A small number of other trials, however, failed to prove that the number of (facultative) pathogens was indeed affected by dietary MOS [25,26]. Singboottra [27] concluded that subject to their chemical structure (bonds and proportion of mannose) the efficiency of mannan products might differ in regard to their potential to moderate the number of *E. coli* and *Salmonella typhimurium*.

The microbiota exists in a dynamic state; increasing the number of any bacterial specie may result in the decrease of another specie. Some studies show that the reduction in the number of pathogenic bacteria in response to dietary MOS supplementation was indeed associated with an increase of beneficial flora, particularly lactobacilli [21,23,28], but this finding is not consistent throughout the literature [24,29,30]. Sims *et al.* [5] found in turkey that dietary MOS supported the growth of *Enterococcus*. These bacteria produce not only short chain fatty acids but also bacteriocin and enterocin and thus enhance the competition and development of beneficial flora [31]. Other results also confirm the positive effect of MOS, since it is reported to reduce the ammonia concentration in the gut [32].

Literature data show that dietary MOS supplementation can efficiently reduce the number of pathogens post-infection; however, it is unable to modify consistently the quantity of harmful species under adequate hygienic conditions. The shift in population of beneficial bacteria is not consistent in the different studies, therefore it can be concluded that although dietary MOS may support the maintenance of eubiosis, it probably has no real prebiotic effect.

4.2. The Effect of Dietary MOS on Gut Morphology

The first studies proving that MOS supplementation has a significant impact on gut morphology were conducted with broilers [33]. Results showed that dietary mannan products increase the villous height: crypt depth ratio in young broilers [33] and turkey [14] and also in weaned piglets [34,35]. In a recent study with nursery pigs, Poeikhampha and Bunchasak [36] found that 3 g MOS/kg in the diet resulted in increased crypt depth in the jejunum. An increased villous height: crypt depth ratio is generally associated with a bigger absorptive surface; this ratio is, however, usually reduced during the initial post-weaning period [37]. There are several hypotheses on the beneficial effect of MOS on intestinal morphology, but not all of them were proven in swine. As discussed earlier, the reduction in the *Enterobacteria* population [24] and/or increase in beneficial flora [21,23,28] enhance the short chain fatty acid production in the intestine, which positively affects the recovery of the epithelia. A number of studies reported that dietary MOS supplementation increases the lactic acid and/or volatile fatty acid production in the hind gut [5,28,36]. The microflora as a whole has a trophic effect on the epithelium, which can lead to a faster turnover rate [20]. In particular, butyric acid has anti-inflammatory property, alleviating the hypersensitive reaction of the gut wall associated with the post-weaning period. In turkey, the increased production of the mucus gel layer [14], and in pigs, a prompt recovery of the intestinal mucosal cells [38], was promoted by MOS. Moreover, enhanced gut maturation was reported in broilers [33,39].

The post-weaning morphological changes in the gut might be alleviated if the pig feed contains MOSs. It seems, however, that the structural changes of the epithelial cells are associated with functional changes of the gut tissue. Kim *et al.* [40] found slightly better apparent ileal amino acid digestibility for valine, isoleucine, leucine, lysine and arginine when the piglet diet was supplemented with 1 g MOS product/kg of feed, but the differences were statistically not significant. Nochta *et al.* [41] reported that supplementation of a mannan containing feed additive significantly improved the apparent ileal digestibility of nutrients, particularly that of indispensable amino acids (Lys, Met, M+C, Thr), Ca and P (Table 1). When the diet was supplemented with MOS at the rate of 2 g/kg, the apparent ileal digestibility of nutrients was similar to that in the treatment containing an antibiotic growth promoter. However, further increase of MOS (4 g/kg) did not further improve the ileal digestibility data. In addition to a lessened erosion of the absorptive surface, the better digestibility might be explained at least partly by the lower pH attributed to the more active fermentation by hind gut bacteria. Beneficial microflora produces short chain fatty acids that reduce the pH of the ileal digesta, which can thus result in higher protein hydrolysis and improved protein and amino acid, as well as Ca and P digestibility [41]. The relatively big improvement in apparent threonine digestibility generally indicates a decreased endogenous threonine excretion.

Table 1. The effect of dietary mannan oligosaccharide (MOS) supplementation on the apparent ileal digestibility of nutrients in weaned pigs (%) [41].

	TREATMENTS*					
	M0	M1	M2	M4	AB	RMSE
Dry matter	76.4 [b]	77.3 [ab]	79.2 [a]	77.9 [ab]	77.2 [ab]	1.9
Crude protein	72.5 [c]	73.6 [bc]	77.5 [a]	76.3 [ac]	77.5 [a]	2.0
Crude fat	93.9 [a]	93.0 [a]	93.8 [a]	93.8 [a]	94.1 [a]	0.8
N-free extract	79.6	80.7	81.2	80.9	79.8	2.3
Calcium	47.1 [b]	55.5 [a]	54.2 [ab]	53.9 [ab]	50.2 [ab]	5.1
Phosphorus	55.1 [b]	62.8 [a]	64.7 [a]	65.2 [a]	61.0 [a]	3.2
Lysine	74.1 [b]	74.5 [b]	79.3 [a]	76.1 [b]	79.5 [a]	1.7
Methionine	84.4 [c]	89.5 [a]	88.0 [ab]	86.9 [b]	82.8 [c]	1.4
Cystine	54.0 [c]	74.8 [a]	72.2 [a]	65.0 [b]	74.1 [a]	3.1
Met+Cys	72.7 [d]	83.8 [a]	81.9 [ab]	78.5 [c]	79.4 [bc]	2.0
Threonine	59.1 [c]	70.4 [ab]	71.4 [a]	66.6 [b]	68.2 [ab]	2.9

* M0: negative control without antibiotic or MOS; M1: supplementation of 1 g AgriMos/kg feed; M2: supplementation of 2 g AgriMos/kg; M4: supplementation of 4 g AgriMos/kg feed; AB: antibiotic growth promoter containing feed with 0.2 g Maxus-200 (40 ppm avilamycin supplementation)/kg feed; RMSE: Root mean square error.

[a,b,c,d] common letters within rows indicate no differences at $P < 0.05$.

Therefore, the data shown in Table 1 suggest that the endogenous protein loss can be reduced by MOS supplementation likely due to a faster recovery of the intestinal mucosal cells [38]. Due to the high threonine content of gut cell and mucus protein, the ileal threonine excretion can be twice as high as the lysine or methionine+cystine excretion [42]. The higher threonine digestibility might be associated with a lower turnover of the gut wall layer and less endogenous N losses. However, there are no data available in the literature that would report a reduced endogenous protein and/or threonine loss when dietary MOS is fed.

In conclusion, dietary mannan products can improve the gut morphology in the post-weaning period, which in turn has a positive impact on nutrient supply and on the first line of defense in nursery pigs.

4.3. The Effect of Dietary MOS on the Immune Response of Weaned Pigs

Dietary MOS has both an indirect and a direct effect on the immune response of farm animals. According to the fact that the microbiota can modulate the local immunity of the host, the MOS induced shift in gut flora may result in the changes of certain immune variables.

More evidence has been reported that addition of dietary mannan products directly enhance the immune competence of pigs, particularly that of sows and weaned pigs. Numerous studies with rats, dogs and chickens show that dietary MOS enhances the secretory IgA in different segments of the intestinal mucosa [43,44,45]. The higher mucosa IgA production is likely attributable to an activation of the local immune defense through the mannose binding receptors located in the gut surface. Davies *et al.* [46] reported that 21 days of feeding 2 g of phosphorylated mannan per kg diet altered the T lymphocyte repertoire of jejunal lamina propria. The local immune function initiates the systemic immune response of the host, which is frequently reported in response to MOS supplementation.

Recent results of the present authors show that the non-specific cellular immune variables are modulated by dietary MOS, and the level of supplementation is determining in this respect (Table 2). Nochta *et al.* [47] found that a lower dose of MOS (1 g/kg) increased the responsiveness to lymphocyte stimulation test (LST) with non-specific mitogens (pokeweed mitogen: PWM, Concavalin A: ConA and phytohaemagglutinin: PHA) in weaned pigs; bigger doses (2 or 4 g/kg), however, had no influence or even impaired it. Although Davis *et al.* [48,49] did not prove that MOS increases the responsiveness of LST with PWM or PHA mitogens, their results with regard to the ratio of CD4+/CD8+ lymphocytes could also indicate that MOS supplementation enhances the establishment of a mature T cell repertoire within the gastrointestinal tract of 3-weeks old weaned pigs. Moreover, in the study of Davies *et al.* [46] the percentage of neutrophils tended to increase and percentage of lymphocytes significantly increased in the peripheral blood when piglets were fed 2 g of mannan product/kg feed. The authors supposed that the alteration in systemic immune function was possibly an indirect response to changes that were occurring in the gastrointestinal immunity [46].

Table 2. Effect of dietary MOS supplementation on non-specific and specific cellular immune response of weaned pigs (lymphocyte stimulation index) [47].

	TREATMENTS*							
Time (Wks)	M0	M1	M2	M4	AB	NI	RMSE	*P*-value **
Con A								
1	5.08	5.10	5.13	5.10	5.08	-	0.054	0.19
2	5.21	5.21	5.20	5.20	5.17	-	0.122	0.96
5	5.21 ^{b}	5.37 ^{a}	5.27 ^{ab}	5.24 ^{ab}	5.20 ^{b}	-	0.105	0.02
PHA								
1	3.73	3.72	3.76	3.69	3.71	-	0.071	0.45
2	3.81	3.86	3.81	3.78	3.80	-	0.146	0.84
5	3.81 ^{ab}	3.98 ^{a}	3.86 ^{ab}	3.78 ^{b}	3.86 ^{ab}	-	0.134	0.02
PWM								
1 †	2.89	2.91	2.90	2.90	2.90	-	0.070	0.96
4	3.06	3.17	3.09	3.12	3.07	-	0.125	0.34
5	3.06 ^{a}	3.27 ^{b}	3.18 ^{ab}	3.13 ^{ab}	3.10 ^{ab}	-	0.148	0.048
Auj								
1	0.88	1.01	1.01	1.00	1.01	1.01	0.135	0.56
2 †	1.28	1.34	1.24	1.28	1.19	1.02	0.274	0.10
3	1.57 ^{ab}	1.95 ^{a}	1.55 ^{ab}	1.56 ^{ab}	1.40 ^{bc}	1.04 ^{c}	0.339	0.0001
4 †	2.62 ^{a}	2.82 ^{a}	2.61 ^{a}	2.50 ^{a}	2.53 ^{a}	1.03 ^{b}	0.736	0.0001
5	2.94 ^{a}	3.15 ^{a}	3.23 ^{a}	3.11 ^{a}	2.78 ^{a}	1.04 ^{b}	0.686	0.0001

RMSE: root mean square error; M0: negative control without antibiotic or MOS; M1: supplementation of 1 g AgriMos/kg feed; M2: supplementation of 2 g AgriMos/kg; M4: supplementation of 4 g AgriMos/kg feed; AB: antibiotic growth promoter containing feed with 0.2 g Maxus-200 (40 ppm avilamycin supplementation)/kg feed; NI: non-immunized group fed no supplementation.

^{a,b,c} common letters within rows indicate no differences at $P < 0.05$.

** There was no effect of interaction between treatment and replication.

† Replication effect was significant ($P < 0.05$).

Recent results suggest that it is worth supplementing dietary mannan in case of an immune challenge. Dietary MOS supplementation (1 g MOS/kg feed) enhanced the specific immune response, particularly the virus neutralization, 2 weeks after immunization with inactivated Aujeszky virus in a study carried out with weaned pigs [47]. In agreement with those results, Franklin *et al.* [50] reported that the specific immunity was enhanced by MOS supplementation as evidenced by greater serum rotavirus neutralization titers in cows supplemented with 10 g MOS daily compared with control cows. The humoral immune response is the result of the activation of the B-cells responsible for the production of antigen specific immunoglobulin. According to a serial study, White *et al.* [21] reported that the serum IgG concentration tended to increase in early weaned piglets fed with mannan containing yeast. Moreover, Shashidhara and Devegowda [51] found significantly higher specific antibody titers in the serum of broiler breeders after vaccination with bursal disease virus when the diet was supplemented with 0.5 MOS g/kg feed.

The earlier and stronger immune response is essential for the livestock in order to moderate or eliminate the antigen attack. There are two potential modes of action of dietary MOS as discussed by Newman [52] and Franklin *et al.* [50]. In the present paper those pathways are summarized briefly. The first underlying mechanism involves the presence of a collectin, a mannose-binding protein in the blood serum that may act as opsonin. Opsonins are molecules that make foreign antigens more susceptible to the action of the phagocytes. Mannose-binding proteins may bind to mannose-containing structures of a number of viruses and bacteria and trigger the complement cascade of the host immune system [52]. Nielsen *et al.* [53] reported increased presence of mannose-binding proteins in chickens during virus infection. It is likely that MOS stimulates the production of mannose-binding proteins resulting in improved phagocytosis, activation of the complement system, and enhancement of the immune response [50]. The other possible mode of action of MOS involves the natural production of antimannan antibodies [50]. The antimannan antibodies are directed against an oligosaccharide-based epitope of the viruses and microbes and these carbohydrate-specific antibodies may be produced during a normal immune response against the intestinal microflora. Dietary MOS probably enhances the production of these antimannan antibodies at the gut level, which in turn may enter the blood stream allowing for an enhanced response to a viral challenge [21,47].

In a recent study with 3-week old weaned pigs, Che *et al.* [54] found that MOS supplementation was associated with rapidly increased numbers of leukocytes, lymphocytes, and neutrophils at the early stage of porcine reproductive and respiratory syndrome (PRRS) virus infection (7 days post-infection). Results of the same study and an earlier trial conducted with mice suggest that dietary MOS has the potential to alleviate inflammation and has an anti-allergic effect, caused by the activation of cellular immunity [55]. Ozaki *et al.* [55] reported lower number of peritoneal acidophils in MOS-fed mice compared to that in control diet-fed ones, moreover MOS treatment reduced interleukin-10 production and tended to suppress ovoalbumin-specific IgE in serum. Che *et al.* [56] supported evidence on the potential of dietary MOS (2 g/kg) to alleviate the hypersensitive reaction post-infection. It has been proved that MOS down-regulates the expression of non-immune and immune genes in pig leukocytes, perhaps providing benefits by enhancing the pig's immune responses to an infection, while preventing over-stimulation of the immune system [56,57]. It also altered the expression of genes regulating pathogen detection in the peripheral blood molecular cells [56] and thus MOS may enhance disease resistance in pigs.

Based on the discussed data it can be concluded that in the most critical period, right after weaning, dietary MOS supplementation may boost the immune response of the pigs and save nutrients for growth in case of infection.

5. THE EFFECT OF DIETARY MOS ON THE GROWTH PERFORMANCE OF WEANED PIGS

Subject to the duration of starvation in the post-weaning period, the growth performance and the immune defense of the piglets suffer to a different extent. Some data show that the growth rate right after weaning has a significant impact on the pig performance in the growing and fattening phases [58]. Although the pig has the ability for compensation, the bigger the stress the less the chance is for recovery. This is certainly true for intensive genotypes. Therefore, any feeding strategy or feed supplement that alleviates the reduction in growth rate post weaning may have a positive effect on the efficiency of pork production.

The real growth promoter effect of MOS in nursery pigs is apparently inconsistent. Some studies report no benefits [25,26,59], while others found an improved rate of daily gain and/or feed efficiency in weaning pigs [24,48,60]. The results appear to be better in the case of younger pigs, especially in a challenged environment [8,61]. Based on the above-mentioned mode of action, the effect of MOS supplementation on piglet performance is affected by different factors, principally by weaning age, health status, duration of feeding and the amount of MOS addition. If the piglets are weaned at 4 weeks of age—as it is commonly done in the European Union, or even earlier, as in the United States—the intestine is less matured and the weaning is associated with a higher rate of gut epithelial atrophy than at a later age. Thus in the case of epithelial atrophy the positive effect of dietary mannan on gut wall repair can be demonstrated. Accordingly, the pig performance, e.g., growth rate and/or feed conversion were reported to be significantly enhanced in studies where the gut cell wall atrophy was reduced by dietary MOS supplementation [34,35,36].

In a meta-analysis involving studies with 54 comparisons, Miguel *et al.* [61] found that dietary MOS supplementation improves the growth rate mainly in the first 2 weeks of nursery, which is in agreement with the results showing that dietary mannans are associated with enhanced gut integrity in the post-weaning period.

The same report suggests that pigs that received 1 or 2 g of the used mannan product (Bio-Mos) per kg of feed had a more pronounced growth response than pigs fed diets with supplementation of 3 or 4 g/kg [61]. This fact is supported by our findings as well; *i.e.*, in contrast to 4 g/kg, 1 or 2 g of a MOS product (AgriMos) per kg of feed increased the ileal digestibility of nutrients [41], as discussed earlier in this paper.

Recent studies conducted on any growth enhancer, in general, are in accordance with the earlier antibiotic growth promoter studies: under higher environmental pressure the treatment results in a better improvement of the zootechnical parameters (such as the average daily gain, feed conversion), however, the growth promoting effect is low if the animal performance is close to the genetic potential [62].

It has to be noted, however, that the activated immune system requires extra nutrient supply, therefore less amino acids and energy is available for growth in case of an immune challenge. Considering that dietary MOS supplementation has the potential to boost immune functions and defense mechanisms of the pig, the fact that the higher immune response is usually not associated with a lower growth rate is a benefit *per se*.

6. CONCLUSIONS

Based on the relevant literature it is likely that mannans help to maintain the intestinal integrity and the digestive and absorptive function of the gut post-weaning. Therefore the malabsorption syndrome associated with this period can be alleviated with dietary MOS supplementation. Recent results suggest that MOS enhances the disease resistance in swine by promoting antigen presentation, thus enhancing the shift from an innate to an adaptive immune response. Accordingly, dietary MOS supplementation may have a growth promoting effect in pigs kept in a poor hygienic environment, while the positive effect of MOS is not observed in healthy pig herds with high hygienic standards, that are able to maintain their high growth rate post-weaning.

In addition to the economic benefits resulting from the maintenance of gut health and support of the defense mechanisms, the use of medication and drugs can be reduced during the post-weaning period with the use of dietary MOS supplementation that enables healthier and safe, and therefore more desirable pork production.

ACKNOWLEDGMENTS

Veronika Halas is supported by the TÁMOP-4.2.1/B-10/1-2010-0002 program, which is gratefully acknowledged.

REFERENCES AND NOTES

1. Gibson, G.R.; Roberfroid, M.B. Dietary modulation of the human colonic microbiota: Introducing the concept of prebiotics. *J. Nutr.* **1995**, *125*, 1401–1412.
2. Adam, C.A. The role of nutricines in health and total nutrition. Available online: http://sydney.edu.au/vetscience/apss/documents/2000/APSS2000-adams-pp17-24.pdf (accessed on 10 May 2012).
3. Ofek, I.; Beechey, E.H. Mannose binding and epithelial cell adherence of *Escherichia coli*. *Infect. Immun.* **1978**, *22*, 247–253.
4. Spring, P.; Wenk, C.; Dawson, K.A.; Newman, K.E. The effects of dietary mannan-oligosaccharides on caecal parameters and the concentration of enteric bacteria of the caeca of salmonella-challenged broiler chicks. *Poult. Sci.***2000**, *79*, 205–211.
5. Sims, M.D.; Dawson, K.A.; Newman, K.E.; Spring, P.; Hoogell, D.M. Effects of dietary mannan oligosaccharide, bacitracin methylene disalicylate, or both on the live performance and intestinal microbiology of turkeys. *Poult. Sci.***2004**, *83*, 1148–1154.
6. Hooge, D.M. Meta-analysis of broiler pen trials evaluating dietary mannan oligosaccharide 1993–2003. *Int. J. Poult. Sci.* **2001**, *3*, 163–174.
7. Hooge, D.M. Turkey pen trials with dietary mannan oligosaccharide: Meta-analysis. *Int. J. Poult. Sci.* **2004**, *3*, 179–188.
8. Pettigrew, J.E. Bio-Mos Effects on Pig Performance: A Review. In Proceedings of Alltech's 16th Annual Symposium; Nottingham University Press: Nottingham, UK, 2000; pp. 31–45.
9. O'Quinn, P.R.; Funderburke, D.W.; Tibbets, G.W. Effects of dietary supplementation with mannan oligosaccharides on sow and litter performance in a commercial production system. *J. Anim. Sci.* **2001**, *79* (Suppl.1), 212.
10. Newman, K.E.; Newman, M.C. Evaluation of mannan-oligosaccharide on the microflora and immunoglobulin status of sows and piglet performance. *J. Anim. Sci.* **2001**, *79* (Suppl.1), 189.
11. Pettigrew, J.E.; Miguel, J.C.; Carter, S. Bio-Mos In sow diet, performance response and economics. In Biotechnology in the Feed Industry: Proceedings of Alltech's 21th Annual Symposium; Lyons, T.P., Jacques, K.A., Eds.; Nottingham University Press: Nottingham, UK, 2005; pp. 213–220.
12. Kogan, G.; Kocher, A. Role of yeast cell wall polysaccharides in pig nutrition and health protection. *Livest. Sci.* **2007**, *109*, 161–165.
13. Spring, P. Mannan-oligosaccharides as an alternative to antibiotic use in Europe. *Zootechnica Int.* **1999**, *22*, 38–41.
14. Ferket, P.R. Use of oligosaccharides and gut modifiers as replacements for dietary antibiotics. In Proceedings of 63rd Minnesota Nutrition Conference, Eagan, MN, USA, 17–18 September 2002; pp. 169–182.
15. Dong, G.Z.; Pluske, J.R. The Low Feed Intake in Newly-weaned Pigs: Problems and Possible Solutions. *Asian-Aust. J. Anim. Sci.* **2007**, *20*, 440–452.
16. Montagne, L.; Boudry, G.; Favier, C.; Le Huerou-Luron, I.; Lalles, J.; Seve, B. Main intestinal markers associated with the changes in gut architecture and function in piglets after weaning. *Br. J. Nutr.* **2010**, *97*, 45–57.
17. Hampson, D.J.; Kidder, D.E. Influence of creep feeding and weaning on brush border enzyme activities in the piglet small intestine. *Res. Vet. Sci.* **1986**, *40*, 24–31.

18. Lindemann, M.D.; Cornelius, S.G.; El Kandelgy, S.M.; Moser, R.L.; Pettigrew, J.E. Effect of Age, Weaning and Diet on Digestive Enzyme Levels in the Piglet. *J. Anim. Sci.* **1987**, *62*, 1298–1307.
19. Burrin, D.; Stoll, B. Intestinal nutrient requirements in weanling pigs. In *The Weaner Pig: Concepts and Consequences*; Pluske, J.R., Verstegen, M.W.A., Le Dividich, H., Eds.; Wageningen Academic Publishers: Wageningen, The Netherlands, 2003; pp. 301–335.
20. Williams, B.A.; Verstegen, M.W.A.; Tamminga, S. Fermentation in the monogastric large intestine: Its relation to animal health. *Nutr. Res. Rev.* **2001**, *14*, 207–227.
21. White, L.A.; Newman, M.C.; Cromwell, G.L.; Lindemann, M.D. Brewers dried yeast as a source of mannan oligosaccharides for weanling pigs. *J. Anim. Sci.* **2002**, *80*, 2619–2628.
22. Burkey, E.T.; Dritz, S.S.; Nietfeld, B.J.; Johnson, J.B.; Minton, J.E. Effect of dietary mannanoligosaccharide and sodium chlorate on the growth performance, acute phase response, and bacterial shedding of weaned pigs challenged with *Salmonella enterica* serotype Typhimurium. *J. Anim. Sci.* **2004**, *82*, 397–404.
23. Liu, P.; Piao, X.S.; Kim, S.W.; Wang, L.; Shen, Y.B.; Lee, H.S.; Li, S.Y. Effects of chito-oligosaccharide supplementation on the growth performance, nutrient digestibility, intestinal morphology, and fecal shedding of *Escherichia coli* and *Lactobacillus* in weaning pigs. *J. Anim. Sci.* **2008**, *86*, 2609–2618.
24. Castillo, M.; Martín-Orúe, S.M.; Taylor-Pickard, J.A.; Pérez, J.F.; Gasa, J. Use of mannanoligosaccharides and zinc chelate as growth promoters and diarrhea preventative in weaning pigs: Effects on microbiota and gut function. *J. Anim. Sci.* **2008**, *86*, 94–101.
25. Hancock, J.D.; Jones, C.L.; Starkey, C.W. Mannanoligosaccharides in diets for nursery pigs. Available online: http://krex.k-state.edu/dspace/bitstream/2097/ 2116/1/ Mannanoligosaccharides%20in%20Diets%20for%20Nursery%20Pigs-%20Swine%20Day%202003.pdf (accessed on 15 May 2012).
26. Van der Peet-Schwering, C.M.C.; Jansman, A.J.M.; Smidt, H.; Yoon, I. Effect of yeast culture on performance, gut integrity, and blood cell composition of weanling pigs. *J. Anim. Sci.* **2007**, *85*, 3099–3109.
27. Singbroota, P. Agglutination of Type-1 Fimbrial Bacteria by Different Yeast Cell Wall Products. Ph.D. Thesis, North Carolina State University, Raleigh, NC, USA, 2005; pp. 41–63.
28. Rekiel, A.; Wiecek, J.; Bielecki, W.; Gajewska, J.; Cichowicz, M.; Kulisiewicz, J.; Batorska, M.; Roszkowski, T.; Beyga, K. Effect of addition of feed antibiotic flavomycin or prebiotic BIO-MOS on production results of fatteners, blood biochemical parameters, morphometric indices of intestine and composition of microflora. *Arch. Tierz. Dummerstorf.* **2007**, *50*, 172–180.
29. Campbell, A.J.; Gardiner, G.E.; Leonard, F.C.; Lynch, P.B.; Stanton, C.; Ross, R.P.; Lawlor, P.G. The effect of dietary supplementation of finishing pigs with organic acids or mannan oligosaccharide on coliform Lactobacillus and Bifidobacterium flora of intestinal contents and faeces. *Pig J.* **2006**, *57*, 22–30.
30. Bonos, E.; Christaki, E.; Abrahim, A.; Soultos, N.; Florou-Paneri, P. Effect of dietary supplementation of mannan oligosaccharides on hydrogen ion concentration of the digestive tract and microbial populations of the ceca of Japanese quail (*Coturnix japonica*). *Turk. J. Vet. Anim. Sci.* **2011**, *35*, 263–269.
31. Leroy, F.; Foulquie-Moreno, M.R.; De Vuyst, L. *Enterococcus faecium* RZS C5, an interesting bacteriocin producer to be used as a co-culture in food fermentation. *Int. J. Food Microbiol.* **2003**, *88*, 235–240.
32. Juskiewicz, J.; Zdunczyk, Z.; Jankowski, J. Effect of adding mannanoligosaccharide to the diet on the performance, weight of digestive tract segments, and cecal digesta parameters in young turkeys. *J. Anim. Feed Sci.* **2003**, *12*, 133–142.
33. Iji, P.A.; Saki, A.A.; Tivey, D.R. Intestinal structure and function of broiler chickens on diets supplemented with a mannan oligosaccharide. *J. Sci. Food Agr.* **2001**, *81*, 1186–1192.
34. Zhao, J.; Harper, A.F.; Estienne, M.J.; Webb, K.E.; McElroy, A.P., Jr.; Denbow, D.M. Growth performance and intestinal morphology responses in early weaned pigs to supplementation of antibiotic-free diets with an organic copper complex and spray-dried

plasma protein in sanitary and nonsanitary environments. *J. Anim. Sci.* **2007**, *85*, 1302–1310.
35. Shen, Y.B.; Piao, X.S.; Kim, S.W.; Wang, L.; Liu, P.; Yoon, I.; Zhen, Y.G. Effects of yeast culture supplementation on growth performance, intestinal health and immune response of nursery pigs. *J. Anim. Sci.* **2009**, *87*, 2614–2624.
36. Poeikhampha, T.; Bunchasak, C. Comparative Effects of Sodium Gluconate, Mannan Oligosaccharide and Potassium Diformate on Growth Performances and Small Intestinal Morphology of Nursery Pigs. *Asian-Aust. J. Anim. Sci.* **2011**, *24*, 844–850.
37. Cera, K.R.; Mahan, D.C.; Cross, R.F.; Reinhart, G.A.; Whitmoyer, R.E. Effect of age, weaning and postweaning diet on small intestinal growth and jejunal morphology in young swine. *J. Anim. Sci.* **1988**, *66*, 574–584.
38. Caine, W.R.; Sauer, C.W.; He, J. Prebiotics, probiotics and egg yolk antibodies: Novel alternatives to antibiotics for improving health of piglets and growing pigs. In Proceedings of 10th International Symposium on Animal Nutrition, Kaposvár, Hungary, 9 October 2001; pp. 33–53.
39. Yang, Y.; Iji, P.A.; Kocher, A.; Mikkelsen, L.L.; Choct, M. Effects of dietary mannanoligosaccharide on growth performance, nutrient digestibility and gut development of broilers given different cereal-based diets. *J. Anim. Physiol. Anim. Nutr.* **2008**, *92*, 650–659.
40. Kim, J.D.; Hyun, Y.; Sohn, K.S.; Woo, H.J.; Kim, T.J.; Han, I.K. Effects of immune stimulators on growth performance and immune response in pigs weaned at 21 days of age. *J. Anim. Sci.* **2000**, *9*, 333–346.
41. Nochta, I.; Halas, V.; Tossenberger, J.; Babinszky, L. Effect of different levels of mannan oligosaccharide supplementation on the apparent ileal digestibility of nutrients, N-balance and growth performance of weaned piglets. *J. Anim. Physiol. Anim. Nutr.* **2010**, *94*, 747–756.
42. Jansman, A.J.M.; Smink, W.; van Leeuwen, P.; Rademacher, M. Evaluation through literature data of the amount and amino acid composition of basal endogenous crude protein at the terminal ileum of pigs. *Anim. Feed Sci. Techn.* **2002**, *98*, 49–60.
43. O'Carra, R. An assessment of the potential of mannan oligosaccharides as immunostimulants. M.S. Thesis, National University of Ireland, Galway, Ireland, 1997.
44. Kudoch, K.; Shimizu, J.; Ishiyama, A.; Wada, M.; Takita, T.; Kanke, Y.; Innami, S. Secretion and excretion of immunoglobulin A to cecum and feces differ with type of indigestible saccharides. *J. Nut. Sci. Vitaminol.* **1999**, *45*, 173–181.
45. Swanson, K.S.; Grieshop, C.M.; Flickinger, E.A.; Bauer, L.L.; Healy, H.P.; Dawson, K.A.; Merchen, N.R.; Fahey, G.C., Jr. Supplemental fructooligosaccharides and mannanoligosaccharides influence immune function, ileal and total tract nutrient digestibilities, microbial populations and concentrations of protein catabolites in the large bowel of dogs. *J. Nutr.* **2002**, *132*, 980–989.
46. Davis, M.E.; Maxwell, C.V.; Erf, G.F.; Wistuba, T.J. Dietary supplementation with phosphorylated mannans improves growth response and modulates immune function of weanling pigs. *J. Anim. Sci.* **2004**, *82*, 1882–1891.
47. Nochta, I.; Tuboly, T.; Halas, V.; Babinszky, L. The effect of different levels of dietary mannan-oligosaccharide on specific cellular and humoral immune response in weaned piglets. *J. Anim. Physiol. Anim. Nutr.* **2009**, *93*, 496–504.
48. Davis, M.E.; Maxwell, C.V.; Brown, D.C.; de Rodas, B.Z.; Johnson, Z.B.; Kegley, E.B.; Hellwig, D.H.; Dvorak, R.A. Effect of dietary mannan oligosaccharides and(or) pharmacological additions of copper sulfate on growth performance and immunocompetence of weanling and growing/finishing pigs. *J. Anim. Sci.* **2002**, *80*, 2887–2894.
49. Davis, M.E.; Brown, D.C.; Maxwell, C.V.; Johnson, Z.B.; Kegley, E.B.; Dvorak, R.A. Effect of phosphorylated mannans and pharmacological additions of zinc oxide on growth and immunocompetence of weanling pigs. *J. Anim. Sci.* **2004**, *82*, 581–587.
50. Franklin, S.T.; Newman, M.C.; Newman, K.E.; Meek, K.I. Immune parameters of dairy cows fed mannan oligosaccharide and subsequent transfer of immunity to calves. *J. Dairy Sci.* **2005**, *88*, 766–775.

51. Shadhishara, R.G.; Devegowda, G. Effect of dietary mannan oligosaccharide on broiler breeder production traits and immunity. *Poult. Sci.* **2003**, *82*, 1319–1325.
52. Newman, K.E. Mannan-oligosaccharides: Natural polymers with significant impact on the gastrointestinal microflora and the immune system. In Biotechnology in the Feed Industry: Proceedings of Alltech's Tenth Annual Symposium; Lyons, T.P., Jacques, K.A., Eds.; Nottingham University Press: Nottingham, UK, 1994; pp. 167–174.
53. Nielsen, O.L.; Jensenius, J.C.; Jorgensen, P.H.; Laursen, S.B. Serum levels of chicken mannan-binding lectin (MBL) during virus infections; Indication that chicken MBL is an acute phase reactant. *Vet. Immunol. Immunopathol.* **1999**, *70*, 309–316.
54. Che, T.M.; Johnson, R.W.; Kelley, K.W.; Van Alstine, W.G.; Dawson, K.A.; Moran, C.A.; Pettigrew, J.E. Mannan oligosaccharide improves immune responses and growth efficiency of nursery pigs experimentally infected with porcine reproductive and respiratory syndrome virus. *J. Anim. Sci.* **2011**, *89*, 2592–2602.
55. Ozaki, K.; Fujii, S.; Hayashi, M. Effect of dietary mannan oligosaccharides on the immune system of ovoalbumine sensitized mice. *J. Health Sci.* **2007**, *53*, 766–770.
56. Che, T.M.; Johnson, R.W.; Kelley, K.W.; Van Alstine, W.G.; Dawson, K.A.; Moran, C.A.; Pettigrew, J.E. Mannan oligosaccharide modulates gene expression profile in pigs experimentally infected with porcine reproductive and respiratory syndrome virus. *J. Anim. Sci.* **2011**, *89*, 3016–3029.
57. Che, T.M.; Johnson, R.W.; Kelley, K.W.; Dawson, K.A.; Moran, C.A.; Pettigrew, J.E. Effects of mannan oligosaccharide on cytokine secretions by porcine alveolar macrophages and serum cytokine concentrations in nursery pigs. *J. Anim. Sci.* **2012**, *90*, 657–668.
58. Pollmann, S.D. Effect of nursery feeding programs in subsequent grower finisher pig performance. In Proceeding of the Fourth Western Nutrition Conference University of Alberta, Edmonton, AB, Canada, 1993; pp. 653–664.
59. White, L.A.; Newman, M.C.; Cromwell, G.I.; Lindemann, M.D. Efficacy of brewers dried yeast as a source of mannan-oligosaccharides without and with organic acids and carbadox on performance and intestinal bacterial population of weanling pigs. *J. Anim. Sci.* **2001**, *79* (Suppl. 2), 78.
60. Le Mieux, F.M.; Naranjo, V.D.; Binder, T.D.; Southern, L.L. Effect of Dried Brewers Yeast on Growth Performance of Nursing and Weanling Pigs. *Profess. Anim. Scient.* **2010**, *26*, 70–75.
61. Miguel, J.C.; Rodriguez-Zas, S.L.; Pettigrew, J.E. Efficacy of Bio-Mos for improving nursery pig performance. *J. Swine Health Prod.* **2004**, *12*, 296–307.
62. Halas, V.; Nochta, I.; Pásti, Z.; Szabó, C.; Tóthi, R.; Tossenberger, J.; Babinszky, L. Cellular Immune Response of Weaned Pigs Fed Diet Supplemented with an Essential Oil. *Agr. Consept. Scient.* **2011**, *76*, 279–282.

CHAPTER 4

Artemia, a New Source of Animal Protein Ingredient in Poultry Nutrition

A. Zarei

Department of Animal Science, College of Agriculture and Natural Resources, Islamic Azad University- Karaj Branch, Karaj, Iran

1. INTRODUCTION

In the nutritional behavior of single stomach animals, the origin of protein is important and its quality varies between different sources and animal origin is better than plant origin [7]. From the standpoint of salmonella contamination and due to high microbial potential, however, some of these proteins such as meat and bone meal ought to be used with caution. If heat treatment is not enough during fish meal processing, thiaminase will remain in fish meal and will cause harmful effects.Some researches indicated that thiamin will be reduced by thiaminase under special storage condition of ingredient before feeding animal [6,7]. Also, severe heating during meat and bone meal and fish meal processing to be confident that poultry by-product is safe, however some amino acids will probably degenerate and their bioavailability will decrease [5,7, 8,9].

Although, meat & bone and fish meal have been used in poultry feeding, exclusively, but artemia biomass is also one of the animal proteins with high nutritional value which can be used in aquaculture and animal nutrition [1,4].

The aim of this study is a survey of biology and characteristics of artemia and possibility of its usage in poultry nutrition as a protein source.

2. WHAT IS ARTEMIA?

Artemia or brine shrimp belongs to the animal kingdom, phylum of *arthropoda*, subphylum of *crustacean*,class of *branchiopoda*,order of *anostraca*,family of *artemidae* and genus of *artemia*. Linnaeus (1758) and Leach (1819) called it "*Cancer salinus*" and "*Artemia salina*", respectively. The latter name is because of the effect of salinity on morphological growth and development of artemia. Two species of artemia in Iran are: *Artemia urmiana* and *Artemia parthenogenetica*. The first is native of Urmia lake and the second was observed in 12 regions of artemia habitats in Iran.

Artemia spreads in tropical and sub tropical regions in saline environments of the world, and over 500 artemia regions are discovered around the globe. Nine species of artemia were recognized in these regions.

More than two million kilograms of dried cysts of artemia with 0.4mm diameter are transacted in world markets every year. It is used as an aquaculture feed for hatched naplius. Uniformity of cysts and embryos with diapose has made artemia a unique source of aquaculture feeds. Artemia cysts can spread by wind and migratory birds.

Artemia contains 40-60% crude protein (dry matter basis) [11].

3. MORPHOLOGY AND ECOLOGY

Morphologically, artemia has fragmented body with leave and wide shaped appearance. It's body consists of three compartments; head, thorax and abdomen, with total length of 8-10 mm and 10-12 mm in male and female respectively (Figure 1). Width of body is 4mm in both sexes. The exoskeleton of artemia is extremely thin (0.3-1μ) and flexible that called" chitin" and it is connected to muscle from inner surface.

The blood circulatory system of artemia is open.

This animal is euryhalin and can tolerate high concentration of salty habitat. There is a glandular organ in back of artemia neck that named;"salt gland" or "neck organ".This organ exudates extra salt from the body to environment. Salty organ extinct at maturity and then this function is performed by exopodits of legs.

Although there are some limitations in the living environment of artemia, like high temperature and high salinity and drought, this animal can tolerate these conditions by producing cysts and going to diapose until the condition become suitable, and then it will continue its living.

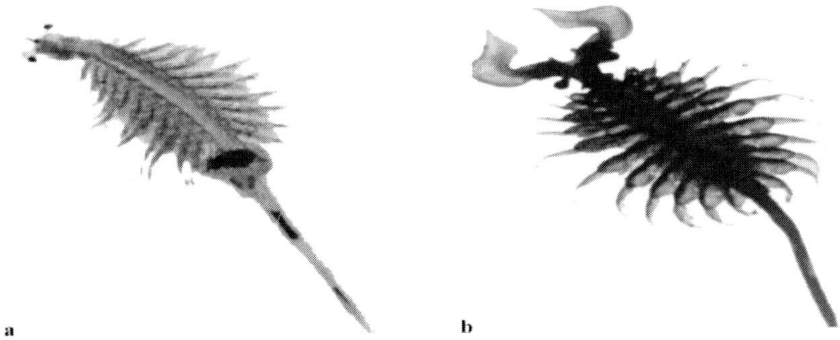

Figure 1. Artemia morphology a) female and b) male

Salinity and temperature are two important factors for growth and survival of artemia.

Artemia can tolerate salinity even more than 250 g/l and suitable range of temperature is 6-35°C.This crustacean has adapted itself with hard environmental conditions. In hypoxia, artemia increases the oxygen carrying capacity through the increase of the amount of hemoglobin. In this situation, body color turns red from original pale brown.

Physiological adaptation of artemia in high salinity is an effective defense method against predators using following mechanisms:

1) A powerful and effective osmoregulation system
2) Overcome high hypoxia in high salinity condition by higher pigmentation (inhalation pigments)
3) Production of embryos in diapose stage in cysts that can tolerate environmentally unfavorable condition.

4. NUTRITION

From nutritional standpoint, artemia is non- selective filter feeder, and eats algae, bacteria, protozoa and yeasts, as long as feed particles diameter is not over 50-70 μm. Artemia feed can be alive or dead in artificial culture system. Artemia can uses bacteria and protozoa as feed sources, which grow in artemia culture medium. This protozoa (such as; *Candida, Rhodotorula*) can also be directly swallowed by artemia. The best algae's for artemia nutrition includes: *Dunaliella salina*, *Spirulina* and *Scenedesmus*. For artemia culture, agricultural products can be utilized such as; rice, corn, wheat, barley flours and their bran.

5. REPRODUCTION

All bisexual species holds 42 chromosomes (2n = 42). *A.persimilis* holds 44 chromosomes (2n = 44) and*Artemia partenogenic* is diploid, triploid, tetraploid and even pantaploid. As a general rule, artemia populations are defined on the basis of the number of their chromosomes. However, contrary to mammalian, female artemia is heterogametic [3]. Artemia is produced by two ways: sexual and parthenogenesis (development of a new individual from an unfertilized egg). The mature female ovulates each 140 hours.

According to strain of artemia or method of living, it selects one of the following conditions; oviparous or ovoviviparous. In suitable situation of rising, reproduction trend is as larvae production (ovoviviparous) and in unsuitable situation of growth (salinity >50gm/lit and oxygen <5mg/lit) oviparous will occur. In the latter condition, growth of embryo will stop and enters diapauses. In suitable saline and nutritional conditions, females can produce 75 naplius each day and over its lift cycle (50 days), it reproduces 10-11 times.

In extreme hypoxia, due to increasing hemoglobin production, the color of artemia will change from light brown to yellow and then red.

Artemia cysts are spread by wind and birds. Earth pond or region of high salty water is suitable for culture and reproduction of artemia.

6. DIFFERENT KINDS OF ARTEMIA FROM THE NUTRITIONAL POINT OF VIEW

From the nutritional point of view,different kinds of artemia are:

- Decapsulated cysts
- Newly hatched nauplii
- Metanauplii and juvenile and adult
- Frozen and freeze – dried artemia

These forms of artemia are commonly used for newly hatched shrimps, sturgeons, trout, aquarium fishes and some crustaceans. Artemia biomass (consist of cysts and different living stages of artemia) is a suitable protein resource for other animals like poultry that consists of different stages of artemia growth.

7. METHODS OF ARTEMIA HARVESTING

According to artemia habitat, different biomass harvesting is utilized. In breeding pools and lake beaches, artemia is collected using a lace net that is fastened to two large floaters from each side (figure 2).

Because of phototropism characteristic, artemia can be collected easily by light source at night.

After harvesting, artemia biomass can be dried and cured under sunlight (Figure 3). Then the dried artemia will be milled before using in poultry diet.

Figure 2. Artemia biomass harvest

8. CHEMICAL COMPOSITION OF ARTEMIA MEAL

The chemical composition of different kinds of artemia meal (dried at 50-60°C as sun cured or oven dried) is shown in Table 1. As shown in table 1, the chemical composition of 3 kinds of artemia meal (collected from different regions of Iran) is not identical. The quality of those,depends on region, species, time of harvest and percentage of artemia mixture (artemia in different stages of living shows different compositions). So prior to using this ingredient, it must be analyzed for main nutrients.

Figure 3. Flaked artemia

Table 1. Chemical composition of three kinds of artemia meal (ULAM[2], EPAM[3], GSLAM[4]) (as g/kg, MJ /kg or mg/kg – DM basis)1

Chemical composition		Kind of Artemia meal		
		ULAM	EPAM	GSLAM
Dry matter	g/kg	928	924	938
Crude Protein	g/kg	401.9	390.8	423.5
Gross Energy	MJ/kg	16.86	16.32	14.98
Crude Fat	g/kg	136	85.5	206.5
Crude Fiber	g/kg	36	18	28
Crude Ash	g/kg	240	287	284
Calcium (Ca)	g/kg	23.4	20.2	26.1
Phosphorus (P)	g/kg	11.1	8.6	14.2
Sodium (Na)	g/kg	12.1	9.6	16.4
Magnesium (Mg)	g/kg	3.3	4.1	3.1
Potassium (K)	g/kg	16.5	20.9	13.9
Iron (Fe)	mg/kg	1147.25	1642.75	437.75
Manganese (Mn)	mg/kg	53.78	132.45	84.08
Copper (Cu)	mg/kg	3.5	3.55	5.05
Zinc (Zn)	mg/kg	52.75	46.75	59

1- Zarei,A (2006) ,2- Urmia Lake Artemia Meal , 3- Earth Pond Artemia Meal , 4- Ghom Salt Lake Artemia Meal

9. METABOLIZABLE ENERGY OF ARTEMIA MEAL

An experiment was designed to determine different classes of metabolizable energy (AME, AMEn, TME, TMEn) in artemia meals [13].For determination of metabolizable energy of artemia meal and comparison with fish meal, samples gathered from 3 regions include : Urmia Lake Artemia Meal(ULAM), Earth Ponds Artemia Meal(beside urmia lake)(EPAM) and Ghom Salt Lake Artemia Meal(GSLAM). Then samples dried, milled and used in a biological experiment with fish meal. 20 Rhode Island Red cockerels with approximately same live weight used in Sibbald assay with completely randomized design with 5 treatments and 4 repetitions for determination of AME, AMEn, TME and TMEn.

Results showed there were significant differences between treatments from standpoints of metabolizable energy ($P<0.05$). ULAM and FM had highest ME and EPAM and GLAM had lowest ME. The highest TME belong to FM and the lowest TME pertained to EPAM. Except to EPAM that had the lowest TMEn, other treatments didn't have any differences between them.

10. PROTEIN AND AMINO ACIDS DIGESTIBILITY OF ARTEMIA MEAL

Result from *in vitro* and *in vivo* experiments showed that this ingredient has high quality of protein and the amount of digestibility was more than 90% [12].

In order to determination of artemia meal's amino acid digestibility, five-week old male broiler chicks were given a semi-purified diet in which artemia meal was the sole source of protein. Apparent amino acid digestibility values of the assay diet, using ileum and excreta contents, were calculated using chromic oxide as indigestible marker. True digestibility values were calculated using endogenous output determined by feeding a nitrogen-free diet. The results showed (Table 2) that in determination of apparent amino acid digestibility of excreta, serine had the lowest (0.80) and methionine had the highest (0.92) digestibility, while glycine had the lowest (0.88) and arginine and leucine had the highest (0.95) apparent ileal digestibility. In measuring true excreta and ileal amino acid digestibility, alanine and glycine had the lowest (0.90 and 0.93) and methionine had the highest (0.96 and 0.99) digestibility, respectively. In general, the site of measurement had no effect on apparent or true amino acid digestibility of artemia meal [2].

11. ARTEMIA MEAL IN BROILER DIETS

In another experiment, different levels of protein from two kinds of artemia meal include artemia meal from Urmia lake and artemia meal from earth ponds beside Urmia lake with levels of 0, 25, 50, 75, 100 percent replaced to prue fish meal protein [12]. The experimental design was completely randomized with factorial method; include 10 treatments and 3 repetitions that in each repetition there were 10 one day-old male broilers from Ross 308 strain. This experiment was performed in 7 weeks and during and end of it, traits that related to broiler performance and carcass, was measured and analyzed. Results showed that effect of kind of artemia meal and effect of level of protein replacement weren't significant for feed intake. But interaction between these two was significant for this trait ($P<0.05$). The highest feed intake belong to Urmia lake artemia meal treatment with 50% level of replacement and the lowest feed intake related to treatment of without artemia meal (contain 5% fish meal). For body weight gain and feed conversion ratio, effect of kind of artemia meal and effect of level of protein replacement and effect of interaction between these two weren't significant.

Table 2. Apparent and true digestibility (coefficients) of artemia meal determined by sampling either excreta or ileum contents

Amino acids	Apparent digestibility				True digestibility			
	Excreta	Ileal	SEM[1]	P[2]	Excreta	Ileal	SEM	P
Methionine	0.92	0.94	0.004	NS	0.96	0.99	0.004	0.09
Lysine	0.88	0.92	0.007	NS	0.92	0.96	0.007	NS
Threonine	0.85	0.90	0.013	NS	0.93	0.98	0.011	NS
Tryptophan	0.88	0.94	0.014	NS	0.90	0.97	0.017	NS
Arginine	0.89	0.95	0.008	0.09	0.93	0.98	0.008	NS
Isoleucine	0.88	0.94	0.011	NS	0.92	0.98	0.011	NS
Leucine	0.89	0.95	0.009	0.06	0.94	0.98	0.009	NS
Valine	0.87	0.93	0.011	NS	0.93	0.98	0.010	NS
Histidine	0.89	0.93	0.007	NS	0.95	0.97	0.007	NS
Phenylalanine	0.87	0.94	0.009	0.09	0.92	0.97	0.009	NS
Glycine	0.81	0.88	0.015	NS	-	0.93	-	-
Serine	0.80	0.89	0.018	NS	0.91	0.97	0.017	NS
Alanine	0.85	0.91	0.014	NS	0.90	0.94	0.014	NS
Aspartic acid	0.86	0.91	0.010	NS	0.91	0.94	0.005	0.09
Glutamic acid	0.87	0.93	0.014	NS	0.93	0.95	0.013	NS
Total	0.85	0.92	0.010	0.09	0.94	0.96	0.011	NS
CP(N×6.25)[3]	0.81	0.89	0.013	NS	0.89	0.94	0.012	NS

NS – Non Significant ; [1]- Standard Error of Mean ; [2] – Probability ; CP- Crude Protein ;N – Nitrogen [3] – The values (protein digestibility) were not corrected for uric acid.

These effects weren't significant for all carcass traits and gastro intestinal parts exception for femur percent that treatment of without artemia meal (contain 5% fish meal) had a lowest percent to comparison with other treatments for this trait.

12. CONCLUSION

Results of this studies revealed that artemia meal can be used as a feedstuff in poultry and other farm animal's diets because it has high level of protein and high protein digestibility. Compared with other animal proteins, artemia does not contain any feather, bone, hair or gastrointestinal tract components. In addition, in artemia production there is no requirement for high pressure and high temperature treatments which can influence protein quality. Artificial culture of artemia is easy and is possible everywhere

REFERENCES

1. Abatzopoulos, TH.J, J.A. Beardmore., J.s.Clegg. and P.Sorgeloos. 2002. Artemia: Basic and Applied Biology. Kluwer Academic Publishers.
2. Aghakhanian, P., A. Zarei, H. Lotfollahian and N. Eila. 2009. Apparent and true amino acid digestibility of artemia meal in broiler chicks.South African Journal of Animal Science.39 (1).
3. Bowen, S.T.1964: The genetics of *Artemia salina*. IV. Hybridization of wild population with mutant stocks. Biol. Bull., 126, 333.
4. Gilbert, V.S.1995.Introduction, biology and ecology of artemia. laboratory of aquaculture and artemia reference center university of Gent, Belgium.
5. Johnson, M.L. and C.M. Parsons., G.C. Fahey, Jr., N.R. Merchen., C.G. Aldrich. 1998. Effects of species raw material source, ash content, and processing temperature on amino acid digestibility of animal by product meals by cecectomized roosters and ileally cannulated dogs. Journal of Animal Science.76:1112-1122.
6. Klasing, K.C. 1998.Comparative Avian Nutrition.CAB International.
7. Leeson. S. J.D.Summers. 2001. Nutrition of the Chicken.4 th edition. university Books.
8. Parsons, C.M.1999. Protein quality and amino acid digestibility. Multi-state. Poultry meeting, May 25-27.
9. Shirley, R.B. and C.M. Parsons. 2000. Effect of pressure processing on amino acid digestibility of meat and bone meal for poultry. Poultry Science.79:1775-1581.
10. Sibbald, I.R., 1976.The true metabolizable energy values of several feedingstuffs measured with roosters, laying hens, turkeys and broiler hens. Poultry Science. 55:1459-1463.
11. Sorgeloos, P.1989. Two strains of Artemia in Urmia lake (Iran), Artemia Newsl., 13, 5.
12. Zarei, A. (1).2006. Use of artemia meal as a protein supplement in broiler diet. International Journal of Poultry Science .5(2).142-148.
13. Zarei, A. (2). 2006. Apparent and true metabolizable energy of artemia meal International Journal of Poultry Science .5(7):627-628.
14. Zarei, A. 2007. Determination of protein digestibility of animal protein feeds under condition of *in vitro* and *in vivo*, BSAS Annual Conference, 2-4 April 2007, Southport, England (oral presentation).

CHAPTER 5

Micro Mineral Concentrations of Dairy Cows in Selected Areas from Slovak Republic

Skalicka m*, Nad P, Hresko Samudovska A and Bujnak I

Institute of Nutrition, Dietetics and Feed Production, University of Veterinary Medicine and Pharmacy in Kosice, Slovak Republic

ABSTRACT

The aim of study was evaluation of the quality of feeds for dairy cattle on the content of copper and zinc from two agro-ecological zones (western Slovakia and eastern Slovakia). Feed samples-corn silage (n=70), lucerne silage (n=32), grass silage (n=25), and total mixed ration for dairy cows-TMR (n=66) came from livestock farms. The Cu and Zn contents were assessed in samples using atomic absorption spectrophotometer (AAS). Higher average values of zinc were found in lucerne silage from eastern Slovakia (75.04 mg/kg), in other silage were the concentration of zinc in range from 51.77 to 58.13 mg/kg. The lowest average values were found of copper in silages-corn, lucerne and grass (9.80; 9.67 and 8.10 mg/kg), respectively. In the TMR was detected the higher average content of Zn in production phases, before calving 100.43 mg/kg, after calving 109.35 mg/kg and on the top of lactation 105.29 mg/kg. A total of 396 blood serum samples were collected from 22 farms of cattle. In western Slovakia, the percentage of animals deficient in serum Zn and Cu were observed in all production phases. The levels of Zn in serum were lower than the critical level 45.07% before calving and 38.03% after calving. The serum concentration of Cu were higher at after calving 33.8% and at top lactation 33.33% of the total number of analyzed samples. In eastern Slovakia, the levels of Zn and Cu in serum samples were lower. Of the total number of samples, the amounts of Cu in serum were reduced in all production phases (before calving 14.20%, after calving 7.81% and top of lactation 16.33%. Similarly, the amount of Zn in serum was reduced (10.65%, 19.17% and 12.78%, respectively).

Keywords: Blood serum; Silage; Total mixed ration (TMR) for dairy cows; Zinc; Copper; Atomic absorption spectrophotometer (AAS)

1. INTRODUCTION

The quality of feed is a complex of several interrelated factors evaluated in several ways. An important indicator is the mineral content. Monitoring the mineral content is important in terms of the ecosystem (cycle minerals) but also in terms of quality feed [1]. Khan et al. [2] reported that mineral contents in forages depend on soil, drainage and pH, forage species and varieties, forage maturity, pasture management, forage yield and climate. A natural source of minerals in the diet of farm animals is feed. Utilization minerals from food can be changed to different combinations of feed and representation grasses, legumes and herbs [3,4]. Forages species vary widely in the range and quantity of trace elements. Fertilizer application may also affect trace elements concentration. On the 16 trace elements currently recognized as being essential for farm livestock, only eight of them have practical significance.

Minerals have a special role in ensuring efficient growth [5], production [6] and immunocompetence in animals [7]. The transitional period (defined as the period from 3 weeks before to 3 weeks after calving) is extremely challenging for the dairy cow [8]. It has been reported that the concentration of some trace minerals are affected during the transitional period, especially around the time of parturition [9,10].

Trace mineral status, especially selenium (Se) and zinc (Zn), affects neutrophil function in postpartum cows [11]. Mineral deficiency remains the top issue in countries with dairy farming.

Furthermore, evidence suggests that herds have increased risks of metritis, mastitis locomotion problems or diarrhea in calves when zinc (Zn) or copper (Cu) status are either marginal or deficient [5]. Dairy cow feeds typically a range of different compounds that possess antioxidant activities, many of which are minerals or are mineraldependent. The key trace elements involved in animal feed are zinc, copper, iron and manganese [7].

Supplementation of zinc in the deficient animals has caused a significant improvement in hormone levels like thyroxin, triiodothyronine, oestrogen, and progesterone [12]. It was observed that marginal or low zinc and copper contents in pasture, soil or animal feed may be responsible for the delayed age of puberty [13]. Inadequate intakes of Cu, Zn, and Fe might produce oxidative stress, growth retardation in young animals, anemia, decrease in follicular growth and fertility, increase in abortion, disturbance in **ANTI-OXIDANT**/ free radical balance, decrease immune system function and enhance the virulence of the infectious agent [14-16].

Copper is a component of a several range of physiologically important metalloenzymes, lipid metabolism, anti-oxidant defense as an integral part of the essential enzyme superoxide dismutase (SOD), immune function and carbohydrate metabolism. Zinc is the second most abundant trace element in mammals and birds, and forms a structural component of over 300 enzymes, where it may also be key to catalytic and regulatory activity. It play and important role in antioxidant defense as an integral part of SOD. These elements have each other both negative and positive relationship within intermediary metabolism [7]. Positive chemical bonds (Mn–Zn) are less than the antagonistic bonds (Cu-Zn). Absorption of Zn and Mn is inhibited also by Ca [2,17,18]. Occurrence of secondary Zn deficiency in animals with excessive content of Ca, P and phytic acid in feed rations are frequent.

In the present study, we intended to monitor the serum concentration of Cu and Zn in dairy cows in different production phases (pre-calving, calving and at the top of lactation) in relation of mineral content in feed.

2. MATERIALS AND METHODS

The study was carried out in two regions of Slovakia, Western and Eastern Slovakia. The monitoring was conducted in two phases; the first phase was the collection of feeds and biological samples – blood serum and the second phase was laboratory analysis of the collected samples. The farms were not in the industrial area. The climate of the locality is the mild climate zone characterized for Central Europe.

Ten **DAIRY CATTLE** farms were located in Western Slovakia and twelve farms in Eastern Slovakia. In each farm, sample was collected from eighteen healthy dairy cows (3-6 years old). A total of 396 blood serum samples were collected from 22 farms of cattle. Feed samples–corn silage (n=70), lucerne silage (n=32), grass silage (n=25) and TMR–total mixed ration (n=66) were used.

Blood samples were collected in the morning via the jugular vein. After proper clotting, the blood samples were centrifuged at 3500 rpm for 15 minutes and the serum samples were stored at -20°C until analyses. Before measuring, serum samples were deproteinized by supplementing trichloracetic acid at a 1:1 ratio. After **CENTRIFUGATION**, the content of Zn and Cu in the supernatant was measured directly by using the flame method of an atomic absorption spectrometer (Unicam Solar, 939, Great Britain).

Feed samples were processed by digestion in the microwave oven (MLS-1200 Mega, Milestone) by using 5 mL HNO_3 and 1 mL HCl per 1 g of sample. The program of digestion was as follow: 1st step-250 W, 2 minutes; 2nd step-0W, 2 minutes; 3rd step-250W, 5 minutes; 4th step-400 W, 5 minutes; 5th step-500 W, 5 minutes; and 6th step-600W, 2 minutes. The digested samples of feeders were analyzed for the presence of Cu and Zn by using the flame method of an atomic absorption spectrometer (Unicam Solar, 939, Great Britain). The flame conditions were those recommended by the instrument manufacturer for Cu and Zn (wavelength 324.8 and 213.9, respectively, band pass 0.5 nm). The content of Cu and Zn in forage and blood were determined according to the methodology used by the Official lists methods and laboratory diagnosis of food and feed [19].

3. RESULTS AND DISCUSSION

3.1 Zinc and copper in blood serum

In the western Slovakia, the mean level of serum Cu was: in production phases (16.80 μmol/L) before calving, after calving (17.07 μmol/L) and in the top of lactation (16.96 μmol/L). The mean zinc content obtained in these phases was: 29.25 μmol/L; 26.90 and 26.55 μmol/L, respectively. The maximum increase of serum concentrations of Zn were found in the period before calving 45.07% and after calving 38.03%. Zn concentration in the serum fluctuates with age, stress, and the reduction of feed.

In the eastern Slovakia, the average concentration of Cu in production phases-before calving, after calving and in the top of lactation, were 17.11 μmol/L; 16.75 μmol/L and 17.69 μmol/L. The zinc in serum in these phases was 25.64 μmol/L; 26.59 μmol/L and 24.10 μmol/L, respectively. These results of zinc concentration in serum were a comparable with serum zinc content from dairy cows in western Slovakia.

The results of study were summarized in **TABLES 1** and **2**. The average serum concentration of trace elements between western and eastern Slovakia did not show any statistical significance. The monitored amounts of zinc and copper serum, in all productions phases of dairy cows, were within the physiological range (Zn-12.2-30.0 μmol/L; Cu-12.6-18.9 μmol/L, respectively). The balanced total mixed ration for dairy cows ensures the supply of nutrients in all phases of lactation and does not occur of some minerals. Although there is no deficiency of micro nutrient in this study, several samples of plants and blood serum have shown deficiency of minerals. The relationship of plant– animal is a complex chain of biochemical events.

Table 1: The concentration of Cu and Zn in blood serum (μmol/L) of dairy cows from western Slovakia.

Element		**Before calving**	**Before calving**	**Top of lactation**
		n=60	**n=60**	**n=60**
Cu	Mean	16.80	17.07	16.96
	SD	3.22	4.64	2.80
Zn	Mean	29.25	26.90	26.55
	SD	10.65	9.44	9.40

Table 2: The concentration of Cu and Zn in blood serum (μmol/L) of dairy cows from eastern Slovakia.

Element		**Before calving**	**Before calving**	**Top of lactation**
		n=72	**n=72**	**n=72**
Cu	Mean	17.11	16.75	17.69
	SD	4.80	2.27	2.13
Zn	Mean	225.64	26.59	24.10
	SD	8.14	7.28	7.60

Singh et al. [20] reported the lower concentration in blood serum of Cu and Zn from **GRAZING** cows in India in compared with results in this study. The average of serum Zn in grazing cows was in range from 0.81 to 1.03 ppm and the concentration of serum Cu was from 0.64 to 0.69 ppm. Deficiency up to 50% has been reported by Ramana et al. [21], similarly, Kawitkar [22] reported lower serum Zn values in lactating animals than pregnant and heifers. Mishra [23] also observed low serum Zn content in 46% of anestrus cows in India. The reason might be the effect of season physiological state of the animals and availability of feeds. The Cu deficiency observed in some grazing **RUMINANTS**, this could be due to poor biological availability of Cu, which was mostly due to increased lignification's in fodders and the susceptibility of Cu to form biologically unavailable complexes (Cu-Fe, Cu-Zn) which is also responsible for high incidence of Cu deficiency syndrome, particularly in grazing ruminants [20]. Trace minerals bound to organic compounds are more readily adsorbent from the digestive tract, and maybe more biologically compared with the inorganic salts to these minerals. Results from study of Hackbart et al. [24] and other studies indicate that trace mineral source can have a positive effect on **MILK PRODUCTION**. Adding inorganic Zn, Mn, Cu, Co with organic forms would increase milk production. Noaman et al. [25] have concluded that Holstein dairy cattle reared under semiindustrial dairy farming were deficient in serum Cu concentrations, especially in summer.

3.2 Zinc and copper in feed

The quality of roughage is dependent on weather conditions, soil potential and the stage of maturity of the **PLANT SPECIES**. TMR (total mixed ration for dairy cows) provides a balance of nutrients for dairy mixing fodder and grain, minerals and vitamins. The analytical characteristic of the nutritional value of TMR is the presence of minerals. The contents of Cu and Zn in the TMR for the **DAIRY COW** are listed in **TABLES 3** and **4**.

The content of Cu in TMR for dairy cows before calving was found 18.02 mg/kg and 20.12 mg/kg, after calving 22.68 and 19.95 mg/kg, and on the top of lactation 22.49 and 19.80 mg/kg, respectively. In TMR from eastern Slovakia, the content of Zn (100.43 mg/kg) was found higher, for cows before calving, in comparison with feed from western Slovakia (70.93 and 84.81 mg/kg, respectively). The slight increase of Zn content (95.10 and 109.35 mg/kg) weas found in TMR, phase after calving, in compared with tolerance specified by the NRC [26].

The Zn content in measured samples was in the tolerance values according to NRC [26] in the TMR intended for dairy cows at top lactation. Skalická et al. [27] found almost the same amounts of Cu in the analyzed TMR in production stages, but found higher zinc content (117.59 mg/kg before calving; after calving 134.79 mg/kg and 132.49 mg/kg at top lactation) compared to western and eastern Slovakia. In other hand, Noaman et al. [25] found low content of Cu (13.00 mg/kg) and Zn (47.30 mg/kg) in total diet for cows in in comparison with the results of our study.

Table 3: The contents of trace elements (mg/kg) in the TMR for dairy cows from western Slovakia.

Element		Before calving	Before calving	Top of lactation
		n=10	n=10	n=10
Cu	Mean	18.02	22.68	22.49
	SD	11.91	17.56	17.93
Zn	Mean	70.93	95.10	95.24
	SD	27.45	18.01	23.05

Table 4: The contents of trace elements (mg/kg) in the TMR for dairy cows from eastern Slovakia.

Element		Before calving	Before calving	Top of lactation
		n=12	n=12	n=12
Cu	Mean	20.12	19.95	19.80
	SD	10.95	10.60	5.14
Zn	Mean	100.43	109.35	105.29
	SD	55.56	47.89	38.92

Tables 5 and **6** shows the average mineral content of the silage. The analyzed levels of minerals finding from western Slovakia in corn silage were for Cu 9.80 mg/kg and Zn 55.99 mg/kg, respectively. In eastern Slovakia, were found the slightly higher contents only in case of Cu (11.02 mg/kg). The monitored average contents of Cu and Zn were higher compared to limits specified by the NRC [26], in corn silage (Cu 6 mg/kg, Zn 24 mg/kg). By contrast in Ireland, Rogers et al. [28] found lower amounts of

Cu (5.11 mg/kg) and Zn (26.61 mg/kg) in corn silage. In eastern Slovakia, were found the higher contents of Cu in corn silage, lucerne silage and grass silage compared from western Slovakia. The content of Zn in Lucerne silage from eastern Slovakia was higher (75.04 mg/kg), in other silage were the concentration of zinc in range from 51.77 to 58.13 mg/kg.

Corn silage is one easily digestible carbohydrate food for dairy cows. Lucerne is widely grown throughout the world as forage for cattle, and is most often harvested as hay, but can also be made into silage and grazed. In terms of the ratio of nutrients is one of the **PROTEIN FEED**. Lucerne silage is a major source of Ca compared with maize silage. The content of trace elements in the feed is governed by the nature of geochemical soil and plant species. Several authors [1,17,29,30] have dealt with the occurrence of trace elements in feed and blood serum of cattle from various weather conditions who similarly found a low Cu content in grass and grass hay.

The micro mineral content of feeds from various climatic conditions was presented by Shahjalal et al. [17]. The mean values obtained in this study were low for Cu (6.12 mg/kg), and higher for (Zn 291.3 mg/kg). The results of Zn were considerable higher than those indicated in our study (**TABLES 5** and **6**). Nearly the same values of micro elements were found in grass by Ramana et al. [30] in India (Cu 27.9 mg/kg and Zn 31.00 mg/kg). In the observed period, we found higher levels of Zn in silage–corn, lucerne and grass. The average detected levels of Cu, Zn in the TMR were in range of tolerance values [26].

Table 5: The content of Cu and Zn (mg/kg) in feed from western Slovakia.

Silage		Corn	lucerne	Grass
Element		n=30	n=11	n=10
Cu	Mean	9.80	9.67	8.10
	SD	4.70	4.39	4.83
Zn	Mean	55.99	57.52	53.60
	SD	34.61	27.51	30.10

Raising cattle is influenced by the quality of feed. Correctly aligned phase nutrition program minimizes metabolic problems of dairy cows. Because of the antagonistic relationship and high body burden of high production dairy cows should be in the diet of dairy production, dosage minerals. In the present study, concerning the serum Cu and Zn concentrations was no statistically significant between monitored farms. The serum concentrations were within the recommended range. It is essential to consider all the dietary nutrient sources including forages, concentrate mineral supplements when evaluating mineral interaction.

Table 6: The content of Cu and Zn (mg/kg) in feed from eastern Slovakia.

Silage		Corn	lucerne	Grass
Element		**n=30**	**n=11**	**n=10**
Cu	Mean	11.02	13.78	13.02
	SD	6.07	6.16	6.88
Zn	Mean	21.77	75.04	58.13
	SD	41.91	49.30	31.68

ACKNOWLEDGEMENT

This work was supported by the Ministry of the education of the Slovak Republic, project VEGA 1/0373/15.

REFERENCES

1. Mandal AB, Yadav PS, Kapoor V (2004) Mineral profile and their retention in lactating cows in relation to soil, fodder and feed in Kamrup district of Assam. Indian J AnimSci 71: 421-429.
2. Khan ZI, Muhammad A, Kafeel A, Ijaz J, Valeem EE (2008) A comparative study on mineral status of blood of small ruminants and pastures in Punjab, Pakistan. Pakistan J Bot 40: 1143-1151.
3. Dorszewski P, Grabowicz M, Mikolajczak J (2006) Usefulness of various biological additions for ensiling green fodder of alfalfa. Polish J Nat Sci 3: 45-52.
4. Illek J (2006) Current trends fault diagnosis health and production of dairy cattle diseases. Proceedings of lectures Days of Nutritions and Veterinary Dietetics VII, Sept.13-14, Košice: 27-30.
5. Enjalbert F, Lebreton P, Salat O (2006) Effects of copper, zinc and selenium status on performance and health in commercial dairy and beef herds: Retrospective study. J AnimPhysiolAnimNutr (Berl) 90: 459-466.
6. Rabiee AR, Lean IJ, Stevenson MA, Socha MT (2010) Effects of feeding organic trace minerals on milk production and reproductive performance in lactating dairy cows: a meta-analysis. J Dairy Sci 93: 4239-4251.

7. Andrieu S (2008) Is there a role for organic trace element supplements in transition cow health. Vet J 176: 77-83.
8. Machado VS, Oikonomou G, Lima SF , Bicalho MLS, Kacar C, et al. (2014) The effect of injectable trace minerals (selenium, copper, zinc, and manganese) on peripheral blood leukocyte activity and serum superoxide dismutase activity of lactating Holstein cows. Vet J 200: 299-304.
9. Goff JP, Stabel JR (1990) Decreased plasma retinol, alpha-tocopherol, and zinc concentration during the periparturient period: effect of milk fever. J Dairy Sci 73: 3195-3199.
10. Xin Z, Waterman DF, Hemken RW, Harmon RJ (1993) Copper status and requirement during the dry period and early lactation in multiparous Holstein cows. J Dairy Sci 76: 2711-2716.
11. Cebra CK, Heidel JR, Crisman RO, Stang BV (2003) The relationship between endogenous cortisol, blood micronutrients, and neutrophil function in postparturient Holstein cows. J Vet Intern Med 17: 902-907.
12. Sharma MC, Joshi C (2005) Therapeutic efficacy of zinc sulphate used in clustered model treatment in alleviating zinc deficiency in cattle and its effect on hormones, vitamins and production parameters. Vet Res Commun 29: 609-628.
13. Ahmed MM, Fadlalla IM, Barri ME (2002) A possible association between dietary intake of copper, zinc and phosphate and delayed puberty in heifers in Sudan. Trop Anim Health Prod 34: 75-80.
14. Noaman V (2013) Assessment of some serum trace elements in Holstein dairy cattle on industrial farms of Istafan Province, Iran. World ApplSci J 21: 1158-1161.
15. Sakhaee E, Kazeminia S (2011) Relationship between liver and blood plasma copper level and abortion in cattle. Comp Clin Path 20: 467-469.
16. Moeini MM, Kiani A, Karami H, Mikaeili E (2011)The effect of selenium administration on the selenium, copper, iron and zinc status of pregnant heifers and their newborn calves. J AgriSci Tech 13: 53-59.
17. Shahjalal M, Khaleduzzaman ABM, Khandaker ZH (2008) Micro mineral profile of cattle in four selected areas of Mymensingh district. Bang J AnimSci 37: 44-52.
18. Wilde D (2006) Influence of macro and micro minerals in the peri-parturient period on fertility in dairy cattle. AnimReprodSci 96: 240-249.
19. Bulletin of the Ministry of Agriculture SR (2004) Listing the Official Methods of Laboratory Diagnosis of Food and Feed, XXXVI. pp. 339.
20. Singh RK, Mishra SK, Swain RK, Dehuri PK, Sahoo G (2011) Mineral profile of feeds, fodders and biochemical profile of animals in west-central table land zone of Odisha. Indian J AnimSci 81: 1148-1153.
21. Ramana JV, Prasad CS, Gowda NKS, Ramachandra KS (2001) Mineral profiles of soil, feed, fodder and blood samples of animals in northern dry and northern transition zones of Karnataka. Indian J Dairy Sci 54: 40-46.
22. Kawitkar SB (2004) Mineral status of Deoni animals in relation to soil feed and fodders in UdgirTaluka of Maharashtra State. Ph.D. thesis submitted to MAFSU, Nagpur.
23. Mishra SK (2006) Effect of supplementation of minerals on the reproductive performance of cows in Dhenkanal district of Orissa. Ph.D. thesis submitted to Orissa University of Agriculture and Technology, Bhubaneswar, Odisha.
24. Hackbart KS, Ferreira RM, Dietsche AA, Socha MT, Shaver RD, et al. (2010) Effect of dietary organic zinc, manganese, copper, and cobalt supplementation on milk production, follicular growth, embryo quality, and tissue mineral concentrations in dairy cows. J AnimSci 88: 3856-3870.
25. Noaman V, Rasti M, Ranjbari AR, Shirvani E (2012) Copper, zinc, and iron concentrations in blood serum and diet of dairy cattle on semi-industrial farms in central Iran. Trop Anim Health Prod 44: 407-411.
26. NRC (2001) Nutrient requirements of dairy cattle, Seventh Revised Edition, National Academy Press, Washington D.C, USA, pp. 408.

27. Skalická M, Maskalová I, Vajda V (2010) Relationship nutritional levels and levels of mineral metabolism in peripartum dairy cows and the production phase. Proceedings of lectures Lazar Days of Nutrition's and Veterinary Dietetics IX. Košice: 73-76.
28. Rogers P, Murphy J, Kavanagh S (2001) Bovine mineral-vitamin balancers for Irish maize silage. Grange Research Centre, Dunsany, Co. Meath, Ireland.
29. Petrikovic P, Sommer A, Cerešnáková Z (2000) The nutritive value of feeds.The issue of Nitra, VÚVZ, pp. 2-35.
30. Ramana JV, Prasad CS, Gowda NKS (2000) Mineral profile of soil, feeds, fodders and blood plasma in southern transition zone of Karnataka. Indian J AnimNutr 17: 179-183.

CHAPTER 6

Vitamin-Like Supplementation in Dairy Ruminants: The Case of Choline

Luciano Pinotti

Department of Health, Animal Science and Food Safety, Università di Milano, Italy

1. INTRODUCTION

Recent developments in nutrition have established that choline is an essential nutrient for mammals when a sufficient supply in methionineand folates are not available in the diet. Vitamin B12 is also involved in this process. The dynamic interactions between these components introduced the concept of choline as vitamin-like compound. Two types of choline functions are known: as choline *per se*, for which the choline moiety is required, and functions as a methyl donor. Choline per se plays a major role in lipid metabolism, particularly in lipid transport, as lipotropic agent. Choline is also an important source of labile methyl groups for the biosynthesis of other methylated compounds. Based on this second function choline and methionine are interchangeable, as sources of methyl groups. Accordingly choline occupies a key position between energy and protein metabolism in mammals.

Choline and methyl group metabolism in ruminants however are different. In adult ruminants, choline is extensively degraded in the rumen; for this reason dietary choline contributes insignificantly to the choline body pool and methyl group metabolism is generally conservative with a relatively low rate of methyl catabolism and an elevated rate of de novo synthesis of methyl groups via the tetrahydrofolate (THF) system. This can be exacerbated in lactating dairy ruminant, in which the dietary availability of choline is nearly non-existent, but the output of methylated compounds in milk is high, while methionine as well as other sources of methyl groups are likely to be in short supply, especially at the onset of lactation. In light of this, the hypothesisthat choline can be a limiting nutrientfor milk production has been formulated and tested in several studies. Accordingly this chapter will focus on the effects of rumen-protected choline (RPC) supplementation to transition and early lactating dairy ruminants (cows and goats) on milk production and on metabolic health. Experimental data will be discussed in a systematic analysis, in order to define possible recommendations for high yielding dairy ruminants.

2. WHAT IS CHOLINE?

In 1849 Adolph Strecker, a German chemist, isolated a compound from pig bile, to which he subsequently (in 1862) applied the name choline (from the Greeke, *chole*, bile). In 1867, Bayer, determined the chemical structure of choline (McDowell, 1989).

Choline is a beta-hydroxyethyltrimethylammonium hydroxide (figure 1), which is widely distributed in nature as free choline, acetylcholine, and more complex phospholipids and their metabolic intermediates. Its role in the body is complex. It is needed for neurotransmitter synthesis (acetylcholine), cell-membrane signaling (phospholipids), lipid transport (lipoproteins), and methyl-group metabolism (homocysteine reduction) (Zeisel & Da Costa, 2009).

2.1. Vitamin or not?

In mammals choline has been classified as one of the B-complex vitamins but it does not satisfy the standard definition of a vitamin (Pinotti et al., 2002): it is synthesised endogenously and there is no evidence that it is an enzyme co-factor; furthermore, unlike other water-soluble vitamins, it is difficult to identify a deficiency syndrome for choline in healthy mammals because of its interrelation with methionine, folic acid, and vitamin B12 (Scott, 1999; Zeisel, 1988). Finally, choline is a vital component of tissues and is required in the diet of non-ruminant species at much higher levels than the water-soluble vitamins (g vs. mg) (Whitehead & Portsmouth, 1989). However, the presence of an endogenous synthetic pathway does not render choline dispensable and deficiency results in several dysfunctions when other nutrients are limiting (Zeisel et al., 1991). For this reason it has been suggested that choline may be an essential nutrient (or vitamin like compound) for mammals when excess methionine and folic acid are not available in the diet (Zeisel & Da Costa, 2009). Consequently choline was officially recognized as an essential nutrient for humans by the Institute of Medicine in 1998 (Food and Nutrition Board, 1998).

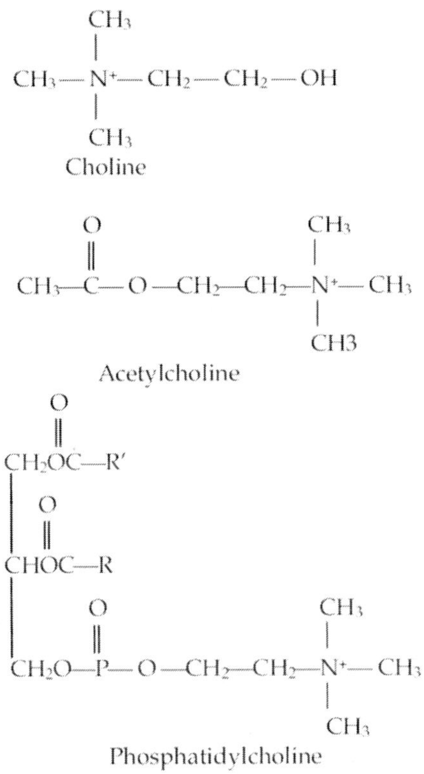

Figure 1. Structural formula for free choline, acetylcholine, and phosphatidylcholine (lecithin). R and R' refers to any fatty acids (adapted from McDowell, 1989).

2.2. How requirement for choline can be met

The metabolic need for choline can be met in two ways: either by dietary choline and via de novo biosynthesis through the methylation of phosphatidylethanolamine to phosphatidylcholine.From the point of view of animal nutrition, relatively rich sources of choline are soybean, soybean meal, rapeseed meal, fish meal and dried yeast, even though its bioavailability is considered "moderate" (Pinotti et al., 2002; Whitehead & Portsmouth, 1989).Dietary choline from a variety of choline-containing feed and food is absorbed by the intestine and uptake is mediated by choline transporters nutrient (McDowell, 1989). *De novo* synthesis of choline occurs by the sequential methylation of phosphatidylethanolamine, the methyl groups being supplied by S-adenosyl-L-methionine (SAM) (Mato et al., 1994).However, *de novo* synthesis of choline alone is not sufficient or rapid enough to satisfy all the animal's need. Methyl groups needed in this pathway may be derived from exogenous sources such as methionine, and betaine but can arise de novo in the body from the tetrahydrofolate (THF) system (Zeisel, 1988; Zeisel, 1992); vitamin B12 is involved in this process (Kennedy et al., 1995). Thus, dietary factors such us methionine, betaine, myo-inositol, folic acid, and vitamin B12 or combination of different levels of fat, carbohydrates, and protein in the diet, as well as the physiological state, all have influence on the "requirements" of choline (McDowell, 1989; Zeisel & Da Costa, 2009).

2.3. Functions of choline

Choline is considered a vitamin-like compound with two main functions: as choline *per se*, for which the choline moiety is required, and as methyl donor, although the two roles overlap.

Choline *per se* is an essential constituent of all cell membranes, where it is required to make the phospholipids phosphatidylcholine, lysophosphatidylcholine, choline plasmalogen, and sphingomyelin. Phosphatidylcholine, one of the most abundant phospholipids in higher plants and animals, is the predominant phospholipid (>50%) in most mammalian cell membranes (Kuksis & Mookerjea, 1978; Ruiz et al., 1983; Zeisel, 1988; Zeisel, 1992). Choline *per se* plays a major role in lipid metabolism, particularly in lipid transport, as lipotropic agentbecause of its ability to prevent or correct excess fat deposition in the liver generally arising as a result of its dietary deficiency (Kuksis & Mookerjea, 1978; Zeisel, 1988). Impaired triglyceride secretion to very low density lipoproteins (VLDL) is considered a major cause of fatty liver in dietary choline deficiency (Zeisel, 1988). In this context, it is noteworthy that bovine VLDL phospholipids are mainly phosphatidylcholine, with smaller proportions of sphingomyelin and phosphatidylethanolamine (Moore & Christie, 1981). Thus, when massive mobilisation of fatty acids – as at the onset of lactation in dairy ruminants- is associated with lipotropic factor deficiency (e.g. choline) triglycerides accumulate in the liver and may lead to the development of fatty liver (Gruffat et al., 1996).Choline *per se* is required to prevent hemorrhagic kidney lesions in rats (Kuksis & Mookerjea, 1978); and together with other nutrients particularly manganese salts, is required to prevent perosis, a bone disease of poultry (Ruiz et al., 1983). In the prevention of perosis, choline is needed as constituent of phospholipids required for normal maturation of the cartilage matrix of the bone, whereas in the prevention ofhemorrhagic kidney lesions choline seems involved in the renal phospholipids turnover (Kuksis & Mookerjea, 1978).

Choline *per se* is also essential for the synthesis of the neurotransmitter acetylcholine (Kuksis & Mookerjea, 1978; Zeisel, 1988). It plays important roles in brain and memory development in the fetus and appears to decrease the risk of the development of neural tube defects (Zeisel & Da Costa, 2009).

As methyl donor, choline, like methionine, is an important source of labile methyl groups for biosynthesis. Actually, the two principal methyl donors in animal metabolism are betaine, a choline metabolite, and S-adenosyl-L-methionine (SAM) a metabolite of methionine (Pinotti et al., 2002 for references). At least 50 SAM-dependent reactions have been identified in mammals, and it is likely that the number is much higher. Such methylation reactions play major roles in biosynthesis of lipids, the regulation of several metabolic pathways, and detoxification in the body (Zeisel & Da Costa, 2009). Accordingly it has been suggested that choline, and methionine are closely interrelated metabolically (Mato et al., 1994); it is also clear that choline has only little capacity to reduce the requirement for methionine, even though choline seems important to spare methionine as a methyl donor.

Although the choline-methionine interrelationship has been studied extensively (see Pinotti et al, 2002 for reference), less emphasis has been placed on the link between choline and its oxidation product, betaine. Betaine serves as an osmoregulator and is a substrate in the betaine–homocysteine (HCy) methyltransferase reaction, which links choline and betaine to the folate-dependent one-carbon metabolism (Kempson & Montrose, 2004; Ueland, 2011). The availability of choline and its metabolite betaine appear to influence the conversion or recycling of the HCy moiety of methionine, through betaine-HCy methyltransferase activity (Stipanuk, 1986).However, betaine fails to prevent fatty livers and haemorrhagic kidney (McDowell, 1989), indicating that the requirement for choline per se must be met as choline, and that betaine can substitute only the methyl donor function of choline, probably because betaine cannot be reduced to choline (Zeisel, 1988). Thus, choline *per se* must provide at least 50% of the total choline requirement, while the remaining portion of the choline requirement can be replaced by betaine (Dilger et al., 2007).

3. CHOLINE IN RUMINANTS

Ruminants differ from most other mammals in regard to their choline and methyl group metabolism. In adult ruminants, choline is extensively degraded in the rumen. Quantitative studies in sheep showed that 76% of [14C]-choline injected into the rumen was expired as methane over 6h, whereas approximately 15% accumulated as trimethylamine. Under such conditions, less than 10% of choline escapes degradation by incorporation, as phosphatidylcholine, into the structural membranes of ciliate protozoa (Neill et al., 1979). Nevertheless, the concentration of phosphatidylcholine in ruminal digesta was higher than in abomasal digesta, suggesting that protozoa are selectively retained in the rumen (Neill et al., 1979). In sheep with a defaunated rumen, the concentration of phosphatidylcholine in the abomasal digesta was higher than in the rumen (Dawson et al., 1981), leading to the suggestion that some of this abomasal Phosphatidylcholine was derived from non-dietary sources (regurgitation of bile from the lower digestive tract) (Robinson et al., 1984). In fact, intravenous injection of labelled choline in sheep indicated that the small amount of phosphatidylcholine present in abomasal digesta is largely (69%) of non-dietary or ruminal origin (Dawson et al., 1981).

The physiological state however, is important (Table 1). In the pre-ruminant lamb the activities of liver Cho-oxidase and betaine-HCy methyltransferase increase markedly after birth, but subsequently decrease as the animals reach the ruminant state (Xue & Snoswell, 1986a). The increasing activities of liver choline oxidase and betaine-HCy methyltransferase in pre-ruminant lambs are probably related to the abundance of choline-containing compounds in the milk (Xue & Snoswell, 1986a- Figure 2). By contrast, in adult ruminants only small quantities of methyl group nutrients are available from the diet and methionine synthase assumes a much more important role (Dawson et al., 1981; Neill at al., 1978; Neill et al., 1979; Kennedy et al., 1995). Methionine synthase is responsible for the de novo synthesis of methionine methyl groups from one-carbon units furnished by tetrahydrofolate (THF) (figure 1).

The enzyme utilises methyl-THF as methyl donor and methylcobalamin as tightly-bound coenzyme. De novo synthesis of methyl groups via this system appears to be minimal when labile methyl group intake is sufficient or excessive (Stipanuk, 1986).

Dietary choline therefore contributes insignificantly to the choline body pool in adult ruminants (Table1; Figure 3). Methyl group metabolism is generally conservative with a relatively low rate of methyl catabolism and an elevated rate of de novo synthesis of methyl groups via the tetrahydrofolate (THF) system (Girard et al., 2010; Henderson et al., 1983;

Robinson et al., 1984; Snoswell & Xue, 1987; Xue & Snoswell, 1986a; Xue & Snoswell, 1986b). In dairy ruminants the situation can be even worse (Table 1): the dietary availability of choline is still low while the output of methylated compounds to milk is particularly high, and methionine, as well as other methyl group sources - one-carbon units for methylneogenesis via the THF system-, is likely to be in short supply, especially at the onset of lactation (Pinotti et al., 2002). In dairy ruminants producing large quantities of milk in fact, methionine is also the first limiting amino acid. This means that the elevated requirement for methioninefor transmethylation reactions and milk protein synthesis may lead to altered methyl group metabolism (Girard et al., 2010; LaCount et al., 1995; Lobley et al., 1996). The effects of lactation on methyl group metabolism have been examined in sheep (Xue & Snoswell, 1985); it was found that the activities of hepatic phospholipid methyltransferase (for phosphatidylcholine synthesis) and methionine synthase were significantly higher (+ 33% and +34%, respectively) in lactating than non-lactating ewes (Xue & Snoswell, 1985).

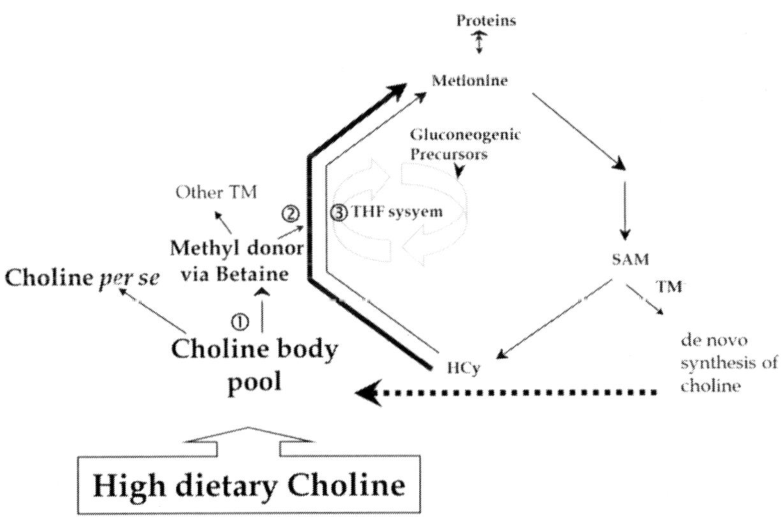

Figure 2. Choline and methyl groups metabolism in pre-ruminants. SAM, S-adenosyl-L-methionine. HCys, Homocysteine. TM, Transamethylation pathway. THF Tetrahydrofolate system. ①Choline oxidase (EC 1.1.3.17). ② Betaine-homocysteine methyltransferase (EC 2.1.1.5). ③ Methionine synthase (EC 2.1.1.13).(adapted from Pinotti et al., 2002).

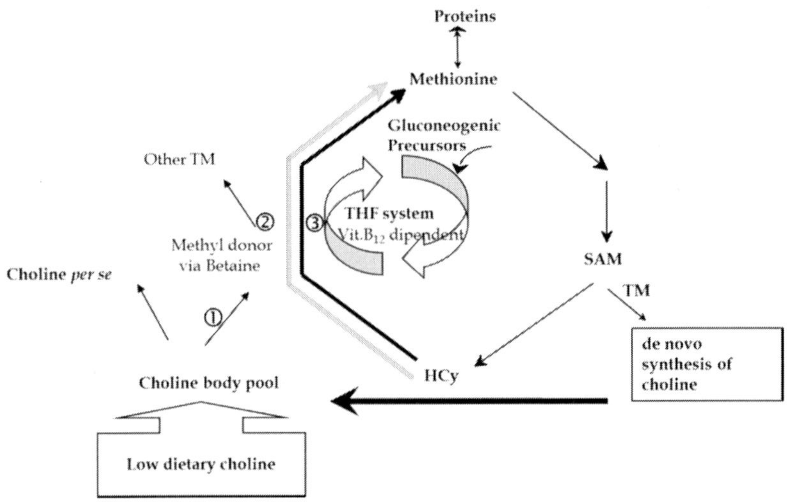

Figure 3. Choline and methyl groups metabolism in adult ruminants in positive energy balance (adapted from Pinotti et al., 2002). SAM, S-adenosyl-L-methionine. HCys, Homocysteine. TM, Transamethylation pathway. THF Tetrahydrofolate system. ①Choline oxidase (EC 1.1.3.17). ② Betaine-homocysteine methyltransferase (EC 2.1.1.5). ③ Methionine synthase (EC 2.1.1.13).(adapted from Pinotti et al., 2002).

Table 1. Evolution of metabolism in methyl groups in ruminants in three physiological stages[*at the onset of lactation, the substrates of this path often become limiting](adapted from Pinotti et al., 2002).

Physiological Stage	Availability	Enzymes Activities
Pre-Ruminants	⇧ choline ⇧ betaine	⇧ Cho-oxidase ⇧Betaine-HCy-methyltransferase
Adult Ruminants	⇩ choline ⇩ betaine ⇧ methyl syntheses via THF	⇧ Methionine synthetase
Lactating Ruminants	⇩ choline ⇩ betaine ⇩methionine ⇩gluconeogenetic precursors ⇧ methyl syntheses via THF*	⇧ Methionine synthetase

4. CHOLINE IN DAIRY COWS NUTRITION

The earliest investigations were interested in determining the effects of added dietary choline on milk production. For this purpose supplementation experiments using unprotected choline (Erman et al., 1984; Atkins et al., 1988; Sharma & Erdman, 1988b), were performed. In these studies using unprotected choline, neither milk production nor milk composition were affected, due, it was suggested, to rapid choline degradation in the rumen (Atkins et al., 1988; Sharma & Erdman, 1988b). In fact the work of Atkins et al. (1988) indicated that choline chloride was more degradable than naturally occurring choline in feed, while increasing choline chloride intake from 23 to 326 g/d only raised duodenal choline flow from 1.2 to 2.5 g/d, so that recovery was low indeed (Sharma & Erdman, 1988b). The mean estimates of rumen degradable choline (measured *in vitro*) actually were 80-98% for common feedstuffs and supplements (Table 2), indicating that use of choline-rich feedstuffs or additives in diets can only marginally increase the post ruminal flow of choline in ruminants (Sharma and Erdman, 1989a).

However, others studies (Sharma & Erdman, 1988a; Sharma & Erdman, 1989b) on the effects of abomasally infused choline, suggested a possible requirement for supplementing choline in lactating dairy cows.For example, Sharma & Erdman (1988a) investigated abomasal infusion of choline or methionine in dairy cows, either alone or with the choline synthesis inhibitor 2-amino-2-methyl-1-propanol (2AMP). They found that abomasal infusion of 30 g/d of choline was more effective than abomasal infusion of 45.6 g/d methionine (the molar equivalent of 30 g choline) in increasing milk yield and milk fat content. Additionally, milk production, milk fat percentage, milk fat yield, milk protein percentage, and milk protein yield of cows infused with 2AMP plus methioninewere all lower than in cows infused with 2AMP plus choline (Sharma & Erdman, 1988a). These data suggest not only that supplemental choline is required to achieve maximal performance, but also that choline formed from methionine (Emmanuel & Kennelly, 1984) may be partially responsible for these stimulatory effects on milk production. Form a technical/feeding point of view these studies also evidenced that choline must be rumen-protected to be effective in dairy cow nutrition.

4.1. Recent advance in choline supplementation in dairy cow

Although the choline requirement of dairy cows is still unknown (National Research Council, 2001), higher choline availability (by feeding rumen-protected choline, RPC) can have a favorable effect on milk production

(Erdman and Sharma 1991; Erdman, 1994;Hartwell et al. 2000; Piepenbrink and Overton 2003; Pinotti et al. 2003; Scheer et al. 2002). In dairy ruminants, in fact choline has been proposed as limiting nutrient, especially at the onset of lactation (Erdman and Sharma 1991; Pinotti et al., 2002). Based on those considerations, the effects of rumen-protected choline (RPC) supplementation to transition cows have been investigated in several studies (figure 4).

Table 2. Rumen stability: In vitro choline degradation (Sharma & Erdman, 1989a). Data obtained during four individual fermentation runs, samples were incubated in vitro with an inoculum mixture containing rumen fluid obtained from a rumen-fistulated dairy cow fed 17.5% corn silage and 28.7% grass silage and 53.8% concentrate diet.

Feedstuffs	Effective degradation, %
Barley	79.4
Cottonseed	84.7
Soybean meal	82.9
Fish-meal	83.8
Supplements	
Choline stearate	98.0
Choline chloride	98.6

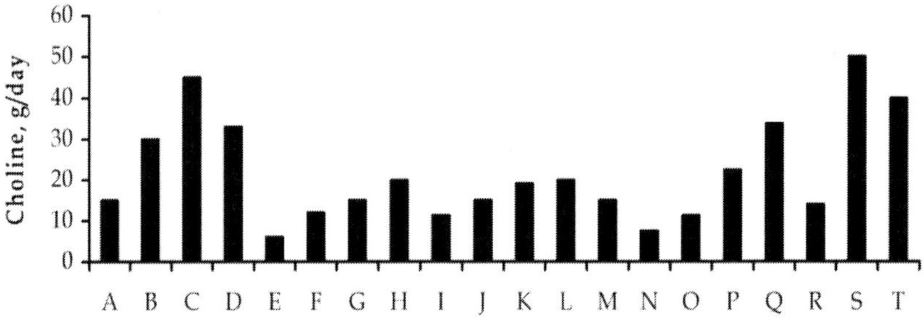

Figure 4. Levels of choline chloride supplementation [provided in a rumen-protected (RP) form] to transition dairy cows in published studies. Citations keys are reported on the x axis: (A)–(C) Erdman & Sharma, 1991; (D) Erdman, 1994; (E)-(F)Hartwell et al., 2000; (G) Scheer et al., 2002; (H) Pinotti et al., 2003; (I)–(K) Piepenbrink & Overton, 2003; (L) Pinotti et al., 2004; (M) Janovick Guretzky at al., 2006; (N)-(Q)Xu et al., 2006; (R) Zahra et al., 2006; (S)Elek et al., 2008; (T) Davidson et al., 2008. Choline chloride supplementation is shown on the y axis.

Findings in transition and early lactating dairy cows suggest that greater choline availability can improve not only milk production (Erdman and Sharma, 1991; Hartwell et al., 2000; Pinotti et al., 2003), but also lipid (Piepenbrink Overton, 2003; Pinotti et al., 2003), methyl group metabolism (Baldi & Pinotti, 2006) and choline secretion in milk (Deuchler et al., 1998; Elek et al., 2008; Pinotti et al., 2003; Pinotti et al, 2004). Furthermore, in the Pinotti et al. (2003) study, although plasma concentrations of vitamin E declined after parturition in choline-supplemented animals, the reduction was less than in controls, suggesting improved vitamin E status. However, the mechanisms of this effect have not been elucidated with certainty, even though an improved fat absorption and transport induced by choline supplementation has been proposed (Pinotti et al., 2003). Janovick Guretzky et al. (2006) found that choline supplementation did not affect milk production in either Holstein or Jersey transition cows, but in discussing their results they noted that calculated methionine balance was negative postpartum, so that supplemental choline might not have spared enough methionine to produce a physiological benefit. This conclusion is in line with another study (Brüsemeister & Südekum, 2006) that, in reviewing several published works on choline supplementation in dairy cows, concluded that negative methionine balance – and therefore a basal diet of adequate quality and composition (Baldi & Pinotti, 2006) – is essential for obtaining a response to rumen-protected choline. A furtherstudy of Zahra et al. (2006), reported that choline supplementation improves milk production (1.2 kg/day) in the first 60 d of lactation, although supplementation was only from 21 d prepartum to 28 days in milk. However, choline's effect on milk yield was attributable mainly to increased milk production (4.4 kg/day) in animals with body condition score ≥4 three weeks before calving, that were also consuming more feed. Supplementation of rumen-protected choline would therefore seem to be essential for optimising high-quality milk production in high-yielding dairy cows, as also reported in subsequent studies (Davidson et al., 2008; Elek et al., 2008; Xu et al., 2006), particularly in animals fed basal diets that limit post-ruminal methionine supply. The magnitude of the production response is likely to be affected by basal diet composition, the dose and mode of administration of the rumen-protected choline, and the stage of lactation, as discussed elsewhere (Baldi & Pinotti, 2006; National Research Council, 2001).

With regard to metabolic health, fewstudies in dairy cow also investigated the effects of choline supplementation on lipid metabolism (Piepenbrink & Overton, 2003; Pinotti et al. 2003; Pinotti et al. 2004), and hepatic lipidosis (Grummer, 2006; van Vuuren et al., 2010). In dairy ruminants it is known that the necessity to increase the lipolysis of adipose fat stores is most critical during the first stage of lactation (Grummer, 1993; McCarthy et al., 1968). During this stage, fat mobilisation leads to increased blood levels of non-esterified fatty acids (NEFA) which are taken up by the liver and oxidised to ketone bodies or carbon dioxide or esterified to triglycerides. The liver normally packages the triglycerides in VLDLs and secretes them, but sudden increases in plasma NEFA may not be adequately processed by the ruminant liver (Gruffat et al., 1996; Pullen et al., 1990). Choline serves as methyl donor in the synthesis of carnitine (Griffith, 1987) which is essential for fatty acid oxidation. The esterification of NEFA to triglycerides, and their secretion to VLDL also involves choline (Gruffat et al., 1996). In this context the NEFA/cholesterol ratio can be considered a measure of the fat retained in or metabolized by the liver and its lowered value observed in RPC cows suggested a lower risk of fatty liver in these animals (Holtenius, 1989). Thus the lowered plasma NEFA levels and the NEFA/cholesterol ratio in RPC-supplemented cows at parturition suggest more efficient liver function and improved lipid metabolism in general (Pinotti et al. 2003).

A subsequent study (Pinotti et al. 2004) provided similar results, in that plasma NEFA was reduced by 18% on day 20 of lactation in supplemented animals. Likewise, Chung et al. (2009) found that RPC have a dose dependent effect on reducing plasma NEFA concentration by 11% and 23% in cows receiving 25g and 50 g of rumen-protected choline. These findings are also consistent with the evidence that RPC supplementation to dairy cows throughout the periparturient period decreased liver accumulation of lipids (stored as intracellular triglycerides- van Vuuren et al., 2010) and increased liver glycogen content (Piepenbrink & Overton, 2003). The effects of choline at liver level have been recently confirmed in transition dairy cows (van Vuuren et al., 2010), in which RPC supplementation resulted in lower TAG concentration, by 36% and 32%compare to untreated animals, in week 1 and 3 post-partum, respectively. Results obtained in these studies indicated that not only hepatic fatty acid metabolism and cow performance are responsive to increasing the supply of choline (Piepenbrink & Overton, 2003), but also that choline supplementation can reduce cellular lipid accumulation in the liver especially during the transition period.

A further aspect investigated in some of these studies was the effect of choline on methyl group status of lactating cows. Studies carried out by Pinotti et al. (2004) showed that the administration of 50 g of (RPC, providing 20 g of choline chloride) to transition dairy cows increased plasma levels of choline-containing phospholipids and folates by the end of the transition period (20 days in milk). These findings indicate that that RPC supplementation is effective both in increasing choline availability and in optimizing methyl group metabolism: the increase in plasma folates may be attributable to a sparing effect of choline on methyl group metabolism. As noted, choline is an important source of labile methyl groups, but when it is in short supply, methyl groups (e.g. for methionine resynthesis from homocysteine) must be synthesized *de novo* by the THF system or derived from other sources. The THF system consumes gluconeogenic precursors, so in periods of glucose imbalance (e.g. early lactation) such precursors can also be limiting (Armentano 1994; Rukkwamsuk et al. 1999). An unexpected finding of studies in this area is that plasma vitamin B12 concentrations in RPC-supplemented cows were lower than in controls (Baldi & Pinotti 2006). Vitamin B12 is considered a marker of methyl group status, and the lower concentrations in treated animals appear inconsistent with the observed increases in plasma folates and choline-containing phospholipids. However, vitamin B12 is required as a coenzyme in the biosynthesis of methionine from homocysteine and in gluconeogenesis from propionate. Increased gluconeogenesis may therefore deplete vitamin B12, as indicated by the finding of a tendency toward higher concentrations of plasma glucose in RPC-supplemented cows. Nevertheless, these results were obtained from a limited number of animals, and therefore they should considered with caution since our knowledge of the interaction of choline with vitamins and other essential nutrients are incomplete.

4.2. Systematic analysis of choline supplementation in dairy cows

In the recent past different models (linear and logistic) were developed to permit the calculation of changes in milk yield and milk composition from amounts of rumen-protected choline in dairy cow diets. For example, Pinotti et al. (2010) investigated the effects of RPC administration on milk production by a regression model based on the results of 11 different studies (42 experimental groups) published between 1991 and 2008 (Davidson et al., 2008; Deuchler et al., 1998; Elek et al., 2008; Erdman et al., 1991; Hartwell et al., 2000; Janovick Guretzky et al., 2006; Pipenbrink and Overton, 2003; Pinotti et al., 2003; Pinotti et al., 2004; Xu et al., 2006; Zahra et al., 2006). The analysis included RPC supplementation (control/RPC) as fixed effect, the variability among experiments as random effect and their interaction. Treatments schedule, dry matter intake and dietary composition, accounted for most of the variability among experiments and were highly correlated in a preliminary data analysis; therefore, these variables were considered as experimental effects and treated as random components in the mixed model, assuming that the results of different experiments are affected by different experimental conditions. For each study, data were also weighted by the reciprocal of the squared standard error of the mean value, in order to consider the unequal variance among studies.

The data reviewed in this analysis were consistent with the fact that choline supplementation significantly increased milk yield in dairy cows, even though, as expected, the dose of choline administered is also relevant (table 3). Considering three supplementation classes, i.e. control (no choline: 0 g/d), low level (<10 g/d of choline chloridein a rumen protected form) and high level (>10 g/d of choline chloridein a rumen protected form), the meta-analysis indicated that a low level (less than 10 g per day) of RPC supplementation tended to increase milk yield and a high level of supplementation of RPC significantly increased milk yield. This results were in line with another meta-analysis, performed using a nonlinear model (Sales et al., 2010), on the effects of rumen-protected choline in dairy cows,in which choline supplementation has a positive effect on milk yield and milk protein content, but not on milk fat content.

Table 3. Results from mixed model when the different amounts of RPC administrated were grouped into three supplementation levels (adapted from Pinotti et al., 2010).

Choline amount administered	Milk yield estimate (kg/d)	RPC effect
Control (0 g/d)	32.35	-
RPC low level (<10 g/d)	33.40	0.15
RPC high level (> 10 g/d)	35.18	<0.001

5. OTHER DAIRY RUMINANT

In the case of small ruminants the effects of choline infusion on transmethylation reactions have been studied in sheep by Lobley et al. (1996), while Emmanuel and Kennely (1984) investigated methionine and choline incorporation into plasma and milk of lactating goats. More recently Banskalieva et al. (2005) assessed the efficacy of rumen-protected choline supplementation in meat goats. However, studies on choline supplementation and its effects on milk production, and lipid and methyl group status in periparturient dairy goats have been published only recently (Pinotti et al., 2008; Savoini et al., 2010). In a production trial (Pinotti et al., 2008) a supplementation of 4g/d of choline in rumen protected form was given to periparturient dairy goats, starting 4 weeks prior to expected kidding and continuing for 5 weeks after parturition. The quantity of choline given was based on experiments in dairy cows (Pinotti et al., 2005) and metabolicBody weight ($BW^{0.75}$) of the goats at the beginning of the experiment. During the first 6 weeks of lactation, milk yield and four percent fat-corrected milk (FCM) yield were increased by 7 and 12%, respectively, in RPC supplemented than non-supplemented goats. Milk fat concentration, fat and protein yield were also increased by RPC treatment.The link between choline supplementation and milk response has been mainly attributed to the metabolic interchangeability of choline and methionine, in the sense that both can furnish labile methyl groups. This interchangeability has been investigated by Emmanuel and Kennelly (1984) in goats, who estimated that 6% of the choline pool was derived from methionine and that approximately 28% of methionine is used for choline synthesis via the pathway for the de novo biosynthesis of the choline moiety involving sequential methylation of phosphatidy-lethanolamine. Reduction of methyl group demands from the methionine pool may therefore offer production benefits to dairy ruminants and might be achieved by nutritional manipulation other than alteration of the methionine supply (Lobley et al., 1996). This was probably the situation in the study by Pinotti and coworkers (2008), in which during the early stages of lactation of dairy goats dietary methionine was probably limiting (goats received about 80% of the optimal level of methionine). These considerations imply that feed composition, mainly protein supply and the availability of methionine (National Research Council, 2001) influence the effects of choline supplementation not only in dairy cows (Baldi & Pinotti, 2006; Brüsemeister & Südekum, 2006), but also in goats.

The milk production response to choline supplementation in dairy goats, was obtained without any detrimental effect on methyl groups status. In fact, folate availability at kidding was higher in choline-supplemented than in non-supplemented goats, while during lactation lack of a drop in plasma folate and vitamin B12 in the choline-supplemented animals, suggests that good methyl group status was maintained in choline-treated goats (Pinotti et al., 2008; Savoini et al., 2010). This results was obtained, even though RPC receiving goats were producing more milk than the choline un-suplemented ones.

In a parallel study (Baldi et al. 2011) the effect of rumen-protected choline administration on metabolic profile, selected liver constituents, and on mRNA expressions of selected enzymes, transcription factors, and nuclear receptors involved in mammary lipid metabolism in dairy goats, have been investigated. It was evident from this study that at comparable milk yield during the first month of lactation, choline supplemented goats showed a lower level of plasma β-hydroxybutyrate, whereas no other differences were observed for other metabolites. The reasons for this difference are uncertain: authors conjecture that it might be due to the role of choline in lipid metabolism as already reported in dairy cows (Baldi & Pinotti, 2006; Chung et al., 2009). With regard to liver variables, it has been reported that total liver lipid/ DNA content was significantly lower (-35%) in RPC than CTR goats. As already reported above, dairy cows receiving choline have also been reported to have reduced hepatocellular lipid accumulation (Cooke et al., 2007; Grummer, 2006).

Within the mammary tissue many selected enzymes [e.g. lipoprotein lipase (LPL)], transcription factors and nuclear receptors that can be affected by choline availability are involved in lipid uptake and milk fat secretion. Furthermore, choline is actively secreted into mammalian milk. The major choline-containing compounds in bovine milk are unesterified choline, phosphatidylcholine, and sphingomyelin (Pinotti et al., 2003). Kinsella (1973) reported that a bovine mammary gland yielding 25 litres milk secretes 10±3g phospholipids per day, corresponding on average to 5% of the phospholipids of the mammary tissue. The phospholipids of the membrane of the milk fat globule constitute the major choline-containing component of bovine milk (McPherson & Kitchen, 1983). The above data suggest that choline is an important metabolite in lactating mammary tissue, that it is used avidly when available (Kinsella, 1973), mainly in the lipid metabolism. Thus, it has been proposed that choline supplementation can have positive effects on lipid trafficking particularly lipid transport toand within extra-hepatic tissues (Baldi & Pinotti, 2006; Cooke et al., 2007; Piepenbrink & Overton, 2003) including the mammary gland. However, at udder level in lactating goats, choline did not affect the expression of several mammary gland transcripts involved in lipid metabolism (Baldi et al. 2011). The effects of nutrition on the expression of mammary lipogenic genes in dairy ruminants (cows, ewes, goats) have been investigated mainly by feeding milk-fat depressing diets, characterized by high levels of concentrates, or conjugated linoleic acids (CLA), or other lipids (fish oil, vegetable oils, full fat seeds, etc.) (Bauman, 2008; Bernard et al., 2006; Harvatine and Bauman 2006; Lock et al., 2008). Absence of significant effects on gene expression observed in goats could be due to the dose or to the type of supplementation. Thus, while in Baldi et al. (2011) were supplied 4 g/d choline (about 0.2% of DMI), other studies used higher levels of supplementation (for fats 3.6-11.2% of DMI) or more potent nutritional factors (e.g. CLAs are potent inhibitors of mammary gland fat synthesis) (see Bernard et al., 2006 for review). Nevertheless, these results were obtained from a limited number of animals, and choline effects and function in lactating mammary tissue deserve further investigation.

6. CONCLUSIONS

It can be concluded from this outline that dietary supply of choline may not always be sufficient to maximize milk production in both dairy cows and goat; and although the requirement for choline can in theory be satisfied by other nutrients, it is unlikely that this happens in practice especially at the onset of lactation. The magnitude of the production response is likely to be affected by basal diet composition, the dose of the rumen-protected choline, and the stage of lactation (Baldi & Pinotti, 2006; National Research Council 2001). In particular with respect to lactation cycleseveral studies provided convincing evidence that protected choline positively affects lactation performance especially when feeding is initiated prior to calving and is continued during early lactation (Grummer, 2011). In fact,assuming that choline is limiting in high yielding dairy ruminants, the greatest demand for choline should occur at lactation onset. It is also at this stage that the availability of other choline-related nutrients (i.e. methionine, folates, and vitamin B12) is often low (Baldi & Pinotti, 2006; Girard et al., 2010). From a metabolic and hephatic point of view, since choline is a lipotropic factor, it may be particularly beneficial at this time in view of the adipose and liver metabolism changes that occur during the transition from late pregnancy to early lactation (Piepenbrink and Overton 2003): choline may optimize the balance between fat retained and fat metabolized by the liver, thereby improving lipid metabolism in general. These roles of choline however, do not exclude that RPC supplementation is also effective in optimizing methyl group metabolism, which might be an added benefit of adequate choline supplementation. With regard to dose, even though an estimate of the minimum choline needed in dairy cows for maintenance functions (based on metabolic body size), is approximately 4 to 6 g/day (Donkin, 2011), production response were abstained when higher doses were fed (Pinotti et al., 2010). Consequently, although insufficient data are available to establish the choline requirement, it can be suggested that supplementation doses in this range are likely to be adequate for transition dairy ruminants.

However, it is also evident that our knowledge of the interaction of choline with vitamins and other essential nutrients is incomplete in the dairy ruminants. For example there are indications that both methyl group precursors (including choline) and appropriate co factors (folate and vitamin B12) are important for the optimal metabolic support of milk production (Pinotti et al., 2008), even though methionine may not always be involved in this scenario (Preynat et al, 2009).These aspects not only reflect the inadequacy of a nutritional approach based on the supply and utilisation of individual nutrients, but also suggest that the requirements of B-complex vitamins in dairy ruminants should be reconsidered.

ACKNOWLEDGMENT

This chapter has been developed in the frame of the COST Action FA0802 Feed for Health. The author wish also to thanks Dr. Gerwin Meijer from the Animal Science Group, Wageningen University, for his advisor in the early literature study on this topic, Prof. Antonella Baldi and Prof. Vittorio Dell'Orto, as well as all the member of the Animal Nutrition Research Group from the Veterinary Medicine Faculty of the Università di Milano (Italy), for their collaboration and support in several experiments carried out in this area.

REFERENCES

1. Armentano L. (1994). Impact of metabolism by extragastrointestinal tissues on secretory rate of milk proteins. *Journal of Dairy Science*, Vol. 77, No. 9, (September 1994), pp. 2809-2820., ISSN: 0022-0302
2. Atkins, K. B.; Erdman, R. A. & Vandersall, J. H. (1988). Dietary choline effects on milk yield and duodenal choline flow in dairy cattle. *Journal of Dairy Science*, Vol. 71, No. 1, January 1988, pp. 109-116, ISSN: 0022-0302
3. Baldi, A.; Bruckmaier, R., D'ambrosio, F., Campagnoli, A., Pecorini, C., Rebucci R., & Pinotti, L. (2011). Rumen-protected choline supplementation in periparturient dairy goats: effects on liver and mammary gland. *Journal of Agricultural Science*, Vol. 150, No.
4. FirstView, Jenuary 2011, pp 1-7,DOI ISSN: 0021-8596 DOI:10.1017/ S0021 859611000104 Baldi, A. & Pinotti, L. (2006). Choline metabolism in high-producing dairy cows; metabolic and nutritional basis. *Canadian Journal of Animal Scicience,* Vol 86, No. 2, June 2006, pp. 207-212. ISSN 1918-1825
5. Banskalieva, V.; Puchala, R., Goetsch, A.L., Luo, J., & Sahlu T. (2005). Effects of ruminally protected betaine and choline on net flux of nutrients across the portal-drained viscera and liver of meat goat wethers consuming diets differing in protein concentration. *Small Ruminant Research*, Vol. 57, No. 2-3, (March 2005), pp. 193-202, ISSN 0921-4488.
6. Bauman, D.E.; Perfield, J.W., Harvatine, K.J. & Baumgard, L.H. (2008). Regulation of Fat Synthesis by Conjugated Linoleic Acid, Lactation and the Ruminant Model. *Journal of Nutrition*, Vol. 138, No. 2 (February, 2008), pp. 403-409, ISSN 1541-6100
7. Bernard, L.; Leroux, C. & Chilliard, Y. (2006). Characterisation and nutritional regulation of the main lipogenic genes in the ruminant lactating mammary gland. In, *Ruminant physiology, digestion, metabolism and impact of nutrition on gene expression, immunology and stress*, Sejrsen, K., Hvelplund, T. & Nielsen, M.O. (Ed.), 295-326, Wageningen Academic Publishers, ISBN 978-953-7619-81-7 Wageningen, The Netherlands.
8. Brüsemeister, F. & Sudekum, K.H. (2006). Rumen-protected choline for dairy cows: the in situ evaluation of a commercial source and literature evaluation of effects on performance and interactions between methionine and choline metabolism. *Animal Research*, Vol. 55, No. 2 (March-April 2006), pp. 93-104, ISSN 1627-3591
9. Chung, Y.-H.; Brown, N. E., Martinez, C. M., Cassidy, T. W. & Varga, G. A. (2009). Effects of rumen-protected choline and dry propylene glycol on feed intake and blood parameters for Holstein dairy cows in early lactation. *Journal of Dairy Science*, Vol. 92, No. 6, (June 2009), pp. 2729–2736, ISSN 0022-0302
10. Cooke, R.F.; Silva Del Río, N., Caraviello, D.Z., Bertics, S.J., Ramos, M.H. & Grummer, R.R. (2007). Supplemental Choline for Prevention and Alleviation of Fatty Liver in Dairy Cattle. *Journal of Dairy Science*, Vol. 90, No. 5, May 2007, pp. 2413-2418, ISSN 0022-0302
11. Davidson, S.; Hopkins, B.A., Odle, J., Brownie, C., Fellner, V. & Whitlow, L.W. (2008). Supplementing limited methionine diets with rumen-protected methionine, betaine, and choline in early lactation holstein cows. *Journal of Dairy Science*, Vol. 91, Vol. 91, No. 4 (April 2008), pp 1552-1559, ISSN 0022-0302
12. Dawson, R. M. C.; Grime, D. W., & Lindsay, D. B. (1981). On the insensitivity of sheep to the almost complete microbial destruction of dietary choline before alimentary-tract absorption. *Biochemical Journal*, Vol. 196, No. 2 (May 1981), pp. 499-504, ISSN 1470-8728
13. Deuchler, K. N.; Piperova, L. S. & Erdman, R. A. (1998). Milk choline secretion as an indirect indicator of postruminal choline supply. *Journal of Dairy Science*, Vol. 81, No. 1, (January 1998), pp. 238-242, ISSN 0022-0302
14. Dilger, R.N.; Garrow, T. A., & Baker, D. H. (2007). Betaine can partially spare choline in chicks but only when added to diets containing a minimal level of choline. *Journal of Nutrition*, Vol. 137, No. 10 (October 2007), pp. 2224-2228, ISSN 0022-3166
15. Donkin, S.S., (July, 2011) Rumen-Protected Choline, 11.07.2011, Available from http://www.extension.org/pages/26158/rumen-protected-choline#

16. Elek, P.; Newbold, J.R., Gaal, T., Wagner, L. & Husveth, F. (2008). Effects of rumen-protected choline supplementation on milk production and choline supply of periparturient dairy cows. *Animal*, Vol. 2, No. 11 (November 2008), pp.1595-1601.
17. Emmanuel, B. & Kennelly, J. J. (1984). Kinetics of methionine and choline and their incorporation into plasma lipids and milk components in lactating goats. Journal of Dairy Science, Vol. 67, No. 9 (September 1984), pp. 1912-1918, ISSN 0022-0302
18. Erdman, R. A. & Sharma, B. K. (1991). Effect of dietary rumen-protected choline in lactating dairy cows. *Journal of Dairy Science*, Vol. 74, No. 5 (May 1991), pp. 1641-1647, ISSN 0022-0302
19. Erdman, R. A. (1994). Production responses in field study herds fed rumen protected choline. *Journal of Dairy Science*, Vol. 77(Suppl. 1), pp 186. (Abstr.) , ISSN 0022-0302
20. Erdman. R. A.; Shaver, R. D., & Vandersall, J. H. (1984). Dietary choline for the lactating cow: possible effects on milk fat synthesis. *Journal of Dairy Science*, Vol. 67, No. 2 (February 1984), pp. 410-415, ISSN 0022-0302
21. Food and Nutrition Board (1998), Institute of Medicine. Choline, In *Dietary Reference Intakes: Thiamin, Riboflavin, Niacin, Vitamin B-6, Vitamin B012, Pantothenic Acid, Biotin, and Choline*, National Academy of Sciences, ISBN 0-39-06554-2, pp 390-422, Washington,
22. D.C., US
23. Girard, C. L.; Preynat, A., Graulet, B & Lapierre, H. (2010). Could 'minor' nutrients such as B vitamins alter major metabolic pathways in lactating dairy cows?, In: Energy and protein metabolism and nutrition, G.M. Crovetto (Ed.), pp. 235-243, Wageningen Academic Publishers (EAAP publication N°127), ISBN978-90-8686-153-8, Wageningen. The Netherlands
24. Griffith O. W. (1987). Mammalian sulfur amino acid metabolism: an overview. *Methods Enzymology*, Vol. 143, pp 366-376, ISSN 0076-6879
25. Gruffat, D.; Durand, D., Graulet, B., & Bauchart, D. (1996). Regulation of VLDL synthesis and secretion in the liver. *Reproduction Nutrition Development*, Vol. 36, No. 4, pp. 375-389. ISSN 1297-9708
26. Grummer, R. R. (1993). Etiology of lipid related metabolic disorders in periparturient dairy cattle. *Journal of Dairy Science*, Volume 76, No. 12 (December 1993), pp. 3882-3896, ISSN 0022-0302
27. Grummer, R.R. (2006). Etiology, pathophysiology of fatty liver in dairy cows. In *Production diseases in farm animals*, Joshi, N.P. & Herdt, T.H. (Ed.), pp. 141-153. Wageningen Academic Publishers, ISBN 9076998574, Wageningen. The Netherlands
28. Grummer, R.R. (July 2011). Managing the Transition Cow-Emphasis on Ketosis and Fatty Liver Syndrome, 15.07.2011, Available from http://www.vtdairy. dasc.vt.edu/ pdf/cow_college/2011/18%20grummer-paper.pdf
29. Hartwell, J. R.; Cecava, M. J. & Donkin, S. S. (2000). Impact of dietary rumen undegradable protein and rumen-protected choline on intake, peripartum liver triacylglyceride, plasma metabolites and milk production in transition dairy cows. *Journal of Dairy Science*, Vol. 83No. 12 (December 2000), pp. 2907-2917, ISSN 0022-0302
30. Harvatine, K.J. & Bauman, D.E. (2006). SREBP1 and thyroid hormone responsive spot 14 (S14) are involved in the regulation of bovine mammary lipid synthesis during dietinduced milk fat depression and treatment with CLA. *Journal of Nutrition* 136, No. 10 (October 2006), pp. 2468–2474.
31. Henderson, G. D.; Xue, G. P. & Snoswell, A. M. (1983). Carnitine and creatine content of tissues of normal and alloxan-diabetic sheep and rats. *Comparative Biochemistry and Physiology Part B: Comparative Biochemistry*, Vol. 76B, No. 2, pp. 295-298, ISSN: 1096-4959
32. Holtenius, P. (1989). Plasma lipids in normal cows around partus and in cows with metabolic disorders with and without fatty liver. *Acta Veterinaria Scandinavica*, Vol. 30, No. 4 (December 1989), pp. 441-445, ISSN: 1751-0147
33. Janovick Guretzky, N.A.; Carlson, D.B., Garrett, J.E. & Drackley, J.K. (2006). Lipid metabolite profiles and milk production for holstein and jersey cows fed rumen-protected choline during the periparturient period. *Journal of Dairy Science*, Vol. 89, No. 1, (January 2006), pp. 188-200, ISSN 0022-0302

34. Kempson, S.A. & Montrose, M.H. (2004). Osmotic regulation of renal betaine transport: transcription and beyond. *Euroopean Journal Physiology*, Vol. 449, No. 3 (December 2004), pp. 227–234 ISSN0 031-6768
35. Kennedy, G. D.; Young, P. B., Kennedy, S., Scott, J. M., Molloy, A. M., Weir, D. G., & Price J. (1995). Cobalt-vitamin B-12 deficiency and the activity of methylmalonyl CoA mutase and methionine synthase in cattle. *International Journal for Vitamin and Nutrition Research*, Vol. 65, No. 4, pp. 241-247, ISSN: 03009831
36. Kinsella, J. E. (1973). Preferential labelling of phosphatidylcholine during phospholipid synthesis by bovine mammary tissue. *Lipids*, Vol. 8, No. 7 (July 1973), pp. 393-400, ISSN 0024-4201
37. Kuksis, A. & Mookerjea, S. (1978). Choline. *Nutrition Reviews*, Vol. 36, No. 7 (July 1978), pp. 201–207, ISSN: 0029-6643
38. LaCount, D. W.; Drackley, J. K., & Weigel, D. J. (1995). Responses of dairy cows during early lactation to ruminal or abomasal administration of L-carnitine. *Journal of Dairy Science*, Vol. 78, No. 8 (August 1995), pp. 1824-1836, ISSN 0022-0302
39. LaCount D. W.; Emmert L. S., & Drackley J. K. (1996). Dose response of dairy cows to abomasal administration of four amounts of L-carnitine. *Journal of Dairy Science*, Vol. 79, No. 4 (April 1996), pp. 591-602, ISSN 0022-0302
40. Lobley, G. E.; Connell, A., & Revell, D. (1996). The importance of transmethylation reactions to methionine metabolism in sheep: effect of supplementation with creatine and choline. *British Journal of Nutrition*, Vol. 75, No. 1 (January 1996), pp. 47-56, ISSN 1475- 2662
41. Lock, A.L.; Rovai, M., Gipson, T.A., De Veth, M. J. & Bauman, D. E. (2008). A conjugated linoleic acid supplement containing Trans-10, Cis-12 Conjugated Linoleic Acid reduces milk fat synthesis in lactating goats. *Journal Dairy Science*, 91, No. 9 (September 2008), pp. 3291-3299, ISSN 0022-0302
42. Mato, J. M.; Alvarez, L., Corrales, F. J., & Pajares, M. A. (1994). S-adenosylmethionine and the liver. In *The Liver: Biology and Pathobiology*, I. M. Arias, J. L. Boyer, N. Fausto, W. B. Jakoby, D. A. Schachter, & D. A. Shafritz (Ed.), pp. 461-470, Raven Press Ltd, ISBN 0- 7817-0133-3, New York, US
43. McCarthy, R. D.; Porter, G. A., & Griel, L. C., Jr. (1968). Bovine ketosis and depressed fat test in milk: a problem of methionine metabolism and serum lipoprotein aberration. *Journal of Dairy Science*, Vol. 51, No. 6 (June 1968), pp. 459-462, ISSN 0022-0302
44. McCarthy, R. D.; Porter, G. A. & Griel Jr., L.C. (1968). Bovine ketosis and depressed fat test in milk: a problem of methionine metabolism and serum lipoprotein aberration. *Journal of Dairy Science*, Vol.51, No.3, (March 1968), pp. 459-462, ISSN 0022-0302
45. McDowell, L.R. (1989). Choline. In: *Vitamins in animal nutrition: comparative aspect to human nutrition*, L.R. McDowell (Ed.), New York: Academic Press, Inc., pp. 347-364, ISBN 0-12-438372-1
46. McPherson, A. V. & Kitchen, B. J. (1983). Reviews of the progress of dairy science: the bovine milk fat globule membrane-its formation, composition, structure and behavior in milk and dairy products. *Journal of Dairy Research*, Vol.50, No.1, pp. 107-133, ISSN: 0022-0299
47. Moore, J. H. & Christie, W. W. (1981). Lipid metabolism in the mammary gland of ruminant animals. In: *Lipid Metabolism in Ruminant Animals.* W.W. Christie (Ed.), Pergamon Press, pp. 227-278, ISBN 0-08-023789-4, Oxford, UK.
48. Neill, A. R.; Grime, D. W. & Dawson, R. M. C. (1978). Conversion of choline methyl groups through trimethylamine into methane in the rumen. *Biochemical Journal*, Vol.170, No. 3, pp. 529-535, ISSN 0264-6021
49. Neill, A. R.; Grime, D. W.; Snoswell, A. M.; Northrop, A. J.; Lindsay, D. B. & Dawson, R. M. C. (1979). The low availability of dietary choline for the nutrition of the sheep. *Biochemical Journal*, Vol.180, No. 3, pp. 559-565, ISSN 0264-6021
50. National Research Council (2001). Vitamins. In: *Nutrient Requirements of Dairy cattle*, Seventh Revised Edition, Subcommittee on Dairy Cattle Nutrition, Committee on Animal
51. Nutrition, Board on Agriculture and Natural Resource, pp. 162-177, National Academy Press, ISBN-10:0-309-06997-1 Washington DC:

52. Preynat A.; Lapierre, H., Thivierge, M.C., Palin, M.F., Matte, J.J., Desrochers, A., &
53. Girard, C.L. (2009). Effects of supplements of folic acid, vitamin B12, and rumenprotected methionine on whole body metabolism of methionine and glucose in lactating dairy cows. *Journal Dairy Science*, Vol. 92, No.2 (January 2009), pp 677-689, ISSN 0022-0302
54. Piepenbrink, M.S. & Overton, T.R. (2003). Liver metabolism and production of cows fed increasing amounts of rumen-protected choline during the periparturient period. *Journal Dairy Science*, Vol.86, No. 5, (May 2003), pp. 1722-1733, ISSN 0022-0302
55. Pinotti, L.; Baldi, A.; Politis, I.; Rebucci, R.; Sangalli, L. & Dell'Orto, V. (2003). Rumen protected choline administration to transition cows: effects on milk production and Vitamin E status. *Journal of Veterinary Medicine Series A*, Vol.50, (February 2003) No.1, pp. 18-21, ISSN 0931–184X
56. Pinotti, L.; Baldi, A. & Dell'Orto, V. (2002). Comparative mammalian choline metabolism with emphasis on role in ruminants, especially the high yielding dairy cow. *Nutrition Research Review*, Vol. 15, No.2, (December 2002), pp. 315-331L, ISSN 0954-4224
57. Pinotti L.; A. Campagnoli, V. Dell'Orto, A. Baldi (2005). Choline: Is there a need in lactating dairy cow? *Livestock Production Science*, Vol. 98, No. 1-2 (1 December 2005), pp 149-152, ISSN: 0301-6226
58. Pinotti, L.; Campagnoli, A.; Sangalli, L., Rebucci, R., Dell'Orto, V. & Baldi, A. (2004). Metabolism of periparturient dairy cows fed rumen-protected choline. *Journal Animal and Feed Science*, Vol.13, Suppl. 1 (August 2004), pp. 551–554, ISSN 1230-1388
59. Pinotti, L.; Polidori, C., Campagnoli, A., Dell'Orto, V. & Baldi, A. (2010). A meta-analysis of the effects of rumen protected choline supplementation on milk production in dairy cows. In: *Energy and protein metabolism and nutrition*, G.M. Crovetto (Ed.), pp. 321-322, Wageningen Academic Publishers (EAAP publication N°127) ISBN978-90-8686-153-8, Wageningen, The Netherlands.
60. Pullen, D.L.; Liesman, J. S. & Emery, R. S. (1990). A species comparison of liver slice synthesis and secretion of triacylglycerol from non-esterified fatty acids in media. *Journal of Animal Science*, Vol.68, No.5, (May 1990), pp. 1395-1399, ISSN 0021-8812
61. Robinson, B. S.; Snoswell, A. M., Runciman, W. B. & Upton, R. N. (1984). Uptake and output of various forms of choline by organs of conscious chronically catherized sheep. *Biochemical Journal*, Vol. 217, No. 2, pp. 399-408, ISSN 0264-6021
62. Ruiz, N.; Miles R. D. & Harms, R. H. (1983). Choline, methionine and sulphate interrelationships in poultry nutrition - A review. *World's Poultry Science Journal*, Vol.39, pp. 185-198, ISSN 0043-9339
63. Rukkwamsuk, T.; Wensing, T. & Geelen, M. J. H. (1999). Effect of fatty liver on hepatic gluconeogenesis in periparturient dairy cows. *Journal of Dairy Science*, Vol. 82, No.3, (March 1999), pp. 500-505, ISSN 0022-0302
64. Sales, J.; Homolkaa, P. & Koukolováa, V. (2010). Effect of dietary rumen-protected choline on milk production of dairy cows: A meta-analysis. *Journal of Dairy Science*, Vol. 93, No. 8, (August 2010), pp. 3746-3754, ISSN 0022-0302
65. Savoini, G.; Agazzi, A., Invernizzi, G., Cattaneo, D., Pinotti, L., & Baldi, A. (2010). Polyunsaturated fatty acids and choline in dairy goats nutrition: production and health benefits. *Small Ruminants Research*, Vol. 88, No. 1-2 (February 2010), 135-144, ISSN 0921-4488
66. Scheer, W.A.; Lucy, M.C., Kerley, M.S. & Spain, J. N. (2002). Effects of feeding soybeans and rumen protected choline during late gestation and early lactation on performance of dairy cows. *Journal of Dairy Science* , Vol. 85, Suppl.1, pp. 276, ISSN 0022-0302
67. Scott, J. M. (1999). Folate and vitamin B12. Proceedings of the Nutrition Society, Vol. 58, No. 2 (June 1999) pp. 441-448, ISSN 0029-6651
68. Sharma, B. K. & Erdman, R. A. (1988a). Abomasal infusion of choline and methionine with or without 2-amino-2-methyl-1-propanol for lactating dairy cows. *Journal of Dairy Science*, Vol. 71, No. 9, (September 1998), pp. 2406-2411, ISSN 0022-0302
69. Sharma, B. K. & Erdman, R. A. (1988b). Effects of high amounts of dietary choline supplementation on duodenal choline flow and production responses of dairy cows. *Journal of Dairy Science*, Vol. 71, No.10, (October 1988), pp. 2670-2676, ISSN 0022-0302

70. Sharma, B. K. & Erdman, R. A. (1989a). In vitro degradation of choline from selected feedstuffs and choline supplements. *Journal of Dairy Science*, Vol. 72, No. 10, (October 1989), pp. 2771-2776, ISSN 0022-0302
71. Sharma, B. K. & Erdman, R. A. (1989b). Effects of dietary and abomasally infused choline on milk production responses of lactating dairy cows. *Journal of Nutrition*, Vol. 119, No. 2, (February 1989), pp. 248-254, ISSN 0022-3166
72. Snoswell, A. M. & Xue, G. P. (1987). Methyl group metabolism in sheep. *Comparative Biochemistry and Physiology Part B: Comparative Biochemistry*, Vol. 88, No. 2, pp. 383-394, ISSN 1096-4959
73. Stipanuk, M. H. (1986). Metabolism of sulfur-containing amino acids. *Annual Reviews Nutrition*, Vol. 6, pp. 179-209, ISSN 0199-9885
74. Ueland, P.M. (2011). Choline and betaine in health and disease. *Journal of Inherited Metabolic Disease*, Vol. 34, No.1 , (February 2011), pp. 3–15, ISSN 0141-8955
75. Vuuren, A. M.; van Goselink, R. M. A., Zom, R., Veth, M. J. de, Baal, J. van (2010). Effect of rumen-protected choline on performance and hepatic triacylglycerol concentrations in early-lactating dairy cattle. In: *Energy and protein metabolism and nutrition*, G.M. Crovetto (Ed.), pp. 493-494, Wageningen Academic Publishers (EAAP publication N°127) ISBN978-90-8686-153-8, Wageningen, The Netherlands.
76. Whitehead, C. C. & Portsmouth, J. I (1989). Vitamin requirements and allowances for poultry. In: *Recent Advances in Animal Nutrition*, Haresign, W. & Cole D. A. J., (Ed.), 35- 86, Butterworths, ISBN 0-407-01165X, London, UK.
77. Xu, G.; Ye, J., Liu, J. & Yu, Y. (2006). Effect of rumen-protected choline addition on milk performance and blood metabolic parameters in transition dairy cows. *Asian-Australasian Journal of Animal Sciences*, Vol. 19, No. 3, pp. 390-395, ISSN 1011-2367
78. Xue, G. P. & Snoswell, A. M. (1985). Regulation of methyl group metabolism in lactating ewes. *Biochemistry International*, Vol. 11, No. 3, (September 1989), pp. 381-385, ISSN 0158-5231
79. Zahra, L.C.; Duffield, T.F., Leslie, K.E., Overton, T.R., Putnam, D. & LeBlanc, S.J. (2006). Effects of Rumen-Protected Choline and Monensin on Milk Production and Metabolism of Periparturient Dairy Cows. *Journal Dairy Science*, Vol. 89, No. 12, (December 2006), pp. 4808-4818, ISSN 1811-9743
80. Zeisel, S. H. (1988). "Vitamin-like" molecules. In: *Modern Nutrition and Health and Disease*, Shils M. & Young V., (Ed.), 440-452, ISBN-10 0781741335, Philadelphia: Lea & Febiger.
81. Zeisel, S. H. (1992). Choline: an important nutrient in brain development, liver function and carcinogenesis. *Journal American College of Nutrition*, Vol. 11, No. 5, pp. 473-481, ISSN 0731-5724
82. Zeisel S. H., & Da Costa K. A. (2009). Choline: An Essential Nutrient for Public Health. *Nutrition Reviews*, Vol. 67, No. 11, (November 2009), pp. 615–623, ISSN 0029-6643
83. Zeisel, S.H.; Da Costa, K.A., Franklin, P.D., Alexander, E.A., Lamont, J.T, Sheard, N.F., and A. Beiser Choline, an essential nutrient for humans. *The FASEB Journal*, Vol. 5, pp. 2093-2098, ISSN 0892-6638
84. Zeisel, S. H.; Mar, M., Zhou, Z. & Da Costa, K. A. (1995). Pregnancy and lactation are associated with diminished concentrations of choline and its metabolites in rat liver. *Journal of Nutrition*, Vol. 125, No. 12, pp. 3049-3054, ISSN 0022-3166

CHAPTER 7

Nutrition, Metabolic Status and Reproductive Efficiency in Dairy Herds

György Gábor[1*], Orsolya Gabriella Balogh[1], László Kern[1], Petra Réka Gábor[2], Hedvig Fébel[1]

[1]*National Agricultural Research Institute—Research Institute for Animal Breeding, Nutrition and Meat Science, Herceghalom, Hungary*

[2]*AndrovetKft, Budapest, Hungary*

ABSTRACT

Nutrition of transition period in dairy cows dramatically increases requirements for energy, glucose, amino acids and other nutrients in dairy cows and, parallel to that, feed intake is depressed. An insufficient or imbalanced nutrient intake might lead to the malfunction of organs, causing reproductive failure. To minimize the metabolic/reproductive damage caused by negative energy balance (NEBAL), feeding of a glucogenic and lipogenic diet containing fat-enriched supplements is recommended. In heat-stressed dairy cows, the reduction of dry matter intake causes an energy deficit, thereby accentuating NEBAL. Heat stress may also have a direct impact on oocyte quality due to the high ovarian temperatures. The intrauterine environment is also compromised in heat- stressed cows, with reduced blood flow to the uterus and increased uterine temperature; this may impair embryonic development, increase early embryonic loss and reduce the proportion of successful inseminations. Postpartum NEBAL suppresses immune function and promotes metabolic disorders, potentially explaining relationships between infectious and noninfectious transition disorders. In postpartum dairy cows, pathologic ovarian phenomena including ovulatory (cystic corpora lutea) and non-ovulatory (follicular and luteal cysts) forms frequently occur causing decreased fertility. The main objective of this review was summarizing the most relevant information regarding nutrition and reproduction in dairy cows.

Keywords: Dairy Cattle, Reproduction, Nutrition, Metabolic Status

1. INTRODUCTION

Efficient reproduction in dairy cattle herds is of great economic importance. Increased milk production together with suboptimal farm management (especially nutrition or reproduction) can reduce fertility. Over the past decades, the calving interval has increased in parallel with milk production in several countries including the US and Hungary (Table 1 and Figure 1).

Extreme selection pressure for milk production in dairy cows has been highly successful. Unfortunately, there are clear negative correlations between milk yield and fertility, presence of mastitis, and other health traits. The decline in fertility is of multifactorial origin and not exclusively associated with increased milk production. To improve the decreased fertility, cow longevity, health and fertility have been used as selection criteria in many breeding programs in numerous countries [4] . Factors such as management and husbandry, nutrition and reproductive diseases determine reproductive performance (number of open days, calving interval, number of AI per pregnancy, pregnancy rate, and early/late embryonic loss). Over the past few years, several trials have been conducted to determine the effect of a more economical dry period length [5] on the postpartum health [6] , production [7] [8] , and reproduction [9] [10] of dairy cows.

2. NUTRIENT INTAKE, METABOLIC AND HORMONAL STATUS AND REPRODUCTIVE FAILURES

The major problem in postpartum cows is the imbalance between the body reserves and milk production. The transition (late pregnancy to early lactation) period dramatically increases requirements for energy, glucose, amino acids and other nutrients in dairy cows and, parallel to that, feed intake is depressed. This is why a negative energy balance (NEBAL) occurs in the background of reduced fertility (Figure 2).

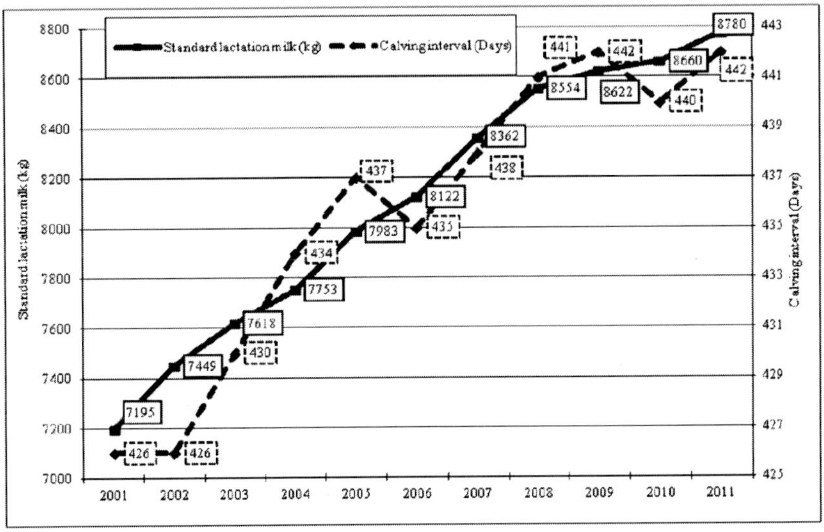

Figure 1. Changes of milk production and calving interval in Hungary between 2001 and 2011 [3] .

Table 1. Change of the calving interval (days) in the Holstein breed in the US.

Parity	Nieuwhof *et al.*, 1989 [1]	Hare *et al.*, 2006 [2]
1	394.2	402.9
2	392.8	402.4
3	394.6	403.4
4	398.1	406.2
5	400.4	408.9
6	404.7	410.8
7		412.9

Since the transition period involves the most important physiological changes of the time around calving, it is a favorable target for reproduction and nutrition physiologists. Both calving and the start of milking act as stress factors for cows, influencing metabolic and hormonal changes (Figure 3).

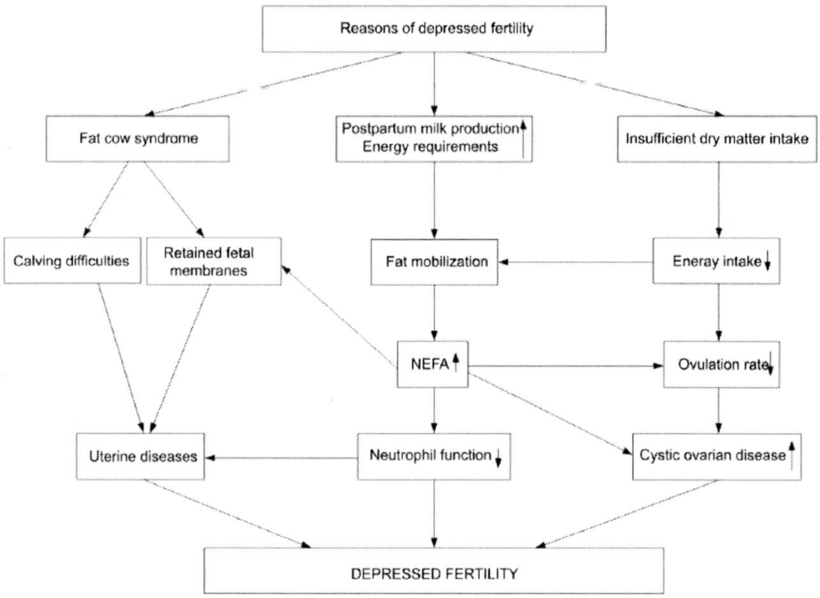

Figure 2. Most common causes of depressed fertility.

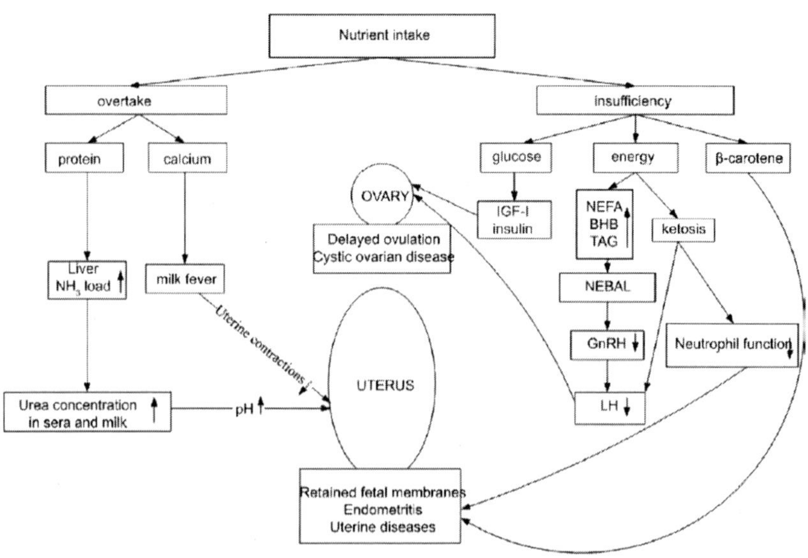

Figure 3. Effect of possible nutritional factors on ovaries and uterus in the transition period.

An insufficient or imbalanced nutrient intake might lead to the malfunction of organs (liver, hypothalamus, hypophysis, ovary, uterus), causing reproductive failure.

3. CALVING AND IMMUNE FUNCTION

It is estimated that approximately 30% - 50% of cows are affected by some form of metabolic and/or infectious diseases in the periparturient period. Calving itself represents a major oxidative stress for cows. Oxidative stress is an imbalance between radical-generating and radical-scavenging activity, resulting in oxidation products and tissue damage. The production of free radicals and reactive metabolites is one of the basic mechanisms underlying reproductive diseases, and antioxidant status is a major determinant of reproductive function in farm animals [11].

Calving disorders have a strong influence on the postpartum period. Retained fetal membranes (RFM) and puerperal metritis are more frequent when calving is assisted. During the postpartum period, rapid and uneventful involution of the uterus, early resumption of normal ovarian activity, and accurately detected estrus are required. Most cows develop a mild non-pathological endometritis during the early puerperal phase.

This physiological condition, together with reduced immunocompetence due to high milk production, NEBAL and poor calving hygiene, can cause puerperal endometritis [12] . Metritis causes endometrial damage, which can delay conception.

4. DIFFERENT DIETARY SOLUTIONS FOR THE PREVENTION OF METABOLIC IMBALANCE AND REPRODUCTIVE DISORDERS

To minimize the metabolic/reproductive damage caused by NEBAL, feeding of a glucogenic and lipogenic diet containing fat-enriched supplements is recommended [12] . This issue has been the focus of an extremely high number of studies, which sometimes brought contradictory results (Table 2).

Gluconeogenic materials (propylene glycol, glycerol, propionic acid) are precursors of glucose in the liver and increase glucose concentration of the serum. Elevated serum glucose concentration results in decreased fat mobilization (lower energy demand and increased insulin level), and indirectly improves reproductive efficiency. Gluconeogenic feed additives are thought to exert their favorable effect through the stimulation of insulin secretion [27].

Propylene glycol seems to be the most effective gluconeogenic substance, as it was found to decrease the plasma non-esterified fatty acid (NEFA) and beta-hydroxybutyrate (BHB) levels and increase the blood insulin and glucose concentrations in several experiments. Propylene glycol and its metabolites (lactate, pyruvate) directly stimulate insulin secretion [28].

Glycerol supplementation seems to have moderate metabolic effects: it did not modify plasma glucose, insulin, NEFA and BHB levels [29] . However, in another trial [30] 1 kg glycerol applied as a drench increased plasma glucose and insulin and decreased plasma NEFA concentration.

Table 2. Literature review of the effect of dietary gluconeogenic and fat supplementation on reproductive efficiency.

Nutritional factor	Product	References	Comment
Gluconeogenic feed additives	Propylene glycol (PGL)	[13]	Despite evidence of improved metabolic status, PGL failed to increase LH pulse frequency, and failed to increase the proportion of first postpartum follicle waves resulting in ovulation
		[14]	Reduced interval from calving to first ovulation was observed and pulsatile release of LH at 2 and 5 weeks was greater in cows fed PGL
		[15]	Cows fed PGL commenced ovarian cyclicity earlier than controls
		[16]	Improved conception rate (33 to 57%) and day of first ovulation from 44.5 Day to 32.3 Day
	PGL + Ca-propionate	[17]	Treatment did not change the proportion of puerperal diseases and reproductive performance
	Glycerol	[18]	No effect on DMI, glucose ↑, NEFA and BHB ↓
		[19]	Conception rate and insulin cc. ↑
	Propionate salts	[20]	Improves reproductive efficiency
Fat supplementation	Ca-long chain fatty acid (LCFA)	[21]	Pregnancy rate per AI increased from 35.5% to 51.1%
	Fish oil	[22]	Reduced PGFM
	Ca-LCFA of palm oil	[23]	From cows fed 800 g fat/day a higher percentage of oocytes developed into blastocysts
	Soybean oil refining by-products (SORB)	[24]	Loss of body condition was less and first ovulation was experienced earlier in cows fed SORB
	Saturated fatty acid (FA), Ca-salts of t-octadecanoic FA, Ca-salts of safflower oil FA	[25]	Mean concentrations of glucose, IGF-1 and P4 were higher in cows fed a safflower oil enriched diet
	Ca salts of palm oil, Ca salts of safflower oil (SO) and fish oil (FO)	[26]	Overall pregnancy rate per AI was higher in cows fed an SO diet and pregnancy loss was reduced in FO-fed cows

The sometimes contradictory results can be attributed to the different experimental conditions (e.g. dosage, duration and method of, application, lactation period, level of NEBAL, and certainly the body condition of cows).

The feed intake of cows is lower in early lactation, and therefore dietary fat supplementation in the early postpartum period alters energy status even though a more energy dense ration is fed. Increasing the caloric density of the ration by fat feeding might improve the reproductive parameters of cows. Improvement of the energy status of dairy cows has been found to influence reproduction by increasing the frequency of LH pulses [31] , modulating the number of larger follicles, altering the size of the dominant follicle, and increasing progesterone concentrations during the luteal phase of the estrous cycle [32] . Supplemental fat improves ovarian follicular dynamics, which is in a positive correlation with pregnancy rate [33] [34] . The use of fat in dairy cow diets usually increases the energy density of the ration and improves reproduction; however, these effects might be mediated by the fatty acid (FA) composition of the fat source. Polyunsaturated fatty acids (PUFA) can influence different levels of the hypothalamo-pituitary-ovarian axis, and they can locally stimulate follicular growth by altering hormone and metabolite levels. Fatty acids are precursors for ovarian steroidogenesis and for the synthesis of endometrial eicosanoids ($PGF_{2\alpha}$, PGE_2). Different PUFAs are important mediators of uterine involution and the ovulatory processes. Feeding sunflower oil rich in linoleic acid (C18:2, n-6) promotes uterine involution. Supplementation of linseed (α-linolenic acid, C18:3, n-3) or the feeding of fish oil containing a high amount of eicosapentaenoic acid (C20:5, n-3) and docosahexaenoic acid (C22:6, n-3) may inhibit endometrial $PGF_{2\alpha}$ synthesis through mechanisms such as the decreased availability of its precursor, arachidonic acid. PUFAs of the n-3 family suppress luteolytic signals around the maternal recognition of pregnancy. This might improve the mechanism of embryo preservation and support embryonic survival in cattle.

The natural defense of postpartum cows is based on both cellular and humoral (local) immunity. An increased level of antioxidants should prevent the depression of neutrophil function caused by free radicals in the postpartum period. It was found that adequate beta-carotene supplementation significantly reduced the incidence of RFM and metritis [35] . Administration of vitamin E and/or selenium has been reported to reduce the incidence of postpartum reproductive disorders such as RFM, metritis and cystic ovaries, and improved fertility in cattle. Circulating NEFA and BHB are known to have a direct negative effect on neutrophil function.

5. ENVIRONMENTAL FACTORS INFLUENCING FERTILITY THROUGH METABOLIC IMBALANCE

Since 2002, two databases [Catus: ultrasound examinations of the cow's reproductive tract; Bopella: laboratory examination of pregnancy and late embryonic loss by pregnancy-specific protein B (PSPB) and P4 ELISA tests] have been developed in our institute for recording the metabolic and reproductive data of dairy herds (Figure 4). These databases provide an opportunity for analyzing relationships between the results of ultrasonic, endocrine and metabolic measurements.

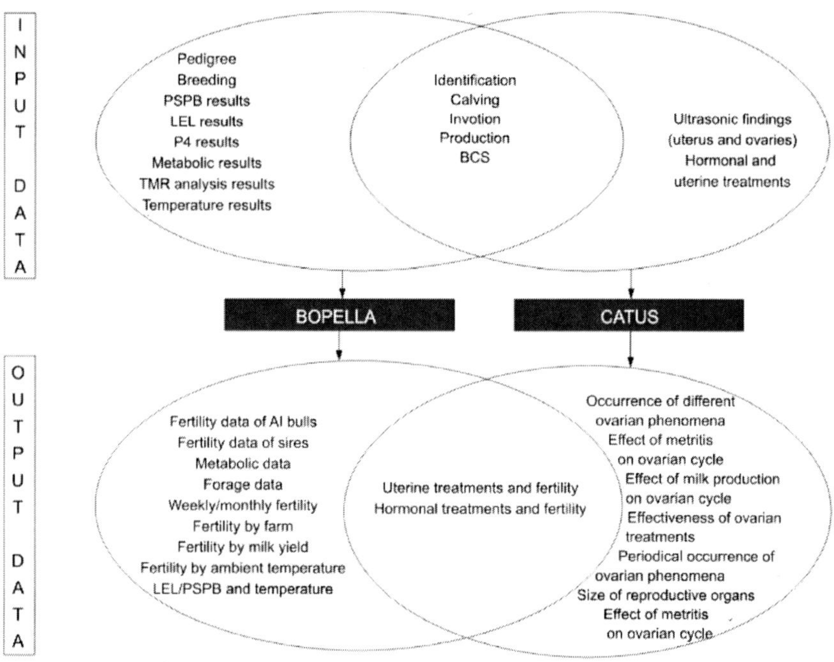

Figure 4. Structure of Bopella and Catus databases.

5.1. Effect of Ambient Temperature on Postpartum Reproductive Performance

Extremely high ambient temperature may cause heat stress, which considerably impacts production and reproduction in dairy cattle. In heat-stressed dairy cows, the reduction of dry matter intake causes an energy deficit, thereby accentuating NEBAL. In addition, reduced rumination and/or absorption of nutrients together with increased maintenance requirements also result in a net decrease of dry material/energy intake. Heat stress may also have a direct impact on oocyte quality due to the high ovarian temperatures. The intrauterine environment is also compromised in heat-stressed cows, with reduced blood flow to the uterus and increased uterine temperature; this may impair embryonic development, increase early embryonic loss and reduce the proportion of successful inseminations. During the hot summer period, poor expression of estrus, a rapid decrease in pregnancy rate and a pronounced increase of late embryonic loss (LEL) are common.

Based on the assessment of ~68,000 samples for serum pregnancy PSPB concentration and re-checking the pregnant cows by transrectal palpation, the rate of pregnant samples (RPS) decreased while LEL increased (Figure 5) parallel to the rise of daily average temperature [36] .

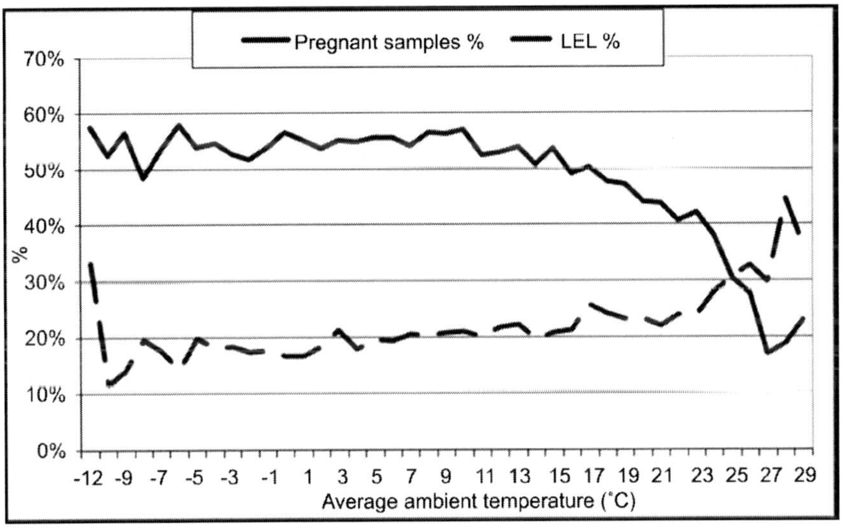

Figure 5. Relationship of the rate of pregnant samples and late embryonic losses (LEL) with ambient temperature [36] .

5.2. Effect of Ambient Temperature on Metabolic Status

The mean plasma NEFA and BHB concentrations were significantly higher (Figure 6 and Figure 7), while serum carotene and plasma IGF-I concentrations were significantly lower in summer than in winter. The high plasma NEFA concentration found in summer appeared to be in association with the lower body condition score (BCS) caused by depressed appetite. Statistical analysis supports the hypothesis that increased plasma NEFA and BHB and decreased plasma IGF-I concentrations may result in reduced fertility in summer. These changes may be associated with the more frequent appearance of cystic ovarian diseases (COD, Table 3) and probably have a negative effect on ovarian function and/or oocyte quality [37].

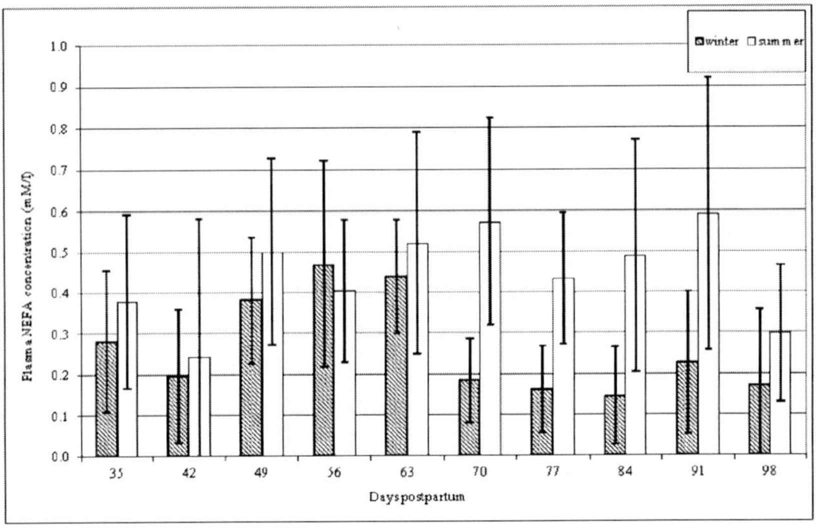

Figure 6. Plasma NEFA concentrations (mean ± SD) in summer and winter [37]

6. ULTRASONIC OVARIAN MORPHOLOGY IN POSTPARTUM COWS

Postpartum NEBAL (indicated by lower BCS as well as higher plasma NEFA and BHB levels) suppresses immune function and promotes metabolic disorders, potentially explaining relationships between infectious and noninfectious transition disorders. In postpartum dairy cows, pathologic ovarian phenomena including COD and ovulatory fluid-filled ovarian phenomena (corpus luteum with cavity and cystic corpus luteum) frequently occur. Some of these phenomena are associated with metritis, but due to the increased plasma concentrations of NEFA the possible effect of NEBAL could not be excluded either. Oocytes can accumulate fatty acids, which may change their lipid content and composition. The accumulation of lipid in oocytes and embryos can reduce their quality and cryotolerance.

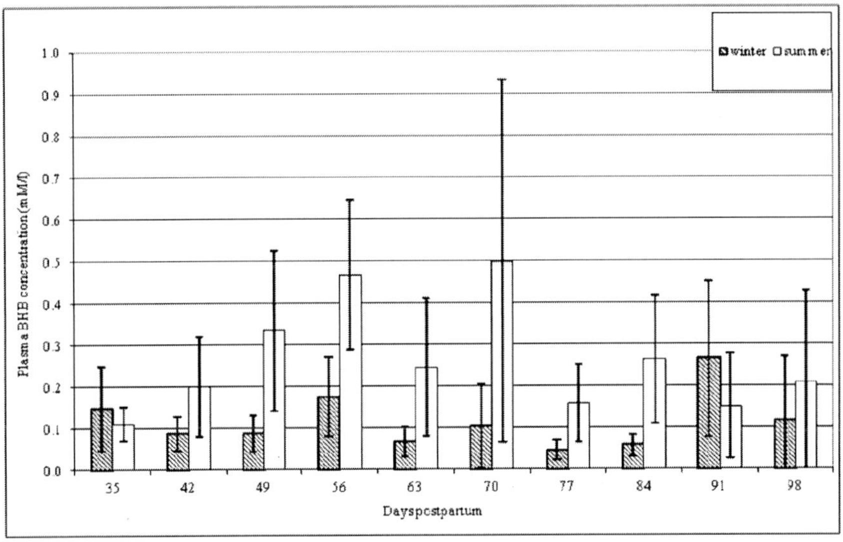

Figure 7. Plasma BHB concentrations (mean ± SD) in summer and winter [37].

Table 3. Frequency and rate of ovulatory (ovulatory fluid-filled ovarian phenomena, OFOP) and non-ovulatory (non-ovula- tory cystic ovarian phenomena, NCOP) ovarian phenomena (source: Catus database, 2008-2011).

	2008	2009	2010	2011
Number of ultrasound examinations	659	547	454	619
Number of detected NCOP	83	84	76	142
NCOP, %	12.6	15.4	16.7	22.9
Number of detected OFOP	119	61	59	70
OFOP, %	18.1	11.2	13.0	11.3
Daily milk production (kg) in NCOP cows	36.9	37.9	37.9	38.2
Daily milk production (kg) in OFOP cows	35.9	36.2	36.1	37.1

Ultrasound examination of dairy cows in the same Hungarian dairy herds between 2008 and 2011 showed a definite increase in the rate of non-ovulatory cystic ovarian phenomena (Table 3) in parallel with a decreased rate of ovulatory fluid-filled ovarian phenomena and increased milk production (Table 3).

7. CONCLUSION

Increased milk production together with suboptimal farm management (especially nutrition or reproduction) can reduce fertility. The major problem in postpartum cows is the imbalance between the body reserves and milk production. This is why a negative energy balance (NEBAL) occurs in the background and causes reduced fertility. An insufficient or imbalanced nutrient intake might lead to the malfunction of organs, causing reproductive failure. Supplementation of gluconeogenic materials (propylene glycol, glycerol, propionic acid) indirectly improves reproductive efficiency. Our recently developed databases (Catus and Bopella) provide an opportunity for analyzing relationships between the results of ultrasonic, endocrine and metabolic measurements.

ACKNOWLEDGEMENTS

This study was supported by NFÜ (National Development Agency) KMR_12-1-2012-0075 project.

REFERENCES

1. Nieuwhof, G.J., Powell, R.L. and Norman, H.D. (1989) Ages at Calving and Calving Intervals for Dairy Cattle in the United States. Journal of Dairy Science, 72, 685-692. http://dx.doi.org/10.3168/jds.S0022-0302(89)79160-8
2. Hare, E., Norman, H.D. and Wright, J.R. (2006) Trends in Calving Ages and Calving Intervals for Dairy Cattle Breeds in the United States. Journal of Dairy Science, 89, 365-370. http://dx.doi.org/10.3168/jds.S0022-0302(06)72102-6
3. Mészáros, G. (2012) Calving Period and Lactation in Hungarian HF Population between 2001 and 2011. Hungarian Dairy Herd Recording Ltd. Newsletter, 2001-2012.
4. Rodriguez-Martinez, H., Hultgren, J., Bage, R., Bergqvist, A.S., Svensson, C., Bergsten, C., et al. (2008) Reproductive Performance in High-Producing Dairy Cows: Can We Sustain It under Current Practice? In: I.V.I.S., Eds., IVIS Reviews in Veterinary Medicine, International Veterinary Information Service, Ithaca NY (www.ivis.org), Last Updated: 11-Dec-2008; R0108.1208.
5. Santschi, D.E., Lefebvre, D.M., Cue, R.I., Girard, C.L. and Pellerin, D. (2011) Economic Effect of Short (35-d) Compared with Conventional (60-d) Dry Period Management in Commercial Canadian Holstein Herds. Journal of Dairy Science, 94, 4734-4743. http://dx.doi.org/10.3168/jds.2010-3596
6. Santschi, D.E., Lefebvre, D.M., Cue, R.I., Girard, C.L. and Pellerin, D. (2011) Incidence of Metabolic Disorders and Reproductive Performance Following a Short (35-d) or

Conventional (60-d) Dry Period Management in Commercial Holstein Herds. Journal of Dairy Science, 94, 3322-3330. http://dx.doi.org/10.3168/jds.2010-3595

7. de Feu, M.A., Evans, A.C.O., Lonergan, P. and Butler, S.T. (2009) The Effect of Dry Period Duration and Dietary Energy Density on Milk Production, Bioenergetic Status, and Postpartum Ovarian Function in Holstein-Friesian Dairy Cows. Journal of Dairy Science, 92, 6011-6022. http://dx.doi.org/10.3168/jds.2009-2374
8. Collier, R.J., Annen-Dawson, E.L. and Pezeshki, A. (2012) Effects of Continuous Lactation and Short Dry Periods on Mammary Function and Animal Health. Animal, 6, 403-414. http://dx.doi.org/10.1017/S1751731111002461
9. Watters, R.D., Wiltbank, M.C., Guenther, J.N., Brickner, A.E., Rastani, R.R., Fricke, P.M., et al. (2009) Effect of Dry Period Length on Reproduction during the Subsequent Lactation. Journal of Dairy Science, 92, 3081-3090. http://dx.doi.org/10.3168/jds.2008-1294
10. Gümen, A., Keskin, A., Yilmazbas-Mecitoglu, G., Karakaya, E. and Wiltbank, M.C. (2011) Dry Period Management and Optimization of Post-Partum Reproductive Management in Dairy Cattle. Reproduction in Domestic Animals, 46, 11-17. http://dx.doi.org/10.1111/j.1439-0531.2011.01851.x
11. Leblanc, S. (2010) Assessing the Association of the Level of Milk Production with Reproductive Performance in Dairy Cattle. Journal of Reproduction and Development, 56, S1-S7. http://dx.doi.org/10.1262/jrd.1056S01
12. Thatcher, W.W., Santos, J.E.P. and Staples, C.R. (2011) Dietary Manipulations to Improve Embryonic Survival in cattle. Theriogenology, 76, 1619-1631. http://dx.doi.org/10.1016/j.theriogenology.2011.06.005
13. Butler, S.T., Pelton, S.H. and Butler, W.R. (2006) Energy Balance, Metabolic Status, and the First Postpartum Ovarian Follicle Wave in Cows Administered Propylene Glycol. Journal of Dairy Science, 89, 2938-2951. http://dx.doi.org/10.3168/jds.S0022-0302(06)72566-8
14. Chagas, L.M., Gore, P.J.S., Meier, S., Macdonald, K.A. and Verkerk, G.A. (2007) Effect of Monopropylene Glycol on Luteinizing Hormone, Metabolites, and Postpartum Anovulatory Intervals in Primiparous Dairy Cows. Journal of Dairy Science, 90, 1168-1175. http://dx.doi.org/10.3168/jds.S0022-0302(07)71603 X
15. Miyoshi, S., Pate, J.L. and Palmquist, D.L. (2001) Effects of Propylene Glycol Drenching on Energy Balance, Plasma Glucose, Plasma Insulin, Ovarian Function and Conception in Dairy Cows. Animal Reproduction Science, 68, 29-48. http://dx.doi.org/10.1016/s0378-4320(01)00137-3
16. Jaskowski, J.M., Nowak, W., Mikula, R., Wlodarek, J., Kostencka, E. and Olechnowicz, J. (2011) Prevention of Negative Energy Balance in the Transition Period-Implications for Plasma Metabolites, Production and Reproduction of Cows. Medycyna Weterynaryjna, 67, 647-652.
17. Melendez, P., Donovan, G.A., Risco, C.A., Littell, R. and Goff, J.P. (2003) Effect of Calcium-Energy Supplements on Calving-Related Disorders, Fertility and Milk Yield during the Transition Period in Cow Fed Anionic Diets. Theriogenology, 60, 843-854. http://dx.doi.org/10.1016/S0093-691X(03)00103-1
18. Wang, C., Liu, Q., Yang, W.Z., Huo, W.J., Dong, K.H., Huang, Y.X., et al. (2009) Effects of Glycerol on Lactation Performance, Energy Balance and Metabolites in Early Lactation Holstein Dairy Cows. Animal Feed Science and Technology, 151, 12-20. http://dx.doi.org/10.1016/j.anifeedsci.2008.10.009
19. Leon, A.O., Ceron, J.H. and Gutierrez, C.G. (2010) Glycerol Supplementation after Artificial Insemination Increases Conception Rate in Holstein Cows. Revista Mexicana de Ciencias Pecuarias, 1, 69-74.
20. Mulliniks, J.T., Kemp, M.E., Cox, S.H., Hawkins, D.E., Cibils, A.F., VanLeeuwen, D.M., et al. (2011) The Effect of Increasing Amount of Glucogenic Precursors on Reproductive Performance in Young Postpartum Range Cows. Journal of Animal Science, 89, 2932-2943. http://dx.doi.org/10.2527/jas.2010-3198
21. McNamara, S., Butler, T., Ryan, D.P., Mee, J.F., Dillon, P., O'Mara, F.P., et al. (2003) Effect of Offering Rumen- Protected Fat Supplements on Fertility and Performance in Spring-Calving Holstein-Friesian Cows. Animal Reproduction Science, 79, 45-56. http://dx.doi.org/10.1016/S0378-4320(03)00111-8

22. Mattos, R., Staples, C.R., Arteche, A., Wiltbank, M.C., Diaz, F.J., Jenkins, T.C., et al. (2004) The Effects of Feeding Fish Oil on Uterine Secretion of PGF2α, Milk Composition, and Metabolic Status of Periparturient Holstein Cows. Journal of Dairy Science, 87, 921-932. http://dx.doi.org/10.3168/jds.S0022-0302(04)73236
23. Fouladi-Nashta, A.A., Gutierrez, C.G., Gomg, J.G., Garnsworthy, P.C. and Webb, R. (2007) Impact of Dietary Fatty Acids on Oocyte Quality and Development in Lactating Dairy Cows. Biology of Reproduction, 77, 9-17.http://dx.doi.org/10.1095/ biolreprod. 106. 058578
24. Boken, S.L., Staples, C.R., Sollenberger, L.E., Jenkins, T.C. and Thatcher, W.W. (2005) Effect of Grazing and Fat Supplementation on Production and Reproduction of Holstein Cows. Journal of Dairy Science, 88, 4258-4272.http://dx.doi.org/10.3168/jds.S0022-0302(05)73112-X
25. Caldari-Torres, C., Lock, A.L., Staples, C.R. and Badinga, L. (2011) Performance, Metabolic, and Endocrine Responses of Periparturient Holstein Cows Fed 3 Sources of Fat. Journal of Dairy Science, 94, 1500-1510.http://dx.doi.org/10.3168/jds.2010-3748
26. Silvestre, F.T., Carvalho, T.S.M., Francisco, N., Santos, J.E.P., Staples, C.R., Jenkins, T.C., et al. (2011) Effects of Differential Supplementation of Fatty Acids during the Peripartum and Breeding Periods of Holstein Cows: I. Uterine and Metabolic Responses, Reproduction, and Lactation. Journal of Dairy Science, 94, 189-204.http://dx.doi.org/10.3168/jds.2010-3370
27. Hayirli, A. (2006) The Role of Exogenous Insulin in the Complex of Hepatic Lipidosis and Ketosis Associated with Insulin Resistance Phenomenon in Postpartum Dairy Cattle. Veterinary Research Communications, 30, 749-774.http://dx.doi.org/10.1007/s11259-006-3320-6
28. Nielsen, N.I. and Ingvartsen, K.L. (2004) Propylene Glycol for Dairy Cows: A Review of the Metabolism of Propylene Glycol and Its Effects on Physiological Parameters, Feed Intake, Milk Production and Risk of Ketosis. Animal Feed Science and Technology, 115, 191-213. http://dx.doi.org/10.1016/j.anifeedsci.2004.03.008
29. Defrain, J.M., Hippen, A.R., Kalscheur, F. and Jardon, P.W. (2004) Feeding Glycerol to Transition Dairy Cows: Effects on Blood Metabolites and Lactation Performance. Journal of Dairy Science, 87, 4195-4206.http://dx.doi.org/10.3168/jds.S0022-0302(04)73564-X
30. Linke, P.L., Defrain, J.M., Hippen, A.R. and Jardon, P.W. (2004) Ruminal and Plasma Responses in Dairy Cows to Drenching or Feeding Glycerol. Journal of Dairy Science, 87, 343.
31. Hightshoe, R.B., Cochran, R.C., Corah, L.R., Kiracofe, G.H., Harmon, D.L. and Perry, R.C. (1991) Effects of Calcium Soaps of Fatty Acids on Postpartum Reproductive Function in Beef Cows. Journal of Animal Science, 69, 4097-4103.
32. Lucy, M.C., Staples, C.R., Michel, F.M. and Thatcher, W.W. (1991) Effect of Feeding Calcium Soaps to Early Postpartum Dairy Cows on Plasma PGF2α, Luteinizing Hormone, and Follicular Growth. Journal of Dairy Science, 74, 483-489. http://dx.doi.org/ 10.3168/jds.S0022-0302(91)78195
33. Sklan, D., Moallem, U. and Folman, Y. (1991) Effect of Feeding Calcium Soaps of Fatty Acids on Production and Reproductive Responses in High Producing Lactating Cows. Journal of Dairy Science, 74, 510-517.http://dx.doi.org/10.3168/jds.S0022-0302(91)78198-8
34. Scott, T.A., Shaver, R.D., Zepeda, L., Yandell, B. and Smith, T.R. (1995) Effects of Rumen-Inert Fat on Lactation, Reproduction, and Health of High Producing Holstein Herds. Journal of Dairy Science, 78, 2435-2451.http://dx.doi.org/10.3168/jds.S0022-0302(95)76872-2
35. Spears, J.W. and Weiss, W.P. (2008) Role of Antioxidants and Trace Elements in Health and Immunity of Transition Dairy Cows. The Veterinary Journal, 176, 70-76. http://dx.doi.org/10.1016/j.tvjl.2007.12.015

36. Gábor, G. and Balogh, O.G. (2010) Factors Influencing Pregnancy Rate in Dairy Cattle. Proceedings of the International Strategic Programs for Conservation of Genetic Resources for Food and Agriculture OECD Workshop, Vancouver, 17 July 2010, 78-83.
37. Balogh, O.G., Febel, H., Huszenicza, G., Kulcsar, M., Abonyi-Toth, Z., Endrodi, T., et al. (2012) Seasonal Fertility Differences in Synchronised Dairy Cows: Ultrasonic, Metabolic and Endocrine Findings. Acta Veterinaria Hungarica, 60, 131-143. http://dx.doi.org/10.1556/AVet.2012.011

CHAPTER 8

Genetically Modified Soybean in Animal Nutrition

Tudisco Raffaella, Calabrò Serena, Cutrignelli Monica Isabella, Piccolo Vincenzo and Infascelli Federico

Department of Animal Science and Food Control; University of Naples; Naples Italy

1. INTRODUCTION

In recent years, genetically modified (GM) plants, whose DNA has been changed using genetic engineering techniques, are mainly used as foods for human and feeds and foods for farm animals. To date, a number of GM products have been approved for human consumption but concerns over safety persist, mainly as regards either the detection of transgenic plant genes and proteins in animal systems or allergenicity and toxicity of GM plants.

Since their commercial release in 1996, the global cultivation area dedicated to the production of GM plants has increased significantly (ISAAA, 2010). The majority of GM crops currently produced, like soybean, corn, cotton and canola, have been engineered to enhance agronomic performance by transformation with genes encoding herbicide tolerance and pest resistance. GM soybean has been rendered tolerant to the glyphosate family of herbicides through expression of transgenic DNA from the CP4 strain of *Agrobacterium tumefaciens* that encodes 5-enolpyruvylshikamate-3-phosphate synthase (CP4 EPSPS). Roundup Ready (RR) soybean have been grown commercially from 1996 and continued to be the principal biotech crop in 2010. Farm animals are currently fed soybean and soybean meal developed from genetic transformation as well as corn and corn products. The European Union imports soybean from USA,

Brazil, and Argentina, the main users of biotech crops globally. About 90% of the compound feed produced in the EU contains GM soybean.

Although regulations with regard to GM plants have been developed primarily from the perspective of human consumption of GM food, it is generally assumed that these criteria are suitable for a risk assessment of the consumption of GM feed by livestock. The protocol for establishing "substantial equivalence" of GM plant compared to isogenic parental lines does not complete a nutritional safety assessment of a GM plant, rather, it provides a starting point for the overall assessment (FAO/WHO, 2000). Based on the European novel food and feed regulation, all foods and feeds containing or derived from approved GM products in amounts greater than a 0.9% threshold are subject to labelling rules (European Commission, 2003). Labelling of feeds containing GM ingredients gives farmers the choice of using such feed for their livestock. However, products such as milk, meat, and eggs, that are derived from livestock fed transgenic feeds are exempt from EU-labelling laws. Several studies have been conducted to evaluate the safety of GM crops, but there is still a debate on the risk of GM consumption and their potential passage into tissues.

Current researches suggest that the passage of plant DNA fragments across the intestinal barrier is a natural event, as demonstrated by the detection of endogenous, high copy number chloroplast genes from plants in several animal tissues and products. Low copy endogenous and transgenic DNA in animal tissues have been detected but to a lesser extent than high copy genes.

For several years, no direct evidence that GM food may represent a possible danger for health has been reported and the scientific literature in this field is still quite poor, especially as to the possible effect of a diet involving a significant amount of GM plants. More recently, a number of papers have been published and controversial results have been obtained. However, some have found significant modifications in some nuclear features in mice fed GM soybean and, more recently, it has been reported that the activity of some enzymes was altered in rabbit and goats fed GM soybean, as confirmed also by histochemistry which showed a widespread distribution of enzyme activity in myocytes, myocardiocytes, epithelial cells of renal tubules and hepatocytes. These observations suggest that the risk of genetically modified crops cannot be ignored and requires further investigations in order to identify possible long-term effects of GM plants on both livestock and human consumption. The main focus of this chapter concerns the genetically engineered soybean, its effects on human and animal health, the productivity of this GM crop and the outcome for environment.

2. AGRONOMIC IMPACT OF GENETICALLY MODIFIED SOYBEAN

Genetic engineering has been widely applied to agriculture to obtain specific plant characteristics which can lead to an improvement in both food quality and yield. Compared with traditional plant breeding methods, such as artificial crossing or hybridization, biotechnology now allows for the introduction of DNA from outside the plant kingdom. Selective inclusion of single or multiple traits can be performed to change the quality of agricultural crops. According to statistics released by the International Service for the Acquisition of Agri-Biotech Applications (ISAAA, 2010), the area of planted transgenic crops was 148 million hectares in 2010, a approximately 87-fold increase from the 1996 level (1.7 million hectares of biotech crops). The number of countries adopting biotech crop cultivation has increased crops consistently from 6 in 1996 to 29 in 2010. The Unites States (US), followed by Brazil, Argentina, India, Canada, and China continued to be the principal adopters of biotech crops globally, with 66.8 million hectares planted in the US.

The majority of genetically modified (GM) crops currently produced have been engineered to enhance agronomic performance by transformation with genes encoding herbicide tolerance or pest resistance. From the first commercialization of biotech crops in 1996, to 2010 herbicide tolerance has consistently been the dominant trait. In 2010, herbicide tolerance deployed in soybean, corn, canola, cotton, sugarbeet and alfalfa, occupied 61% or 89.3 million hectares of the global biotech area. In 2010, the stacked double and triple traits occupied a larger area (22% or 32.3 million hectares) than insect resistant varieties (26.3 million hectares) at 17%. The insect resistance trait products were the fastest growing trait group between 2009 and 2010 at 21% growth, compared with 13% for stacked traits and 7% for herbicide tolerance.

Biotech herbicide tolerant soybean continued to be the principal biotech crop in 2010, occupying 73.3 million hectares or 50% of global biotech area, followed by biotech corn (46.8 million hectares at 31%), biotech cotton (21.0 million hectares at 14%) and biotech canola (7.0 million hectares at 5%) of the global biotech crop area.

Farm animals are currently fed soybean and soybean meal developed from genetic transformation as well as corn and corn products such as corn gluten feed and meal. Europe is strongly dependent upon the American continent for its protein requirements amounting up to 90 to 95% for soybean, 40 to 60% for corn derivatives and partly for canola grain or meal (Aumaitre, 2004).

3. GENETIC MODIFICATION OF SOYBEAN

Traditionally, plants with desirable characteristics were chosen for food of the next generation. The desirable characteristics arose from naturally occurring variations in the genetic make-up of individual plants. Unlike conventional genetic modification that is carried out through time-tested conventional breeding of plants as combining genes from different organisms is known as recombinant DNA technology and the resulting organism is said to be genetically modified, or genetically engineered or transgenic (Pandey et al., 2010).

Transgenic plant is one that has received a segment of DNA or genes from another organism (known as heterologous or foreign DNA) using recombinant DNA techniques. The foreign DNA is integrated through natural systems present in plant cells into the plant's genome. The newly introduced genes are subsequently inherited in a normal Mendelian manner through pollen and egg cells. The mainly process of introducing DNA into plants (called transformation) uses the *Agrobacterium* mediated method and it can be achieved both in monocotyledonous plants such as wheat, barley and rice and in dicotyledonous plants such as soyabean, potato and tomato.

The soil bacterium *Agrobacterium tumefaciens* causes crown gall disease on some plants, in particular in dicotyledonous species. In causing crown gall disease *A. tumefaciens* transfers DNA (the transferred DNA or T-DNA) from the bacterium to the plant. In nature the transferred bacterial DNA cause the symptoms associated with crown gall disease. In the early 1980s scientists removed the disease causing genes from this bacterium and the TDNA is now routinely used to transport foreign genes into plants. *Agrobacterium* cells carrying the foreign genes of interest are incubated with cultured cells of the recipient crop plant and transgenic plants are regenerated from them. Not all cells subjected to this process are successfully modified so it may be necessary to identify the modified cells using marker genes which are closely linked to the genetic material that is transferred. These selectable marker genes usually confer resistance to an antibiotic such as kanamycin or resistance to an herbicide.

The genetically modified soybean (named RoundUp Ready, RR) has been rendered tolerant to the glyphosate family of herbicides through expression of transgenic DNA from *Agrobacterium tumefaciens* sp. strain CP4 that encodes 5-enolpyruvylshikamate-3-phosphate synthase (CP4 EPSPS). The CP4 EPSPS protein expressed in GE glyphosate tolerant plants is functionally equivalent to endogenous plant EPSPS enzymes with the exception that CP4 EPSPS displays reduced affinity for glyphosate (Franz et al. 1997). This soybean is, also, composed of a 35S promoter from cauliflower mosaic virus (CMV) and a NOS-terminator, a terminator of nopaline synthase gene.

According to the Center for Environmental Risk Assessment (2010), the 5-enolpyruvylshikimate-3-phosphate synthase (EPSPS: EC 2.5.1.19) family of enzymes is ubiquitous in plants and microorganisms. EPSPS enzymes have been isolated from both sources, and their properties have been extensively studied. EPSPS proteins catalyze thetransfer of the enolpyruvyl group from phosphoenol pyruvate (PEP) to the 5-hydroxyl of shikimate-3-phosphate (S3P), thereby yielding inorganic phosphate and 5-enolpyruvylshikimate-3-phosphate (Alibhai and Stallings, 2001). Shikimic acid obtained is a substrate for the biosynthesis of aromatic amino acids (phenylalanine, tryptophan and tyrosine) as well as many secondary metabolites, such as tetrahydrofolate, ubiquinone, and vitamin K. Importantly, the shikimate pathway and, hence, EPSPS proteins, are absent in mammals, fish, birds, reptiles and insects (Alibhai and Stallings, 2001). In contrast, it has been estimated that aromatic molecules, all of which are derived from shikimic acid, represent 35% or more of the dry weight of a plant (Franz et al. 1997).

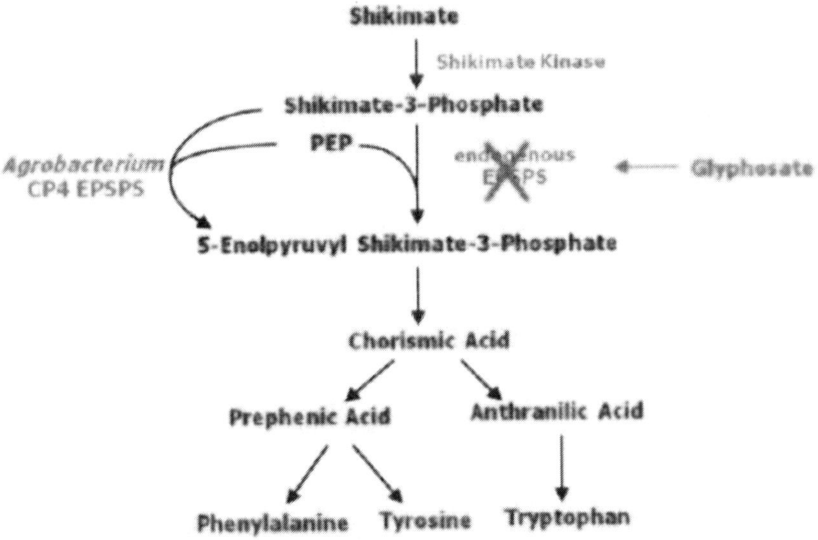

Figure 1. Schematic representation of glyphosate mode of action and mechanism of CP4 EPSPS mediated tolerance (Center for Environmental Risk Assessment, 2010).

4. NUTRITIONAL ASSESSMENT OF GENETICALLY MODIFIED SOYBEAN

In animal nutrition, many studies with GM plants were carried out in target species using the substantial equivalence method. The application of this method to animal studies led to the development of the concept of nutritional equivalence which implies specific measurements regarding animal production. The European Commission has a combined safety approach that requires an assessment of risks for humans, animals, and the environment prior to approval of importation or cultivation of a novel crop (European Commission, 2001). An integral part of the safety evaluation of GM plants is to test for "substantial equivalence". The concept of substantial equivalence is the starting point and guiding concept for safety assessment (Food and Agriculture Organization/World Health Organization, 2000). It is not the conclusion, but it is part of the safety assessment (Konig et al., 2004; Kuiper & Kleter, 2003). The aim of such a test is to determine whether a transgenic plant is substantially equivalent to its conventional counterpart at a chemical and nutritional level. While the parameters to be measured have not been formally defined, minimal analyses performed should determine whether the major nutritional components (i.e., lipids, carbohydrates, proteins, vitamins, minerals, trace elements) and known antinutrients and toxins of transgenic plants are equivalent to those in conventional varieties that have a history of safe use. Guidelines have been established by several organizations regarding assessment of the allergenic risk of each novel protein expressed in a GM plant, prior to market approval (FAO/WHO, 2000; Konig et al., 2004; Martens, 2000). These typically include comparison of amino acid sequence homology of the novel protein to known allergens and digestion of the protein in simulated gastric environments. While allergic reactions are primarily a concern for human consumption of GM foods, certain proteins in soybean have been shown to elicit allergenic reactions in calves and piglets. The assessment of the safety of GM organisms addresses both intentional and unintentional effects that may result as a consequence of genetic engineering of the food source. Future transgenic crops are expected to contain fewer or no marker genes in the final products since marker free insertion techniques or methods to eliminate marker genes from transgenic plants are being improved. The assessment of safety measures are a lengthy and tedious process (Figure 2). The nutritional aspects, risk characterization and exposure assessment are preliminary steps being taken. Before hitting the market, all GM products have to pass all the allergic tests and provide the details. Only those products find as possessing no harmful or allergic effects are only recommended.

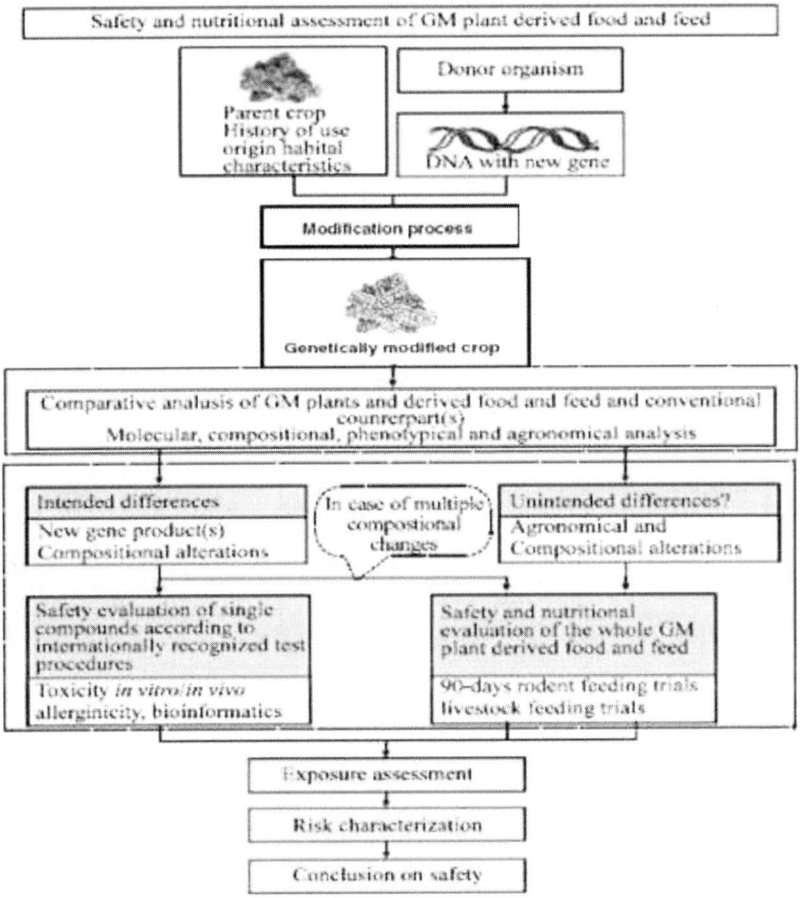

Figure 2. Pre-market safety and nutritional testing of genetically modified plant derived food and fed (EFSA, 2008).

In most countries authorities and agencies involved in feed/food safety assessment have based their safety assessment strategies and guidelines on this approach. To provide consumers the opportunity for choice, in 2004 the European Union (EU) has extended regulations concerning GM foods to include animal feeds and feed additives. According to Regulation (EC) No. 1829/2003 (European Commission, 2003), all foods and feeds containing or derived from approved GM products in amounts greater than a 0.9% threshold are subject to labelling rules. Labelling of feeds containing GM ingredients informs farmers and gives them the choice of using such feed for their livestock. However, products such as milk, meat, and eggs, that are derived from livestock fed transgenic feeds are exempt from EU-labelling laws. One of the controversies, important for safety aspects of GM feeds, is a potential possibility of transfer of the transgenic DNA to animal tissues, and in consequence its negative effect on consumers of such products originating from animals fed diets containing GM plants. Detailed studies of the feeding qualities of GM plants for livestock and their nutritional evaluation have been reviewed previously (Aumaitre et al., 2002; Flachowsky et al., 2005a; Flachowsky & Aulrich, 2001). A lot of animal studies with GM plants aimed to evaluate the compositional and nutritional equivalency of transgenic feeds and their conventional counterparts.

Accordingly to the substantial equivalence theory, feed chemical analyses were performed to determine whether the macronutrients, vitamins, minerals and even trace elements were found at the same level as in the conventional or isogenic plants. The composition studies confirmed the substantial equivalence of genetically modified (GM) soybean to conventional counterpart (Table 1) (Cromwell et al., 2001; Padgette et al., 1996).

Table 1. Chemical composition (% dry matter) of genetically modified soyabean (RoundUp Ready, RR) and its conventional counterpart (Adapted from Cromwell et al., 2001). NDF: neutral detergent fiber.

	Conventional soybean	GM soybean
Dry Matter	90.30	91.00
Crude Protein	51.50	51.20
NDF	4.95	4.85
Lysine	3.16	3.09
Methionine+Cysteine	1.47	1.51

The concerns that have been raised with respect to the potential risk associated with the use of GM plant products in animal feed are related to the possible unintended effects of inserting novel DNA into the plant by biotechnology. The possible "side effects" of the genetically modification is often termed unintended effects and may result from the random integration in the genome of the novel DNA which may result in an over-expression in the plant of inherently toxic substances such as anti-nutritional factors (ANFs), silencing of endogenous plant genes (e.g essential nutrients), or alterations in host metabolic pathways (Novak & Hasleberger 2000; Saxena & Stotzky, 2001). The possible combination of an unexpected increase in expression of endogenous ANFs and the presence of new exogenous toxicants have been of particular concern as these could compromise the quality of feedstuffs and may affect animal health and nutrition (Francis et al., 2001). Studies also include any known anti-nutritional factors (ANFs), such as trypsin inhibitors in soybean, interfering with nutrient absorption or natural toxicants typically present in the genus (Padgette et al., 1996). Trypsin inhibitors are similarly destroyed during heating associated with oil extraction and preparation of soybean meal. In addition, the activity of urease measured by the variation in pH is efficiently reduced by heat treatment whatever the genotype of the soybean kernel (Table 2).

Table 2. Main anti-nutritional factors in raw soybean and soybean meal (parental vs. RR): absence of effect of genetic modification for glyphosate resistance (Adapted from Padgette etal.1996).

	Parental soybean		RR	
	Raw	Meal	Raw	Meal
Lectins (HU/mg sample)	1.2	-	1.0	-
Trypsin inibithor (TIU/mg sample)	22.6	3.4	23.7	3.3
Urease (pH)	2.18	0.03	2.17	0.04

Additional nutritional data can be of importance in the case of oil seeds or oil-rich cereals such as corn because they can markedly affect the composition of fatty tissues when fed to farm animals. Data obtained from the analysis of corn kernels demonstrated similar proportions of fatty acids in oil of kernels of insect resistant and herbicide-resistant corn. Similarly, and in the majority of cases, it has also been observed that insect resistant and herbicide-tolerant corn kernels contain similar proportions of amino acids. The introduction of Bt and herbicide tolerance genes in corn has never been found to create starch modifications expressed as the proportion of amylose and amylopectin. Thus similar proportions of 21.5 and 21.0% of amylose have been found in Bt and herbicide-tolerant modified corn, respectively, compared to 22.4 and 22.7% of amylose in starch of the isogenic varieties, respectively (Benetrix, 2000). Data from the literature have many times corroborated the substantial equivalence in major nutrients and minerals and trace elements in corn and kernels of GM compared to isogenic control corn (Brake & Vlachos, 1998; Sidhu et al., 2000). The whole modified corn plant (Clarke & Ipharraguerre, 2001; Faust, 2000) has also been found to be substantially equivalent in composition to isogenic plants. All these data suggest a similar nutritional value for the feed material derived from the modified plants (Aumaitre et al., 2002).

In order to evaluate the nutritive value of feeds for ruminants, nowadays the in vitro gas production technique (Theodorou et al. 1994) (IVGPT) is commonly used. IVGPT is based on the assumption that the accumulated gas production by a substrate, incubated in with rumen liquor, is proportional to the amount of digestible carbohydrates, and thus highly correlated to the energy value of feeds. In addition, IVGPT allows to study also the fermentation kinetics of feeds. Tudisco et al. (2004) in a research, aimed to compare the fermentation kinetics of Roundup Ready defatted soybean to its conventional counterpart by the IVGPT, found that the genetic modification, although did not affect the chemical composition, led to a significantly lower cumulative gas production and volume per gram of incubated organic matter. It could be hypothesized that the genetic modification may lead to pleiotropic effects (effect of a single gene on multiple phenotypic traits) that may alter the starch and/or protein structure. Alternatively, the results could be explained with a plant DNA transfer to ruminal bacteria which may modify their fermentation activity.

In any case, the results of this research arouse concerns in term of food safety, because other unpredictable metabolic effects, such as metabolic interferences, or direct or indirect insertional mutagenesis cannot be excluded. With this regards Seralini et al. (2011) report that by insertion of the transgene in varieties producing Cry1Ab toxin caused a complex recombination event, leading to the synthesis of new RNA products encoding unknown proteins, or/and to metabolic pathways variations which caused up to 50% changes in measured osmolytes and branched aminoacids.

4.1 Nutritional testing of GM feed with GM soybean in target animal species

Many studies with GM plants were carried out in target species to assess the nutritive value of the feed and their performance potential and were revealed no significant differences in performance indices and quality parameters of meat, eggs or milk, when farm animals were fed diets containing GM or conventional feeds (Aumaitre et al., 2004; Flachowsky et al., 2005a; Świątkiewicz S. & Świątkiewicz M., 2009). These studies have focused on livestock nutrition, in order to confirm nutritional equivalence and to obtain further information concerning the safety of animal products.

As regards monogastric livestock, many feeding studies with 1-day-old broiler chicks have been reported (EFSA, 2008), including GM lines of corn, soybean, canola and wheat and appropriate counterparts. Few experiments are available with laying hens (Aulrich et al., 2001; Halle et al., 2006) where GMOs were compared with near isogenic counterparts. McNaughton et al. (2011) compared the nutritional performance of laying hens fed corn grain (event DP-Ø9814Ø-6) and processed soybean meal (event DP-356Ø43-5), individually or in combination, with the performance of hens fed diets containing conventional corn and soybean meal. The performance (body weight, feed intake and egg production) and egg quality of hens fed GM feeds was comparable with that of hens fed diets formulated with conventional feed. In each study, chemical composition and nutritional value of GM lines and the near-isogenic non-GM lines were found to be comparable without biologically significant differences in the production parameters measured. Experiments with growing and laying quails were carried out to test diets with isogenic or GM Bt 176 corn (Flachowsky et al., 2005b; Halle et al., 2006). Health, hatchability and performance of quails and the quality of meat and eggs were unaffected by the diets. Improvements in livestock performance were noted significant with the diet containing Bt corn compared to diets containing conventional corn grain (Piva et al., 2001a, 2001b). The authors attributed the results to the fact that the use of Bt lines reduced secondary fungal infection and, as a consequence, reduced mycotoxin contamination. Research conducted with growing and finishing pigs (EFSA, 2008) including GM corn grain, sugar beet, soybean meal, rapeseed meal, rice and wheat, showed that when compositional analyses of GM lines and the nearisogenic non-GM line and commercial varieties were comparable, then nutritional equivalence was also established.

As regards ruminants, comparable performance of beef cattle fed corn grain, corn silage or stover from GM plants or from conventional plants is reported, and in dairy cows the inclusion in the diet of feed ingredients derived from a wide range of GM plants unaffected feed intake, milk yield and composition (EFSA, 2008). Milk quality is generally measured as the fat, protein and lactose concentration and as such there is no evidence to suggest that the inclusion of GM feed ingredients affects milk quality. As with other livestock species, studies with lactating dairy cows also showed that once compositional equivalence was demonstrated then nutritional equivalence occurred.

Table 3. Some experiments carried out to establish the qualitative and quantitative performance of monogastric or ruminant livestock fed with GM soybean. BWG: Body Weight Gain; ADG: Average Daily Gain; FCR: Feed Conversion Ratio.

	Parameters	Results	References
Poultry	BWG, ADG, FCR, muscle weight	No significant differences	(Hammond et al., 1996)
Pig	BWG, DMI, FCR, carcass quality, sensory score of meat	No significant differences	(Cromwell et al., 2001)
Cow	Milk yield and composition, DMI, nitrogen balance, rumen VFA composition	No significant differences	(Hammond et al., 1996)

Finally, the production studies carried out with fish provided similar conclusions to those drawn from studies conducted with other livestock species (EFSA, 2008).

5. FATE OF TRANSGENIC DNA AND NEW PROTEIN IN ANIMAL ORGANS AND PRODUCTS

One of the most important questions about the use of GM products in animal nutrition is the possibility that modified DNA could be transferred from plants to animal products or to bacteria, with harmful consequences (FAO, 2004). Other problems regard the ability of transgenic proteins to provoke food intolerance or allergic reaction in susceptible people. Hence, it is necessary to consider the destiny of these molecules within the animal organism (Alexander et al., 2007).

The gastrointestinal tract is constantly exposed to DNA that is released from partially or completely digested food, ingested microbes, and DNA from intestinal microflora. Ingested food is mechanically disrupted and the released DNA, although poorly digested, is cleaved through acid hydrolysis and enzymatic digestion into small DNA fragments. Eventually some of these fragments are converted to single nucleotides. Acid hydrolysis in the gastrointestinal tract is expected to depurinate most adensine and guanine nucleotides of the food DNA (Klinedinst & Drinkwater, 1992). The presence of various phosphatases and deaminases continue to destroy the structural integrity of any free DNA. The breakdown products of DNA are absorbed for using at the cellular level for synthetic processes as they may be found in blood and tissues (McAllan, 1982). All though there were conflicting reports on the fate of GM DNA in the biological system it was observed that DNA could pass through the gut wall into the blood stream and taken up by cells in the blood, liver, spleen and passed through the placenta to the cells of the feoetus and the newborn one (Doerfler & Schubert, 1998).

In ruminants, experimental evidence suggests that more than 80% of DNA is completely disrupted after 2 hours (Wiedemann et al., 2006). However, this degradation is not complete and not immediate (van den Eede et al., 2004).

In animal tissues some fragments of chloroplast DNA have been found. The reason why chloroplast DNA is more frequently detected in animal products is the number of the genes involved and the sensitivity of the PCR method. In transgenic plants, every cell contains hundreds chloroplast genes, but only one transgenic gene (Aumaitre et al., 2002).

Research on the fate of foreign DNA in the mammalian organism showed that PCR products specific to foreign DNA could be detected therein. It was concluded that DNA fragments from the gastrointestinal tract could reach the bloodstream and be transported through the epithelium of the gut and the cells of the Peyer's patches to spleen and liver cells. Such DNA fragments are probably retained for a short while and then digested (Schubbert et al., 1994, 1997, 1998). While the intestinal tract does not seem an absolute barrier against the uptake of macromolecules or even of microorganisms, the mechanism of foreign DNA uptake by the intestinal wall epithelia is unknown. In addition, not much is known about the degradation and integration of the DNA. There is some evidence that fragments of foreign DNA are not digested in the gut and might enter the organism or become incorporated into cells lining the gut wall (Doerfler, 2000; Tony et al., 2003). Transfer of the plant DNA to bacteria needs several steps, and the expectation seems to be extremely low (Kuiper et al., 2003; Sharma e al., 2004). Bacterial resistance to antibiotics is not a specific problem of genetic engineering. According to Directive 2001/18/EC, use of GMOs containing antibiotic-resistant genes will be forbidden starting from 01/01/2009 (European Commission, 2001).

Data are also available on the fate of recombinant plant DNA in the gastrointestinal tract of humans. By in vitro simulation of human digestion, 80% of the transgene in naked GM soybean DNA was degraded in the gastric simulations, while no degradation of the transgenes contained within GM soybean and corn was observed in these acidic conditions (Martin-Orúe et al., 2002). In the small intestinal simulations, transgenes in naked soybean DNA were degraded. In contrast, the corn nucleic acid was hydrolysed in the small intestinal simulations in a biphasic process in which approximately 85% was rapidly degraded, while the rest of the DNA was cleaved at a low rate of degradation.

The number of transgene copies passing to the small intestine of human ileostomists consuming GM soya were successful quantified, and up to 3.7% of the transgene could survive passage (Neterwood et., 2004).

Finally, another factor that will directly affect gene persistence throughout the digestive tract and therefore indirectly affect the chance of passage across the GIT epithelium is the digestibility of the ingested plant species. Feedstuffs with relatively greater digestibility, such as soybean meal, are likely to have their DNA degraded more rapidly, decreasing the chance of absorption.

A low copy endogenous (soybean lectin) and recombinant (CP4 epsps) gene in longissimus dorsi muscle samples from pigs fed herbicide-tolerant soybean meal, in the grower and finisher phases, respectively were attempted to detect (Jennings et al., 2003). The same results about the fate of the CP4 epsps gene in other species were reported (Klotz & Einspanier, 1998).

The CP4 epsps transgenic gene was not found in muscle and liver of chicken fed herbicidetolerant soybean up to 7 weeks after ingestion (Khumnirdpetch et al., 2001). According to the authors, GM soybean fragment were degraded in the gastrointestinal tract.

As regards ruminants, chloroplast gene fragments were found in the leucocytes of dairy cows fed small quantities of transgenic soybean meal, while no fragments of the transgenic DNA were found in any tissue examined and in milk (Klotz & Einspanier, 1998). Similarly, high copy chloroplast "rubisco" gene fragments were found in the blood of cattle fed GM and soybean meal, but transgenic sequences were never detected (Phipps et al., 2003).

In fish, soybean meal is used as dietary source of protein in their diet, however, because of the presence of anti-nutritional factors (ANFs) their inclusion levels should be kept low (Olli et al., 1994). The fate of ingested GM soybean DNA fragments (120 and 195 bp) and a 180 bp fragment of the lectin gene of soybean in Atlantic salmon and their survival through the gastrointestinal tract (GIT) were investigated and the DNA was traced in a variety of fish tissues (liver, muscle and brain) (Sanden et al., 2004). Only the smaller GM DNA fragment (120 bp) was amplified from the content of the stomach, pyloric region, mid intestine and distal intestine, while no transgenic or conventional soybean DNA fragments were detected. The uptake of dietary DNA into blood, kidney and liver of salmon was investigated also by other authors (Nielsen et al., 2005) which determined the DNA fragment size if dietary DNA was detected. Most of the feed (partially digested) was found in the pyloric region, mid intestine, and distal intestine at 4, 8, and 16 h after force-feeding, while the highest concentrations of dietary DNA in liver and kidney were found 8 h after force-feeding, and blood up to 64 h. Finally, the cauliflower mosaic virus 35S promoter fragment (220 bp) of the GM defatted soybean meal was detected in the muscle of rainbow trout receiving both levels of GM soybean (approximately 15 and 30%) diet by nested PCR, but the frequency of detection was greater at the higher inclusion level (Chainark et al., 2006). Additionally, the promoter fragment was not detected by the fifth day after changing the diet to non-GM soybean. Conversely, the promoter fragment was not detected from fish fed with the non-GM SBM diet. Successively, Chainark et al. (2008) traced foreign DNA fragments from genetically modified defatted soybean meal (GM SEM) in rainbow trout by nested polymerase chain reaction (PCR) and located by in situ hybridization. Either a GM or non-GM SBM formulated diet (42% protein) was fed to fish (average weight 50.5 g) for 2 weeks. The degradation results showed that the cauliflower mosaic virus 35S promoter (220 bp) fragment was detected in the contents of digestive system only in fish fed the GM SBM diet, and it was not detected on the third day after changing the diet to the non-GM SBM diet. For the possible transferral results, the promoter fragment was detected in the leukocyte, head kidney and muscle only of fish fed the GM SBM diet; it was not detected on the fifth day after changing the diet to the non-GM SBM diet. These results suggest that a foreign DNA fragment was not completely degraded and might be taken up into organs through the gastrointestinal tract. However, foreign DNA was not detected after the withdrawal period. Thus, the data show that uptake of DNA from GM SBM might not remain in the tissues of fish fed GM SBM diet. Similarly, Ran et al. (2009) found DNA fragments from RR soybean in different tissues and organs of tilapias (*Oreochromis niloticus*, GIFT strain).

Tudisco et al. (2006) in order to evaluate the presence of plant DNA fragments in tissues to follow the fate of plant fed, carried out a research on twenty weaned 30-day-old New Zealand rabbits (10 males and 10 females), individually caged, which were equally assigned to control (C) and treated (T) groups. The animals were given a diet containing soybean meal (solved extracted) which was from conventional or Roundup Ready beans, for group C and T, respectively. The presence of chloroplast DNA was found in tissues and blood from both control and treated groups. The percentage of positive samples were: 50% (blood), 70% (muscle), 80% (heart), 70% (liver) and 80% (kidney). By contrast specific fragments of soybean were not detected in all samples but only in the plant samples. Similarly transgenic fragments gave undetectable results.

Subsequently, the same authors (Tudisco et al. 2010) investigated the presence of DNA fragments in blood and milk from goats fed conventional (control) or Roundup Ready soybean and in blood, skeletal muscle and organs from their offspring. Transgenic target DNA sequences (35S and CP4 EPSPS) were detected in blood and milk from goats that received a diet containing transgenic soybean as well as from some samples of their offspring, not in the control group. Those findings show plant DNA fragments are likely to survive digestive processes to some extent (Duggan et al., 2003; Einspanier et al., 2004), as well as their transfer to blood and milk. In addition, the detection of plant DNA in tissues and organs of nursed kids could support the hypothesis of a gene transfer through milk.

6. EFFECTS ON ANIMAL HEALTH OF GM SOYBEAN

In different experiments food and feed derived from GM plants, mixed in animal diets have been fed to rats, mice or other animal species during different periods of administration, and parameters such as body weight, feed consumption, blood chemistry, organ weights, histopathology, etc., have been measured. With respect to recent studies on safety assessment of GM soybeans, the scientific literature shows rather contradictory results.

No immunotoxic activity or an increase in the IgE in serum and histopathological abnormalities were found in the mucosa of the small intestine of rats and mice fed heattreated GM soybean meal containing the cp4-epsps (Teshima et al., 2000).

In Sprague–Dawley rats, Appenzeller et al. (2008) conducted a subchronic feeding study with the herbicide-tolerant soybean DP-356Ø43-5 (356043). Diets were fed to young adult animals for at least 93 days. Compared with rats fed with the isoline control or conventional reference diets, no biologically-relevant, adverse effects were observed in rats fed diets containing 356043 soybean with respect to body weight/gain, food consumption/efficiency, clinical signs, mortality, ophthalmology, neurobehavioral assessments (sensory response, grip strength and motor activity), clinical pathology (hematology, coagulation, serum chemistry and urinalysis), organ weights, and gross and microscopic pathology. Similarly, Delaney et al. (2008) carried out in Sprague–Dawley rats a subchronic feeding study of high oleic acid soybeans (Event DP-3Ø5423-1). DP-3Ø5423-1 (305423) is a GM soybean produced by biolistic insertion of a gm-fad2-1 gene fragment and the gm-hra gene into the germline of soybean seeds. Compared with rats fed the non-GM control diet, no biologically-relevant differences were observed in animals fed the 305423 diet with respect to body weight/gain, food consumption/efficiency, mortality, clinical signs of toxicity, or ophthalmologic observations. In addition, no diet-related effects were noted on neurobehavioral assessment, organ weights, or clinical or anatomic pathology. Based on the results of these studies, the authors concluded that 356043 and 305423 soybeans were as safe and nutritious as conventional non-GM soybeans. Sakamoto et al. (2007; 2008) conducted 52-week and 104-week feeding studies of genetically modified soybeans in F344 rats. Although in both studies several differences in animal growth, food intake, serum biochemical parameters and histological findings were observed between rats fed the GM (glyphosate-tolerant) soybeans and those fed a commercial diet, body weight and food intake were similar for the rats fed the GM and non-GM soybeans. Gross necropsy findings, hematological and serum biochemical parameters, organ weights, and pathological findings showed no meaningful differences between rats fed the GM and non-GM soybeans. These results indicate that longterm intake (54 and 104 weeks) of GM soybeans at the level of 30% in the diet had no apparent adverse effect in rats.

In a 42-day feeding trial study conducted in broiler chickens (McNaughton et al., 2008), it was also concluded that 356043 soybean was nutritionally equivalent to non-transgenic control soybean with a comparable genetic background.

Finally, also related to GM soybeans, Mathesius et al. (2009) assessed the safety of a modified acetolactate synthase protein (GM-HRA) used as a selectable marker in GM soybeans. The authors did not find adverse effects in mice following acute oral exposure to GM-HRA at a dose of at least 436 mg/kg of body weight, or in a 28-day repeated dose dietary toxicity study at doses up to 1247 mg/kg of body weight/day. It was concluded that GM-HRA protein is safe when used in agricultural biotechnology.

In contrast to the above results, in a long-term study on female mice fed with a GM modified soybean (insertion of the bacterial CP4 EPSPS gene to confer a high level of tolerance to glyphosate), focused on assessing the effects of this diet on liver of old animals (until 24 months of age) and to elucidate possible interference with aging, Malatesta et al. (2008a) found that GM soybean intake could influence the liver morpho-functional features during the physiological process of aging. Several proteins belonging to hepatocyte metabolism, stress response, calcium signaling and mitochondria were differentially expressed in GM-fed mice, indicating a more marked expression of senescence markers in comparison to controls. Moreover, hepatocytes of GM-fed mice showed mitochondrial and nuclear modifications indicative of reduced metabolic rate. In previous studies on hepatocytes from young and adult (2–8 months of age) female mice fed GM soybeans, nuclear modifications involving structural constituents of the transcription and splicing properties pathways were seen (Malatesta et al., 2002a). Although the cause(s) of the observed alterations could not be conclusively established, it was noted that these modifications disappeared when GM soybean was replaced by a non-GM one in the diet (Malatesta et al., 2005). Since the GM soybean used was tolerant to glyphosate and was treated with the glyphosate-containing herbicide Roundup, the effects observed might be due to herbicide residues. Accordingly, and aiming to verify this hypothesis, Malatesta et al. (2008b) treated rat hepatoma tissue culture (HTC) cells with 1–10 mM Roundup and analyzed cellular features by flow cytometry, fluorescence, and electron microscopy. Under these experimental conditions, the death rate and the general morphology of HTC cells were not affected, as well as most of the cytoplasmic organelles. However, in HTC-treated cells, lysosome density increased and mitochondrial membranes were modified indicating a decline in the respiratory activity. In addition to the above, nuclei underwent morphofunctional modifications suggesting a decreased transcriptional/splicing activity. The authors did not exclude that factors other than the presence of the herbicide residues could be responsible for the cellular modifications described in GM-fed mice. However, they indicated that the concordance of the effects induced by low concentrations of Roundup on HTC cells suggested that the presence of Roundup residues could be one of the factors interfering with multiple metabolic pathways.

Cisterna et al. (2008) investigated the ultrastructural and immunocytochemical features of pre-implantation embryos from mice fed either GM or non-GM soybean in order to verify whether the parental diet could affect the morpho-functional development of the embryonic ribonucleoprotein structural constituents involved in pre-mRNA pathways. Morphological observations revealed that the general aspect of embryo nuclear components were similar in the GM and non-GM soybean-exposed groups. However, immunocytochemical and in situ hybridization results suggested a temporary decrease of pre-mRNA transcription and splicing in 2-cell embryos and a resumption in 4–8-cell embryos from mice fed GM soybean. In addition, pre-mRNA maturation seemed to be less efficient in both 2-cell and 4–8-cell embryos from GM-fed mice than in non-GM-fed animals.

Battistelli et al. (2010) investigated the duodenum and colon of mice fed on genetically modified (GM) soybean during their whole life span (1-24 months) by focusing their attention on the histological and ultrastructural characteristics of the epithelium, the histochemical pattern of goblet cell mucins, and the growth profile of the coliform population. Even if the GM soybean-containing diet did not induce structural alterations in duodenal and colonic epithelium or in coliform population, the histochemical approach revealed significant diet-related changes in mucin amounts in the duodenum. In particular, the percentage of villous area occupied by acidic and sulphomucin granules decreased from controls to GM-fed animals.

In a previous ultrastructural analysis of testes from mice fed GM soybean conducted by the same research group (Vecchio et al., 2004), it was found that the immunolabelling for Sm antigen, hnRNPs, SC35 and RNA Polymerase II was decreased in 2 and 5 month-old GMfed mice, and was restored to normal at 8 months. In GM-fed mice of all ages considered, the number of perichromatin granules was higher and the nuclear pore density lower. Moreover, enlargements in the smooth endoplasmic reticulum in GM-fed mice Sertoli cells were also observed. Consequently, all these studies at the microscopic and ultramicroscopic levels showed cellular changes attributable to GM soybean intake.

Magaña-Gómez et al. (2008) conducted a study in Wistar rats, in which the hypothesis was that the intake of GM (SUPRO 500E) soybean could induce pancreatic stress or injury by analyzing the expression of pancreatitis-associated protein (PAP) and trypsinogens by qRTPCR in rats fed GM soy protein for 30 days. The hypothesis was based on the results of previous investigations showing that mice chronically fed since gestation with GM had problems in synthesis and processing of zymogens by pancreatic acinar cells and reduced nucleoplasmic and nucleolar and perichromatin granule accumulation on pancreatic acinar cell nuclei (Malatesta et al., 2002b; 2003). Magaña-Gómez et al. (2008) did not find differences in nutritional performance among rats fed non-GM and GM diets. The GM diet induced significant zymogen-granule depletion after 15 days feeding, returning to normal levels after 30 days. Acinar disorganization started as early as 5 days after initiation of the GM diet and it recovered after 30 days. Levels of PAP mRNA significantly increased in the GM diet between day 1 and day 3 and decreased to the basal level by day 15. In turn, trypsinogen mRNA peaked at two different times: at day 1 and at day 15, decreasing to basal levels after 30 days, while plasma amylase levels remained unchanged at all times. The authors indicated that GM soy protein intake affected pancreas function, evidenced by the early acute PAP mRNA increased levels and pancreas cellular changes followed by recuperation of acinar cells after 30 days.

Evaluating the GM soybean in Atlantic salmon diet, enlarged spleen and possible impaired spleen function as the number of smaller-sized red blood cells simultaneously increased were indicated (Hemre et al., 2005).

The same authors (Sagstad et al., 2008) reported lower plasma triacylglycerol levels and a significantly larger spleen somatic index in fish groups fed GM soybean compared to groups fed non-GM soybean.

Ermakova (2006) examined the effect of glyphosate-resistant (RR) GM soybean seeds fed to pregnant female rats on the number and weight of pups delivered The study was originally published in Russian, and was heavily criticised for using coated seeds ready for planting instead of beans suitable for feed. The control non-GM soybean was not the isogenic parent line, either. However, because of the possible serious implications of the results of this study for humans and animals it should have been repeated and possibly verified by other scientists with the correct GM soybean diets. Indeed, she has repeatedly pleaded for this but no one dared to try to reproduce her experiments. In this study rats were fed with laboratory rat chow and this diet was complemented with GM or conventional soybean for two weeks before mating, during the pregnancy and during suckling and the body mass and the number of pups were observed. The data indicated that on the GM soybeansupplemented rat chow significantly fewer pups were born, and with smaller body mass, than on the control non-GM soybeans.

In order to evaluate the possible health effects of a GM diet, Tudisco et al (2006) studied the activity of organ specific enzymes in two groups of New Zealand rabbits, given a diet containing soybean meal which was from conventional or Roundup Ready beans.

Statistical differences were detected in kidney for alanine aminotransferase (ALT), gamma glutamyltransferase (GGT) and lactic dehydrogenase (LDH) (higher activity for group fed GM soybean) whereas in the heart such result was seen only for LDH. No statistical differences were found for serum, liver and skeletal muscle. Significant differences between groups were detected for heart LDH1 and LDH2 and for kidney LDH1, thus confirming the significant increase of the enzyme in these tissues. Moreover, despite no significant differences were found for LDH total activity in liver, a significant increase (LDH1) and decrease (LDH4) were found also in this organ.

Brasil et al. (2009) found that rats fed on GM soy showed altered morphology of the uterus and the ovaries: had greater volume density of endometrial glanular epithelium, reduced follicle number and increased corpus luteum numbers (a tendency to abort or less of a chance to get pregnant). Although the GM diet was not supplemented with cysteine as the other diets, and it is difficult to assess if the results were due to consumption of the transgenic soy itself or were due to the presence of glyphosate (and/or AMPA), always present in GM seeds, the findings are disturbing and warrant further studies.

A recent study found that the hamsters fed with GM Soybean showed the growth of hairs inside the pouches of the mouth and the number of hairy mouthed hamsters was much higher in the third generation of GM soy fed animals than in others (Baranov et al., 2010). According to the authors, it remains unclear why these hair structures appear in the oral cavity of mammals. We may only speculate on the origin of this phenomenon. The gingival pouches may result from paradontitis and paradontosis caused by feeding on compound food in the vivarium, i.e., by a suboptimal diet. This pathology may be exacerbated by elements of the food that are absent in natural food, such as genetically modified (GM) ingredients (GM soybean or corn meal) or contaminants (pesticides, mycotoxins, heavy metals, etc.). Probably, hair growth in the gingival pouches is a protective reaction of the body suppressing the progress of gingival pathology, because the hair bundles are so dense that they prevent food from getting into the pouches and the resultant inevitable inflammation. Hair grows in the parts of the mucosa that, being affected by mechanical factors, acquire the capacity for keratinization.

More recently, Tudisco et al (2010) studied the possible effects on cell metabolism, by determination of several specific enzymes in serum of goats fed conventional or Roundup Ready soybean and in heart, skeletal muscle, liver and kidney of their offspring. Aspartate aminotransferase (AST) and ALT enzyme activity were significantly lower in serum from goats fed GM soybean but enzyme levels were in the normal range. Statistical differences were detected in kid's kidney for GGT and LDH, whereas in the heart and skeletal muscle this result was seen only for LDH. The increase in LDH activity was confirmed by histochemistry. In addition, significant differences between control and treated animals were detected for heart, kidney, muscles and liver LDH isoenzyme distribution, particularly concerning the LDH1, as found previously in the rabbits. Since LDH1 is known to be involved in cell metabolism by favouring the reaction of lactate to pyruvate (Van Hall, 2000), these results could indicate a general increase in cell metabolism.

A summary of experimental studies concerning health of animals fed genetically modified soybean is reported in Table 4.

Table 4. A summary of experimental studies concerning health of animals fed with genetically modified soybean.

Animal species	Length of study	Main effects	Reference
BN rats and B10A mice	15 weeks	No immunotoxic activity. No histopathological abnormalities.	Teshima et al. (2000)
Sprague-Dawley rats	> 93 days	No adverse effects on body weight/gain, food consumption, clinical signs, mortality, ophthalmology, neurobehavioral assessment, clinical pathology, organ weights and gross and microscopic pathology	Appenzeller et al. (2008)
Sprague-Dawley rats	-	No adverse effects on body weight/gain, food consumption, and mortality, clinical signs of toxicity or ophthalmological observations, neurobehavioral assessments, organ weights or clinical and anatomic pathology	Delaney et al. (2008)
Mice	28 days	No adverse effects	Mathesius et al. (2009)
F344 rats	52 weeks	No adverse effect in gross necropsy findings, hematological and serum biochemical parameters, organ weights and pathological findings	Sakamoto et al. (2007)
Broilers	42 days	No adverse effects were found. It was concluded that GM 356043 was nutritionally equivalent to non-GM soybean with comparable genetic background	McNaughton et al. (2008)
Mice	-	Enlargements in the smooth endoplasmic reticulum of Sertoli cells	Vecchio et al. (2004)
Mice	-	Several proteins belonging to hepatocyte metabolism, stress response, calcium signaling and mitochondria were differentially expressed in GM-fed mice indicating a more marked expression of senescence markers in comparison to controls. GM-fed mice showed mitochondrial and nuclear modifications indicative of reduced metabolic rate	Malatesta et al. (2008a)

Mice	-	No morphological differences in embryos of GM and non-Gm soybean-exposed groups. Microscopic and ultramicroscopic cellular changes attributed to GM soybean intake	Cisterna et al. (2008)
Wistar rats	30 days	No adverse effects in nutritional performance. Altered pancreas function evidenced by the early acute PAP mRNA increased levels and pancreas cellular changes	(Malatesta et al., 2002a) and (Malatesta et al., 2002b)
Wistar rats	30 days	Significant zymogen-granule depletion	Magaña-Gómez et al. (2008)
Pregnant Rats	2 weeks	Fewer pups born with smaller body mass	Ermakova et al. (2006)
Rabbits	60 days	Significant alteration of kidney ALT, GGT and LDH activity. Significant alteration of heart LDH activity	Tudisco et al. (2006)
Hamsters	-	Growth of hairs inside the pouches of the mouth and the number of hairy mouthed	Baranov et al. (2010)
Kids	60 days	Significant alteration of kidney GGT and LDH activity. Significant alteration of LDH isoenzyme distribution in heart, kidney, muscle and liver	Tudisco et al. (2010)

7. HUMAN/ANIMAL SAFETY OF GLYPHOSATE

Glyphosate is not a genetically modified product but because its use in agriculture is inseparable from the cultivation of herbicide-tolerant GM crops in a particular technology package, its effects on health need to be examined also with that of the glyphosate-resistant GM crops. Although the declared aim of the introduction of glyphosate resistant GM crops was that with these crops the amount of herbicide sprayed on the land should decrease, due to the ever increasing area of cultivation of glyphosate-resistant Roundup Ready GM crops, the use of glyphosate has in fact increased (Benbrook, 2004; 2009). The glyphosatecontaining sprays destroy all weeds but the growth of the glyphosate-resistant GM crop is protected regardless of how much glyphosate is sprayed on to the land. To make sure that all weeds are destroyed the use of glyphosate and consequently the glyphosate load of the land has been substantially increasing after the first few years of a slight reduction (Benbrook, 2004; 2009). This has happened despite the ever-increasing number of publications showing that glyphosate has many serious and detrimental effects on the environment and biodiversity (Relyea 2005) with the development of herbicide-resistant weeds (Duke, 2005; Owen & Zelaya 2005; Warwick et al., 2008; Zelaya et al., 2007). There is also an urgent need to consider the potentially seriously damaging effects of this total herbicide on human/animal health, particularly as it is used in large amounts. Indeed, there are a number of recently published papers that all indicate possible damaging effects of glyphosate on health and reproduction which need to be taken seriously. On previous work the findings of Marc et al. (2005) have confirmed and extended their previous results by showing that the main ingredient of commercial Roundup formulations, glyphosate, in a mM concentration range, particularly when used together with the obligatory polyoxyethylene amine surfactant, inhibited the transcription of one of the enzymes involved in hatching of sea urchin embryos and therefore significantly delayed their hatching. When it is considered that farm workers inhale commercial herbicide sprays in which the active ingredient concentration exceeds by about 25 times of that used in the transcription inhibition studies by Marc et al. (2005), health concerns due to the use of glyphosate must be acute. In another study it was shown that in the oral treatment of Wistar rats with increasing concentrations of the herbicide Glyphosate-Biocarb, a formulation used in many countries such as Brazil, the number of Kupffer cells in hepatic sinusoids increased, followed by large deposition of reticulin fibres and the leakage of hepatic aspartate aminotransferase and alanine aminotransferase into the circulation, indicating hepatic damage in these animals (Benedetti et al., 2004). Successively, Richard et al. (2005) and Benachour et al. (2007) showed that glyphosate, particularly as used together with polyoxyethylene amine surfactant in Roundup Ready formulations, was toxic to human placental JEG3 cells at concentrations lower than that used in agricultural practices. Even at subtoxic concentrations RR was an endocrine disruptor on aromatase activity and its mRNA level as glyphosate interacted with the active site of the purified enzyme. It is possible that the pregnancy problems in agricultural workers using Roundup may be traced back to the exposure to this herbicide (Savitz et al., 2000).

Recently, Gasnier et al. (2009) exposed human liver HepG2 cells, a well-known model to study xenobiotic toxicity, to four different formulations and to glyphosate, which is usually tested alone in chronic *in vivo* regulatory studies. They measured cytotoxicity with three assays (Alamar Blue®, MTT, ToxiLight®), plus genotoxicity (comet assay), anti-estrogenic (on ER_, ER_) and anti-androgenic effects (on AR) using gene reporter tests. They also checked androgen to estrogen conversion by aromatase activity and mRNA. All parameters were disrupted at sub-agricultural doses with all formulations within 24 h. These effects were more dependent on the formulation than on the glyphosate concentration. First, the observed a human cell endocrine disruption from 0.5ppm on the androgen receptor in MDA-MB453-kb2 cells for the most active formulation (R400), then from 2ppm the transcriptional activities on both estrogen receptors were also inhibited on HepG2. Aromatase transcription and activity were disrupted from 10 ppm. Cytotoxic effects started at 10ppm with Alamar Blue assay (the most sensitive), and DNA damages at 5 ppm. According to the authors, the direct G action is most probably amplified by vesicles formed by adjuvants or detergent-like substances that allow cell penetration, stability, and probably change its bioavailability and thus metabolism. These detergents can also be present in rivers as polluting contaminants. The type of formulation should then be identified precisely in epidemiological studies of G-based herbicides effects. Of course to drive hypotheses on *in vivo* effects, not only dilution in the body, elimination, metabolism, but also bioaccumulation and time-amplified effects should be taken into account. These herbicides mixtures also present endocrine effects on human cells, at doses far below agricultural dilutions and toxic levels on mitochondrial activities and membrane integrity.

All these findings indicate that there is an urgent need to carry out systematic and direct studies, independent of the biotech industry, on the short- and long-term effects on animal (and human) health of exposure to glyphosate and its more effective commercial formulations alone and/or preferably in combination with the appropriate GM crop. With the presently cultivated huge areas of Roundup Ready crops and the anticipated even larger future extensions of this glyphosate-dependent GM crop technology the potential danger for animal/human health needs to be dealt with in advance and not if or when it occurs. If we consider that RR soybeans may in themselves damage reproduction, a combination of the similar, possibly synergistic effects of the GM crop and glyphosate could be a potential disaster waiting to happen.

8. CONCLUSIONS

The debate on the safety of genetically modified organisms (GMOs) used for food and feed is still very lively throughout the world, more than 15 years after their first commercial release. Huge social, economical, and political issues have been raised. Unfortunately, although some stakeholders claim that a history of safe use of GMOs can be upheld, there are no human or animal epidemiological studies to support such a claim as yet, in particular because of the lack of labeling and traceability in GMO-producing countries. As a matter of fact, 97% of edible GMOs among cultivated GMOs (soy, corn and oilseed rape or canola, excluding cotton) are grown in South and North America, where GMOs are not labeled. All these plants have been modified to tolerate and/or produce one or more pesticides, and contain therefore such residues at various levels. Most are Roundup residues (it is a major herbicide used worldwide and tolerated by about 80% of GMOs).

As stated by the EFSA (2008), several aspects have to be investigated when considering whether or not recombinant DNA from GM plants, or the derived proteins can end up in animal tissues and products. These include (i) the fate of the recombinant DNA and protein during feed processing and ensiling; (ii) the fate of the recombinant DNA and protein in the gastrointestinal tract of animals fed with GM feed; (iii) the potential absorption of the digested pieces of DNA or protein into animal tissues/products and (iv) the potential of biological functionality of absorbed DNA and protein fragments.

The mere detection of recombinant DNA fragments in animal organs and tissues could not justify, by itself, public concerns regarding human consumption of products from farm animals fed transgenic crops. However, the persistence of DNA after dietary exposure is one aspect of risk assessment for novel food. Indeed, as concerns the hypothetical horizontal gene transfer of recombinant DNA from GM crop-derived feeds to animal and human gut microflora, Netherwood et al. (2004), found that a small proportion of feed DNA survives passage through the human upper gastrointestinal tract and a very small proportion of the small intestinal microflora containing transgenic feed. According to the authors, even if this result does not indicate a complete transgenic transfer to the prokaryotes, the survival of transgenic DNA during the passage through the small intestine should be considered in future safety assessments of GM foods. In addition, any alteration in cell metabolism should be taken into account in this field. For instance, the modification in LDH synthesis suggests an increase in cell metabolism. Therefore, possible long-term effects of such an alteration need to be elucidated.

In conclusion, taking into account the potential risks related to GMP impact, further researches are needed in this area, including studies to determine DNA transport or entry mechanisms/processes across the epithelial layer of the gastro-intestinal tract into the bloodstream, as well as degradation or accumulation of foreign DNA in blood or other organs of animal species. In any case, the traceability of products from animals fed on GMOs is crucial.

REFERENCES

1. Alexander, T.W., Reuter, T., Aulrich, K., Sharma, R., Okine, E.K., Dixon, W.T., & McAllister, T.A. (2007). A review of the detection and fate of novel plant molecules derived from biotechnology in livestock production. *Animal Feed Science and Technology*, 133, pp. 31-62.
2. Alibhai, M.F. & Stallings, W.C. (2001). Closing down on glyphosate inhibition – with a new structure for drug discovery. *Proceedings of the National Academy of Sciences*, 98, pp. 2944-2946.
3. Appenzeller, L.M., Munley, S.M., Hoban, D., Sykes, G.P., Malley, L.A., Delaney, B. (2008). Subchronic feeding study of herbicide-tolerant soybean DP-356Ø43-5 in Sprague–Dawley rats. *Food Chem Toxicol*, 46, pp. 2201–13.

4. Aulrich, K., Böhme, H., Daenicke, R., Halle, I., Flachowsky, G. (2001). Genetically modified feeds in animal nutrition. 1st Com.: Bacillus thuringiensis (Bt) corn in poultry, pig and ruminant nutrition. *Archives of Animal Nutrition*, 54, pp. 183-195.
5. Aumaitre, A., Aulrich, K., Chesson, A., Flachowsky, G., Piva, G. (2002). New feed from genetically modified plants: substantial equivalence, nutritional equivalence, digestibility, and safety for animals and the food chain. *Livestock Production Science*, 74, pp. 223-238.
6. Aumaitre, A. (2004). Safety assessment and feeding value for pigs, poultry and ruminant animals of pest protected (Bt) plants and herbicide tolerant (glyphosate, glufosinate) plants: interpretation of experimental results observed worldwide on GM plants. *Italian Journal of Animal Science*, 3, pp. 107-121.
7. Baranov, A.S., Chernova, O.F., Feoktistova, N., Yu, Surov, A.V. (2010). A New Example of Ectopia: Oral Hair in Some Rodent Species. *Doklady Akademii Nauk*, 431, pp. 559–562.
8. Battistelli, S., Citterio, B., Baldelli, B., Parlani, C., Malatesta, M. (2010). Histochemical and morpho-metrical study of mouse intestine epithelium after a long term diet containing genetically modified soybean. *European Journal of Histochemistry,* 54, pp. 154-157.
9. Benachour, N., Sipahutar, H., Moslemi, S., Gasnier, C., Travert, C., Séralini, G.E. (2007). Time- and dose-dependent effects of roundup on human embryonic and placental cells. *Archives of Environmental Contaminant Toxicology*, 53, pp. 126-133.
10. Benbrook, C.M. (2004). Genetically engineered crops and pesticide use in the United States. The first nine years. *BioTech InfoNet Technical*, 7.
11. Benbrook, C. (2009). Impacts of genetically engineered crops on pesticide use: The first thirteen years. The Organic Center, *Critical issues*, pp.1-47.
12. Benedetti, A.L., Lourdes Vituri, C., de Trentin, A.G., Domingues, M.A.C., Alvarez-Silva, M. (2004). The effects of sub-chronic exposure of Wistar rats to the herbicide Glyphosate-Biocarb. Toxicolology Letters, 153, pp. 227-232.
13. Benetrix, F. (2000). Evaluation de la valeur technologique, alimentaire et zootechnique de produits et de sous produits issus de plantes genetiquement modifiees, *personnal communication*.
14. Brake, J. & Vlachos, D. (1998). Evaluation of Transgenic Event 176 “Bt” Corn in Broiler Chickens. *Poultry Science*, 77, pp. 648-653.
15. Brasil, F.B., Soares, L.L., Faria,T.S., Boaventura, G.T., Barcellos Sampaio, F.J., Ramos, C.F. (2009). The Impact of Dietary Organic and Transgenic Soy on the Reproductive System of Female Adult Rat. *The Anatomical Record,* 292, pp. 587–594.
16. Center for Environmental Risk Assessment (2010). A Review of the Environmental Safety of the CP4 EPSPS Protein. ILSI Research Foundation. Washington D.C., USA, pp. 1-16.
17. Chainark, P., Satoh, S., Hino, T., Kiron, V., Hirono, I., Aoki, T. (2006). Availability of genetically modified soybean meal in rainbow trout *Oncorhynchus mykiss* diets. *Fisheries Science*, 72, pp. 1072–1078.
18. Chainark, P., (2008). Availability of genetically modified feed ingredient II: investigations of ingested foreign DNA in rainbow trout *Oncorhynchus mykiss. Fisheries Science*, 74, pp. 380-390.
19. Cisterna, B., Flach, F., Vecchio, L., Barbino, S.M.L., Battistelli, S., Martin, T.E., Malatesa, M., Biggiogera, M. (2008). Can a genetically modified organism-containing diet influence embryo development? A preliminary study on pre-implantation mouse embryos. European Journal of Histochemistry, 52, pp. 263–267.
20. Clarke, J.H. & Ipharraguerre, I.R. (2001). Livestock performance: feeding biotech-crops. *Journal of Dairy Science*, 84, pp. E9–E18.
21. Cromwell, G.L., Lindemann, M.D., Randolph, J.H., Stanisieweski, E.P., Hartnell, G.F. (2001). Soybean meal from Roundup Ready or conventional soybeans in diets for growingfinishing pigs. *J. Anim. Sci.,* 79, pp. 1318.
22. Delaney, B., Appenzeller, L.M., Munley, S.M., Hoban, D., Sykes, G.P., Malley, L.A. , Sanders, C. (2008). Subchronic feeding study of high oleic acid soybeans (event DP-3Ø5423- 1) in Sprague–Dawley rats. *Food and Chemical Toxicology,* 46, pp.3808–3817.
23. Doerfler, W. (2000). Foreign DNA in Mammalian Systems. Wiley-VCH Verlag GmbH, Weinheim. Doerfler, W., Schubert, R. (1998). Uptake of foreign DNA from the

environment: the gastrointestinal tract and the placenta as portals of entry. *Wien Klin Wochenschr,* 110, pp. 40-44.

24. Duggan, P.S., Chambers, P.A., Heritage, J., Forbes, J.M. (2003). Fate of genetically modified corn DNA in the oral cavity and rumen of sheep. *British Journal of Nutrition,* 89, pp. 159–166.
25. Duke, S.O. (2005). Taking stock of herbicide-resistant crops ten years after introduction. *Pest Management Science,* 61, pp. 211-218.
26. Einspanier, R., Lutz, B., Rief, S., Berezina, O., Zverlov, V., Scwarz, W., Mayer, J. (2004). Tracing residual recombinant feed molecules during digestion and rumen bacterial diversity in cattle fed transgene corn. *European Food Research and Technology,* 218, pp. 269–273.
27. Ermakova, I. (2006). Genetically modified soy leads to the decrease of weight and high mortality in rat pups of first generation: preliminary studies. *EcosInfo.*, 1, pp. 4-10.
28. European Commission (2001). Directive 2001/18/EC of the European Parliament and of the Council of 12 March 2001 on the deliberate release into the environment of genetically modified organism and repealing Council Directive 90/220/EEC. *Official Journal of Communities*, L106, 1–39
29. European Commission (2003). Regulation (EC) 1829/2003 of the European Parliament and of the Council of 22 September 2003 on genetically modified food and feed. *Off J Eur Communities*, L268, 1–23
30. European Food Safety Authority (EFSA) - GMO Panel Working Group on Animal Feeding Trials (2008). Safety and nutritional assessment of GM plants and derived food and feed: The role of animal feeding trials. Food Chemistry and Toxicology, 46, pp. S2- S70.
31. FAO/WHO (2000). Safety Aspects of Genetically Modified Foods of Plant Origin. Report of a Joint FAO/WHO Expert Consultation on Foods Derived from Biotechnology. World Health Organization, Geneva.
32. FAO (2004). The state of food and agriculture 2003-2004. http://www.fao.org/docrep/006/y5160e/y5160e10.html
33. Faust, M. (2000). Livestock products composition and detection of transgenic DNA/proteins. *Proceeding of Annual Meeting ADAS-ASAS Agricultural and Biotechnology Market*, pp. 1-48, Baltimore MD, USA.
34. Flachowsky, G., Aulrich, K., (2001). Nutritional assessment of feeds from genetically modified organism (GMO). *Journal of Animal Feed Science*, 10, pp. 181–194.
35. Flachowsky, A., Chesson, A., Aulrich, K. (2005a). Animal nutrition with feeds from genetically modified plants. *Archive of Animal Nutrition*, 59, pp. 1-40.
36. Flachowsky, G., Halle, I., Aulrich, K. (2005b). Long term feeding of Bt-corn—a 10 generation study with quails. *Archives of Animal Nutrition*, 59, pp. 449–451.
37. Francis, G., Makkar, H.P.S. & Becker, K. (2001). Anti-nutritional factors present in plantderived alternate fish feed ingredients and their effects in fish. *Aquaculture*, 199, pp. 197–227.
38. Franz, J.E., Mao, M.K. and Sikorski, J.A. (1997). Glyphosate: A Unique Global Herbicide. In *ACS Monograph* American Chemical Society (Eds.), Washington D.C., pp. 27-64. Gasnier, C., Dumont, C., Benachour, N., Clair, E., Chagnon, M.C., Séralini, G.E. (2009). Glyphosate-based herbicides are toxic and endocrine disruptors in human cell lines. *Toxicology,* 262, pp. 184–191
39. Halle, I., Aulrich, K., Flachowsky, G. (2006), Four generations feeding GMO-corn to laying hens. *Proc Soc Nutr Physiol*, 15, 114 (Abstract).
40. Hammond, B.G., Vicini, J.L., Hartnell, G.F., Naylor, M.W., Knight, C.D., Robinson, E.H., Fuchs, R.L. and Padgette, SR. (1996). The feeding value of soybeans fed to rats, chickens, catfish and dairy cattle is not altered by genetic incorporation of glyphosate tolerance. *Journal of Nutrition*, 126, pp. 717–727.
41. Hemre, G.I., Sanden, M., Bakke-McKellep, A.M., Sagstad, A., Krogdhal, A. (2005). Growth, feed utilization and health of Atlantic salmon Salmo salar L. fed genetically modified compared to non-modified commercial hybrid soybeans. *Aquaculture Nutrition,* 11, pp.157-167.

42. Jennings, J.C., Kolwyck, D.C., Kays, S.B., Whetsell, A.J., Surber, J. B., Cromwell, G.L., Lirette, R.P., Glenn, K.C. (2003). Determining whether transgenic and endogenous plant DNA and transgenic protein are detectable in muscle from swine fed Roundup Ready soybean meal. *Journal of Animal Science*, 81, pp. 1447–1455.
43. Khumnirdpetch, V., Intarachote, U., Treemanee, S., Tragoonroong, S., Thummabood, S. (2001). Detection of GMOs in the broilers that utilized genetically modified soybean meals as a feed ingredient. In *Plant & Animal Genome IX Conference*, Town and Country Hotel, San Diego, San Diego, California, USA.
44. Klinedinst, D. & Drinkwater, N. (1992). Mutagenesis by apurinic sites in normal and ataxia telangiectasia human lymphoblastoid cells. *Molecular Carcinogenesis*, 6, pp. 32-42.
45. Klotz, A. & Einspanier, R. (1998). Detection of "novel-feed" in animals? Injury of consumers of meat or milk is not expected. *Mais*, 3, pp. 109–111.
46. Konig, A., Cockburn, A., Crevel, R.W.R., Debruyne, E., Grafstroem, R., Hammerling, U., Kimber, I., Knudsen, I., Kuiper, H.A., Peijnenburg, A.A.C.M., Penninks, A.H., Poulsen, M., Schauzu, M.,Wal, J.M. (2004). Assessment of the safety of foods derived from genetically modified (GM) crops. *Food Chem. Toxicol.*, 42, pp. 1047–1088.
47. Kuiper, H.A., Kleter, G.A. (2003). The scientific basis for risk assessment and regulation of genetically modified foods. *Trends of Food Science Technology*, 14, pp. 277-293.
48. ISAAA (International Service for the Acquisition of Agri-biotech Applications) (2010). Global Status of commercialised Biotech/GM crops: 2010. http://www.isaaa.org/resources/publications
49. Magaña-Gómez, J.A., López Cervantes, G., Yepiz-Plascencia, G., Calderón De La Barca, A.M. (2008). Pancreatic response of rats fed genetically modified soybean. *J Appl Toxicol*, 28, pp. 217–226.
50. Malatesta, M., Caporaloni, C., Rossi, L., Battistelli, S., Rocchi, M.B.L., Tonucci, F., Gazzanelli, G. (2002a). Ultrastructural analysis of pancreatic acinar cells from mice fed on genetically modified soybean. *Journal of Anatomy*, 201, 409–415.
51. Malatesta, M., Caporaloni, C., Gavaudan, S., Rocchi, M.B.L., Serafini, S., Tiberi, C., Gazzanelli, C. (2002b). Ultrastructural morphometrical and immunocytochemical analyses of hepatocyte nuclei from mice fed on genetically modified soybean. *Cell Structur and Function*, 27, pp. 173–180.
52. Malatesta, M., Biggiogera, M., Manuali, E., Rocchi, M.B.L., Balzelli, B., Gazzanelli, G. (2003). Fine structural analyses of pancreatic acinar cell nuclei from mice fed on genetically modified soybean. *European Journal Histochemical*, 47, p. 385–8.
53. Malatesta M., Tiberi C., Baldelli B., Battistelli S., Manuali E., Biggiogera M. (2005). Reversibility of hepatocyte nuclear modifications in mice fed on genetically modified soybean. *European Journal Histochemical*, 49, pp. 237–242.
54. Malatesta, M., Boraldi, F., Annovi, G., Balzelli, B., Battistelli, S., Biggiogera, M., Quaglino, D. (2008a). A long-term study on female mice fed on a genetically modified soybean: effects on liver ageing. *Histochemistry Cell Biology*, 130, pp. 967–977.
55. Malatesta, M., Perdoni, F., Santin, G., Battistelli, S., Muller, S., Biggiogera, M. (2008b). Hepatoma tissue culture (HTC) cells as a model for investigating the effects of low concentrations of herbicide on cell structure and function. Toxicology *In Vitro*, 22, pp. 1853–1860.
56. Marc, J., Le Breton, M., Cormier, P., Morales, J., Bellé, R., Mulner-Lorillon, O. (2005). A glyphosate-based pesticide impinges on transcription. *Toxicol. Appl. Pharmacol.*, 203, pp. 1-8.
57. Martens, M.A. (2000). Safety evaluation of genetically modified foods. *Int. Arch. Occup. Environ. Health*, 73, pp. S14–S18.
58. Martin-Orúe, S.M., O'Donnell, A.G., Arino, J., Netherwood, T., Gilbert, H.J., Mathers, J.C. (2002). Degradation of transgenic DNA from genetically modified soya and maize in human intestinal simulations. *British Journal of Nutrition*, 87, pp. 533–542.
59. Mathesius, C.A., Barnett, J.F., Cressman, R.F., Ding, J., Carpenter, C., Ladics, G.S., Schmidt, J., Layton, R.J., Zhang, J.X.Q. , Appenzeller, L.M., Carlson, G., Ballou, S., Delaney, B. (2009). Safety assessment of a modified acetolactate synthase protein (GM-HRA) used as a

selectable marker in genetically modified soybeans. *Regul Toxicol Pharmacol*, 55, pp. 309–320.
60. McAllan, A.B. (1982). The fate of nucleic acids in ruminants. *Proc Nutr Soc*, 41, pp. 309-317.
61. McNaughton, J., Roberts, M., Smith, B., Rice, D., Hinds, M., Sanders, C., Layton, R., Lamb, I., Delaney B. (2008). Comparison of broiler performance when fed diets containing event DP-3Ø5423-1, nontransgenic near-isoline control, or commercial reference soybean meal, hulls, and oil. *Poultry Science*, 87, pp. 2549–2561.
62. McNaughton, J., Roberts, M., Rice, D., Smith, B., Hinds, M., Delaney, B., Iiams, C., Sauber, T. (2011). Nutritional equivalency evaluation of transgenic maize grain from event DP-098140-6 and transgenic soybeans containing event DP-356043-5: laying hen performance and egg quality measures. *Poultry Science*, 90 (2), pp. 377-389.
63. Neterwood, T., Martin-Orúe, S.M., O'Donnell, A.G., Gockling, S., Graham, J., Mathers, J.C., Gilbert, H.J. (2004). Assessing the survival of transgenic plant DNA in the human gastrointestinal tract. *Nature Biotechnology*, 22, pp. 204-209.
64. Nielsen, C.R., Berdal, K.J., Bakke-McKellep, A.M., Holst-Jensen, A. (2005). Dietary DNA in blood and organs of Atlantic salmon (Salmo salar L.). *European Food Research and Technology*, 221, pp. 1–8.
65. Novak, W.K. & Hasleberger, A.G. (2000). Substantial equivalence of anti-nutrients and inherent plant toxins in genetically modified novel foods. *Food Chemistry and Toxicology*, 38, pp. 473–483.
66. Olli, J.J., Krogdahl, A., van den Ingh, T.S.G.A.M. & Brattas, L.E. (1994). Nutritive value of four soy products in diets for Atlantic salmon (*Salmo salar* L.). *Acta Agriculture Scandinave A Animal Science*, 44, pp. 50–60.
67. Owen, M.D.K., Zelaya, I.A. (2005). Herbicide-resistant crops and weed resistance to herbicides. *Pest Management Science*, 61, pp. 301-311.
68. Padgette, S.R., Biest Taylor, N., Nida, D.L., Bailey, M.R., MacDonald, J., Holden, L.R., Fuchs, R.L. (1996). The composition of glyphosate-tolerant soybean seeds is equivalent to that of conventional soybean. *Journal of Nutrition*, 126, pp. 702–716.
69. Pandey, A., Kamle, M., Yadava, L.P., Muthukumar, M., Kumar, P., Gupta, V., Ashfaque, M., Pandey, B.K. (2010). Genetically modified food: Its uses, future prospects and safety assessments. *Biotechnology*, 9, pp. 444-458.
70. Phipps, R.H., Deaville, E.R., Maddison, B.C. (2003). Detection of transgenic and endogenous plant DNA in rumen fluid, duodenal digesta, milk, blood, and feces of lactating dairy cows. *Journal of Dairy Science*, 86, pp. 4070–4078.
71. Piva, G., Morlacchini, M., Pietri, A., Piva, A., Casadei, G., (2001a). Performance of weaned piglets fed insect-protected (MON 810) or near isogenic corn. *Journal of Animal Science*, 79, pp. 106(Abstract).
72. Piva, G., Morlacchini, M., Pietri, A., Rossi, F., Grandini, A., (2001b). Growth performance of broilers fed insect-protected (MON 810) or near isogenic control corn. *Poultry Science*, 80 (Suppl. 1), 320 (Abstract).
73. Ran, T., Mei, L., Lei, W., Aihua, L., Ru, H., Jie, S. (2009). Detection of transgenic DNA in tilapias (*Oreochromis niloticus*, GIFT strain) fed genetically modified soybeans (Roundup Ready). *Aquaculture Research*, 40, pp. 1350-1357.
74. Relyea, A.A. (2005). The impact of insecticides and herbicides on the biodiversity and productivity of aquatic communities. *Ecological Applications*, 15, pp. 618-627.
75. Richard, S., Moslemi, S., Sipahutar, H., Benachour, N., Seralini, G.-E. (2005). Differential effects of glyphosate and Roundup on human placental cells and aromatase. Environmental Health Perspective, 113, pp. 716-720.
76. Sagstad, A., Sanden, M., Krogdahl, A., Bakke-McKellep, A.M., Froystad, M., Hemre, G.-I. (2008). Organs development, gene expression and health of Atlantic salmon (Salmo salar L.) fed genetically modified soybeans compared to the near-isogenic nonmodified parental line. *Aquaculture Nutrition*, 14, pp. 556-572.
77. Sakamoto, Y., Tada, Y., Fukumori, N., Tayama, K., Ando, H., Takahashi, H. (2007). A 52-week feeding study of genetically modified soybeans in F344 rats. *Journal Food Hygiene Society of Japan*, 48, pp. 41–50.

78. Sakamoto, Y., Tada, Y., Fukumori, N., Tayama, K., Ando, H., Takahashi, H. (2008). A 104-week feeding study of genetically modified soybeans in F344 rats. *Journal Food Hygiene Society of Japan*, 49, pp. 272–282.
79. Sanden, M., Bruce, I.J., Rahman, M.A., Hemre, G.-I. (2004). The fate of transgenic sequences present in genetically modified plant products in fish feed, investigating the survival of GM soybean DNA fragments during feeding trials in Atlantic salmon, *Salmo salar* L. *Aquaculture*, 237, pp. 391–405.
80. Savitz, D.A., Arbuckle, T., Kaczor, D., Curtis, K.M. (2000). Male pesticide exposure and pregnancy outcome. *Am. J. Epidemiol,*. 146, pp. 1025-1036.
81. Saxena, D. & Stotzky, G. (2001). Bt corn has a higher lignin content than non-Bt corn. *Am. J. Bot.*, 88, pp. 1704–1706.
82. Schubbert, R., Lettmann, C., Doerfler, W. (1994). Ingested foreign (phage M 13) DNA survives transiently in the gastrointestinal tract and enters the bloodstream of mice. *Molecular Genetics*, 242, pp. 495-504.
83. Schubbert, R., Renz, D., Schmitz, B., Doerfler, W. (1997). Foreign (M13) DNA ingested by mice reaches peripheral leukocytes, spleen, and liver via the intestinal wall mucosa and can be covalently linked to mouse DNA. *Proc Natl Acad Sci USA,* 94, pp. 961-966.
84. Schubbert, R., Hohlweg, U., Renz, D., Doerfler, W. (1998). On the fate of orally ingested foreign DNA in mice: chromosomal association and placental transmission to the fetus. *Molecular Genetics*, 259, pp. 569–576.
85. Seralini, G.E., Mesnage, R., Clair, E., Gress, S., Spiroux de Vendomois, J., Cellier, D. (2011), Genetically modified crops safety assessments: present limits and possible improvements. *Environmental Sciences Europe,* 23, pp. 10. doi:10.1186/2190-4715-23-10
86. Sharma, R., Alexander, T.W., John, S.J., Dugan, M.E. R., Aalhus, J.L., Stanford, K., Mcallister, T.A. (2004). Relative stability of transgene DNA fragments from GM rapeseed in mixed ruminal cultures. *British Journal of Nutrition*, 91, pp. 673-681.
87. Sidhu, R.S., Hammond, B.G., Fuchs, R.L., Mutz, J.-N., Holden, L.R., George, B. & Olson, T. (2000). Glyphosate-tolerant corn: The composition and feeding value of grain from glyphosate-tolerant corn is equivalent to that of conventional corn (Zea mays L.). *Journal of Agricultural and Food Chemistry*, 48, pp. 2305-2312.
88. Świątkiewicz S., Świątkiewicz M. (2009). Second generation of genetically modified plants in animal nutrition. *Medycyna Wet*, 65, pp. 460-465.
89. Teshima, R., Akiyama, H., Okunuki, H., Sakushima, J.-I., Goda, Y., Onodera, H,, Sawada, J.-I. and Toyoda, M. (2000). Effect of GM and non-GM soybeans on the immune system of BN rats and B10a mice. *Journal of Food Hygiene Society Japan*, 41,pp. 188–193.
90. Theodorou, M.K., Williams, B.A., Dhanoa, M.S., McAllan, A.B., France, J. (1994). A simple gas production method using a pressure transducer to determine the fermentation kinetics of ruminant feeds. *Animal Feed Science and Technology*, 48, pp. 185-197.
91. Tony, M.A., Butschke, A., Broll, A., Grohman, L., Zagon, J., Halle, I., Danicke, S., Schauzu, M., Hafez, H.M., Flachowsky, G. (2003). Safety assessment of Bt 176 maize on broiler nutrition: degradation of maize-DNA and its metabolic fate. *Archive of Animal Nutrition*, 57, pp. 235–252.
92. Tudisco, R., Calabro', S., Terzi, V., Cutrignelli, M.I., Bovera, F., Piccolo, V., Zicarelli, F., Piccolo, G., Infascelli, F. (2004). In vitro fermentation kinetics of some genetically modified feeds. *Proceeding of 2nd International Meeting on Veterinary morpho-functional Biotechnologies,* Naples, Italy, Giugno 2004.
93. Tudisco, R., Lombardi, P., Bovera, F., d'Angelo, D., Cutrignelli, M.I., Ma stellone, V., Terzi, V., Avallone, L., Infascelli, F. (2006). Genetically modified soybean in rabbit feeding: detection of DNA fragments and evaluation of metabolic effects by enzymatic analysis. *Animal Science*, 82, pp. 193-199.
94. Tudisco, R., Ma stellone, V., Cutrignelli, M.I., Lombardi, P., Bovera, F., Mirabella, N., Piccolo, G., Calabrò, S., Avallone, L., Infascelli, F. (2010). Fate of transgenic DNA and evaluation of metabolic effects in goats fed genetically modified soybean and in their offsprings. *Animal*, 4, pp. 1662-1671.
95. Van den Eede, G., Aarts, H., Buhk, H.J., Corthier, G., Flint, H.J., Hammes, W., Jacobsen, B., Midtvedt, T., van der Vossen, J., von Wright, A., Wackernagel, W., Wilcks A. (2004). The

relevance of gene transfer to the safety of food and feed derived from genetically modified (GM) plants. *Food Chemistry and Toxicology*, 42, pp. 1127-1156.

96. Van Hall, G. (2000). Lactate as a fuel for mitochondrial respiration. *Acta Physiologica Scandinava*, 168, pp. 643-656.
97. Vecchio, L., Cisterna, B., Malatesta, M., Martin, T.E., Biggiogera, M. (2994). Ultrastructural analysis of testes from mice fed on genetically modified soybean. *European Journal of Histochemistry*, 48, pp. 448–454.
98. Wiedemann, S., Lutz, B., Kurtz, H. F., Schwarz, J., Albrecht, C. (2006). In situ studies on the time-dependent degradation of recombinant corn DNA and protein in the bovine rumen. *Journal of Animal Science*, 84, pp. 135-144.
99. Warwick, S.I., Legere, A., Simard, M.J., James, T. (2008). Do escaped transgenes persist in nature? The case of an herbicide resistance transgene in a weedy Brassica napa population. *Molecular Ecology*, 17, pp. 1387-1395.
100. Zelaya, I.A., Owen, M.D.K., VanGessel, M.J. (2007). Transfer of glyphosate resistance: Evidence of hybridisation in Conyza (Asteraceae). *Amer. J. Botany*, 94, pp. 660-673.

CHAPTER 9

Relationship between Starch and Amino Acid Levels of Broiler Diets on Growth Performance and Feed Efficiency

NA Hartono[1]*, Nahrowi[2] and Sumiati[2]

[1]*Student, Post Graduate School, Bogor Agricultural University, Indonesia*

[2]*Nutrition and Feed Science, Animal Husbandry Faculty, Indonesia*

ABSTRACT

This study was conducted to investigate the relationship between starch and amino acid levels of broiler diets on growth performance and feed efficiency. In this study, 240 one-day old chicks were distributed to 6 treatments using a completely randomized design (CRD) with four replications. Three treatments were fed with high starch and three others were fed with low starch. Both type of diets were formulated as isocaloric with three levels of total lysine, varying from 1.34%, 1.43%, 1.48% for starter period (0-18 days) and 1.18%,1.3%,1.35% for finisher period (19-35 days). The result showed that high starch level increased body weight gain ($p < 0.01$) and feed intake at 35 days ($p < 0.05$). Feed conversion ratio was better for the birds were fed a high starch level diet (1.66) compare with the low starch level (1.71). No interaction between starch level and amino acid level was observed for body weight gain, feed intake and feed conversion ratio at 35 days. No differences for blood parameter glucose, uric acid and insulin. Significant interaction between starch level and amino acid was observed for uric acid. It was concluded that feeding high starch level based on corn soya diet increased body weight gain and feed intake. Starch level has relationship with blood glucose, insulin and utilization of protein and energy in broiler.

Keywords: Starch; Amino acid; Broiler performance

1. INTRODUCTION

Feed efficiency aims to achieve production results that match genetic potential with low feed costs and reduce the potential environmental pollution caused by excess nutrients which are not digested by livestock, especially proteins. One technique to improve feed efficiency is to use the concept of balance energy and protein in broiler feed formulation. In addition, protein is composed of amino acids with a certain ratio of the amino acid lysine (ideal protein concept model). Starch is a part of carbohydrate as important energy source for broiler chickens. The amount of starch is still not taken into account in predicting the energy content of a material. Starch provides a more than 50% of energy requirement in broiler feed which based on corn and soybean meal. Is generally assumed to be very high starch digestibility close to 100%, but the rate of degradation in the intestine varies [1].

Feed with the starch content gradually broken down in the small intestine, will provide continuous glucose into the bloodstream. Glucose in the blood will be responded by the release of the hormone insulin. Insulin plays an important role in the transport of glucose and protein into muscle deposition during growth [2]. Continuous availability of glucose to the posterior part of the small intestine could prevent the use of amino acids as an energy source for the intestinal wall [3].

This study focused on investigate concept of balance energyamino acid related to the difference in the speed of starch digestibility in the small intestine Availability of different amounts of starch in the isocaloric feed in intestine, is expected to give different growth responses in broilers.

2. MATERIALS AND METHODS

2.1 Animal and Housing

In this study, 240 birds (DOC) placed in to 24 pens. Six types of feed with four replication each replication contained 10 birds. Age 0-18 days use the same feed that kind of starter feed. Age 19-35 days used grower feed.

2.2 Diets

There are 6 types of feed formulations were used, which are 2 of the content of starch (high starch 36% and lower starch 32%), which combined with 3 levels of the amino acid Lysine-based Total 1.34%, 1:43%, 1:48% for the starter phase, and 1.18%, 1.3%, 1.35% for the grower phase. Other amino acids will follow the ratio of the total Lysine according to nutrient requirement ROSS 308. Prediction systems for nutrient content of raw materials using SFR (Schothorst Feed Research) system with Format International Formulation Software from England.

The feed is given in the form of crumble, with a minimum temperature of 80°C conditioner, die size of 3.2 mm, minimum 90% PDI. Composition and nutrient content of research feed as in **Tables 1** and **2**.

2.3 Feed quality and blood parameters analysis

- Proximate analysis: Quality feed were analyzed using full proximate AOAC method (1980).
- Blood parameters: Blood glucose levels (Glucose Hexokinase Kit, Bayer, USA), levels of the hormone insulin (Coat-A-Kits COUNTA Insulin, DPC, LA, CA) and blood uric acid (Biosystems kit), be read with a spectrophotometer.
- Starch Analysis: Polarimeter methods read with a spectrophotometer

2.4 Starch and protein digestibility

- Ileal Starch Digestibility (%): calculated by subtracting the total amount of starch consumed with remaining starch in the small intestine digesta, divided by the total starch consumed multiplied by 100%.

Table 1: Composition and nutrient content of pre starter feed (0-18 days).

Raw Material	Treatment					
	R0	R1	R2	R3	R4	R5
Corn	57.118	57.449	58.671	46.497	46.075	47.468
Ricebran	3.502	2.838	0.000	13.809	15.000	11.321
Soybean meal 47%	27.764	26.970	27.792	26.966	24.757	27.445
Fish meal 60%	5.000	5.000	5.000	5.000	5.000	5.000
Crude Palm Oil	3.756	3.331	3.195	5.000	4.508	4.638
Meat Bone Meal 50%	0.000	2.300	3.400	0.000	1.050	2.050
Poultry Meat Meal	0.000	0.000	0.000	0.000	1.400	0.000
L-Lysine 78%	0.195	0.270	0.285	0.190	0.290	0.300
L-Threonine 95%	0.062	0.113	0.122	0.063	0.105	0.112
DL-methionine 99%	0.294	0.353	0.370	0.288	0.352	0.370
Tryptopan 100%	0.000	0.000	0.003	0.000	0.000	0.000
Limestone Ca 38%	0.400	0.000	0.000	0.500	0.200	0.000
Dicalcium Phosphate 18%	1.250	0.827	0.629	1.120	0.723	0.748
Salt NaCl	0.145	0.118	0.106	0.135	0.101	0.112
Mineral Premix	0.100	0.100	0.100	0.100	0.100	0.100
Vitamin Premix	0.100	0.100	0.100	0.100	0.100	0.100
Choline chloride 60%	0.100	0.100	0.100	0.101	0.101	0.096
Antimold propionic acid	0.050	0.050	0.050	0.050	0.050	0.050
Antioxidant	0.020	0.020	0.020	0.020	0.020	0.020
Anticoccydium	0.060	0.060	0.060	0.060	0.060	0.060
Nutrient Content						
Crude Protein (%)	21.34	22.17	22.88	21.04	22.00	22.60
ME (Kkal/kg)	3100.0	3100.8	3100.1	3100.2	3100.6	3100.0
Ca (%)	0.95	0.95	1.04	0.95	0.95	0.91
P (%)	0.75	0.79	0.78	0.86	0.89	0.87
Crude Fiber (%)	2.51	2.46	2.25	3.29	3.37	3.12
Crude Fat (%)	7.55	7.26	6.94	9.60	9.47	9.19
Starch (%)	**36.00**	**35.98**	**36.01**	**32.09**	**32.13**	**32.05**
Lysine (%)	**1.34**	**1.43**	**1.48**	**1.34**	**1.43**	**1.48**
Cl(%)	0.278	0.289	0.29	0.284	0.284	0.286
Na (%)	0.16	0.16	0.16	0.16	0.16	0.16
Methionine (%)	0.664	0.732	0.757	0.661	0.732	0.755
Methionine + Cystine (%)	1.00	1.07	1.11	1.00	1.07	1.11

R0: starch 36%- lysine 1.34%, R1: starch 36%- lysine 1.43%, R2 : starch 36%- lysine 1.48%, R3: starch 32% -lysine 1.34%, R4 : starch 32% - lysine 1.43%, R5: starch 32%- lysine 1.48%

Table 2: Composition and nutrient content of starter feed (19-35 days).

Raw Material	Treatment					
	R0	R1	R2	R3	R4	R5
Corn	59.021	54.482	55.154	45.787	45.799	44.655
Ricebran	10.000	10.000	8.407	15.000	15.000	15.000
Soybean meal 47%	18.962	22.121	21.513	17.836	17.195	16.178
Fish meal 60%	5.000	5.000	5.000	5.000	5.000	5.000
Crude Palm Oil	3.226	4.024	3.753	5.000	5.000	6.000
Meat Bone Meal 50%	1.600	2.300	4.450	0.450	4.600	2.400
Poultry Meat Meal	0.000	0.000	0.000	4.500	1.050	0.000
L-Lysine 78%	0.220	0.245	0.275	0.235	0.335	0.240
L-Threonine 95%	0.085	0.110	0.123	0.088	0.140	0.088
DL-methionine 99%	0.285	0.341	0.362	0.328	0.368	0.283
Wheat Pollard	0.000	0.000	0.000	0.000	0.000	4.050
Copra Expeller	0.000	0.000	0.000	5.000	5.000	5.000
Dicalcium Phosphat P 18%	1.014	0.816	0.428	0.265	0.000	0.541
Salt NaCl	0.121	0.111	0.086	0.062	0.063	0.104
Mineral Premix	0.100	0.100	0.100	0.100	0.100	0.100
Vitamin Premix	0.100	0.100	0.100	0.100	0.100	0.100
Choline chloride 60%	0.136	0.120	0.119	0.119	0.120	0.130
Antimold propionic acid	0.050	0.050	0.050	0.050	0.050	0.050
Antioxidant	0.020	0.020	0.020	0.020	0.020	0.020
Anticoccydium	0.060	0.060	0.060	0.060	0.060	0.060
Nutrient Content						
Crude Protein (%)	18.949	20.509	21.182	21.262	21.072	19.205
ME (Kkal/kg)	3200.7	3200.9	3200.0	3200.8	3200.8	3200.6
Ca (%)	0.90	0.94	1.09	0.87	1.06	0.88
P (%)	0.84	0.85	0.88	0.84	0.91	0.89
Crude Fiber (%)	2.90	2.93	2.80	3.78	3.74	4.01
Crude Fat (%)	7.95	8.60	8.40	10.74	10.79	11.54
Starch (%)	36.77	36.00	36.00	31.98	31.99	32.02
Lysine (%)	1.18	1.30	1.35	1.18	1.30	1.35
Cl(%)	0.284	0.284	0.286	0.26	0.284	0.27
Na (%)	0.16	0.16	0.16	0.16	0.16	0.16
Methionine (%)	0.629	0.702	0.730	0.701	0.735	0.629
Methionine + Cystine (%)	0.930	1.020	1.050	1.024	1.05	0.931

- Total protein digestibility (%): calculated by subtracting the total protein consumed by the amount of protein remaining in the feces, divided by the total protein consumed multiplied by 100%.

2.5 Chicken growth

- Feed intake (g/bird): measured by weighing the amount of feed consumed during the study.
- Average weight gain (g/bird): calculated by calculating the weight by the end of the study reduced the initial weight.
- Feed Conversion Ratio: calculated by comparing feed consumption with weight gain.

2.6 Statistical Analysis

The experimental design used in this study is completely randomized design (CRD) with 4 replications 2x3 factorial. The data obtained were analyzed using analysis of variance (ANOVA), the differences among mean treatments were analyzed using Tukey test [4].

3. RESULTS AND DISCUSSION

3.1 Weight gain

Based on **Table 3**, the average weight gain of birds aged 0-35 days in all treatments ranged from 1659.82-1891.54 gram. Potential genetic of modern broiler strain ROSS for weight gain in 35 days old is 1979 gram. Best result in this study was 95.5% of genetic potential.

Different starch content in feed between 36% and 32% gave a significant influence on the weight of the chicken 0-35 days ($P<0.01$). Weight gain was not influenced by difference levels of amino acids. The results showed no interaction between level of starch and amino acids on chicken weight gain. Feed with starch content of 36% led to higher feed intake than the diet feed containing 32% of starch ($P<0.05$). Difference in feed intake is causing different starches consumption. Starch is part of energy source carbohydrate which is essential for broiler. Starch broke down into glucose in small intestines. Pancreatic α-amylase secreted into the lumen of the small intestine, especially in the jejunum to break the bonds α-1,4 starch.

Table 3: Average weight gain, feed consumption and conversion ratio age 0-35 days.

Item	R0	R1	R2	R3	R4	R5
Gain (g/bird)	1891.54±110.18[a]	1795.36±41.37[a]	1811.02±143.30[a]	1659.30±90.93[b]	1665.82±99.48[b]	1676.43±96.65[b]
Consumption (g/bird)	3199.32±99.06[a]	2971.80±144.8[a]	3003.95±295.25[a]	2829.07±60.98[b]	2795.76±128.3[b]	2931.84±70.02[b]
FCR	1.69±0.1	1.65±0.06	1.657±0.11	1.70±0.06	1.68±0.03	1.75±0.08

Means in the same row with different superscript differ significantly ($P<0.05$).

R0: starch 36%- lysine 1.34%, R1: starch 36%- lysine 1.43%, R2 : starch 36%- lysine 1.48%, R3: starch 32% -lysine 1.34%, R4 : starch 32% - lysine 1.43%, R5: starch 32%- lysine 1.48%

Glucose is absorbed through the small intestinal wall, through sodium and potassium pump mechanism. Glucose is needed as a precursor in the citric acid cycle (citric acid cycle). Oxidation of glucose produces energy which is necessary for the synthesis of glycogen, amino acids, fatty acids and other metabolites. Feed that containing 36% starch produces more glucose compared with the diet containing 32% starch. This evidenced by the digestibility of starch in the small intestine (ileal Digestibility starch) were significantly different ($P < 0.05$), ie 99.46% and 99.26% between diets with a high starch compared with lower starch diet. Glucose produced of feed containing different starches will be determined by the amount of starch entering the small intestine multiplied by it digestibility level.

Feed with adequate starch content, will provide continuous glucose into the bloodstream. Glucose in the blood will be responded by the release of the insulin hormone [5]. Insulin plays an important role in the transport of glucose and deposition protein into muscle during growth [2]. Black et al. [6] stated that the unequal rate of starch digestibility and protein affect protein deposition and broiler performance.

3.2 Consumption and Feed Conversion Ratio

Based on **Table 3**, the average feed consumption in chickens aged 0-35 days all treatments ranged from 2795.76 -3199.32 gram. Genetic potential of modern broiler strain ROSS for feed consumption to 35 days old is 3248 gram. Average feed intake in this study was 90.9% of genetic potential. Starch content in different feed between 36% and 32% gave significant effect on feed consumption of chicken 0-35 days age ($P>0.05$). Feed consumption is not influenced by differences in the levels of amino acids in the diet. The results showed there is no interaction between level of starch and amino acids on broiler feed consumption. Differences on starch content in the diet cause differences gelatinization process at the pellet-making process heating steam temperature above 80 oC, leading to differences resistance pellet. Pellet quality was also influenced by the feed composition. A good quality pellet and not easily destroyed is one of the things that led to increased feed intake [1].

Feed conversion 0-35 day old chicks in all treatments ranged between 1.65-1.75. Genetic potential of modern broiler strain ROSS for feed conversion for age 35 days is 1.607. Average feed conversion in this study was 95% of the genetic potential. Starch content in different feed between 36% and 32% had no real effect on 0-35 days age chicken feed conversion ($P<0.05$). Feed conversion is not influenced by differences in the levels of amino acids in the diet. The results showed no interaction between level of starch and amino acids to the broiler feed conversion.

Table 4: Average intestine starch digestibility and total protein digestibility age 0-35 days.

Item	Treatment					
	R0	R1	R2	R3	R4	R5
Protein Dig.	68.98±2.93[c]	74.36±4.24[c]	82.70±2.49[b]	77.70±5.13[b]	81.68±5.7[b]	84.22±2.26[a]
Starch Dig.	99.42±0.02[a]	99.48±0.11[a]	99.47±0.11[a]	99.09±0.45[b]	99.39±0.3[b]	99.26±0.29[b]

Means in the same row with different superscript differ significantly ($P<0.05$).
R0: starch 36%- lysine 1.34%, R1: starch 36%- lysine 1.43%, R2 : starch 36%- lysine 1.48%, R3: starch 32% -lysine 1.34%, R4 : starch 32% - lysine 1.43%, R5: starch 32%- lysine 1.48%

Feed containing higher starch tends to provide a better feed conversion ration compared with lower starch content is 1.66 and 1.69.

3.3 Ileal starch digestibility

Based on **Table 4**, the average digestibility of starch in the small intestine 35 days old chicken on all treatments ranged from 99.09 - 99.48%. Starch content in different feed between 36% and 32% gave significant effect on digestibility of starch in the small intestine of chicken 0-35 days (P<0:05). Digestibility of starch in the small intestine is not influenced by differences in the levels of amino acids in the ration. The results showed no interaction between level of starch and level of amino acids for starch digestibility to the small intestine for 35 day old chickens. Feed containing higher starch provides a higher digestibility of starch in the intestine than the lower starch diet containing (99.46% and 99.26%).

3.4 Total protein digestibility

Based on **Table 4**, the average of total feed protein digestibility of chickens aged 35 days on all treatments ranged from 68.98 - 84.22%. Starch content in different feed between 36% and 32% gave real effect on protein digestibility chicken feed 0-35 (P<0:05). Protein digestibility is influenced by differences in the levels of amino acids in the diet (P < 0:05). The results showed no interaction between level of starch and level of amino acids for protein digestibility. Feed containing higher starch gives a lower protein digestibility than the diet containing lower starch (75.35% and 84.22%). Results of this study indicate that although feed with low starch content have higher protein digestibility, but did not give weight gain better than the feed that contain more starch. Selle et al. [7] states that the rate of absorption of amino acids associated with the rate of starch digestibility and absorption of glucose that had effect on the growth and deposition protein. Nitrogen Retention not correlated with the rate of nitrogen digestibility.

3.5 Content of Glucose, Uric Acid and Insulin in Chicken Blood

Blood glucose levels ranged between 200.14 mg/dl - 223.68 mg/ dl. Hazelwood [8] stated that the levels of glucose in broilers is higher compared with other mammals, the range 180-250 mg/dL. Chickens in a state of hypoglycemia when blood glucose levels 137 mg/dL and hyperglicemia on blood glucose levels 363 mg/dL [9]. Results showed no significant differences in all treatments.

Tabel 5: Average Glucose, Uric Acid and Insulin in blood.

Parameter	Treatment					
	R0	R1	R2	R3	R4	R5
Glukosa (mg/dL)	223.68±14.17	212.14±10.71	209.09±16.90	209.28±7.21	209.22±19.31	200.14±14.45
Uric Acid (mg/dL)	2.32±0.02	2.82±0.26	3.02±0.11	2.92±0.03	2.87±0.18	3.12±0.05
Insulin (µIU/mL)*	2.52±0.25a	2.16±0.18c	2.72±0.20a	2.39±0.20b	1.73±0.05d	2.34±0.20b

Means in the same row with different superscript differ significantly (P<0.05).
R0: starch 36%- lysine 1.34%, R1: starch 36%- lysine 1.43%, R2 : starch 36%- lysine 1.48%, R3: starch 32% -lysine 1.34%, R4 : starch 32% - lysine 1.43%, R5: starch 32%- lysine 1.48%

Feed containing higher starch tend to have higher blood glucose levels compared with a diet containing a low starch content (214.97 mg/dL and 206.21 mg/dL). Blood glucose always regulated in relatively constant levels through glucoregulation mechanism which is controlled by several metabolic hormones such as insulin, glucagon, pancreatic polypeptide, corticosteron and thyroxin [9]. Levels of insulin in the blood of chickens significantly influenced by the levels of starch and amino acids in the diet ($p < 0.05$). Interaction between starch content and amino acids in the diet did not affect the levels of insulin. It showed that glucose results from breakdown of starch in diet stimulating secretion of insulin hormone by the pancreas [5]. Difference in starch content in feed will also result in different amounts of glucose. Grizard et al. [10] stated that insulin plays a role in protein deposition in the body. This is thought to cause better weight gain on diet containing more starch. Adding of the amino acid from Lysine 1:34% to 1:43% did not respond higher insulin levels, both on the high starch and lower starch diet. Uric acid level in the blood can be used as one indicator to the final result nitrogen metabolism Hartman et al. [11].

Donsbough et al. [12] stated levels of uric acid in broilers in the range of 2:02 mg/dL - 5:13 mg/dL. The result of this study shows that differences in the levels of starch and amino acids in the diet treatment did not affect the levels of uric acid in the chicken blood. The result of this study similar to that expressed by Donsbough et al. [12]. Levels of uric acid had a tendency to increase at each addition of acid amino. Xie et al. [13] suggest that increased levels of uric acid by the addition of amino acids methionine in diet. Uric acid as an antioxidant stated by Hare and Johnson [14], therefore its existence is also associated with a state of oxidative stress [15]. Uric acid levels increase when the nitrogen intake increased Hevia and Clifford [16].

4. CONCLUSIONS

Diet containing the same type starch but with different numbers gives different of weight gain and feed consumption in broilers ages 35 days. Feed containing high starch tends to have more levels of glucose and insulin in the blood. Total digestibility of proteins not directly describes the growth performance of chickens on feed which the starch content distinguished.

RECOMMENDATIONS

Broiler feed formulation needs to pay attention to the content of starch in the ration, because the process of starch digestion and glucose absorption associated with protein deposition.

REFERENCES

1. Weurding RE, Veldman A, Veen WA, van der Aar PJ, Verstegen MW (2001) In vitro starch digestion correlates well with rate and extent of starch digestion in broiler chickens. J Nutr 131: 2336-2342
2. Fox SI (1996) Human Physiology page 588. Wm.C Brown Publishers, Dubugue.
3. Weurding RE, Enting H, Verstegen MW (2003) The relation between starch digestion rate and amino acid level for broiler chickens. PoultSci 82: 279-284
4. Steel RGD, Torrie JH (1995) Prinsipdanprosedurstatistika :suatupendekatanbiometrik. Gramedia, Jakarta.
5. Henquin JC (2000) Triggering and amplifying pathways of regulation of insulin secretion by glucose. Diabetes 49: 1751-1760
6. Black JL, Hughes RJ, Nielsen SG, Tredrea AM, MacAlpine R, et al. (2005)Proceedings, Australian Poultry Science Symposium 16: 21-29
7. SellePH, Liu SY, Cai J, Cowieson AJ(2012) Animal Production Science.
8. Hazelwood RL (1984) Pancreatic hormones, insulin/glucagon molar ratios, and somatostatin as determinants of avian carbohydrate metabolism. J ExpZool 232: 647-652.
9. Goodwin, Denise B, John B, Mc Murray L, William JR et al. (1994) Blood Glucose value and definitions of Hypoglicemia and Hyperglycemia. Georgia Poultry Laboratory. 38: 861-865
10. Grizard J, Picard B, Dardevet D, Rochon BMC (1999)Protein metabolism and Nutrition,EAAP Publication no 96.Wageningen Pers. Pp 184-187
11. Hartman S, Taleb SA, Geng T, Gyenai K, Guan X, et al. (2006) Comparison of plasma uric acid levels in five varieties of the domestic turkey, Meleagrisgallopavo. PoultSci 85: 1791-1794.
12. Donsbough AL, Powell S, Waguespack A, Bidner TD, Southern LL (2010) Uric acid, urea, and ammonia concentrations in serum and uric acid concentration in excreta as indicators of amino acid utilization in diets for broilers. PoultSci 89: 287-294.
13. Xie M, Hou SS, Huang W, Zhao L, Yu JY, et al. (2004) Interrelationship between methionine and cystine of early Peking ducklings. PoultSci 83: 1703-1708.
14. Hare JM, Johnson RJ (2003) Uric acid predicts clinical outcomes in heart failure: insights regarding the role of xanthine oxidase and uric acid in disease pathophysiology. Circulation 107: 1951-1953.
15. Cutler RG (1984) Urate and ascorbate: their possible roles as antioxidants in determining longevity of mammalian species. Arch GerontolGeriatr 3: 321-348.
16. Hevia P, Clifford AJ (1977) Protein intake, uric acid metabolism and protein efficiency ratio in growing chicks. J Nutr 107: 959-964.

CHAPTER 10

Effect of Rumen-Inert Fat Enriched with High Levels of Poly-Unsaturated Fatty Acids on the Productive and Reproductive Response in Ruminants

Alejandro Salvador[1], Rolando Hernandez[2], Thais Diaz[3] and Ricardo Betancourt[2]

1 Unidad Experimental de Producción Caprina, Facultad de Ciencias Veterinarias, Universidad Central de Venezuela, Maracay, Venezuela

2 Cátedra de Nutrición Animal, Facultad de Ciencias Veterinarias, Universidad Central de Venezuela, Maracay, Venezuela

3 Instituto de Reproducción Animal "Dr. Abraham Hernandez Prado", Facultad de Ciencias Veterinarias, Universidad Central de Venezuela, Maracay, Venezuela

1. INTRODUCTION

The chapter will discuss the effect of rumen-inert fat enriched with high levels of poly-unsaturated fatty acids (PUFA) on the productive and reproductive responses in ruminants. The discussion will include characteristics of the milk and its derivatives such as goat cheese. The expertise presented in the chapter will cover the authors' experience about the interrelationship between nutrition and reproduction in ruminants from the College of Veterinary Sciences at the Central University of Venezuela.

The success of milk and meat production systems with ruminants depends largely on an efficient reproductive performance of the herd. Thus, numerous researchers have tried to identify the different factors affecting the productive and reproductive behavior of these herds, being the nutritional component one of the factors that has the highest impact on tropical livestock. Marked shortcomings of forage quality and quantity generate undernourished animals, originating low production parameters in our livestock, far below the desired levels. In countries located in the tropical region of the world the reproductive efficiency in beef and dairy cattle is low, no more than 40-45% (Diaz, 2009). In addition, the pregnancy rate in first calf cows is around 30%, post-partum anestrous period for cattle ranges from 150 to 210 days (Montaño and Ruiz, 2005) and early embryonic mortality in lactating dairy cows, is around 30% (Bach, 2003).

Furthermore, during the early postpartum period the lactating cow (dairy, beef and dual purpose cows), as well as the dairy goats, show a gap between the consumption and the amount of energy needed, especially during the time of maximum requirements, when the cow and the goat do not meet their nutritional requirements, which leads to the massive mobilization of its body reserves, particularly energy, generating a state known as the Negative Energy Balance (NEB), which can cause deleterious effects on milk production, health and reproductive performance.

A practical way to control this NEB is through the increase of the food consumption by animals and/or the increase of the energy density of the diet. Thus, various food technologies have been developed for this purpose: energy banks, the use of strategic crops such as sugar cane, liquid diets with high levels of energy, multi-nutritional blocks and energy supplementation with rumen-inert fat. Increasing the fat content of the ration is a way to decrease the NEB, because of fats have double the energy than sugars. However, there are limitations on their use, particularly if the fat is not a rumen-inert fat.

If rumen-inert fat is used to feed ruminants, some undesirable effects of active fats at ruminal level, would be avoided, and higher amount of fat could be used.

On the other hand, if the fat is rich in polyunsaturated faty acids (PUFA; linoleic, linolenic, docosahexanoic [DHA], or eicosapentaenoic acid [EPA]), the rumen-inert fat enriched not only supplies energy, but allows the incorporation of the beneficial effects that these fatty acids have on the animal body and in the productive and reproductive response, therefore having a nutraceutical effect, i.e., producing meat and milk with high levels of these fatty acids, which may have a beneficial effect on human health.

2. THE FAT AND THE FEEDING OF RUMINANTS

Fats are important in ruminant nutrition because of its high energy content. In this regard, the complete combustion of one gram of fat produces around 9.45 Kcal of net energy, while a typical carbohydrate generates around 4.4 Kcal. So, lipids, in general, provide 2.25 times more energy than traditional sources. Also, it is important to consider the intake of fat-soluble vitamins and essential fatty acids (Mayes, 1988).

Fats are part of a group of organic molecules called lipids, which accomplish many functions in the animal body: structural functions (part of cell membranes), hormonal functions (some hormones are of lipidic nature: estradiol, progesterone, testosterone, among others) and immunological functions. Some vitamins (A, D, E and K) are of lipid nature.

Fatty acids (FA) are fundamental molecules in the structure of a lipid, with hydrogenated carbon chains which contain an acidic or carboxyl group in one end and a methyl group in the other end (Jenkins, 2004). The length of the chains of the FA goes from 2 to 24 or more carbon atoms. It is common to name them according to the number of atoms and the number of double bonds, which serves to classify them as saturated or unsaturated fatty acids.

Animal tissues do not synthesized linoleic and linolenic acids (Omega 6 and Omega 3, respectively); they should therefore be incorporated within the diet (Jenkins, 2004). Both FA are known as essential in animal nutrition because they are required for many metabolic processes. Omega fatty acids belong to one of the three Omega families (ω): ω-6, ω-3 and ω-9 (Table 1). Each family has a parental fatty acid, which can be converted into other biologically active acids within the same ω family (Jenkins, 2004).

The sources of lipids in ruminant feeding systems are forages, cereals, oleaginous seeds, by-products of the industry, such as the tallow, yellow fats, mixtures of vegetable and animal fats, hydrogenated fat, oil palm and calcium soaps. Tropical forages are relatively low in their content of lipids; the concentration of fatty acids in forages rarely exceeds 1.5% of the dry matter of the diet.

Both oils and fats, belong to the group of lipids, but differ in that the first are liquids at ambient temperature, while fats are solid. Another difference is that fats are usually from animal origin (tallow) while most of the oils are from vegetables (excluding fish oil). Fats have high levels of saturated fatty acids, while the oils contain more PUFA.

Table 1. Parental fatty acids and its main metabolites within each Omega family.

ω–Family	Parental fatty acid	Mayor Metabolite
ω – 9	C 18:1 ω-9; Oleic acid	C 20:3 ω-9; Eicosatrienoic acid
ω – 6	C 18: 2 ω-6; Linoleic acid	C 20:4 ω-6, Arachidonic acid
ω – 3	C 18:3 ω-3; Linolenic acid	C 20:5 ω-3, Eicosapentaenoic acid C 22:6 ω-3, Docosahexaenoic acid

(Adapted from Jenkins, 2004.)

Fats and oils have limitations to be incorporated in the feeding of ruminants. It has been reported that levels >5% of dry matter produced a decrease in food consumption of the animal. In this regards, Jenkins (1993) and Palmquist (1996) mention some of the possible ways of how fats can affect the microorganism action in the rumen:

- Rumen microorganisms cannot use the fiber because the formation of a film on the surface of the fiber, thus preventing the enzymatic and bacterial attack and disturbing the fermentation process.
- Reduction of microbial activity by adsorption of fat to the surface of the bacterial membrane.
- Possible formation of calcium and magnesium soaps in the rumen, which reduce the availability of essential minerals for the fermentative activity in the rumen.
- Elimination of a fraction of the microbial population due to possible toxic effects of some PUFA, especially on celulolitic bacteria.

The above generates a reduction in the ruminal microbial growth, which translates into a change in the production of volatile fatty acids in the rumen, with consequences to the acetic:propionic ratio and a decrease in the amount of acetic acid available for the production of fat in the mammary gland. In the particular case of unsaturated fatty acids, once they are free in the rumen, they suffer a massive hydrogenation process known as biohydrogenation (Jenkins, 1993), which consists in the addition of hydrogen atoms on the double bonds, thus transforming the unsaturated fatty acids into saturated fatty acids. Thus, PUFA as oleic (C18:1), linoleic (C18, ω-6) and linolenic acid (C18, ω-3), are transformed into the stearic FA (C18:0). Eicosapentaenoic acid (C20:5, ω-3) and docosahexaenoic acid (C22:6, ω-3) undergo very little hydrogenation in the rumen (Mattos et al., 2000). These two fatty acids are commonly found in oil and fishmeal.

When we use a source of fat, not protected or not a rumen-inert fat, with high levels of PUFA, most is lost due to the biohydrogenation, which is particularly important for PUFA ω-6 and ω-3, which are considered as essential in the diet and which have an important role on reproductive, immunological, metabolic and hormonal functions. Therefore, this type of susceptible fat to interact in the rumen, are known as active fat and its use is limited in ruminants.

New technologies have generated chemically modified fats that allow its use at higher levels and with a lower level ruminal interaction, which reduces the deleterious effects of lipids on the ruminal bacteria activity. This kind of fats is known as inert fat, by-pass fat, or protected fat. In this regard, Jenkins (2004) defines the inert greases as those fats that have been specifically designed to have very little, or no negative effect on the digestibility of foods in ruminants. Rumen-inert fats are often carboxylate calcium salts (calcium soaps), saturated fatty acids or hydrogenated fats.

The use of calcium soaps allows the incorporation of a higher level of unsaturated fatty acids in the diet of ruminants. This is particularly important in the case of essential fatty acids (ω-6 and ω-3) which not only provide an energy effect *per se*, but may have specific effects on the metabolism of tissues and organs (Staples et al., 1998). The melting point of rumen-inert fats, is usually above 100°C and solubility occurs at pH levels below 5.5. These values of temperature and pH are present in the rumen, which allows the fat to by-pass the rumen. However, at the level of the abomasum and first portion of the duodenum pH levels are much lower, allowings the dissociation of the carboxylate salt, leaving the fatty acids available for absorption.

Table 2. Estimated amount of linoleic acid that reaches the small intestine in cows, depending on the source of fat.

Source of fat	Fat source fed (Kg/day)	Linoleic acid fed (g/day)	Linoleic acid appearing in small intestine (g/day)
Whole cottonseeds	2.8	300	30 to 120
Whole soybeans	2.8	300	30 to 120
Yellow grease	0.45	77	8 to 31
Tallow	0.45	23	2 to 9
Calcium soaps (8,5% ω-6)	0.45	38	25 to 34

(Adapted from Staples *et al.*, 1998)

Therefore, it can be concluded that supplementation of rumen-inert fats to ruminants generates an increase in the availability of unsaturated fatty acids at the intestinal level, and therefore may increase the absorption and their incorporation into tissues (Table 2).

3. RUMEN-INERT FAT AND THE PRODUCTIVE AND REPRODUCTIVE RESPONSES IN RUMINANTS

When the rumen-inert fat is included into the diet it incorporates more PUFA, which is not only an energy source, but also generates non-energetic benefits related to the impact of these FA on the metabolism, hormonal and immune responses. The energetic effect is related to the greater amount of energy that lipids have, which helps to reduce the NEB during the postpartum period, which translates into increased production of luteinizing (LH) and follicle-stimulating hormone (FSH) by the pituitary gland, generating more growth and follicular development and promoting the ovulation (Diaz et al., 2009).

Non-energetic effects are associated with the type of FA presents in the supplied fat, these effects are related with the increase in cholesterol (particularly the HDL fraction or "good cholesterol") levels, direct effects on ovaries and uterus, increase the levels of progesterone (P_4) and the modulation of the prostaglandins production (especially $PGF_{2\alpha}$). In addition, rumen-inert fat has direct effects on hormones and growth factors involved with the reproductive and productive activity (insulin, IGF-I, among others). Most of these non-energy effects are favored when PUFA (ω-6 and ω-3) are included in the diet. Figure 1 shows various of the non-energetic proposed mechanisms, that can generate this type of FA.

The impact of this type of fat is associated with the energy balance of the animal, particularly in the cow during early lactation, decreasing the deleterious effects of the NEB and improving the reproductive activity during the early postpartum. On this regard, Staples et al. (1998) mentioned that 11 out of 18 studies reviewing the effect of fat in cow's reproduction, reported an increase in reproductive performance, either because it improves the conception rate at first service or by increasing in the overall conception or pregnancy rates.

In addition, the supplementation with high levels of PUFA can generate a favorable response in milk production and composition, and an improvement in the fatty acid profile of the milk, particularly the levels of conjugated linoleic acid (CLA). Angulo et al. (2005) have described that some anticancer and antidiabetogenic properties of the CLA as well as the prevention of atheroma formation, the potentiation of the immune response and the improvement of bone mineralization. It may also increase levels of essential fatty acids (ω-6 and ω-3) in milk, which can cause the milk produced by cows or goats fed with this type of fat, would be a nutraceutical food.

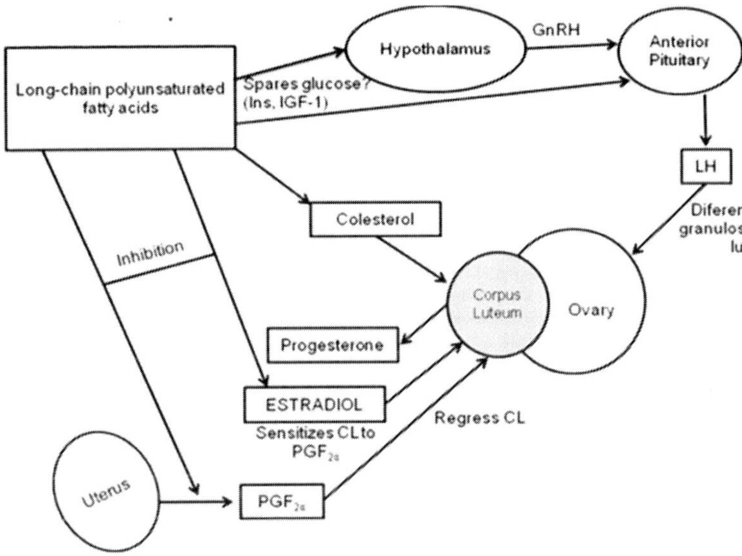

Figure 1. Proposed mechanisms of action by which PUFA supplementation can affect reproductive function. (Adapted from Staples et al., 1998).

4. EXPERIENCES IN VENEZUELA

Experiences in Venezuela with the use of rumen-inert fat enriched with high levels of PUFA are promissory, either dairy and beef cattle, and small ruminants such as goats. Since 2007 the research group in the area of Nutrition - Reproduction interaction, in conjunction with researchers from the area of Animal Production at the College of Veterinary Sciences, Universidad Central de Venezuela, have been working on the effect of the rumen-inert fat enriched with high levels of polyunsaturated fatty acids on the productive and reproductive response in ruminants.

The research has been carried out in different ecological areas, with different types of production, levels of intensification, both in experimental stations and commercial farms, using the supplementation with rumen-inert fat given in different ways: multinutritional blocks (MNB); top-dress supplementation (fat placed on the concentrate supplement), in addition to the use of calcium soaps and mixed with mineral or concentrated supplement.

Regarding the effect of rumen-inert fat enriched with high levels of PUFA on the production of milk in a commercial trial, conducted with multiparous Carora breed cows, we evaluated the milk production during six months, obtaining a higher production per day and per lactation (20% more) in cows supplemented with 400 g of rumen-inert fat enriched with high levels of PUFA (Energras ®) compared to cows in the control group. Results are shown in Table 3.

In crossbred Canary goats under tropical conditions, Salvador et al. (2009) evaluated the effect of by-pass fat supplementation (80 g Energras ® / day) on milk production and composition. They reported a beneficial effect on the duration of lactation (+44 d), the milk production (+29.4%) and the content of milk components (41% more fat, and 31.2% more protein), without altering the physical characteristic (WMT, acidity, pH, cryoscopy) of the milk the supplemented goats (Table 4).

Table 3. Effect of supplementation with rumen-inert fat on milk production of Carora cows.

Parameters	Treatment group (400 g Energras® /day)	Control group
N° animals	8	8
Liters/day	13.8	11.7
Total milk yield (Liters)	16,852	14,055
Diference (Liters/day)	+ 2.1 (17.9%)	
Diference in total milk yield (Liters)	+2,797 (19.9%)	

Otero (2007; unpublished data)

On the other hand, the milk lipid profile can be manipulated when supplementation with rumen-inert fat is given. Milk tends to have low levels of unsaturated fatty acids, being oleic acid (C18:1), a monounsaturated fatty acid, the most abundant (20% approx.). The proportion of unsaturated fatty acids is less than the saturated ones. The formation of the milk fat may come from the *novo* synthesis fatty acids in the mammary gland or from the incorporation of fatty acids from the diet or body reserves. Regarding the long chain fatty acids (including ω-6 and ω-3 fatty acids) which are incorporated in the milk, about 40-45% come from the diet (Palmquist, 1996). Therefore, manipulating the diet by the incorporation of higher levels of fatty acids from rumen-inert fat, could be an strategy to change the milk fat composition in ruminants.

Table 4. Effect of rumen-inert fat on milk production, days on lactation and composition of crossbred Canary goats

Parameter	Treatment group (80 g Energras®/day)	Control group	Probability
Milk yield (kg)	274.4 ± 18.5	193.6 ± 17.9	**
Days of milk	272.6 ± 15.9	228.8 ± 15.4	NS
Fat (kg/lactation)	13.83 ± 1.02	8.16 ±1.05	**
Protein (kg/lactation)	11.29 ± 0.67	7.76 ± 0.69	**
Casein (kg/lactation)	7.59 ± 0.44	5.19 ± 0.46	**
Ashes (kg/lactation)	2.12 ± 0.14	1.49 ± 0.14	**
Lactose (kg/lactation)	12.44 ± 0.80	8.99 ± 0.83	**
Non fatty Solid (kg/lactation)	25.86 ± ±1.59	18.25 ± 1.65	**
Total solid (kg/lactation)	39.70 ± 2.56	26.41 ± 2.65	**

** P < 0.01; NS= not significant

(Adapted from Salvador *et al.*, 2009)

In recent years, the manipulation of milk fat content has taken great importance, with the aim of increasing the concentration of CLA in dairy products, which may contribute significantly to generate milk with nutraceutical properties of essential fatty acids that would contribute to the human health. In this regard, Zamora (2010), supplementing crossbred Canary goats with rumen-inert fat enriched with high levels of fatty acids (45 g/day ENERGRAS®), found significant differences (P<0.01) on oleic and linolenic acids concentrations between cheeses from treated and control goats: (331.34 ± 63.25 vs. 28.02 ±67.29 mg/g and 9.78 ±1, 22 vs. 5.41 ±1.30 mg/g respectively), when compared to the cheeses made from milk of goats supplemented (treatment vs. control) respectively. In addition, there was a trend in linoleic acid concentrations to be higher in chese from treated goats: (30.26 ±9.74 vs. 19.26 ±10.37 mg/g), not affecting the quality and organoleptic properties of fresh cheeses (Table 5).

In addition, the performance of milk in terms of kilograms of cheese showed no significant differences (P>0.05) between cheese made from milk of goats supplemented with rumen-inert fat and cheese made from milk of goats without fat supplementation, (at 0 h and 48 h), after the removal from the mould. Nevertheless, the cheeses made from milk of the treatment group needed less milk to produce 1 kg of fresh cheese (140g and 400g less at 0 and 48 hours after the removal from the mould, respectively). At the moment of curd preparation there was not difference in terms of content of water retained (48.1% and 48.4% for treatment and control groups respectively). This could be due to, the protein content of the serum that could affect the production of cheese (cheese making performance and the drainage of whey, especially when milk from treated goats was heated; Raynal-Ljutovac *et al.,* 2008). However, it is important to mention there were not statistical differences in terms of kg of cheese obtained from treated and control goats, the performance improved for processed cheese made from milk of goats that consumed rumen-inert fat, although they had a higher milk production, there were 47.95 kg of fresh cheese from the treated group and 25.37 kg of fresh cheese in the control group, with a difference of 47.09% for the treatment group.

Table 5. Effect of rumen-inert fat on performance, characteristics and composition of fresh pasteurized goat's cheese.

VARIABLE	Treatment group (80 g Energras®/day)	Control group	Probability
Performance			
Performance to 0 H	3.9 ± 0.1	4.0 ± 0.1	NS
Performance to 48 H	4.8 ± 0.1	5.2 ± 0.1	NS
Characteristic:			
Acidity	13.1 ± 1.7	13.5 ± 1.7	NS
Clorures	2.9 ± 0.3	2.9 ± 0.3	NS
pH	5.6 ± 0.1	5.4 ± 0.1	NS
Components:			
fat (%)	24.5 ± 1.2	21.9 ± 1.2	NS
Protein (%)	10.2 ± 0.4	10.7 ± 0.4	NS
Humidity (%)	48.1 ± 0.9	48.4 ± 0.9	NS
Ash (%)	4.3 ± 0.2	4.1 ± 0.2	NS
Solids (%)	51.8 ± 0.9	51.5 ± 0.9	NS
Solids no fat (%)	27.3 ± 1.3	29.6 ± 1.3	NS
Lactose (%)	12.7 ± 1.0	14.6 ± 1.0	NS
Oleic C18:1 (mg/g)	331.3 ± 63.2	28.0 ± 67.2	**
Linoleic C18:2 (ω-3) (mg/g)	30.2 ± 9.7	19.2 ± 10.3	NS
Linolenic C18:3 (ω-6) (mg/g)	9.7 ±1.2	5.4 ± 1.3	*

NS: No significant; *: $P<0.05$; **: $P<0.01$ (Adapted from Zamora, 2010).

It is important to mention that the rumen-inert fat did not affect the characteristics of the cheese, nor the chemical composition neither of its components, except FA, concentrations and minerals in the treatment group, which had a trend to be higher. Some additional results showed that the effect of the consumption of rumen-inert fat on the proportion of *cis* and *trans* FA found in fresh goat cheese, was not significant (P>0.05). However, the proportion was higher in milk from treated goats than in milk from control goats (95.97 ± 4.31 and 85.73 ± 4.31 respectively) which aggregates a value to the cheese, having a product with unsaturated *cis* FA, which has benefits to human health, because *trans* FA from of partially hydrogenated oils have been linked to deaths due to cardiovascular diseases (Giacopini, 2008).

Regarding to reproductive responses, the use of rumen-inert fat enriched with high levels of PUFA has been evaluated with promising results, particularly in first-calf Brahman cows in areas of well drained savannas in Cojedes States (Farms A and B) and Guárico States (Farm C) with acidic soils, in conditions of limited range and forage quality. In these studies Diaz et al. (2008; 2009) used multinutritional blocks that were formulated with the aim of providing non protein nitrogen, rumen-inert protein, minerals (macro and microelements) and rumen-inert fat (with 17% of linoleic acid [ω-6]). In the case of Farm A, the breeding season was during the rainy season, while for the two other farms (B and C) breeding seasons ware during the dry period. Data of pregnancy rate are shown in Figure 2. Results showed higher pregnancy rates than those reported for the same type of animals under tropical conditions (30%). Pregnancy rate for first calf Brahman cows in Farm A during the previous year was 46% much lower than the present results.

Diaz et al. (2009) and Hernandez (2010) also reported a beneficial effect of the supplementation with rumen-inert fat enriched with high levels of PUFA, during the postpartum period on ovarian activity of first-calf Brahman cows. Their results showed that supplementation with 150 g of fat/cow/d increased the accumulated number of Class 3 follicles (≥ 10 mm) during the first 90 d post-partum ($P<0.06$; Figure 3). In this regard Diaz (2009) suggested that a greater accumulated number of Class 3 follicles could be an indicator of a greater likelihood of having preovulatory follicles, so cows would have a greater chance of ovulation during the first 90 d postpartum. On the other hand, it is an indirect indicator of the onset of LH secretion after parturition, as this hormone is responsible for the final maturation of preovulatory follicles and subsequent ovulation.

Also, Hernandez (2010) reported that independent of significant difference in reproductive efficiency parameters between cows supplemented with rumen-inert fat and cows in the control group, there were numeric differences of 14 percentage points in conception rate, in favor of the supplemented cows (91% vs. 77%) and 12 d shorter interval from calving to conception (days open: 90 vs. 102 d in treatment and control group respectively). Those numbers would translate into an increase in reproductive efficiency and therefore the profitability herd.

In another experiment (cows received 250 g of rumen-inert fat/cow/d) we found a 20% reduction in the interval between calving and first estrus (control cows: 98 d *vs.* 78 d in supplemented cows). In this case the fat used as a supplement contained 5% ω-3 and 17% ω-6, (Unpublished data). It is important to note that the cows supplemented with ω-6 presented problems of consumption at the beginning of the trial.

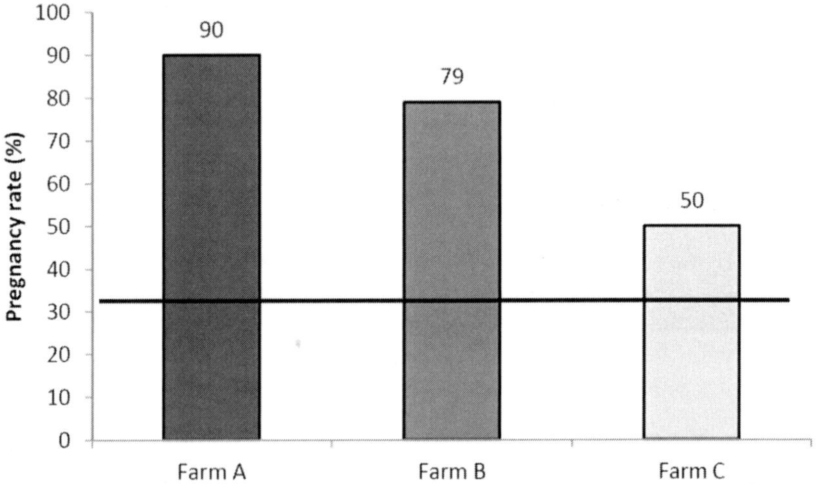

Figure 2. Pregnancy rate for first calf Brahman cows supplemented with multinutritional blocks enriched with rumen-inert fat. The line represents the pregnancy rate for first calf Brahman cows in Venezuela (Adapted from Diaz et al., 2008; 2009).

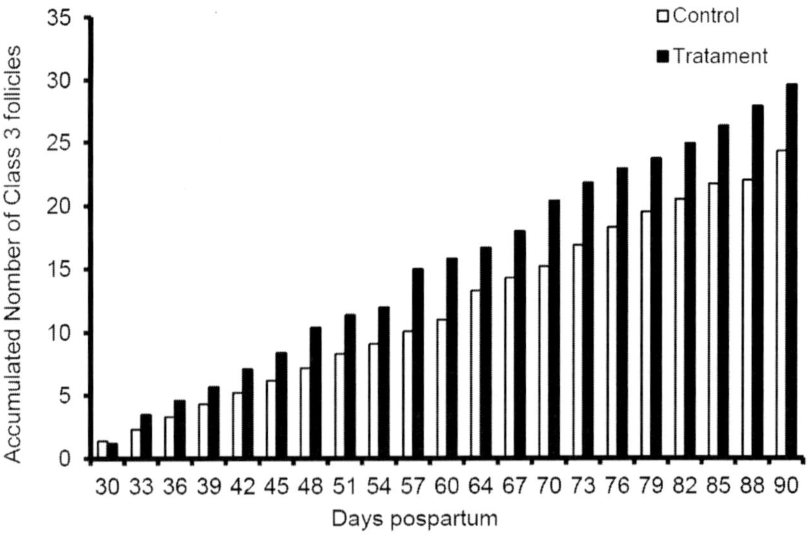

Figure 3. Accumulated number of Class 3 follicles (≥ 10 mm) between 30 and 90 d postpartum in first-calf Brahman cows supplemented or non with-rumen-inert fat.

Diaz et al. (2009) considered that the ω-6 fatty acids should be provided to stimulate cyclic ovarian activity early in the post-partum, while the ω-3 fatty acid (linolenic acid, DHA and EPA) should be supplied during early pregnancy. In this regard, several authors (Petit et al., 2002;) Mattos et al., 2003) indicate that ω-6 fatty acids stimulate synthesis of $PGF_{2\alpha}$, as well as ω-3 stimulates the synthesis of series 3 prostaglandins, which can block the synthesis of the former, specifically the $PGF_{2\alpha}$ by competitive inhibition of enzymes that regulate their synthesis.

However, there are contrasting evidences regarding the role of the ω-6 fatty acids and the synthesis of $PGF_{2\alpha}$. To this respect, Staples et al. (2002), indicate that linoleic acid has inhibitory effects both *in vitro* and *in vivo*, probably dose related, as the excess of this fatty acid might reduce the synthesis of prostaglandins in series 1 and 2 (i.e., $PGF_{2\alpha}$). These same authors point out, that the inhibition may be due to a competitive effect with arachidonic acid by the prostaglandin endoperoxide synthase enzyme (PGSH). Therefore, a rich source of ω-6 fatty acids could reduce the synthesis of $PGF2\alpha$, prolonging the lifespan of the corpus luteum, allowing the implantation of the embryo and decreasing the early embryonic mortality.

On the other hand, Diaz et al. (2009) argue that both the maintenance of pregnancy, and the onset of postpartum ovarian activity may occur with the single supply of linoleic acid. Therefore, the use of rumen-inert fat enriched with high levels of ω-6 and/or ω-3 could contribute positively on uterine involution, the resumption of postpartum ovarian activity and reduction of days open in beef and dairy cows.

5. EFFECT OF RUMEN-INERT FAT SUPPLEMENTATION ON THE VOLUNTARY CONSUMPTION

Salvador et al. (2009) reported that there were not significant differences between groups in terms of voluntary consumption of food. Both groups consumed the whole dose of concentrate (200 g with 18% of crude protein) across the lactation period. Also, there were no differences ($P > 0.05$) in terms of consumption of hay of Bermuda grass (*Cynodon dactylon*) offered ad libitum between groups (0.85 ± 0.02 kg and 0.83 ± 0.02 kg for treatment and control group respectively). These results are consistent with those reported by Teh et al. (1994), González and Bas (2002) and Sanz Sampelayo et al. (2002a).

Figure 4 shows that there was a significant effect ($P < 0.05$) of days in milk on the consumption of hay, showing that hay consumption increased immediately after the peak of lactation and decreased as the milk production reduces, coinciding with results reported by Jimeno et al. (2003) who pointes out that the lowest level of consumption is just one week before calving. On the other hand, the voluntary consumption increased after calving up to the maximum consumption between 6 to 10 weeks; However, it was not linear increase. Then, subsequent to the peak of milk the consumption declined linearly as the milk production decreased.

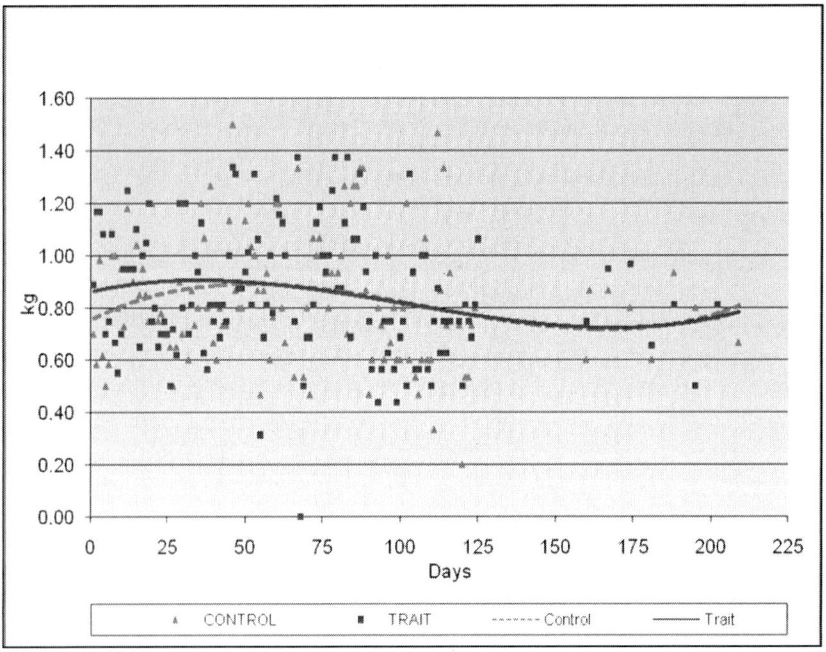

Figure 4. Effect of rumen-inert fat on the voluntary consumption of hay in crossbreed Canarian goats during lactation.

Regarding to the consumption of the of rumen-inert fat (80 g) by goats in the treated group, there was no rejection of the fat, consuming all the by-pass offered.

Results showed that the addition of this type of rumen-inert fat has no a substitute effect of the main component of the diet of ruminants (grass) but works as a supplement, thereby improving the quality of the diet, reducing the NEB, and improving the productive and reproductive parameters.

6. FINAL CONSIDERATIONS

Supplementation with rumen-inert fat enriched with high levels of PUFA, should start enough time before parturition, to promote energy reserves of the animal (to improve body condition score) and to maintain body condition during the early postpartum period, in order to reduce the negative effects that has NEB on the production and reproduction in ruminants. Diaz et al. (2009) recommend that this type of fat supplementation, should begin between 21-40 days before calving, with doses between 100 and 400 g/cow/day, according to the type of cow (dairy or dual purpose cows). The time that the supplementation will be given varies according to the farm, the basal diet, the level of production, the type of animal, etc., but in general it is recommended to keep it during 90 days postpartum, period in which requirements of the cows are higher because they reach the peak of milk production.

Salvador et al. (2009) and Zamora (2010) recommend in dairy goats, that supplementation with this type of fat, would be given during the same way and time but with lower levels (45 and 80 g/goat/d) depending on the weight and the level of production of the goat.

Although the rumen-inert fat is not the solution to all problems of our farms, it is a food resource of great potential, that contains high levels of polyunsaturated fatty acids that allows the incorporation of beneficial effects that essential, fatty acids (ω-6 and ω-3) may have on the reproductive and productive activity of our herds and at the same time generating meat, milk and other derivatives with better nutritional quality.

REFERENCES

1. Angulo, J., Machecha, L., Giraldo, C. and Olivera M. 2005. Prostaglandinas y grasa de la leche: síntesis a partir de ácidos grasos poliinsaturados, en bovinos. In: M. Pabón y J. Ossa (Eds.) Bioquímica, Nutrición y Alimentación de la Vaca. Fondo Editorial Biogénesis. pp. 111-135. Medellín, Colombia.
2. Bach A. 2003. La reproducción del vacuno lechero: nutrición y fisiología. VII Curso de Especialización FEDNA La reproducción del vacuno lechero: nutrición y fisiología. Purina, España. In: http://www.etsia.upm.es/fedna/capitulos/2001CAPV.pdf
3. Diaz T. 2009. Aspectos de la fisiología reproductiva de la hembra Brahman. Trabajo de ascenso. Facultad de Ciencias Veterinarias, Universidad Central de Venezuela. pp. 93.
4. Diaz, T., Betancourt, R., Hernandez, R. and Gallo, J. 2009. El efecto de las grasas omega 3 y omega 6 sobre la reproducción de vacas de primera lactancia pastoreando en los llanos venezolanos. In: E. Chacón y F. Espinoza (Eds). Memorias III Simposium Sobre Recursos y Tecnologías Alimentarias Apropiadas para la Producción Bovina a Pastoreo en Condiciones Tropicales. Pasteurizadora Táchira C.A. (PASTCA). San Cristóbal. Pp. 1-28 CD-ROM.

5. Diaz, T., Betancourt, R., Hernandez, R., Romero, C., Gallo, J. and Cardona, M. 2008. Effects of by-pass fat feeding on the reproductive performance of first-calf Brahman cows under tropical Savannah conditions. Reproduction in Domestic Animals. 43:37. (Abstr.).
6. Giacopini, M. 2008. Efecto de los ácidos grasos *trans* sobre las lipoproteínas del plasma. Archivos Venezolanos de Farmacología y Terapéutica. 27(1): 19-21.
7. González F., and Bas, F. 2002. Efecto de la suplementación con un aceite hidrogenado de pescado sobre la producción de leche en vacas Holstein Friesan. Cien. Invest. Agrop. 29(2): 73-82.
8. Hernandez, R. 2010. Efectos de la suplementación con grasa sobrepasante sobre parámetros productivos y reproductivos en vacas Brahman de primer parto a pastoreo. Trabajo de ascenso. Facultad de Ciencias Veterinarias, Universidad Central de Venezuela. pp. 97.
9. Jenkins T. 2004. Challenges of meeting cow demands for omega fatty acids. Florida Ruminant Nutrition Symposium. In: http://dairy.ifas.ufl.edu/files/rns/2004 /Jenkins.pdf
10. Jenkins T. 1993. Lipid metabolism in the rumen. Journal of Animal Science. 76: 3851-3863.
11. Jimeno V., Rebollar, P., and Castro, T. 2003. Nutrición y alimentación del caprino de leche en sistemas intensivos de explotación. XIX Curso de Especialización FEDNA. Madrid, España.
12. Mattos, R., Guzeloglu, A., Badinga, L., Staples C. R., and Thatcher. W. W. 2003. Polyunsaturated fatty acids and bovine interferon-? modify phorbol ester-induced secretion of prostaglandin F2a and expression of prostaglandin endoperoxide synthase- 2 and phospholipase-A2 in bovine endometrial cells. Biology of Reproduction. 69: 780–787.
13. Mattos R., Staples C.R. and W.W. Thatcher. 2000. Effects of dietary fatty acids on reproduction in ruminants. Reviews of Reproduction. 5:38-45.
14. Mayes P.A. 1988. Lípidos de importancia fisiológica. En: Bioquímica de Harper. 11a Ed, Editorial Manual Moderno, México, D.F. Pp. 129-141.
15. Montaño E.L. and Ruiz Z.T. 2005. ¿Por qué no ovulan los primeros folículos dominantes de las vacas cebú posparto en el trópico colombiano? Rev Col Cienc Pec Vol 18:2 127-135
16. Palmquist DL. 1996. Utilización de lípidos en dietas de rumiantes. En: http://www.etsia.upm.es/fedna/capitulos/96capituloIII.pdf
17. Petit H., R. Dewhurst, N. Scollan, J. Prolux, M. Khalid, W. Haresing, H. Twagiramungu and G. Mann. 2002. Milk production and composition, ovarian function, and prostaglandin secretion of dairy cows fed omega 3 fats. Journal of Dairy Science. 85:889-899.
18. Raynal-Ljutovaca, K., Lagriffoul, K., Paccard, G., Guillet, and P., Chilliard. 2008. Composition of goat and sheep milk products: An update. Rev. ScienceDirect. Small Ruminant Research. 79 : 57–72.
19. Salvador A., Alvarado C., Contreras I., Betancourt R., Gallo J. and A. Caigua. 2009. Efecto de la alimentación con grasa sobrepasante sobre la producción y composición de la leche de cabra en condiciones tropicales. Zoot. Trop. 27:285-298.
20. Sanz Sampelayo M., Martin, J., Pérez, L., Gil, F., and Boza, J. 2002. Effects of concentrates with different content of protected fat rich in PUFAs on the performance lactating Granadina Goats. 1. Feed intake, nutrient digestibility, N and energy utilization for milk production. Small Rum. Res., 43(3): 133-139.
21. Staples C., R. Mattos, S. Boken, L. Sollenberger, W. W. Thatcher and T. Jenkins. 2002. Feeding fatty acids for fertility? Proceedings of the 13th Annual Florida Ruminant Nutrition Symposium. Pp 71-85. En: http://dairy.ifas.ufl.edu/rns/2002/staples.pdf
22. Staples C.R., J.M. Burke and W.W. Thatcher. 1998. Influence of supplemental fats on reproductive tissues and performance of lactating cows. Journal of Dairy Science. 81:856-871.
23. Teh T., Trung, L., Jia, Z., and Gipson, T. 1994. Varying amounts of rumen-inert fat for high producing goats in early lactation. J. Dairy Sci., 77: 253-258.
24. Zamora R. 2010. Producción y composición de la leche y queso fresco pasteurizado de cabras suplementadas con grasa sobrepasante con altos niveles de ácidos grasos poliinsaturados. Tesis de grado para obtener el título de Ingeniero Agrónomo. Facultad de Agronomía. UCV. Maracay, Venezuela.

CHAPTER 11

Aflatoxins Importance on Animal Nutrition

Vivian Feddern[1], Giniani C. Dors[2], Fernando de C. Tavernari[1], Helenice Mazzuco[1], Anildo Cunha [1], Everton L. Krabbe[1] and Gerson N. Scheuermann[1]

[1] *Embrapa Suínos e Aves, Concórdia, SC, Brazil*

[2] *Instituto Federal Catarinense (IFC) – Campus Concórdia, Concórdia, SC, Brazil*

1. INTRODUCTION

Mycotoxins are secondary metabolites produced by several fungal species, toxic to humans, animals and plants. Their ingestion, inhalation or dermal absorption may cause different diseases and even death. These compounds have been described for many decades, however in the beginning of the 1960s, they have been chemically characterized due to the discovery of aflatoxins [1]. The word aflatoxin is the combination of 3 other words "a" for *Aspergillus* genus, "fla" for the species *flavus* and toxin, meaning poison [2].

Aflatoxin is the mycotoxin generating the greatest losses and the highest management costs due to its extremely high toxicity on a unit basis, and its long history of stringent regulation. The costs are inversely related to the regulatory level that must be met, and lower concentration allowances will increase the costs of crop management. Several effective ways for the management of mycotoxin contamination in agriculture have been stressed. One strategy to manage mycotoxin contamination and decrease health risks and economic costs is to instruct food producers and handlers on how to minimize mycotoxin contamination, and to encourage the adoption of process-based guidelines such as Good Agricultural Practices (GAPs) before harvest and good manufacturing practices (GMPs) after harvest. These actions would minimize risk throughout the production, handling, and processing chain, and can complement product standards [3]. To control the presence of aflatoxins in foods, many countries established maximum tolerated concentrations through legislation (Van Egmond, 1989a cited [4]).

This chapter focuses on properties of aflatoxins and their occurrence in feeds and animal products as meat, eggs, liver, kidneys and milk. Topics regarding mycotoxins absorbents and legislation in feed ingredients and feeds are also covered.

2. AFLATOXIN PROPERTIES

All mycotoxins are low-molecular-weight natural products (i.e., small molecules) (Bennett, 1987 cited [3]). The hyphal structure of filamentous fungi has evolved to utilize solid substrates efficiently by growing over surfaces and penetrating into solid matrices. Moulds are able to secrete enzymes to break down complex macromolecular compounds and utilize them for growth and metabolism. They can absorb low molecular weight nutrients, produce and secrete secondary metabolites, which are also relatively low molecular weight compounds but not associated with the process of growth and primary metabolism (Bushell, 1989 cited by [4]).

Some aflatoxin derivatives are products of animal metabolism following ingestion of the mould metabolites, which are divided into the B and G groups based on their blue or green fluorescence under UV light when absorbed to solid substrates. *A. parasiticus* is the most toxigenic species, the majority of strains producing both B and G toxins (Van Egmond, 1989a cited by [4]).

Aflatoxins are crystalline substances, freely soluble in moderately polar solvents such as chloroform, methanol, dimethyl sulfoxide; they dissolve in water to the extent of 10-20 mg L^{-1}. Some important physical and chemical properties of aflatoxins are given in Table 1 [5].

Due to the important structural diversity (Figure 1) of mycotoxins and the variations in their metabolism, it is difficult to edit general rules. Thus, each toxin and respective metabolites have to be investigated as a particular case. The chemical and physical properties of aflatoxins are described as follows [6]:

1) *Description*: Colorless to pale-yellow crystals. Intensely fluorescent in ultraviolet light, emitting blue (aflatoxins B_1 and B_2) or green (aflatoxin G_1) and green–blue (aflatoxin G_2) fluorescence, from which the

designations B and G were derived, or blue–violet fluorescence (aflatoxin M_1)

2) *Melting-points*: see Table 1.
3) *Absorption spectroscopy*: see Table 1.
4) *Solubility*: Very slightly soluble in water (10–30 μg mL^{-1}); insoluble in non-polar solvents; freely soluble in moderately polar organic solvents (e.g., chloroform and methanol) and especially in dimethyl sulfoxide
5) *Stability*: Unstable to ultraviolet light in the presence of oxygen, to pH extremes (< 3, > 10) and to oxidizing agents
6) *Reactivity*: The lactone ring is susceptible to alkaline hydrolysis. Aflatoxins are also degraded by reaction with ammonia or sodium hypochlorite.

Table 1. Physical and chemical properties of aflatoxins.

				UV absorption max (ε(L mol^{-1} cm^{-1})), methanol	
Aflatoxin	***Molecular formula***	***Molecular weight***	**Melting Point**	***265 nm***	***360-362 nm***
B_1	$C_{17}H_{12}O_6$	312	268-269	12,400	21,800
B_2	$C_{17}H_{14}O_6$	314	286-289	12,100	24,000
G_1	$C_{17}H_{12}O_7$	328	244-246	9,600	17,700
G_2	$C_{17}H_{14}O_7$	330	237-240	8,200	17,100

Aflatoxins are quite stable in many foods and are fairly resistant to degradation. The effectiveness of some processes in reducing concentrations of aflatoxins in food can be affected by many factors, such as the presence of protein, pH, temperature and length of treatment. Commercial processing of raw commodities using cleaning regimes including the removal of broken particles, milling and sorting can reduce aflatoxin concentration considerably [5].

Naturally occurring aflatoxins (as a group) and other 107 agents were evaluated as carcinogenic to humans (Group 1). There is sufficient evidence in humans for the carcinogenicity of aflatoxins, being liver cancer (hepatocellular carcinoma) the main effect. Aflatoxin M_1, the metabolite of aflatoxin B_1found in milk of lactating mammals was classified in Group 2B as possibly carcinogenic to humans [8]. Carcinogenicity of naturally occurring mixtures of aflatoxins B_1, G_1 and M_1 is also demonstrated in experimental animals. The intake of these toxins over a long period of time in very low concentrations may be highly dangerous. These compounds can enter the food chain, mainly, by ingestion through the diet of humans and animals (Miraglia et al., 1996 cited by [11]).

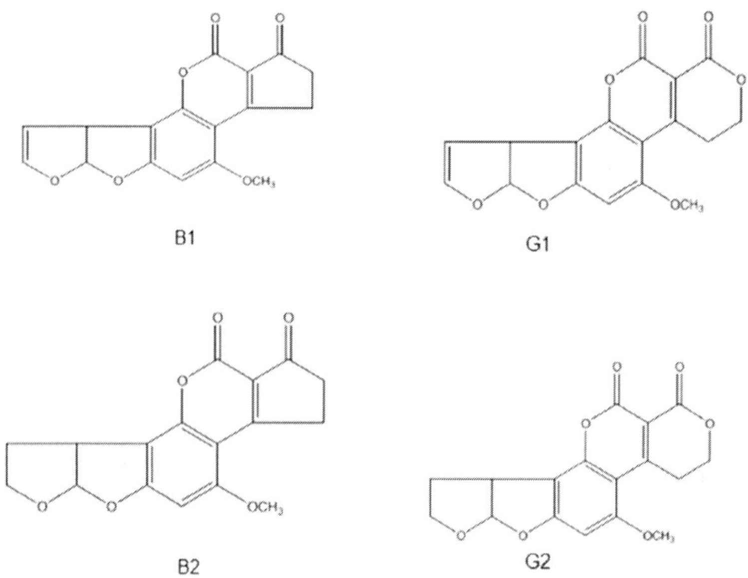

Figure 1. Chemical structures of aflatoxins [7].

There is strong evidence that the carcinogenicity of aflatoxins operates by a genotoxic mechanism of action that involves metabolic activation to a genotoxic epoxide metabolite, formation of DNA adducts, and modification of the TP53 gene. In humans, hepatocellular carcinomas from areas of high exposure to aflatoxins, up to 50% of tumors have been shown to harbor a specific point mutation in the TP53 tumor suppressor gene [9]. Table 2 shows aflatoxin main producing species and toxic effects, pointed by the International Agency for Research on Cancer (IARC).

$AFB_1$is the most potent carcinogenic substance naturally produced by*Aspergillus*species. Indeed, AFB_1 is classified by IARC as Group 1 carcinogen [10]. This compound is certainly acutely toxic to humans, is probably responsible for liver necrosis following chronic exposure, and may be involved in the epidemiology of human liver cancer in some parts of the world perhaps synergistically with hepatitis B virus (Van Egmond, 1989a cited by [4]).

Table 2. Main producing species and effects of aflatoxins [6,10].

Aflatoxins	*Main producing species*	*CAS No.*	*Toxic effect*
B_1	*A. flavus, A. parasiticus, A. nomius, A. bombycis*	1162-65-8	Hepatotoxic, genotoxic, carcinogenic, immunomodulation
B_2	*A. flavus, A. parasiticus, A. nomius, A. bombycis*	7220-81-7	Limited evidence for carcinogenicity
G_1	*A. parasiticus, A. nomius, A. bombycis*	1165-39-5	Sufficient evidence of carcinogenicity
G_2	*A. parasiticus, A. nomius, A. bombycis*	7241-98-7	Inadequate evidence for carcinogenicity

After ingestion, aflatoxin B is metabolized by enzymes to generate a reactive 8,9-epoxide metabolite that can be bound to DNA as well as to serum albumin forming aflatoxin-N-7 guanine and lysine adducts, respectively. Covalent binding to DNA is considered to be a critical step in aflatoxin hepatocarcinogenesis [11].

Determination of these metabolites was solved by developing enzyme linked immunosorbent assay (ELISA) methods (Vidyasagar et al., 1997 and Nayak, et al., 2001 cited by [11]). the biosynthesis of aflatoxins is induced by sugars. The induction is associated with the transcriptional activation of the pathway genes and the pathway regulatory gene, aflR. The regulation of aflatoxin biosynthesis had been examined by manipulating the transcription of aflR. Studies concerning this topic showed that constitutive overexpression of the pathway transcriptional regulatory gene aflR led to higher transcript accumulation of pathway genes and increased aflatoxin production (Flaherty and Payne, 1997 cited by [11]).

Aflatoxins are metabolized in ruminants by the liver and are excreted in the bile. AFB_1 increases the apparent protein requirement of cattle. When significant quantities are consumed, the metabolite M_1appears in milk within 12 hours. Research suggests M_1 is not as carcinogenic or mutagenic as B_1, but it does appear to be as toxic as its parent compound [12]. When an animal ingests food contaminated with an AFB_1, from 0.5 to 5% of the toxin ingested is biotransformed in the liver into AFM_1 (Hussein and Brasel, 2001 cited by [13].

On a worldwide basis about 35% of *A. flavus* strains produce aflatoxins and only the B group. These molds occur in warmer parts of the world and aflatoxins may be produced in a wide range of tropical and subtropical food commodities (Van Egmond, 1989a cited by [4]). The presence of fungi does not necessarily imply the presence of toxins. The fungi species can produce aflatoxins on commodities in the field under stress conditions or in storage when high moisture and warm temperature propitiate their growth (Schuster et al. 1993 cited by [11]). The minimum moisture content of foods that allows the growth of *A. flavus* is around 85% relative humidity (0.85 water activity), and temperature of 25-30 ºC. In cereals with high starch conten as rice, maize, sorghum, wheat, barley, the moisture content in the grain is 18-18.5%. The subsequent drying does not affect the existing level of aflatoxin because it resists drying and roasting temperatures (Pitted, 1998; Sabino, 1996; Wilson and Payne, 1994 cited by [14]). Because of the weather, aflatoxins are commonly found in South America, Africa, Asia and Australia [15].

3. AFLATOXINS OCCURRENCE IN FEEDS

Time of harvest has been shown to be important in influencing the occurrence and levels of aflatoxin because *Aspergillus* does not compete well with other molds when corn presents more than 20% moisture. Harvesting corn when moisture content is above 20% followed by rapid drying to at least 14% moisture content within 24 to 48 hours of harvest can inhibit *Aspergillus* growth and toxin production. Contaminated grains and their byproducts are the most common sources of aflatoxin. Corn silage may also be a source of aflatoxins, because the ensiling process does not destroy toxins already present in silage [12].

On the farm, more than one mold or toxin may be present in the contaminated feed, which often makes definitive diagnosis of aflatoxicosis difficult. The prognosis of aflatoxicosis depends upon the severity of liver damage. Once overt symptoms are noticed the prognosis is poor. Treatment should be directed at the severely affected animals in the herd and further poisoning prevented. Aflatoxicosis is typically a herd rather than an individual cow problem. If aflatoxicosis is suspected, feed should be analyzed immediately. If aflatoxins are present, the source should be eliminated immediately. Levels of protein in feed and vitamins A, D, E, K and B should be increased as the toxin binds vitamins and affects protein synthesis. Good management practices to alleviate stress are essential to reduce the risk of secondary infections which must receive immediate attention and treatment [12].

Importantly, it has been demonstrated that simple measures can significantly reduce the risk of mycotoxin exposure on farm. Storage of grain at appropriate moisture content (below 130 g kg^{-1}), inspection of grain regularly for temperature, insects and wet spots will limit the possibility of fungal development in feeds and feedstuffs as discussed before. The risk of feed contamination will be reduced in animal units with rapid turnover of feed because there will be less time for fungal growth and toxin production [17]. Aflatoxin is just one of many mycotoxins that can adversely affect animal health and productivity. Care regarding animal feed must be extended not only to the nutritional and economic value, but also to food quality [13].

Decades of animal studies have demonstrated that chronic exposure to aflatoxins in animals can also cause growth inhibition and immune suppression [18]. Nursing animals may be affected, and AFM_1may be excreted in the milk of dairy cattle and other dairy animals. This in turn poses potential health risks to both animals and humans that consume that milk. Chronic aflatoxin exposure in animals can result in impaired reproductive efficiency, reduced feed conversion efficiency, increased mortality rates, reduced weight gain, anemia, and jaundice. In the case of laying hens, aflatoxicosis causes an enlarged fatty liver and lowered egg production [19].

Sex and age of animals have also an influence on AFB_1 susceptibility. For instance, males are more susceptible than females and young animals of all species are more susceptible than mature animals to the effects of aflatoxin [12,16]. Feed refusal, reduced growth rate and decreased feed efficiency are the predominant signs of chronic aflatoxin poisoning. In addition, listlessness, weight loss, rough hair coat and mild diarrhea may occur. Anemia along with bruises and subcutaneous hemorrhage are also symptoms of aflatoxicosis. The disease may also impair reproductive efficiency, including abnormal estrous cycles (too short and too long) and abortions. Other symptoms include impaired immune system response, increased susceptibility to disease, and rectal prolapse [12].

A study identified and quantified aflatoxins (AFB_1, AFB_2, AFG_1 and AFG_2) from poultry feed and their recovery, together with their metabolites (AFM_1, AFM_2, AFP_1 and aflatoxicol) in litter. Hens were divided in 3 groups and fed with 2 AFB_1 concentrations: 30 and 500 ppb, besides the control group. Feed samples of the 3 groups presented significant difference with AFB_2 and AFG_2, whereas in litter samples, there were significant differences for AFG_2 in the 500 ppb group. Poultry litter had traces of AFM_1, AFM_2, AFP_1 (can be considered as a demethylated AFB_1) and aflatoxicol with no significant differences among treatments [20].

The presence of molds in foodstuffs causes the appearance of flavors and odors that reduce palatability and affect feed consumption by animals as well as reduce the nutritional value of foods. Mycotoxins, in turn, affect the digestion and metabolism of nutrients in animal production, resulting in nutritional and physiological disorders, besides a negative effect on the immune system [21].

It was reported main effects caused by aflatoxins during swine growth and termination phases. When feed was contaminated with 10-100 ppb, productivity losses without noticeable clinical signs were observed. When this level was 200-400 ppb, reduced growth and feed efficiency occurred. At 400-800 ppb of aflatoxins in feed, there were liver diseases (friable or yellow-tan liver). After 800-1200 ppb of aflatoxins administration in feed, reduction of food intake and growth was observed. Finally, at 1200-2000 ppb, jaundice, coagulopathy, anorexia and even mortality may happen. Not only swine is affected by aflatoxins but all species, being the main clinical signs and lesions reported as decreased weight gain, digestive disorders, liver disease, anorexia, ataxia, tremors and death [22].

A total of 480 poultry feed samples from Rio de Janeiro state were collected monthly during one year and analyzed, being the main fungal species found *P. citrinum* (35% of the samples) followed by *A. flavus* (25%) which is the main aflatoxin producer microorganism. AFB_1 levels ranged from 1.2 to 17.5 ppb. There were no significant differences ($P<0.001$) between all months tested except February and March when the highest and lowest AFB_1 production was found [23]. In Pakistan, a total of 216 samples of poultry feed ingredients were assayed, being found maximum 191.65 ppb for AFB_1, 86.85 ppb for AFB_2, 89.80 ppb for AFG_2 and 167.82 ppb for AFG_1. Minimum aflatoxins were produced in the winter season. The temperature varies from 10 to 45 ºC in this country, favorable to *Aspergillus* growth [24].

Recently [25], a survey reported the association of mycotoxins with hematological and biochemical profiles in broilers. The authors performed meta-analysis using data from 98 articles, totaling 37,371 broilers. Some conclusions of this review were that mycotoxins reduced ($P<0.05$) the hematocrit (−5%), hemoglobin (−15%), leukocytes (−25%), heterophils (−2%), lymphocytes (−2%), uric acid (−31%), creatine kinase (−27%), creatinine (−23%), triglycerides (−39%), albumin (−17%), globulin (−1%), total cholesterol (−14%), calcium (−5%), and inorganic phosphorus (−12%). Mycotoxins also altered ($P<0.05$) the concentrations of alkaline phosphatase, alanine aminotransferase and aspartate aminotransferase. The total protein concentration in blood was 18% lower ($P< 0.05$) in broilers challenged by aflatoxins compared with that of the unchallenged ones. The inclusion of antimycotoxin additives in diets with aflatoxins altered ($P<0.05$) some variables (uric acid, creatinine, alkaline phosphatase, alanine aminotransferase, aspartate aminotransferase, and γ-glutamyl transferase) in relation to the group that received diets with the mycotoxin and without the additive.

Another recent study assumes that aflatoxins could compromise the macrophages functions; in particular, co-exposure to AFB_1, AFB_2, AFM_1 and AFM_2 may exert interactions which can significantly affect immunoreactivity [26].

4. AFLATOXINS OCCURRENCE IN ANIMAL PRODUCTS

When focusing on how mycotoxins play a role in food safety, attention should be limited to mycotoxins that are known to be transferred from feed to food of animal origin, as this food represents a significant route of exposure for humans [27]. Apart from their toxicological effects in affected animals, the carry-over through animal derived products, such as meat, milk and eggs into the human food chains is an important aspect of mycotoxin contamination. FAO has estimated that up to 25% of the world's food crops and a higher percentage of the world's animal feedstuffs are significantly contaminated by mycotoxins.

Aflatoxin or ochratoxin residues in meat are uncommon and rarely found [28]. However, it's more common in organs especially liver. This organ may have its lipid content increased over three fold when 20 ppm aflatoxin is incorporated in broiler feed [29].

The problem in the egg production is that the long-term or short-term hen's exposure, via dietary sources, to low concentrations of certain mycotoxins causes contamination of eggs. This is the case of aflatoxins, which have a high impact in both, human and animal health, causing significant losses in the egg industry, considering the deleterious effect on egg production and quality.

In laboratory studies it was proved that aflatoxin can decrease egg production and increase liver fat (fatty liver syndrome). This classical study established the typical symptoms associated with acute or chronic aflatoxicosis, observed until today in field conditions [30].

A distinctive sequence of events during acute aflatoxicosis in laying hens (30 weeks-old) in a four week experiment with increasing aflatoxin doses in the diet of 0; 1.25; 2.5; 5.0 and 10.0 $\mu g\ g^{-1}$ [31]. Results indicated that egg production was decreased by about 70% from the control value at 10$\mu g\ g^{-1}$concentration in the diet and the liver size was increased significantly by 5 and 10 $\mu g\ g^{-1}$ dietary concentrations of aflatoxin and the liver lipid increasing dramatically by a smaller dose of 2.5 $\mu g\ g^{-1}$. Table 3 shows the dramatic effect of aflatoxin in the liver function [31]. The obtained data suggest that plasma and yolk lipids respond to the inhibition of lipid synthesis and transport from the liver during aflatoxicosis induced by the dietary treatments. The liver malfunction results in an increase in its fat content and a decrease in the levels of plasma lipids.

Table 3. Response of liver lipid and plasma lipid during aflatoxicosis in laying hens.

Dose (μg g^{-1})	***Liver lipids (%)***	***Plasma lipids (g 100 g^{-1})***
0.0	21.2±1.1	2.6±0.3
1.25	24.4±1.8	2.6±0.4
2.5	32.7±1.8	2.1±0.2
5.0	35.6±4.9	1.9±0.1
10.0	46.5±4.7	1.7±0.2

[i] - Adapted from ref. [31]. Values are means ± Standard error of the mean

Zaghini et al. [32] supported the previous finding showing the effects of AFB_1 on egg quality and chemical parameters. In the study, 44 weeks-old laying hens were submitted to a diet containing 2.5 ppm of AFB_1 and by the end of the second and third weeks of the trial, changes were observed like decreased egg weight (from 73.76 g to 72.5 g, week 0-4th, respectively) and reduced shell weight, as indicated by the decline in the percent shell of eggs laid by the hens fed the AFB_1 contaminated diet from 10.49% to 10.19%. In the same study, aflatoxin also influenced color parameters, which were probably related to interference of AFB_1 with lipid metabolism and pigmentary substances deposition in yolk. Additionally, all livers collected from the hens administered the mycotoxin group were positive for AFB_1.

Astonishingly, as little as 0.2 mg kg^{-1} (or 0.2 ppm) of the metabolite AFB_1 has been documented to reduce egg production and egg mass in laying hens from 22 to 40 weeks of age [33]. The most important economic effect of poultry ingesting AF-contaminated feed would be the increase of the mortality index [34] and in addition, aflatoxins intake can decrease productivity due to hepatic [35]; immunological [36] and renal damages [37].

In a detailed study focusing on the effects of aflatoxin chronic intoxication in renal function of laying hens (13 weeks-old), aflatoxins were evident at 17 weeks of the intoxication period. Final concentrations were 0, 0.46, 0.98 and 1.53 mg of aflatoxins per kg of feed, respectively and birds ingested the contaminated diets during 17 and 42 weeks. Body weight of intoxicated hens, showed a tendency to decrease, being significant in 1.0 and 1.5 mg kg^{-1} of feed concentrations in both times of the intoxication period. Evidence of tubular damage in kidney was found as a result of a decreased concentration of Ca^{++} and $PO_4{}^{+3}$ in plasma or even a decreased Ca^{++} absorption from the gut. Also, microscopic lesions of glomerular and tubular structures like inflammatory and degenerative processes of the renal structures in hens kidneys were found. Additionally, the authors pointed out that the renal lesions occurred more frequently in larger doses of AF and over a long period of exposure to the toxin (42 weeks intoxication period) [38].

Other authors concluded that aflatoxins may have direct or indirect effect or both, on functionality of the gastrointestinal tract. Results indicated that specific activity of the intestinal maltase and disaccharidase increased quadratically, by feeding up to 1.2 mg kg^{-1} aflatoxins and declined at 2.5 mg kg^{-1} concentration in the study and the intestinal crypt depth (but not villus length) increased linearly with increasing the level of aflatoxins in the experimental diets [39].

Hens were fed three levels of aflatoxin that might approximate contamination under field conditions [40]. Pure AFB_1 was prepared and mixed in the diet as follows: 0.1 ppm for 10 days; 0.2 ppm for 12 days and 0.4 ppm for 15 days. Results confirmed that AFB_1 fed to hens was transmitted into eggs in measurable amounts at all levels and was found in both, albumen and yolks. The average amounts of aflatoxin distributed between albumen and yolk were 2.2 and 3.6 ppb, respectively. Even at the concentration of 0.1 ppm of AFB_1 in the layer diet, the transmission into eggs occurred as an average of 0.23 ppb.

Mainly aflatoxins and ochratoxin A may be found as residues at significant levels in muscles and muscle foods when contaminated feed is distributed to farm animals. Meat contamination may also result from toxigenic mold development during ripening and ageing. In muscles, only low levels are found, often below detection limits of the methods used, even after exposure of the animals to high doses of AFB_1. In ruminants, many studies evaluated aflatoxin transfer into the milk of lactating cows. However, as for other species, residues can be found in liver and kidney that are edible parts of these animals [41].

It was reported [41] that processing conditions during ageing of hams may allow aflatoxin synthesis. Thus, is important to conduct research evaluating the production of AFB_1 during meat processing and ageing. Studies show that frequency of processed meat contamination with AFB_1 was low and the toxin level within meat was usually <10 ng g^{-1} (ppb). It is not clear whether AFB_1 was produced during meat processing or was present before at the residual level in muscles. The contamination of spices and additives added during meat processing may also represent a source of mycotoxin. Besides, spice addition may lead to a secondary contamination of the final product with aflatoxins.

Table 4 shows residues of one or more aflatoxins in different hen tissues.

Table 4. Residues of aflatoxin in animal products combined or not with other mycotoxin

Animal species	***Dose or exposure time***	***Tissues***	***Residues (µg kg⁻¹)***	***Metabolites***	***Reference***
Laying hens	2.5 ppm AFB_1 in feed for 4 weeks	Muscle Eggs Liver Eggs	0.08±0.03 0.24±0.07 and 0.25±0.09 4.13±1.95 <0.5 and <0.01		[32]
Layer breeder hens	OTA+AFB_1 mg kg^{-1} 0 + 0 0 + 5 3 + 5 5 + 5	Liver/Muscle/Kidney Muscle Kidney Liver Muscle Kidney Liver Muscle Kidney Liver	ND + ND ND + 0.03 ND + 0.25 ND + 1.44 0.34 + 0.02 2.80 + 0.27 1.98 + 0.26 0.51 + 0.02 2.81 + 0.27 2.21 + 0.11	OTA+AFB_1	[42]

Feeding one mycotoxin alone (OTA) resulted in significantly higher residue levels in liver, kidney and breast muscles of hens than their counterpart birds kept on a diet concurrently contaminated with both OTA and AFB_1 [42]. In this study, a total of 72 White Leghorn layer breeder hens at 45 weeks of age were submitted to diets containing different combinations of these mycotoxins (some are shown in Table 4, focusing mainly on AFB_1 residues). In liver of hens fed OTA alone at 5 mg kg^{-1} (ppm) feed, residues level was 22.54±1.48 (mean±SD) ppb, as compared to significantly lower residual concentration of 2.21±0.42, in the same levels of OTA when administered in combination with AFB_1. Residues of OTA were significantly higher in liver than in kidneys of the hens fed OTA alone, in all experimental groups. However, feeding OTA in combination with AFB_1 resulted in higher deposition of OTA in kidneys than in livers. Residues of AFB_1 were significantly higher in liver and breast muscles of the birds kept on AFB_1 contaminated feed compared with those fed OTA and AFB_1 concurrently. When the maximum dosage (5 ppm) was administered, residues of OTA and AFB_1 were also the maximum in the liver, i.e. 22.54±1.48 and 1.44±0.21 ppb, respectively, while a minimum concentration of residues of both mycotoxins was found in the breast muscles of the laying hens. Residues of AFB_1 in the eggs appeared at day 5 of toxin feeding and disappeared at day 6 of withdrawal of AFB_1contaminated diet. As in case of tissues, residues of OTA and AFB_1 are significantly lower in eggs obtained from hens fed both toxins in combination, compared to those fed each mycotoxin alone, possibly due to their protein binding potentials.

Another study shows that concentration of AFB_1 residues in liver and muscles increased with toxin ingestion time and were at its highest levels (6.97 ppb in liver and 3.27 ppb in muscle) on the last day (7th) of feeding AFB_1 contaminated ration. Broiler chicks of 7, 14 and 28 days of age fed same level of AFB_1 showed lower tissue residues in older birds compared with younger ones. Birds given 1600 and 3200 ppb AFB_1 for 7 days at 28 days of age had no detectable AFB_1 levels in liver and muscles after 3 and 8 days of withdrawal of contaminated feed. A rapid decrease in AFB_1 residues below the tolerance limits from muscles and liver within 3 and 7 days of withdrawal of dietary AFB_1 in this study confirm the rapid metabolism of aflatoxins in the body of chicken and that it may not become a significant human health risk. However, in areas with no regulatory limits on AFB_1 levels of poultry feed, the secondary exposure to aflatoxins through consumption of chicken liver and meat derived from the poultry fed contaminated feed may pose a risk to consumers health [43].

Table 5. Mycotoxin levels in vegetable and animal products.

Product	***Aflatoxin***	***Positive/total of samples***	***Range (ppb)***	***Method***
Corn	AFs	76 / 246	2-906	TLC or ELISA
Corn	AFB_2	33 / 292	1-17	TLC
Feed	AFB_1	14 / 96	11-287	TLC
Eggs	AFB_1	2 / 210	2-5	TLC
Eggs	AFM_1	0 / 210	-	TLC
Swine liver	AFB_1	1 / 43	27	TLC
Chicken liver	AFB_1	3 / 6	1.2-3.2	TLC/HPLC

Adapted from ref. [45]. See this review to obtain the references of original publication.

The impact of subchronic exposure of AFB_1 on the tissue residues of enrofloxacin and its metabolite ciprofloxacin was examined in broilers. Broiler chickens given either normal or AFB_1 (750 ppb diet) supplemented diets for 6 weeks received enrofloxacin (10 mg kg^{-1} day^{-1}, p.o.) for 4 days and thereafter, residue levels were determined at 1, 5 and 10 days after the last treatment. In AFB_1-unexposed broiler chickens, enrofloxacin was detected in all the tissues. After 24 h of treatment cessation, concentrations of enrofloxacin were up to 0.85 μg g^{-1} in the following order: liver>skin+fat>muscle>kidney. The parent drug was not found in any of the tissues except liver 10 days after the last dose of enrofloxacin. Ciprofloxacin was not detectable in any tissue. In AFB_1-exposed broiler chickens, higher concentrations of enrofloxacin and ciprofloxacin were found in different tissues, compared with tissues of control broiler chickens. After 24 h of the last dose of enrofloxacin, concentrations up to 4.53 μg g^{-1}were found of the parent drug in the order skin+fat>liver>kidney>muscle. The parent drug persisted in all the tissues except muscle for 10 days. Ciprofloxacin was detected in muscle and skin plus fat 24 h after termination of enrofloxacin administration and it persisted only in muscle for 10 days. The metabolite was not detectable in kidney [44].

A review carried out in Brazil [45] showed high variability among the results (Table 5). For instance, corn contamination with aflatoxins reached 906 ppb, above those levels allowed by legislation (20 ppb). This fact indicates the need for quality control in the reception of this ingredient in the feed mill with the use of rapid tests for mycotoxins. Regarding products of animal origin, major problems were not observed in eggs and tissues of swine and poultry (Table 5). However, among chicken liver samples 50% tested positive but with relatively low levels. Anyway, attention should be paid with liver consumption when there are evidences of corn contamination. From Table 5 data, it can be noted that contaminated feed samples achieved up to 287 ppb AFB_1, above values allowed by legislation.

A survey with hens fed AFB_1 via moldy rice powder feed showed residues in eggs and tissues (kidneys, liver, muscle, blood, and ova) [46]. Hens were fed for 7 days with a contaminated diet (8 μg g^{-1}) followed by additional 7 days on an aflatoxin-free diet. Eggs were collected over the entire 14-day period. The study showed that aflatoxicol (R0), a carcinogenic metabolite of AFB_1, was found in all samples but blood (Table 6). Levels of R0 and AFB_1 were approximately the same in eggs, ova, kidneys, and liver. In eggs, the levels of R0 and AFB_1 (0.02 to 0.2 ng g^{-1}) increased steadily for 4 or 5 days until reaching a plateau and then decreased after B_1 withdrawal at the same rate as they increased. After 7 days of withdrawal, only trace amounts of R0 (0.01 ng g^{-1}) remained in eggs. All samples from hens sacrificed immediately before aflatoxin withdrawal contained R0 or R0+AFB_1. R0 was the only aflatoxin detected in muscle. Seven days after aflatoxin withdrawal, B_1 (0.08 ng g^{-1}) was found in one of nine livers and R0 (0.01-0.04 ng g^{-1}) in eight of nine muscles analyzed, but no aflatoxins were found in any other tissues. Interestingly, the transfer of aflatoxins into eggs is right after administration, since B_1 (0.03 ng g^{-1}) and R0 (0.02 ng g^{-1}) residues were found in eggs laid 1 day after contaminated feed was administered. This indicates that toxins penetrate the egg through eggwhite since yolk was already formed before this period. Aflatoxin apparently can enter the egg at any stage of its development. This is because it takes 7 to 8 days for each oocyte to develop into a mature ovum (yolk) and 24 hours for the egg oviposition.

Table 6. AFB_1 and its metabolites aflatoxicol (R0) and AFM_1 (ng g^{-1}) after contaminated diet ingestion with AFB_1 (8 μg g^{-1}). Values in ppb; ND = Not detected [**46**].

Tissue	*R0*	*AFB_1*	*AFM_1*
Ova	0.25	0.24	ND
Kidneys	0.10	0.25	0.05
Liver	0.20	0.46	ND
Muscle	0.08	ND	ND
Blood	ND	0.05	0.10

5. LEGISLATION IN FEED AND FEED INGREDIENTS

In the last decades, only aflatoxins and, to a lesser extent, ochratoxin A were regulated in foods from animal origin. For other toxins, the risk management was based on the control of the contamination of food from vegetal origin intended for both human and animal consumption. Nowadays, other mycotoxins are included. Regulatory values or recommendations are mainly built on available knowledge on toxicity and potential carryover of these molecules in animal. Therefore, by limiting animal exposure through feed ingestion, one can guarantee against the presence of residues of mycotoxins in animal-derived products. However, accidental high levels of contamination may lead to a sporadic contamination of products coming from exposed animals [41].

Tolerance levels of mycotoxins in foods are needed to ensure product quality and consumer health. The limits differ among countries, i.e., depending on the product and the country there are different tolerance levels for each mycotoxin, but it is certain that their presence in foods has been widely researched and new standards were required over the years, in the last decade. Table 7 shows an average of mycotoxin variation depending on the type of food, required as maximum standard in different countries.

Table 7. Variation among different countries regarding maximum tolerance limits of mycotoxins.

*Mycotoxins**	*Feed[1]*	*Corn[2]*	*Soybean[3]*
Aflatoxin B_1, ppb	1.5 - 50	1 – 50	30 - 50
Aflatoxin B_1, B_2, G_1 and G_2, ppb	0 - 75	5 – 50	20 - 50
Deoxynivalenol, ppb	5 - 1500	-	-
Toxin HT2, ppb	25 - 100	-	-
Ochratoxin, ppb	5 - 300	50 – 300	-
Zearalenone, ppb	-	0.5 – 200	-

[1] Relative to feed and concentrates for all categories and phases of animal

[2] Corn and byproducts

[3] Soybean and byproducts

* Source: Adapted from Resolution RDC Nº7 [47] and EUR-LEX [48].

In Brazil, the most recent resolution on mycotoxins in food is the RDC 07/2011 [47] which establishes maximum tolerated levels for aflatoxins (AFB_1+AFB_2+AFG_1+AFG_2 and AFM_1), ochratoxin A, deoxynivalenol, fumonisins (FB_1 + FB_2), patulin and zearalenone, admissible in ready-to-eat foods and raw materials. To adapt to the new standard required in 2011, the producers of 14 food categories should meet the requirements until 2016. Table 8 shows standard values set for corn, which is the main ingredient added to feed in the country.

Brazil, like different countries, also follows the recommendation to keep mycotoxin levels as low as possible. For that, better practices and technologies in the production, handling, storage, processing and packaging should be accomplished in order to prevent that contaminated food is sold or consumed.

Table 8. Maximum tolerated levels for mycotoxins according to Resolution RDC 07/2011 [47].*The maximum tolerated levels refer to results obtained by methodologies that comply with the performance criteria established by Codex Alimentarius.

Mycotoxins	***Corn and byproducts****
Aflatoxins B_1, B_2, G_1 and G_2, ppb	20
Deoxynivalenol, ppb	3000
Fumonisin B_1+B_2, ppb	5000
Zearalenone, ppb	400

6. AFLATOXIN BINDERS AND STRATEGIES TO REDUCE TOXICITY TO FARM ANIMALS

Adsorbents are necessary and important and may have great impact on improving animal production and health, providing greater security to consumers of animal products, due to the reduction and/or removal of mycotoxins in these products.

Considering that aflatoxins were the first discovered mycotoxins, there are many data available searching for binders and other methods to reduce toxicity in animals. However, due to methodologies used for evaluation, there is certain degree of variation in results.

The most common additives used in animal diets are aluminosilicates, produced synthetically or extracted from clay mines. There are also other alternatives to reduce aflatoxin toxicity, as presented.

6.1 Clay derived sorbents

This type of binder is basically composed by single or blended type of clay. The most common clay is hydrated sodium calcium aluminosilicate (HSCAS). However, there are other sort of clays which can be used as toxin binders, like sodium or calcium bentonites and zeolites. Not often, any particular varying sort of clay can be used as well. It has to be considered that such materials can be synthesized industrially or obtained from mines around the world. In the case of natural sources (mines) it has to be considered that each source may present specific particularities in terms of composition, which can impact the binding capacity, and even clays obtained from the same place, can vary from batch to batch, that has to be well controlled throughout quality control.

There is much information available in the literature comparing those different clays [49]. These authors for example, compared zeolite, bentonite and HSCAS for AFB_1 binding capacity using *in vitro* method, simulating gastrointestinal fluids. This method seems to be the most frequent technology adopted for such assays, including a double condition of pH (3.0 and 7.0). Those researchers find that zeolite and bentonite aflatoxin binding capacity varied according to the pH used for the assay, and both clays were less effective than HSCAS (Figure 2).

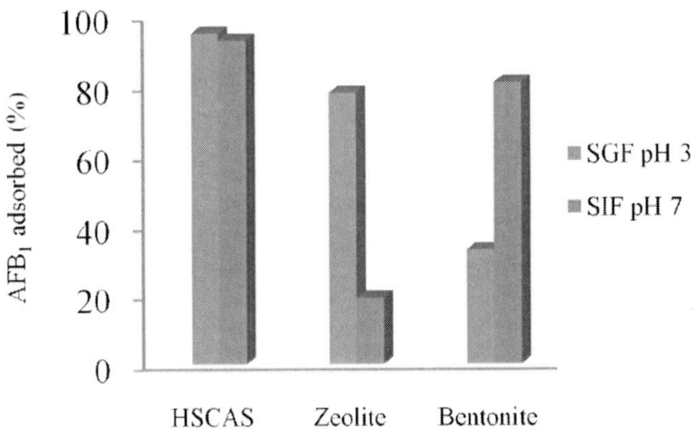

Figure 2. Adsorption percentage of AFB_1 (8 μg/ml) to sorbents (0.5% w/v) in simulated gastrointestinal fluid at pH 3 and pH 7 [49].

It is important to establish the correct inclusion rate to animal diets in order to optimize the binding response. In Figure 3, it can be observed how those three types of clays perform under the same pH (7.0) when increasing doses are included.

It has been shown that montmorillonite (0.5%) added to the diet containing 5 ppm of aflatoxin has proven its effectiveness in preventing the effects of aflatoxicosis in broilers [50].

Based on the data presented in Figures 2 and 3, it is clear that assays condition (especially pH) and toxin:sorbent dosing rate are extremely important. These conditions should be considered when product performance reports are compared. However, when evaluating [51] nine different toxin binders (4 activated charcoals, 3 sodium bentonites, 1 calcium bentonite and 1 esterified glucomannan) all products presented adsorption above 95% of AFB_1, regardless of the pH used (3.0, 7.0, 10.0 and the original pH of each product).

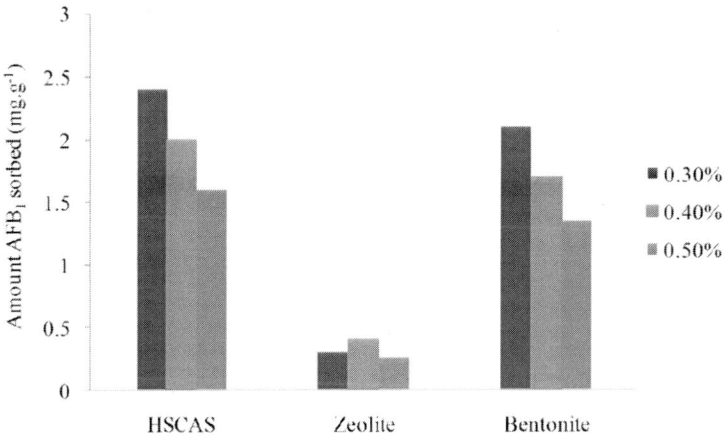

Figure 3. Amount of aflatoxin B_1 adsorbed on sorbents at different concentrations of the adsorbents in simulated intestinal fluid at pH 7 [**50**].

Other methods can be used to evaluate toxin adsorbents, as *in vivo* trials. In this case, the most frequent inconsistency when comparing research reports is related to the source used to obtain aflatoxin (synthetic crystalline vs natural aflatoxin obtained by fermentation) and also the aflatoxin level used in the specific assay. When comparing analytical reports with field accepted levels of aflatoxins, there is much difference. One reason for that is the difference between the experimental conditions (well controlled) where animals are not submitted to stress situation in comparison to the real farm condition.

Another additional evaluation that should be performed is the presence of aflatoxin in specific organs, like liver. Low level of AFB_1 (50 ppb) on broiler performance was studied on biochemical parameters and aflatoxin presence in liver tissue, when monensin and sodium bentonite were added to the feed, from 18 up to 46 days of age. The authors concluded that monensin and AFB_1 compete for adsorption sites on sodium bentonites, indicating a non-selective adsorption capacity of this particular binder. The researchers comment as well that different substances, such as coccidiostats, vitamins, minerals, aminoacids or other dietary components, could affect the ability of the adsorbent to bind low levels of aflatoxin. In addition, significant levels of AFB_1 in livers indicate that this determination is important not only for diagnosis of aflatoxicosis in broilers, but also for quality control of avian products [52].

6.2 Organic sorbents

The most well-known natural toxin binders are yeast based products. Glucans are yeast cell wall constituents. Those compounds have been submitted to esterification process generating a new additive with toxin binding capacity, called esterified glucomannan (EGM). Efficacy of EGM was tested against mycotoxins naturally present in broiler feed [53], being 0.05% EGM efficient to counteract the adverse effects of mycotoxins (Table 9).

Table 9. Efficacy of esterified glucomannans (EGM) on broiler live weight (LW), feed intake (FI) and feed conversion ratio (FCR) fed with a mycotoxin contaminated diet, from one up to 35 days

*Mycotoxin**	*EGM (%)*	*LW (g)*	*FI (g)*	*FCR ($g\ g^{-1}$)*
----	----	1,391.2 ^{b}	3,017.6 ^{b}	2.17 ^{b}
----	0.05	1,441.4 ^{c}	2,994.0 ^{b}	2.07 ^{a}
+++	----	1,258.8 ^{a}	2,803.4 ^{a}	2.22 ^{c}
+++	0.05	1,381.0 ^{b}	2,952.6 ^{b}	2.15 ^{b}
SEM		7.25	20.01	0.015

* Aflatoxin 168 ppb, ochratoxin 8.4 ppb, zearalenone 54 ppb and T2-Toxin 32 ppb [53].

6.3 Other strategies

The use of mechanisms that improve animal health and physiology can be helpful. One example is the use of probiotics which have been used to ameliorate mycotoxicosis. Activated charcoal has been used to prevent animal intoxication by several compounds, including mycotoxins. Plant extracts with specific mode of action, like liver protection, have been used as well to reduce the toxicity of some mycotoxins, specially aflatoxin.

Milk thistle (*Silybum marianum*), which is a medicinal herb found in Pakistan, has been used to treat liver diseases. This herb was tested in poultry feed contaminated with AFB_1 (80 ppb for the first week and 520 ppb from the second until the fifth) at a dose of 1%. The results indicated that milk thistle is effective as hepatoprotectant and growth promoter in the presence of AFB_1 in the feed [54]. Protection against the negative effects of aflatoxin on performance of broiler chickens was observed when broilers were fed during 4 weeks with 0.05% ETE (ethanolic turmeric extract, *Curcuma longa*) plus 3 ppm aflatoxins [55].

Another report concluded that dietary citric acid supplementation can be used as an additive to degrade aflatoxins in the ration as well as to promote growth performance in young broiler chickens. Results showed that aflatoxins in the diet, at a concentration of 39 ppb were almost degraded (92%) by the acidification procedure (up to 50 g kg^{-1}) [56].

Other alternatives to degrade aflatoxins have been tested, like the use of microorganism. Bacteria (*Nocardia corynebacteroides*, NC) showed ability to degrade AFB_1 [57]. In a trial performed with broiler fed AFB_1 (800-1,200 ppb) NC was safe to the birds and showed protection to the animal, indicating that it can be used as a tool to detoxify feed contaminated with AFB_1 at high levels [58].

Humic acid, generated during matter decomposition, has binding capacity for many molecules. The use of oxihumate was evaluated as AFB_1 binder, *in vitro* and *in vivo* [35]. Oxihumate showed a high *in vitro* affinity for AFB_1. *In vivo* trial showed that oxihumate decreased adverse effects caused by AFB_1 on broiler body weight and also protective effect against liver damage, stomach and heart hyperplasia, acting positively preserving standard blood parameters. Enzyme degradation of aflatoxin has been tested as well. Data suggest that lactoperoxidase can be used to hydrolyze aflatoxin [59].

6.4 Other aspects

There are many contradictory data available in the main scientific journals. One and probably the main reason for that is the way the trials have been performed, differing in terms of toxin levels, and environmental condition of the trials. Under a real field condition, the challenges the animals suffer are far stronger then under experimental situation. Toxin binder, especially clays, may affect the cation binding capacity of feeds and consequently influencing water intake and feed consumption. Also, the effects of none, medium (1 g kg^{-1} feed) and high (2.5 g kg^{-1} feed) inclusion levels of HSCAS was evaluated in broiler mycotoxin free diets [60]. The data suggest that increasing HSCAS to diets may modify performance, internal organ weights, gastrointestinal and biochemical parameters. However, other authors [61] did not see effect as the consequence of toxin binders (EGM) presence on broiler body weight and feed efficiency.

Table 10. Mycotoxin detoxifying agents tested in vivo in poultry. Adapted from ref. [62].

Detoxifying agents	**Mycotoxin[3]**								
	AFB_1	**CPA**	**DAS**	**DON**	**Fusaric acid**	**NIV**	**OTA**	**T2 Toxin**	**ZEA**
HSCAS	+	-					-	-	
Clinoptilolite	+/-	-							
Modified nanomontmorilonite	+								
Mg K aluminosilicate	+/-								
Sodium bentonite	+								
Ca montmorillonite	+								
Synthetic crystalline aluminosilicate	+/-								
Acidic phyllosilicate		-							
Zeolite	+		-			-		-	
Diatomaceous earth							+		
Charcoal							-		
Superactivated charcoal	+/-							+/-	
BHT[1]	+								
Cell wall *Saccharomyces cerevisiae*	+								
Yeast glucomannans	+								
Esterified glucomannans	+			+/-	+		+	-	+
Xylanase				-					
Live yeast culture residue	+								

Nocardia corynebacteroides	+/-								
Eubacterium			+	+				+	
Yeast *Trichosporon mycotoxinivorans*							+		-
Saccharomyces cerevisiae	+								
Ammonia	+								
Calcium propionate	+/-								
PVPP[2]	+/-			-					

[1]BHT = butylhydroxytoluene; [2]PVPP = Polyvinylpolypyrrolidone; [3]Mycotoxins AFB_1 = aflatoxin B_1, CPA = cyclopiazonic acid; DAS = diacetoxyscirpenol, DON = deoxynivalenol; NIV = nivalenol, OTA = ochratoxin A, ZEA = zearalenone. Cells highlighted in dark gray indicate that the product has shown positive effects in counteracting deleterious effects of mycotoxins, while light gray color depicts that the product was not effective.

+: positive effect of the mycotoxin-detoxifying agent;

- : negative effect of the mycotoxin-detoxifying agent;

+/-: positive effect of the mycotoxin-detoxifying agent on some parameters, no effect on other parameters.

Additionally, different regions across the world have been dealing with mycotoxin subject in different ways. In US for instance, no toxin binders are officially registered, as a consequence of the control quality assumed for feedstuffs. In EU, since 2009, toxin sequestrants have been considered as a sort of feed additive, and a scientific group of specialists, namely European Food Safety Authority [62] have been working on re-evaluation of the efficacy and biological effects of detoxifying agents in animals. In Tables 10 and 11, a summary of the outcome of that technical group is presented.

In other regions, like South America, due to the climate and grain production conditions, mycotoxin has been a significant challenge along the past decades. This particular situation has been forcing the development of research groups which are involved with commercial sequestrants evaluation. As a consequence, the maximum acceptable aflatoxin limit has been established for different raw materials and feed, as well the specificity and inclusion levels of toxin binders in animal feed. However there are many different criteria for toxin adsorbents registration in different countries, some demanding an extend documentation about product efficacy and others are less restrictive.

Table 11. Mycotoxin detoxifying agents tested in vivo in pigs. Adapted from ref. [62].

Detoxifying agents	AFB_1	DON-NIV	Fumonisin	OTA	T2 Toxin	ZEA
HSCAS	+	-				
Montmorilonite		-				-
Sodium bentonite	+	-				
Calcium bentonite	+					
Zeolite	+					+
Sepiolite	+					
Palygorskite	+					
Ammonium carbonate		-				
Charcoal			+/-			
Polyvinylpolypyrrolidone		-				
Yeast glucomannans	+	+/-			-	+
Apple pommace		+				
Alfafa						+
Content of large intestine of hens		+				
Eubacterium		+				
Combination of Eubacterium BBSH 797 with dried yeast and clays		-				

Cells highlighted in dark gray indicate that the product has shown positive effects in counteracting deleterious effects of mycotoxins, while light gray color depicts that the product was not effective.

+: positive effect of the mycotoxin-detoxifying agent; -: negative effect of the mycotoxin-detoxifying agent;

+/-: positive effect of the mycotoxin-detoxifying agent on some parameters, no effect on other parameters

7. CONCLUSIONS

There is a risk of contamination of meat products, eggs and milk with mycotoxins, although literature shows great variability due to the ingredients contamination which are included in feed. Ingredients should be analyzed before their entrance into the silo.

Mycotoxin contamination of ingredients used in feeds for broilers and laying hens is a reality in Brazil, since there are deficiencies in storage, handling and harvesting. Effects on animal production may vary depending on dose and time of administration and/or combination among mycotoxins which are not fully elucidated yet.

In general, mycotoxin residues tend to decrease rapidly after removal of the contaminated diet, which allows "to clean" broilers when feeding a diet free of mycotoxins few days before slaughter. This management option is not viable for laying hens and dairy cattle, since eggs and milk are generated continuously.

Apart from their toxicological effects in affected animals, the carry-over through animal derived products, such as eggs into the human food chains is an important aspect of aflatoxin contamination. Aflatoxins have a high impact in both, human and animal health, causing significant losses in the egg industry, considering the deleterious effect on egg production and quality. There is scarce literature in Brazil regarding egg contamination by aflatoxins.

After mycotoxin contamination of raw materials and feeds, effects can be minimized by using adsorbents which inhibit intestinal absorption of mycotoxins and can thereby prevent their deleterious effects in poultry production.

REFERENCES

1. Ramos Girona, A. J., Sillué, S. M., & Almenar, V. S. (2011). Micotoxinas. Introducción histórica. Ramos Girona AJ (ed.). *Micotoxinas y micotoxicosis*, Madrid, AMV Ediciones, 1-18.
2. Bakırdere, S., Bora, S., Bakırdere, E., Aydın, F., Arslan, Y., Komesli, O., et al. (2012).Aflatoxin species: their health effects and determination methods in different foodstuffs.*Central European Journal of Chemistry*, 10(3), 675-85.
3. Atanda, S. A., Aina, J. A., Agoda, S. A., & U. OE, P. PO. (2012). Mycotoxin Managementin Agriculture: a Review.*Journal of Animal Science Advances*, 2(Suppl. 3.1),250-60.
4. Moss, M. O. (1996). Mycotoxins. *Mycological Research*, 100(5), 513-23.
5. European Mycotoxins Awareness Network. (2012). The Aflatoxins. Available from, http://services.leatherheadfood.com/eman/Factsheet.aspx?ID=6, accessed 08 oct. 2012.
6. IARC Monographs on the Evaluation of Carcinogenic Risks to Humans. (2002)
7. Kussak, A., Andersson, B., & Andersson, K. (1995). Determination of aflatoxins in airborne dust from feed factories by automated immunoaffinity column clean-up and liquid chromatography.*Journal of Chromatography A*, 708(1), 55-60.
8. IARC Monographs on the Evaluation of Carcinogenic Risks to Humans. (2012).*Agents Classified by the IARC Monographs*, 1-105, Available from, http://monographs.iarc.fr/ENG/Classification/, accessed 28 June 2012.

9. IARC Monographs on the Evaluation of Carcinogenic Risks to Humans. (2012). *A Reviewof Human Carcinogens: Chemical Agents and Related Occupations*, 100F, Availablefrom, http://monographs.iarc.fr/ENG/Monographs/vol100F/mono100F-23.pdf, accessed28 June 2012.
10. IARC Monographs on the Evaluation of Carcinogenic Risks to Humans. (1993). SomeNaturally Occurring Substances: Food Items and Constituents. *Heterocyclic AromaticAmines and Mycotoxins*, 56.
11. Papp, E., -Otta, K. H., Záray, G., & Mincsovics, E. (2002). Liquid chromatographic determination of aflatoxins. *Microchemical Journal*, 73(1-2), 39-46.
12. Cassel, E. K. , Campbell, B., Draper, M., & Epperson, B. (2012). Aflatoxins hazards ingrain/ Aflatoxicosis and livestock. South Dakota State University (SDSU). CooperativeExtension Service / College of Agriculture & Biological Sciences / USDA. *Textadapted with permission from University of Maryland Cooperative Extension Service FactSheet* [444 & 445], FS 907: PDF by CES. October 2001, http://igrow.org/up/ resources/02-2033-2012.pdf, accessed 08 oct. 2012.
13. Gonçalez, E., Pinto, M. M., Manginelli, S., & Felicio, J. D. (2004). Dairy cows poisonedwith cottonseed meal naturally contaminated with aflatoxins. *Ciência Rural*, 34(1),171-4.
14. Maia, P. P., Bastos, M. E. P., & de Siqueira, . (2007). Aflatoxins in pet foods- A Review.*Revista da FZVA*, 14(1), 235-257.
15. Akande, K., Abubakar, M., Adegbola, T., & Bogoro, S. (2006). Nutritional and healthimplications of mycotoxins in animal feeds: A review. *Pakistan Journal of Nutrition*,5(5), 398-403.
16. Arias, S. L., Mary, V. S., Theumer, M. G., & Rubinstein, H. R. (2011). Micotoxicosis enanimales de laboratorio. *In: Ramos Girona AJ (ed.). Micotoxinas y micotoxicosis*, Madrid,AMV Ediciones, 299-328.
17. Bryden, W. L. (2012). Mycotoxin contamination of the feed supply chain: Implicationsfor animal productivity and feed security. *Animal Feed Science and Technology*,173(1-2), 134-58.
18. Khlangwiset, P., Shephard, G. S., & Wu, F. (2011). Aflatoxins and growth impairment:A review. *Critical Reviews in Toxicology*, 41(9), 740-55.
19. Wu, F., Narrod, C., Tiongco, M., & Liu, Y. (2011). The health economics of aflatoxin:global burden of disease. *Working Paper 4*, February Available from, http://www.ifpri.org/ sites/default/files/publications/aflacontrol_wp04.pdf, accessed 04 May2012.
20. Cortés, G., Carvajal, M., Méndez-Ramírez, I., Ávila-González, E., Galván, C., Castillo-Urueta, P., et al. (2010). Identification and quantification of aflatoxins and aflatoxicolfrom poultry feed and their recovery in poultry litter. *Poultry Science*, 89, 993-1001.
21. Bünzen, S., & Haese, D. (2006). Controle de micotoxinas na alimentação de aves e suínos.*Revista Eletrônica Nutritime*, 3(1), 299-304.
22. Mallmann, C. A., Tyska, D., Mallmann, A. O., Mallmann, B. A., & Dilkin, P. (2008).Interação das micotoxinas e seus efeitos nos suínos. *In: Simpósio Brasil Sul de Suinocultura(1.: Chapecó, SC). Anais do I Simpósio Brasil Sul de Suinocultura*, Concórdia: EmbrapaSuínos e Aves, 82-85.
23. Oliveira, G., Ribeiro, J., Fraga, M., Cavaglieri, L., Direito, G., Keller, K., et al. (2006).Mycobiota in poultry feeds and natural occurrence of aflatoxins, fumonisins andzearalenone in the Rio de Janeiro State, Brazil. *Mycopathologia*, 162(5), 355-62.
24. Alam, S., Shah, H., Khan, H., & Magan, N. (2012). The Effect of Substrate, Season,and Agroecological Zone on Mycoflora and Aflatoxin Contamination of Poultry Feedfrom Khyber Pakhtunkhwa, Pakistan. *Mycopathologia*, 174(4), 341-9.
25. Andretta, I., Kipper, M., Lehnen, C. R., & Lovatto, P. A. (2012). Meta-analysis of therelationship of mycotoxins with biochemical and hematological parameters in broilers.*Poultry Science*, 91(2), 376-82.
26. Bianco, G., Russo, R., Marzocco, S., Velotto, S., Autore, G., & Severino, L. (2012).modulation of macrophage activity by aflatoxins B1 and B2 and their metabolites aflatoxinsM1 and M2. *Toxicon*, 59(6), 644-50.

27. FAO and IFIF. (2010). Good practices of the feed industry implementing the CodexAlimentarius Code of Practice in Good Animal Feeding. FAO Animal production and Health Manual n.9. Rome. Section 1. Health Hazards associated with animal feed.
28. Biswas, A. K., Kondaiah, N., Anjaneyulu, A. S. R., & Mandal, P. K. (2010). Food SafetyConcerns of Pesticides, Veterinary Drug Residues and Mycotoxins in Meat and Meat Products. *Asian Journal of Animal Sciences*, 4, 46-55.
29. Wyatt, R. D., Briggs, D. M., & Hamilton, P. B. (1973). The Effect of Dietary Aflatoxinon Mature Broiler Breeder Males. *Poultry Science*, 52(3), 1119-23.
30. Hamilton, P. B., & Gralich, J. D. (1971). Aflatoxin as possible cause of fatty liver syndromein laying hens. *Poultry Science*, 50, 800-4.
31. Huff, W. E., Wyatt, R. D., & Hamilton, P. B. (1975). Effects of dietary aflatoxin on certainegg yolk parameters. *Poultry Science*, 54, 2014-8.
32. Zaghini, A., Martelli, G., Roncada, P., Simioli, M., & Rizzi, L. (2005). Mannanoligosaccharidesand aflatoxin B1 in feed for laying hens: effects on egg quality, aflatoxins B1 and M1 residues in eggs, and aflatoxin B1 levels in liver. *Poultry Science*, 84(6), 825-32.
33. Azzam, A. H., & Gabal, M. A. (1998). Aflatoxin and immunity in layer hens. *AvianPathology*, 27, 570-7.
34. Jand, S. K., Kaur, P., & Sharma, N. S. (2005). Mycoses and mycotoxicosis in poultry:A review. *Indian Journal of Animal Science*, 75, 465-76.
35. Jansen Van Rensburg, C., Van Rensburg, C. E. J., Van Ryssen, J. B. J., Casey, N. H., &Rottinghaus, G. E. (2006). In vitro and in vivo assessment of humic acid as an aflatoxinbinder in broiler chickens. *Poultry Science*, 85, 1576-83.
36. Tessari, E. N., Oliveira, C. A., Cardoso, A. L., Ledoux, D. R., & Rottinghaus, G. E.(2006). Effects of aflatoxin B1 and fumonisin B1 on body weight, antibody titres andhistology of broiler chicks. *British Poultry Science*, 47, 357-64.
37. Glahn, R. P., Van Campen, D., & Dousa, T. P. (1994). Aflatoxin B1 reduces Na(+)-P(i)co-transport in proximal renal epithelium: Studies in opossum kidney (OK) cells.*Toxicology*, 92, 91-100.
38. Martinez-de-Anda, A., Valdivia, A. G., Jaramillo-Juarez, F., Reys, J. L., Ortiz, R., Quezada,T., et al. (2010). Effects of aflatoxin chronic intoxication in renal function of layinghens. *Poultry Science*, 89, 1622-8.
39. Applegate, T. J., Schatzmayer, G., Pricket, K., Trache, C., & Jiang, Z. (2009). Effect ofaflatoxin culture on intestinal function and nutrient loss in laying hens. *Poultry Science*,88, 1235-41.
40. Jacobson, W. C., & Wiseman, H. G. (1974). The transmission of aflatoxin B1 into eggs.*Poultry Science*, 53, 1743-5.
41. Bailly, J. D., & Guerre, P. (2009). Mycotoxins in meat and processed meat products.*In: Toldrá F (ed.). Safety of Meat and Processed Meat, Food Microbiology and Food Safety*,New York, Springer, 83-124, Available from, http://www.springerlink. com/content/m0562t7598299h5j/, accessed 16 April 2012.
42. Hassan, Z. U., Khan, M. Z., Khan, A., Javed, I., & Hussain, Z. (2012). Effects of individualand combined administration of ochratoxin A and aflatoxin B1 in tissues and eggs of White Leghorn breeder hens.*Journal of the Science of Food and Agriculture*, 92,1540-4.
43. Hussain, Z., Khan, M. Z., Khan, A., Javed, I., Saleemi, M. K., Mahmood, S., & Asi, M.R. (2010). Residues of aflatoxin B1 in broiler meat: Effect of age and dietary aflatoxinB1 levels. *Food and Chemical Toxicology*, 48(12), 3304-7.
44. Kalpana, S., Aggarwal, M., Srinivasa Rao, G., & Malik, J. K. (2012). Effects of aflatoxinB1 on tissue residues of enrofloxacin and its metabolite ciprofloxacin in broiler chickens. *Environmental Toxicology and Pharmacology*, 33, 121-6.
45. Rodriguez-Amaya, D. B., & Sabino, M. (2002). Mycotoxin research in Brazil: the lastdecade in review. *Brazilian Journal of Microbiology*, 33, 1-11.
46. Trucksess, M. W., Stoloff, L., & Young, K. (1983). Aflatoxicol and aflatoxins B1 and M1in egg and tissues of laying hens consuming aflatoxin-contaminated feed. *Poultry Science*,62, 2176-82.

47. Resolution RDC Nº 7. (2011). Technical Regulation on maximum tolerated levels formycotoxins in food. from February 18, http://www.brasilsus.com.br/ legislacoes/ anvisa/107378-7.html , accessed 28 June 2012.
48. EUR LEX. (2012). http://eur-lex.europa.eu/pt/index.htm, accessed 25 June 2012.
49. Thieu, N. Q., & Pettersson, H. (2008). In vitro evaluation of the capacity of zeolite andbentonite to adsorb aflatoxin B1 in simulated gastrointestinal fluids. *Mycotoxin Research*,24, 124-9.
50. Batina, P. N., Lopes, S. A., Santurio, J. M., de Souza, C., & Martins, D. B. (2005). Theeffects of the addition of sodic montmorilonite on the feeding diet on the biochemicalprofile of broiler chicken intoxicated by aflatoxin. *Ciência Rural*, 35(4), 826-31.
51. Diaz, D. E., Hagler, W. M., Jr Hopkins, B. A., & Whitlow, L. W. (2003). AflatoxinBinders I: In vitro binding assay for aflatoxin B1 by several potential sequesteringagents. *Mycopathologia*, 156(3), 223-6.
52. Magnoli, C. E., Astoreca, A. L., da Rocha Rosa, C. A., & Dalcero, A. M. (2011). Micotoxicosisen aves. *Ramos Girona AJ (ed.). Micotoxinas y micotoxicosis*, Madrid, AMVEdiciones, 221-239.
53. Aravind, K. L., Patil, V. S., Devegowda, G., Umakantha, B., & Ganpule, S. P. (2003).Efficacy of esterified glucomannan to counteract mycotoxicosis in naturally contaminatedfeed on performance and serum biochemical and hematological parameters inbroilers. *Poultry Science*, 82(4), 571-6.
54. Muhammad, D., Chand, N., Khan, S., Sultan, A., Mushtaq, M., & Rafiullah, . (2012).Hepatoprotective role of milk thistle (*Silybum marianum*) in meat type chicken fedaflatoxin B1 contaminated feed. *Pakistan Veterinary Journal*, 32(3), 443-6.
55. Rangsaz, N., & Ahangaran, M. G. (2011). Evaluation of turmeric extract on performanceindices impressed by induced aflatoxicosis in broiler chickens. *Toxicology and IndustrialHealth*, 27(10), 956-60.
56. Salgado-Tránsito, L., Arjona-Román, J. L., Moreno-Martínez, E., & Méndez-Albores,A. (2011). Effect of citric acid supplemented diets on aflatoxin degradation, growthperformance and serum parameters in broiler chickens. *Archivos de Medicina Veterinaria*,43(3), 215-22.
57. Ciegler, A., Lillehoj, E. B., Peterson, R. E., & Hall, H. L. (1966). Microbial detoxificationof mycotoxins. *Applied Microbiology*, 14, 934-9.
58. Tejada-Castañeda, Z. I., Avila-Gonzalez, E., Casaubon-Huguenin, M. T., Cervantes-Olivares, R. A., Vásquez-Peláez, C., Hernández-Baumgarten, E. M., et al. (2008). Biodetoxificationof aflatoxin-contaminated chick feed. *Poultry Science*, 87(8), 1569-76.
59. Doyle, M. P., & Marth, E. H. (1978). Degradation of aflatoxin by lactoperoxidase.*Zeitschrift für Lebensmitteluntersuchung und Forschung A*, 166(5), 271-3.
60. Basalan, M., Gungor, T., Aydogan, I., Hismiogullari, S. E., Erat, S., & Erdem, E. (2006). Effects of Feeding Mycotoxin Binder (HSCAS) on Gastrointestinal Environment and Metabolism in Broilers at Later Ages. *Archiva Zootechnica*, 9, 5-9.
61. Girish, C. K., & Devegowda, G. (2004, 8-10 September). Efficacy of modified glucomannan(Mycosorb) and HSCAS to alleviate the individual and combined toxicity ofaflatoxin and T-2 toxin in broiler chickens. Istanbul, Tukey. *XXII'nd World PoultryCongress*.
62. EFSA, European Food Safety Authority (2009). Review of mycotoxin-detoxifyingagents used as feed additives: mode of action, efficacy and feed/food safety: Scientific report submitted to EFSA by AFSSA, CODA-CERVA, INRA Clermont-Ferrand, INRA Toulouse, IRTA, ISPA; 192p. Grant number: CFP/EFSA/FEEDAP/2009/01. Available from http:// www.efsa.europa.eu/en/scdocs/doc/22e.pdf

CHAPTER 12

The Interaction Between Dietary Valine and Tryptophan Content and Their Effect on the Performance of Piglets

Sam Millet

Animal Sciences Unit, Institute for Agricultural and Fisheries Research (ILVO), Scheldeweg 68, 9090 Melle, Belgium

SIMPLE SUMMARY

To optimize animal performance, pig diets need the right amount of all essential amino acids. Usually, the ideal amino acid profile is used: If the dietary concentration of one particular amino acid is too low, other amino acids will not be used efficiently. In this experiment it was shown that amino acids that stimulate the feed intake may improve the performance of pigs, even if other amino acids are not at the optimal level.

ABSTRACT

Four experimental diets for newly weaned pigs were formulated: (1) low valine and low tryptophan; (2) low valine and high tryptophan; (3) high valine and low tryptophan and (4) high valine and high tryptophan.

Dietary standardized ileal digestible (SID) lysine content was 1.06 g/kg. The SID valine to SID lysine ratio was 0.58 and 0.67 for the low and high valine diets, respectively, and SID tryptophan to SID lysine ratios were 0.19 and 0.22 for the low and high tryptophan diets, respectively. In total, 64 pens of 6 pigs (3 barrows and 3 gilts) were divided over the four experimental treatments. No interaction between dietary supply of valine and tryptophan was observed ($P > 0.1$ for all parameters). Increasing the dietary valine content increased the daily feed intake, daily gain and gain:feed ($P < 0.001$ for all three parameters). Increasing the dietary tryptophan content improved gain:feed during the first 2 weeks ($P < 0.05$) and overall ($P < 0.05$). Valine supply had a greater effect on performance results than tryptophan supply. It may thus be beneficial to provide a diet with an optimal dietary concentration of valine even if other amino acids are at suboptimal dietary levels.

Keywords: feed intake; performance; piglet; tryptophan; valine

1. INTRODUCTION

The nutrient content of a diet may affect the voluntary feed intake of piglets. Feed intake depends largely on energy density [1,2], so that the daily energy intake remains relatively constant across diets with different energy densities, as long as the density is not too low. On the other hand, feed intake is depressed both by a severe deficiency in the limiting amino acid or an excessive supply of some essential amino acids [2].

When pigs consume energy above maintenance, this will be used for protein and fat deposition [3]. With increasing energy intake, the protein deposition will increase until a maximum (PDmax). Once energy intake leads to PDmax, this protein deposition will stay at that absolute maximum and a further increase in energy intake will not affect protein deposition [4]. In young pigs, feed intake capacity appears to be the main factor limiting protein deposition [4]. Still, increased energy intake will only lead to protein deposition if the composing amino acids are not limiting. While some amino acids can be synthesized by the pigs, other amino acids need to be supplied by the diet. Protein synthesis requires that all necessary amino acids are available to the tissues [5]. For this, the dietary intake of essential and non-essential amino acids needs to meet the requirements. The ideal protein concept is commonly used in feed formulation for pig diets. This concept is defined as the perfect ratio among individual essential amino acids and N required for optimal performance [6]. Amino acid requirements are determined by both the need for maintenance and for protein accretion [6]. It is assumed that, with a sufficient supply of energy and nitrogen, increasing the dietary concentration of a deficient amino acid will only improve performance if the dietary concentration of no other amino acid is limiting growth. If one or more amino acid limits growth, other essential amino acids that are in surplus of the ideal pattern will be catabolized and used as sources of energy and nitrogen for non-essential amino acids [5]. In practice, amino acids are expressed relative to lysine.

Feed conversion is affected by the growth rate and the composition of this growth [7]. Since lean is composed of protein plus water and some lipids, its energetic costs are lower than the costs for fat deposition [8]. With a faster growth rate, the energetic cost for maintenance needs per kg of growth will be lower. Thus, faster growing animals will have a higher feed efficiency and better performance, as long as this faster growth does not lead to fatter animals.

Tryptophan supplementation to a diet that was limiting in tryptophan has been shown to increase piglets' feed intake [9,10,11]. Also, in studies on valine requirements, suboptimal levels were associated with a lower feed intake [12,13]. Therefore, the hypothesis was tested that adding either valine or tryptophan to a diet short in both valine and tryptophan will lead to an increased feed intake and improved performance.

2. EXPERIMENTAL SECTION

2.1. Experimental Design

In 4 batches, 96 pigs (Piétrain boar × hybrid sow) were divided over 16 pens of 6 pigs (3 barrows and 3 gilts per pen) at weaning. Half of the experiment (2 batches) was conducted in spring and the other half in autumn. Each pen was randomly assigned to one of four treatments, with a total of 16 pens per treatment. One pen measured 1 m × 1.8 m (1.8 m^2), with 0.3 m^2 per piglet. The slatted floor was covered with a synthetic coating. The temperature in the compartments ranged from 26 °C at the start to 22 °C at the end of the experiments; the light schedule was natural daylight and supplemental artificial light between 7 h 30 and 15 h 30. The pigs had free access to water and were fed *ad libitum.* The pigs received a standard feed (11.5 g lys/kg, 9.8 MJ net energy per kg) the first week after weaning (at 4 weeks of age) and the experimental feeds between 5 and 9 weeks of age. Four experimental diets were formulated: (1) low valine and low tryptophan (LL); (2) low valine and high tryptophan (LH); (3) high valine and low tryptophan (HL); (4) high valine and high tryptophan

(HH). The standardized ileal digestible (SID) lysine concentration was 1.06 g/kg. The dietary lysine concentration was slightly limiting according to the nutrient requirements of pigs from this genotype [14], so that lysine would be limiting and the other amino acids could be formulated relative to lysine. Dietary SID methionine + cystine, threonine, isoleucine, and leucine contents expressed as a percentage of SID lysine were 57, 62, 51, and 96%, respectively (Table 1 and Table 2). The intended SID tryptophan: lysine ratio and SID valine: lysine ratio were 22 and 70, respectively. These were based on optima determined during ILVO research concerning dietary valine concentration (unpublished results) and recent literature on dietary tryptophan content [8]. The intended degree of limitation in the low tryptophan and valine diets were 15% for both amino acids. For valine, a preliminary test at our institute (unpublished data) showed that this degree of restriction would lead to a measurable difference in performance results. For tryptophan, a similar degree of restriction was taken. First, the LL diet was produced and then free L-valine or L-tryptophan were added to yield the intended amount for the different experimental diets. Based on amino acid analysis (Table 2), the eventual ratio of SID tryptophan to SID lysine was 19% in the low tryptophan diets and 22% in the high tryptophan diets. The ratio of SID valine to SID lysine was 58% in the low valine diets and 67% in the high valine diets.

2.2. Measurements

All diets were subject to proximate analysis [15,16,17,18]. Analysis of amino acids in the diets was performed by Ajinomoto Eurolysine SAS using a JLC-500/V AminoTac AA Analyzer (Jeol, Croissy-sur-Seine, France) after hydrolysis with 6 N HCl at 110 °C for 23 h under reflux. The Met and Cys contents in diets were determined after performic oxidation before hydrolysis. The Trp content in diets was determined after hydrolysis at 120 °C for 16 h with barium hydroxide and separation by reverse-phase HPLC and fluorometric detection.

All pigs were weighed individually three times: at the beginning of the experiment, 2 weeks into the experiment, and at the end. Average daily gain was calculated as weight gain divided by the number of days. For further analysis, the average per pen was used. Feed consumption per pen was recorded and average daily feed intake was calculated per pen for each period. Feed efficiency per pen was calculated as the total amount of feed consumed by the pen divided by the total gain of all pigs in the pen.

2.3. Statistical Analysis

Performance results were analyzed with variance analysis (Statistica 9.0, Statsoft, Tulsa, OK, USA), with tryptophan content, valine content and their interaction as fixed factors and batch as random factor. Results were considered significant if $P < 0.05$. The pen was the experimental unit.

Table 1. Ingredient composition of experimental diets (%, as-fed basis).

Item	Low Valine		High Valine	
	Low tryptophan	High tryptophan	Low tryptophan	High tryptophan
Wheat	25.00	24.99	24.97	24.96
Barley	19.76	19.75	19.74	19.73
Corn	18.00	17.99	17.98	17.97
Full fat soybeans	16.00	15.99	15.98	15.98
Soybean meal	5.76	5.76	5.75	5.75
Rapeseed meal	1.91	1.91	1.91	1.91
Beet molasses	4.00	4.00	4.00	3.99
Soy oil	0.20	0.20	0.20	0.20
Mono calcium phosphate	0.78	0.78	0.78	0.78
Salt	0.36	0.36	0.36	0.36
Limestone	0.30	0.30	0.30	0.30
Premix[1]	6.00	6.00	5.99	5.99
Nutrisure[2]	1.00	1.00	1.00	1.00
L-Lysine HCL, 78%	0.49	0.49	0.49	0.49
Dl-methionine, 99%	0.21	0.21	0.21	0.21
L-threonine, 98%	0.20	0.20	0.20	0.20
L-tryptophan, 98%	0.03	0.07	0.03	0.07
L-Valine, 96.5%	–	–	0.11	0.11
Phytase (5,000 IU/g)	0.01	0.01	0.01	0.01

1.The premix contained 80% dairy products and 20% vitamin and mineral premix, providing the following

quantities of vitamins and minerals per kilogram of basal diet: vitamin A, 15,000 IU; vitamin D3, 2,000 IU;

vitamin E, 100 mg; vitamin K, 2 mg; vitamin B1, 2.5 mg; vitamin B2, 7.5 mg; vitamin B5, 20 mg;

vitamin B6, 5 mg; vitamin B12, 0.04 mg; vitamin C, 100 mg; vitamin PP, 30 mg; choline, 324 mg; folic acid, 3 mg;

biotine, 0.15 mg; Ca, 516 mg; P, 419 mg; Mg, 165 mg; Na, 353 mg; Cl, 1,375 mg; K, 1,227 mg; S, 234 mg;

Fe, 100 mg; Cu, 160 mg; Mn, 60 mg; Zn, 100 mg; I, 2 mg; Se, 0.4 mg.

2. DSM nutritonal products: A mixture of calcium salts of the following organic acids: Lactic acid, Formic acid, Citric acid monohydrate, Orthophosphoric acid, Propionic acid.

Table 2. Analyzed and calculated nutritional composition of the basal diet (as-fed basis) *.

Net energy, MJ/kg [1]	9.8
Dry matter, %	89.77
Crude ash, %	5.01
Crude fibre, %	3.32
Crude protein, %	17.50
Crude fat, %	4.97
Starches, %	35.09
Sugars, %	8.62
Lactose, %	3.35
ADF, %	4.47
NDF, %	10.81
ADL, %	0.93
Ca,%	0.60
P, %	0.60
Digestible P, %	0.40
Cu, mg/kg	165
Lysine, %	1.19
Methionine + cysteine, %	0.68
Methionine, %	0.43
Threonine, %	0.76
Tryptophan, %	0.24
Isoleucine, %	0.65
Leucine, %	1.22
Valine, %	0.75
Arginine, %	0.95
Histidine, %	0.39
Phenylalanine, %	0.77
SID LYS [1,2]	1.06
SID MET [1,2]	0.39
SID M+C [1,2]	0.60
SID THR [1,2]	0.65
SID TRP [1,2]	0.20
SID ILE [1,2]	0.55
SID LEU [1,2]	1.02
SID VAL [1,2]	0.62

3. RESULTS AND DISCUSSION

It has been shown that amino acids may affect feed intake. This increased feed intake will lead to an extra amount of amino acids available for protein deposition, even though they are not perfectly balanced. Therefore, the hypothesis was tested that adding either valine or tryptophan to a diet short in both valine and tryptophan will lead to increased feed intake and improved performance.

No interactions between dietary valine and tryptophan content could be observed ($P > 0.05$ for all parameters, Table 3). However, while valine had a strong effect on feed intake and consequently daily gain ($P < 0.001$ for all parameters), an effect of tryptophan content on daily feed intake or daily gain was not observed ($P > 0.05$).

Table 3. Influence of valine and tryptophan concentration on performances of the piglets [1,2].

	Low Valine		High Valine			P		
	Low Tryptophan	High Tryptophan	Low Tryptophan	High Tryptophan	SEM	SID VAL	SID TRP	SID VAL x SID TRP
Bodyweight, kg								
5 weeks of age	9.45	9.48	9.40	9.44	0.04	0.514	0.616	0.993
7 weeks of age	14.09	14.41	15.29	15.35	0.12	<0.001	0.185	0.374
9 weeks of age	20.88	21.47	23.51	23.58	0.21	<0.001	0.157	0.267
Daily feed intake, g								
5–7 weeks	565	571	648	636	9	<0.001	0.776	0.430
7–9 weeks	897	899	1004	1008	13	<0.001	0.871	0.952
5–9 weeks	731	735	826	822	11	<0.001	0.871	0.952
Daily weight gain, g								
5–7 weeks	332	352	421	423	60	<0.001	0.194	0.287
7–9 weeks	485	504	587	588	8	<0.001	0.326	0.382
5–9 weeks	408	428	504	505	7	<0.001	0.136	0.215
Feed efficiency, g/g								
5–7 weeks	0.587	0.619	0.652	0.665	0.007	<0.001	0.034	0.377
7–9 weeks	0.544	0.563	0.585	0.584	0.005	0.001	0.313	0.266
5–9 weeks	0.560	0.585	0.610	0.616	0.004	<0.001	0.031	0.159

1 Data are means of 16 pens per treatment.

2 SID = standardized ileal digestible, VAL= valine, TRP = tryptopan.

The pigs consuming the low valine diets showed a lower feed intake and daily gain. While this has also been observed in some other studies [9,10,20,21], no clear reason has been given. Recent research [22] on mice showed that somatostatin may play a role in anorexia induced by a valine-deficient diet. Central administration of ghrelin and neuropeptide Y restored feed intake in rats on a valine-deficient diet [23]. Systemic administration of ghrelin did not affect the feed intake in these rats [23].

For tryptophan, the stimulating effect on feed intake has been reported [9,10,11]. Tryptophan competes with large neutral amino acids (LNAA) such as leucine, isoleucine and valine for its passage through the blood-brain barrier by sharing a common transport system [24]. Therefore, the ratio of tryptophan to LNAA affects the feed intake [9]. Henry *et al.* [9] assumed that the negative effect of LNAA on feed intake could be explained by a decrease in the tryptophan and consequently serotonin concentration in the brain. However, it cannot be excluded that the assumed positive effect of tryptophan on feed intake is mediated by an inhibition of the L-Leucine uptake to the brain and consequently diminished mTOR-signaling. L-Leucine depresses feed intake through a stimulation of mTOR-signaling [25]. This might explain why an effect of dietary tryptophan on feed intake was not obvious in the present experiment, despite being clear in the experiments of Jansman *et al.* [11] and Henry *et al.* [9]. In their experiments, the leucine: lysine ratio was above 1.5, while in the present experiment, this was around 1. In addition to this, in the study of Henry *et al.* [9], it is possible that the effect they observed was solely caused by the tryptophan:leucine ratio instead of the tryptophan:LNAA concentration, as the leucine concentration varied together with the LNAA concentration.

Dietary tryptophan content did affect the feed efficiency between 5 and 7 weeks of age and between 5 and 9 weeks of age ($P < 0.05$). Also, between 5 and 7 weeks as well as between 7 and 9 weeks of age, the feed efficiency was worse in the groups receiving the low valine diets ($P \leq 0.001$ for all parameters). This shows that the LL diet was indeed limiting for both tryptophan and valine.

Despite the absence of an effect of tryptophan on feed intake, it was possible to confirm the hypothesis that the addition of either valine or tryptophan to a diet limiting in both valine and tryptophan may improve performance, even if the other amino acid was limiting growth. Indeed, feed efficiency was improved either by adding tryptophan or by adding valine to the LL diet.

Feed intake may have a role in this. When considering the ideal protein concept, one can assume that the first limiting amino acid determines the amount of protein deposition when sufficient energy is available. The excess amino acids will be catabolized. Therefore, one would expect that only one diet (either HL or LH) would be able to elicit positive growth performance, until the level that the second amino acid becomes limiting. With addition of this second limiting amino acid (the HH diet), further improvement can be expected. However, as valine stimulated the uptake of an extra amount of feed, this provided extra amino acids (including tryptophan) that could be retained as protein tissue. Therefore, it is logical to assume that the energy needs for maintenance were relatively smaller per kg of body weight gain and the feed efficiency improved in the HL *versus* the LL diet, although tryptophan was still limiting maximal protein deposition.

The effect of valine addition to the diet on performance was more pronounced than the effect of tryptophan addition. The levels for the HH feed were based on optima determined during ILVO research concerning dietary valine concentration (unpublished results) and recent literature on dietary tryptophan content [11]. However, the analyzed valine concentration of the basal diet was lower than originally intended, and therefore, the high valine diets might still have been slightly limiting in valine (SID valine: SID lysine was 0.67 instead of 0.70 in the HH diet). This may explain why the 16% increase in valine concentration had a larger effect than the 16% increase in tryptophan concentration. A clear effect of tryptophan addition on feed efficiency was seen in the first 2 weeks of the experiment, but not in the following 2 weeks. It is possible that the pigs responded to the lower tryptophan content through an adaptation mechanism. This may be an improved absorption or a lower catabolism rate of the amino acids. It has been shown that pigs use amino acids more efficiently during a period of amino acid deficiency [26].

The results of this experiment have a direct practical consequence. Even if other amino acids are formulated to yield suboptimal performance (e.g., because of the limited benefit in comparison to the cost), it may be beneficial to provide the diet with an optimal amount of valine to improve feed intake and performance.

4. CONCLUSIONS

It can be concluded that either supplementing valine or tryptophan to a diet that was limiting in both amino acids improved performance of newly weaned piglets, without an interaction between the dietary supply of valine and tryptophan. Valine supply had a greater effect on performance results than tryptophan supply; providing the diet with an optimal amount of valine even if other amino acids are at suboptimal dietary levels may improve performance.

ACKNOWLEDGMENTS

Thanks to Myriam Audenaert, Bart de Bock, Kristof Dierkens, Roland Limpens, Jan Staels, and Hans Uitterhaeghen for all their practical support. Thanks to Miriam Levenson for language correction.

REFERENCES

1. Noblet, J.; Van Milgen, J. Energy Value of Pig Feeds: Effect of Pig Body Weight and Energy Evaluation System. *J. Anim. Sci.* **2004**, *82*, E229–E238.
2. Henry, Y. Dietary Factors Involved in Feed Intake Regulation in Growing Pigs: A Review. *Livest. Prod. Sci.* **1985**, *12*, 339–354.
3. Bikker, P.; Karabinas, V.; Verstegen, M.W.A.; Campbell, R.G. Protein and Lipid Accretion in Body Components of Growing Gilts (20 to 45 Kilograms) as Affected by Energy-Intake. *J. Anim. Sci.* **1995**, *73*, 2355–2363.
4. Van Milgen, J.; Quiniou, N.; Noblet, J. Modelling the Relation between Energy Intake and Protein and Lipid Deposition in Growing Pigs. *Anim. Sci.* **2000**, *71*, 119–130.
5. Boisen, S.; Hvelplund, T.; Weisbjerg, M.R. Ideal Amino Acid Profiles as a Basis for Feed Protein Evaluation. *Livest. Prod. Sci.* **2000**, *64*, 239–251.
6. Fuller, M.F.; Mcwilliam, R.; Wang, T.C.; Giles, L.R. The Optimum Dietary Amino Acid Pattern for Growing Pigs. 2. Requirements for Maintenance and for Tissue Protein Accretion. *Br. J. Nutr.* **1989**, *62*, 255–267.
7. Millet, S.; Gielkens, K.; De Brabander, D.; Janssens, G.P.J. Considerations on the Performance of Immunocastrated Male Pigs. *Animal* **2011**, *5*, 1119–1123.
8. National Research Council. *Nutrient Requirements of Swine*, 11th ed.; National Academy Press: Washington, DC, USA, 1998.
9. Henry, Y.; Seve, B.; Colleaux, Y.; Ganier, P.; Saligaut, C.; Jego, P. Interactive Effects of Dietary Levels of Tryptophan and Protein on Voluntary Feed Intake and Growth Performance in Pigs, in Relation to Plasma Free Amino Acids and Hypothalamic Serotonin. *J. Anim. Sci.* **1992**, *70*, 1873–1887.
10. Eder, K.; Peganova, S.; Kluge, H. Studies on the Tryptophan Requirement of Piglets. *Archiv fur Tierernaehrung* **2001**, *55*, 281–297.
11. Jansman, A.J.M.; van Diepen, J.T.M.; Melchior, D. The Effect of Diet Composition on Tryptophan Requirement of Young Piglets. *J. Anim. Sci.* **2010**, *88*, 1017–1027.
12. Theil, P.K.; Fernández, J.A.; Danielsen, V. Valine Requirement for Maximal Growth Rate in Weaned Pigs. *Livest. Prod. Sci.* **2004**, *88*, 99–106.
13. Mavromichalis, I.; Kerr, B.J.; Parr, T.M.; Albin, D.M.; Gabert, V.M.; Baker, D.H. Valine Requirement of Nursery Pigs. *J. Anim. Sci.* **2001**, *79*, 1223–1229.
14. Warnants, N.; Van Oeckel, M.J.; De Paepe, M. Response of Growing Pigs to Different Levels of Ileal Standardised Digestible Lysine Using Diets Balanced in Threonine, Methionine and Tryptophan. *Livest. Prod. Sci.* **2003**, *82*, 201–209.
15. EEC Commission. Establishing Community Methods of Analysis for the Official Control of Feeding Stuffs (EEC Commission Directive 71/250/EEC). Official J. Eur. Communit., 1971.
16. EEC Commission. Establishing Community Methods of Analysis for the Official Control of Feeding Stuffs (Directive 171/393/EEC). Official J. Eur. Communit., 1971.
17. EEC Commission. Amending Annex I to Fourth Directive 73/46/EEC Establishing Community Methods of Analysis for the Official Control of Feeding Stuffs (Directive 92/89/EEC). *Official J. Eur. Communit.* **1992**, *L344*, 36–37.
18. EEC Commission. Amending Annex I to the Third Directive 72/199/EEC Establishing Community Methods of Analysis for the Official Control of Feedingstuffs (Directive 93/28/EEC). *Official J. Eur. Communit.* **1993**, *L179*, 8–10.

19. CVB Table on Feed Ingredients; Information about Chemical Composition, Digestibility and Feeding Value, Centraal Veevoederbureau: Lelystad, The Netherlands, 2007.
20. Lordelo, M.M.; Gaspar, A.M.; Le Bellego, L.; Freire, J.P.B. Isoleucine and Valine Supplementation of a Low-Protein Corn-Wheat-Soybean Meal-Based Diet for Piglets: Growth Performance and Nitrogen Balance. *J. Anim. Sci.* **2008**, *86*, 2936–2941.
21. Wiltafsky, M.K.; Pfaffl, M.W.; Roth, F.X. The Effects of Branched-Chain Amino Acid Interactions on Growth Performance, Blood Metabolites, Enzyme Kinetics and Transcriptomics in Weaned Pigs. *Br. J. Nutr.* **2010**, *103*, 964–976.
22. Nakahara, K.; Takata, S.; Ishii, A.; Nagao, K.; Bannai, M.; Takahashi, M.; Murakami, N. Somatostatin is Involved in Anorexia in Mice Fed a Valine-Deficient Diet. *Amino Acids* **2011**.
23. Goto, S.; Nagao, K.; Bannai, M.; Takahashi, M.; Nakahara, K.; Kangawa, K.; Muraikami, N. Anorexia in Rats Caused by A Valine-Deficient Diet Is Not Ameliorated by Systemic Ghrelin Treatment. *Neuroscience* **2010**, *166*, 333–340.
24. Fernstrom, J.D.; Wurtman, R.J. Brain Serotonin Content: Physiological Regulation by Plasma Neutral Amino Acids. *Science* **1972**, *178*, 414–416.
25. Cota, D.; Proulx, K.; Smith, K.A.B.; Kozma, S.C.; Thomas, G.; Woods, S.C.; Seeley, R.J. Hypothalamic MTOR Signaling Regulates Food Intake. *Science* **2006**, *312*, 927–930.
26. Millet, S.; Langendries, K.; Aluwé, M.; De Brabander, D.L. Effect of Amino Acid Level in the Pig Diet during Growing and Early Finishing on Growth Response during the Late Finishing Phase of Lean Meat Type Gilts. *J. Sci. Food Agric.* **2011**, *91*, 1254–1258.

CHAPTER 13

A Potential Role for Pro-Inflammatory Cytokines in the Development of Insulin Resistance in Horses

Jessica K. Suagee [1,*], Benjamin A. Corl [1] and Raymond J. Geor [2]

[1]*Department of Dairy Science, Virginia Tech, Blacksburg, VA 24071, USA*

[2]*Department of Large Animal Clinical Sciences, Michigan State University, East Lansing, MI 48824, USA*

SIMPLE SUMMARY

Insulin resistance is a metabolic condition involving reduced sensitivity of insulin-sensitive tissues to insulin-induced glucose disposal, including adipose tissue, skeletal muscle, and liver. Insulin resistance occurs in overweight and obese horses, and may increase risk for the development of laminitis. The development of insulin resistance is thought to occur in response to increased production of pro-inflammatory cytokines by adipose tissue in obesity, that then have an inhibitory effect on insulin signaling pathways in multiple tissues. This article reviews current knowledge of the involvement of pro-inflammatory cytokines in the development of insulin resistance in horses and uses data from other species to provide context.

ABSTRACT

Understanding the mechanisms involved in the development of insulin resistance in horses should enable development of effective treatment and prevention strategies. Current knowledge of these mechanisms is based upon research in obese humans and rodents, in which there is evidence that the increased production of pro-inflammatory cytokines by adipose tissue negatively influences insulin signaling in insulin-responsive tissues. In horses, plasma concentrations of the cytokine, tumor necrosis factor-α, have been positively correlated with body fatness and insulin resistance, leading to the hypothesis that inflammation may reduce insuli

n sensitivity in horses. However, little evidence has documented a tissue site of production and a direct link between inflammation and induction of insulin resistance has not been established. Several mechanisms are reviewed in this article, including the potential for macrophage infiltration, hyperinsulinemia, hypoxia, and lipopolysaccharide to increase pro-inflammatory cytokine production by adipose tissue of obese horses. Clearly defining the role of cytokines in reduced insulin sensitivity of horses will be a very important step in determining how obesity and insulin resistance are related.

Keywords: adipose tissue; horses; inflammation; obesity; pro-inflammatory cytokines; skeletal muscle

1. INTRODUCTION

There is information supporting a positive relationship between obesity and the risk of developing laminitis in horses and ponies [1,2,3]. The mechanism explaining this relationship was suggested as early as the 1980's to be insulin resistance when Jeffcott and Field described reduced tolerance to oral glucose administration in obese, previously-laminitic ponies compared to lean, healthy ponies and horses [1]. Current evidence for a relationship between obesity, laminitis, and insulin resistance has been reviewed elsewhere [4,5]. The purpose of this review is to highlight research describing the hypothesis that increased adipose tissue production of pro-inflammatory cytokines during equine obesity leads to the development of insulin resistance. It is relevant to examine the relationship between inflammation and obesity in horses because in other species cytokines directly impair insulin signaling in tissues that have potential to alter whole body glucose metabolism, such as adipose, skeletal muscle, and liver tissue. These data suggest that increased pro-inflammatory cytokines directly contribute to the development of insulin resistance, and this raises two very important questions in equine science: are cytokines involved in the pathology of insulin resistance, and which tissues produce potentially insulin-resistance-inducing cytokines?

In human medicine, increased circulating concentrations of inflammatory proteins have demonstrated utility for the prediction of future metabolic disease. For example, elevated concentrations of C-reactive protein (CRP), an acute-phase inflammatory protein, indicate increased risk for obesity-associated disorders [6]. This finding has potential clinical application to horses if researchers can identify an inflammatory protein similar in its ability to predict the risk of metabolic disease. Therefore, our overall goal was to integrate available knowledge about inflammatory proteins in horses into the larger body of mechanistic information that has been described in other species.

2. PRODUCTION OF PRO-INFLAMMATORY CYTOKINES

Chronic low-grade inflammation is a primary component of obesity-associated-metabolic conditions, such as insulin resistance and diabetes. In humans, low-grade inflammation is caused by increased circulating concentrations of pro-inflammatory cytokines (interleukin [IL]-1β, IL-6, tumor necrosis factor-α [TNF]) and acute phase proteins (serum amyloid A [SAA] and CRP) [7-14]. Because laminitis is an obesity-associated disease in horses, new information from human obesity research led to investigations to determine whether increased systemic inflammation was present in obese horses [1]. In 2007, Vick *et al.* described a positive correlation between obesity and circulating concentrations of TNF in Thoroughbred mares [15]. A comparable, later investigation in pony mares demonstrated that obese, insulin-resistant pony mares had higher serum TNF concentrations than those that were leaner and more insulin-sensitive [3]. These two investigations have provided supporting epidemiological evidence that obesity could be associated with increased inflammation in horses. However, as suggested by two recent studies, it is possible that a relationship between obesity and inflammation does not exist in horses. Suagee *et al.* reported that serum TNF concentrations were not altered in gelded horses while they were being fed to increase in adiposity over several months [16]. In that study, horses did not reach true obesity, and neither were they evaluated after being maintained in the overweight state. In a second study, serum TNF concentrations were found to be lower in obese, hyperinsulinemic horses than normal, lean controls [17]. Horses used for this investigation were aged similar to other studies; however, the investigation did not restrict gender or control diet. Therefore, we suggest that future investigations evaluating a potential influence of obesity on inflammation control for variables such as age, gender, and diet.

2.1. Role of Adipose Tissue

If obesity is correlated to increased concentrations of pro-inflammatory mediators, what is the source of these proteins? In humans, both adipose tissue and skeletal muscle produce pro-inflammatory cytokines. While the majority of research has focused on the contribution of adipose tissues to circulating cytokine concentrations, it is possible that production in skeletal muscle could also be very important to systemic glucose metabolism.

Adipose tissue is present throughout the body in different storage locations, or depots, including intra-abdominal (mesenteric and omental) and subcutaneous, that require consideration. Intriguingly, recent research indicates that abdominal depots may be more “inflammatory” in humans than subcutaneous depots [18]. This potential for ‘depot-specific’ inflammation may be relevant to equine scientists because of the association between excess subcutaneous adipose deposited along the crest of the neck (regional adiposity) and elevated laminitis risk [2]. Furthermore, a recent investigation of adipose-depot-associated inflammation in overweight, but not obese, horses revealed that IL-1β and IL-6 mRNA are higher in subcutaneous adipose deposited along the crest of the neck than other depots including omental, retroperitoneal, mesenteric, and subcutaneous adipose from the tailhead [19]. These are important findings to keep in mind when reviewing equine data because subcutaneous adipose is the primary adipose depot investigated due to ease of sampling.

Within adipose tissue, several cell types are capable of producing pro-inflammatory cytokines, including stromal-vascular cells (preadipocytes, fibroblasts, and non-differentiated mesenchymal cells), differentiated adipocytes, and infiltrated macrophages. When cell types are separated, TNF mRNA abundance is lower in differentiated adipocytes than the non-adipocyte fraction of adipose tissue from humans [20] and mice [21]. However, when differentiated mouse adipocytes were co-cultured with macrophages, TNF and IL-6 production by adipocytes was increased [22]. These results support the ability of differentiated adipocytes to produce pro-inflammatory cytokines, and further, that the up-regulation seen in obesity may require a signal from another cell type. In horses, the basal expression of cytokines from different cell types has not been thoroughly investigated; however, lipopolysaccharide (LPS) stimulation, which directly activates transcription of pro-inflammatory cytokines through the transcription factor nuclear factor-κB, has been used. In rump fat from two horses of unreported body condition, both adipocytes and pre-adipocytes had cytokine mRNA responses to LPS stimulation, although a greater percentage of pre-adipocytes stained for TNF than mature adipocytes [23]. These results could indicate that inflammatory responses of adipose tissue cell types are similar between horses and other species. In order to assimilate this new information with existing knowledge, further investigations and future reports need to include information on body condition and adipose depot. In conclusion, it appears that many cell types are capable of producing cytokines, and may do so constitutively at a low level, even in lean, healthy individuals. This raises an important question: what specific changes occur in the adipose tissue of obese animals that instigates low-grade, systemic, inflammation?

2.2. Role of Skeletal Muscle

As skeletal muscle is the major site of glucose disposal, myocyte production of cytokines with insulin desensitizing action could also contribute to the pathology of equine insulin resistance [24]. In humans, obesity increases pro-inflammatory proteins in skeletal muscle, similar to adipose tissue [25]. Skeletal muscle TNF and IL-6 mRNA are greater in diabetic humans, and TNF protein secretion from muscle is greater in cultured cells derived from diabetic humans [26,27]. It is possible that mechanisms for increased myocyte pro-inflammatory cytokine expression are similar to those mechanisms involved in adipose tissue. This includes evidence for mechanisms of increased macrophage infiltration, and hyperinsulinemia-induced cytokine production [28-32].

In horses, limited data on skeletal muscle inflammation are available. Following lipopolysaccharide (LPS) infusion, neither IL-1β nor TNF mRNA were increased, and IL-6 mRNA was undetectable 24 h following LPS administration in skeletal muscle [33]. As earlier time points have not been investigated, it is difficult to speculate about the role of skeletal muscle in the production of cytokines, either acting locally, or contributing to systemic concentrations. In opposition to the LPS study findings, a 6 h insulin infusion that induced acute, supraphysiological hyperinsulinemia also increased skeletal muscle IL-1β about 1.75 fold above baseline measurements [34]. Thus, neither the local production of cytokines by skeletal muscle nor the effects of cytokines on skeletal muscle have been characterized.

2.3. Mechanisms of Increased Pro-Inflammatory Cytokine Production

The majority of research into potential mechanisms stimulating increased circulating pro-inflammatory cytokine concentrations has been examined in adipose tissue. For this reason, the discussion below focuses on these mechanisms, including macrophage infiltration into adipose tissue, hyperinsulinemia, hypoxia of adipose tissue, and increased LPS. As previously mentioned, however, similar mechanisms may also occur in skeletal muscle.

2.3.1. Macrophage Infiltration into Tissues

Adipose tissue macrophages are a potentially important source of cytokines [21,35], and also have the ability to increase cytokine production by other cell types. Although resident macrophages are present in all tissues as part of the innate immune system, an increased number of macrophages residing in the adipose tissue of obese humans is observed, and preliminary data suggest a similar relationship in horses [21-37]. Data derived from human studies indicate that adipocytes produce chemokines that signal monocyte and macrophage infiltration into adipose tissue. One of these chemokines, with limited data in horses, is monocyte chemoattractant protein-1 (MCP-1), which is responsible for attracting monocytes out of the bloodstream and into adipose tissue [38,39]. Serum concentrations of MCP-1 are significantly correlated to body mass index, waist circumference, and other inflammatory markers in humans, mRNA is increased in adipose tissue of obese mice, and an *intravenous* infusion of MCP-1 increases macrophage infiltration into mouse adipose tissue [39-44]. Thus, evidence indicates that MCP-1 is involved in recruiting macrophages to adipose tissue in obese humans and rodents; however, in non-obese, over-conditioned horses, MCP-1 mRNA is not different across adipose depots or altered by insulin sensitivity status [19]. This could indicate species-specific differences, or that an increase in MCP-1 is not seen until onset of obesity or following a prolonged period of obesity. Of further interest is the knowledge that MCP-1 expression is stimulated by TNF [45-47]; however, 24 h following the LPS infusion that increased plasma TNF, MCP-1 mRNA is not increased in subcutaneous adipose tissue of horses [33]. Adipose tissue MCP-1 expression might be increased earlier than 24 h, as serum TNF concentrations peak 1 h following LPS and return to baseline by 4 h. Alternatively, a localized inflammatory state in adipose tissue that is found with obesity was not induced by LPS infusion. The induction of obesity-associated adipose tissue inflammation in horses, akin to that observed in other species, might require more complex conditions than systemic inflammation affords.

In addition to assessing the potential for increased inflammation of adipose tissue macrophages, research has assessed inflammatory gene expression of circulating white blood cells. Intriguingly, Vick *et al.* reported a positive correlation between adiposity and IL-1β and TNF white blood cell mRNA; however, IL-6 mRNA was not related to adiposity [15]. This could indicate that IL-6 mRNA is not increased by obesity, or that circulating white blood cells in general, and more specifically, white blood cell IL-6 mRNA, are not indicators of adipose tissue inflammation. The capacity to assess adipose tissue inflammation through a blood sample would improve animal welfare by decreasing the need for adipose tissue biopsies; however, the validity of this approach requires further investigation.

2.3.2. Hyperinsulinemia

Anecdotal observation suggests that high glycemic diets are commonly fed to horses and scientific evidence has demonstrated that these diets induce a postprandial increase in plasma insulin concentrations [48]. In humans, evidence exists for the ability of insulin to promote inflammation during acute, experimentally-induced hyperinsulinemia, independent of obesity, increasing inflammatory cytokine concentrations in subcutaneous adipose tissue interstitial fluid as well as circulating IL-6 concentrations [49,50]. These findings that suggest insulin somehow activates the production of cytokines are relevant to equine nutrition because of the demonstrated relationship between high glycemic diets and lower insulin sensitivity in obese horses [51]. Given this association, and the potential relationship between insulin, inflammation, and insulin sensitivity, Suagee et al. demonstrated that geldings consuming a high glycemic diet have increased serum TNF concentrations compared to geldings consuming a low glycemic diet [16]. As these horses were not obese, the higher TNF concentrations suggest high glycemic diets might promote inflammation independent of obesity. Using a more direct approach, hyperinsulinemia, achieved through a six-h insulin infusion, increased circulating TNF and IL-6 concentrations in non-obese mares [34]. However, even though white blood cell IL-6 mRNA was elevated, neck subcutaneous adipose tissue mRNA levels were not altered. Investigation of multiple adipose tissue depot responses to insulin is required. Other factors that could be involved in the insulin-inflammation axis include level of adiposity or degree of macrophage infiltration. Adipose tissue used to investigate insulin-induced inflammation was collected from obese humans, and this could account for some of the observed species differences.

These *in vivo* data provide a good starting point for asking questions such as: which tissues are involved in insulin-induced inflammation in horses, and within those tissues, which cell types. As stated previously, acute hyperinsulinemia did not seem to create a pro-inflammatory state in neck subcutaneous adipose of horses; however, this effect is apparent in differentiated mouse and human adipocytes [52,53]. In these cells, the addition of insulin to culture media increased mRNA abundance and secretion of IL-6 and MCP-1. An exact mechanism describing the ability of insulin to promote inflammation is not yet described for all cytokines. However, the IL-6 promoter contains a response element for cAMP-response element binding (CREB) protein [54]. Insulin activates CREB via cGMP in human adipocytes, thereby increasing IL-6 transcript abundance [55]. Insulin also increased IL-6 protein synthesis and secretion through a MAPK dependent pathway [55]. The functionality of these pathways has not been investigated in horses.

2.3.3. Hypoxia of Adipose Tissue

The possibility that adipose tissue hypoxia leads to increased inflammation involves the theory that expansion of adipose tissue mass may not be paralleled by increased vasculature. Adipose tissue hypoxia was first recognized in obese surgical patients, who had a greater risk of surgical site infections in association with low tissue oxygen tension [56]. It is possible that the greater metabolic activity and oxygen demand of infiltrating immune cells plays a role in adipocyte hypoxia, or that hypertrophied adipocytes have a larger diameter than the O_2 diffusion distance, a fact that would increase the propensity that these cells exist under hypoxic conditions [57,58,59]. The physiological evidence for adipose tissue hypoxia includes impaired postprandial blood flow to adipose of non-obese insulin-resistant humans and hypoxia in adipose tissue of genetically obese (ob/ob) and diet-induced obese mice confirmed by use of interstitial O_2 sensors [60,61,62]. *In vitro* hypoxia increases pro-inflammatory cytokine responses in human stromal-vascular cells as well as differentiated mouse adipocytes, and it is possible that a similar phenomenon could occur in equine adipose tissue [61,63].

The cellular response to hypoxia is highly conserved and serves to adapt cells to preservation of critical metabolic functions. These adaptations are primarily facilitated through activation and maintenance of the transcription factor, hypoxia-inducible factor-1 (HIF-1) [64]. Normal oxygen levels stabilize the enzymes that degrade the α-subunit of HIF-1, thus, only in low oxygen environments is HIF-1α capable of binding HIF-1β (constitutively expressed and stable) to form HIF-1 [65,66]. Stable HIF-1 binds to hypoxia response elements in the promoter region of several genes, and increases mRNA abundance and secretion of IL-1β, IL-6, TNF, and other pro-inflammatory mediators in adipose tissue [61,62,67-69]. Further, abundance of HIF-1α mRNA is increased in adipose tissue from obese humans [70]. Hypoxia also activates NFκB in adipocytes, which then promotes transcription of pro-inflammatory cytokines [61,71]. The activation of NFκB in response to hypoxia is possibly due, in part, to activation of the IκB kinase complex, similar to the mechanism of LPS stimulation, but could involve other mechanisms such as tyrosine phosphorylation of IκBα [72,73]. While it is likely that HIF-1 activation is a stronger response to hypoxia than NFκB, it is possible that NFκB may play a direct role in HIF-1 expression as HIF-1α contains an NFκB binding site in its promoter [73-75]. Thus, several feedback and feedforward signals appear to exist to promote and control cellular responses to hypoxia, but how these signals are moderated or influenced by obesity remains only partly explained. Hypoxia-induced inflammation is a viable theory that, to our knowledge, is uninvestigated in relation to equine obesity.

2.3.4. Increased Uptake of Gastrointestinal-Tract-Derived Lipopolysaccharide

Yet another potential theory explaining the relationship between obesity and inflammation involves a potential capacity for increased production and systemic absorption of hindgut microbial products, including the endotoxin, LPS. The gastrointestinal tract is designed to resist the crossing of bacterial products into the bloodstream. However, it is possible that high starch and sugar diets reduce the structural integrity of the intestinal epithelium, as a percentage of starch escapes the small intestine undigested and is fermented in the cecum and colon. In the cecum and colon, starch is fermented to lactic acid, which reduces cecal, colonic, and fecal pH values [76,77]. This situation is somewhat similar to that observed in ruminants with high starch diet-induced rumen acidosis. In these animals, epithelial tight junction integrity is compromised and increased plasma LPS concentrations are observed [78,79]. In horses, nasogastric administration of starch at values 7 to 10-fold greater than the recommended dietary maximums is a model used to induce laminitis [80-84]. Although this is an extreme scenario, the cecal pH values of horses subjected to carbohydrate overload were decreased from 7.0 to 4.7, and there was degeneration of the tight junctions between epithelial cells of the cecum [85]. In a similar experiment using elevated oligofructose administration, increased plasma LPS concentrations were reported [86]. It does not appear that LPS concentrations have been reported in horses consuming high starch and sugar concentrates at a level that would not directly induce laminitis. However, if consumption of these diets increases plasma LPS concentrations, then it might be possible that this is a mechanism relating these diets to inflammation and insulin resistance.

The reason LPS production and uptake from the gastrointestinal tract is interesting is because LPS induces transient insulin resistance in horses

24 h after intravenous infusion [23,33,87], and also increases inflammatory protein production. In LPS-treated horses, white blood cell IL-1β, IL-6, and TNF mRNA and plasma TNF concentrations are increased within 1 h of infusion, while plasma SAA concentrations are increased at 3 h and remain elevated for at least 24 h [23,33,88]. Of further interest is the finding that IL-6 mRNA is elevated in adipose tissue at 24 h post infusion: the time point corresponding to insulin resistance. If increased uptake of LPS from the gastrointestinal tract occurs, then its potential influence on insulin resistance should be investigated. However, little equine research has focused on either the uptake or function of LPS outside of extreme models. For instance, while a bolus intravenous LPS infusion induces profound, transient insulin resistance, it is unknown how chronic, low-level plasma LPS concentrations might influence insulin resistance in obese horses.

3. EFFECTS OF PRO-INFLAMMATORY CYTOKINES ON EQUINE METABOLISM

During sickness, increased pro-inflammatory cytokine production promotes physiological responses that facilitate the re-prioritization of energy expenditures to host defenses [89,90]. These responses include fever [91], fatigue, and loss of appetite [89,90]. As decreased energy intake occurs simultaneously to the increased energy expenditure of elevated body temperatures [92], fuel re-partitioning must occur. Glucose utilization by the immune system is also increased [93,94], requiring that non-immune cells reduce their use of available glucose. Non-immune tissues increase energy usage from non-glucose sources, and to support this, plasma-free fatty acids are increased during infection [95]. The altered tissue metabolism of sickness that increases energy availability to immune cells is due to the actions of pro-inflammatory cytokines. The response to illness is transient, and normal metabolism is restored once the animal recovers from infection. The chronic low grade systemic inflammation resulting from obesity may, however, cause more long-term alterations in metabolism, including both dyslipidemia and diabetes [14,96-104].

Pro-inflammatory cytokines impair glucose disposal by insulin sensitive tissues, thereby mediating insulin resistance. In adipose tissue and adipocytes, TNF, IL-6 and IL-1β have all been shown to have negative effects on insulin signaling. These include the *in vitro* effect of TNF to decrease mRNA and protein abundance of the insulin receptor, insulin receptor substrate (IRS)-1, and GLUT4, all of which result in reduced GLUT4 plasma membrane translocation [22-107]. Further, TNF increases serine phosphorylation of IRS-1, leading to proteosomal degradation and decreased ability of IRS-1 to promote insulin signaling [107,108]. In addition to the actions of TNF, IL-1β also reduces insulin-stimulated glucose uptake in adipocytes by decreasing GLUT4 protein abundance, IRS-1 protein abundance, insulin-stimulated phosphorylation of IRS-1, and insulin-induced membrane translocation [109,110]. In mice, IL-6 inhibits insulin-stimulated tyrosine phosphorylation of IRS-1 and IRS-2 [111]. In adipocytes, chronic culture with IL-6 decreases transcription of IRS-1 and GLUT4, while *in vivo*, plasma IL-6 concentration is inversely proportional to insulin-stimulated glucose transport into adipose tissue [112-114]. These data demonstrate the profoundly negative consequences of adipose tissue pro-inflammatory cytokine production on insulin signaling in adipose tissue. As inflammatory proteins also act in an endocrine manner in addition to their paracrine and autocrine effects [115], cytokines produced by and released from adipose tissue could act systemically on tissues such as skeletal muscle.

Skeletal muscle insulin responsiveness is the largest component of peripheral insulin sensitivity, and any impairment of glucose utilization in skeletal muscle will have significant impacts on whole body glucose metabolism [24]. The pro-inflammatory cytokines, TNF, IL-6 and IL-1β have the ability to reduce insulin-stimulated glucose disposal into skeletal muscle, although less attention has been paid to this than to the metabolic effects of cytokines on adipose tissue. *In vivo*, TNF reduces glucose uptake into human skeletal muscle [116]. Use of rat-derived muscle cell lines showed this reduced glucose uptake to be due to inhibition of insulin-stimulated activation of the insulin signaling cascade, partly due to altered phosphorylation of several signaling intermediates [117-119]. In a mouse muscle cell line, TNF, but not IL-1β, reduced both glucose uptake and Protein Kinase B (PKB) levels [120]. Similar to TNF, IL-6 treatment also reduces skeletal muscle glucose utilization *in vivo* in mice, and treating a rat muscle cell line with IL-6 decreased insulin stimulated Akt phosphorylation [121,122].

In horses, the role of cytokines to directly and specifically alter metabolism in either adipose or muscle tissue have not yet been investigated. This information could be determined by culturing adipose tissue and muscle with equine cytokines, or through intravenous infusion of equine cytokines. This information could lend support for further research into strategies to prevent and treat insulin resistance by enabling scientists to focus on specific inflammatory markers.

4. CONCLUSIONS

Insulin resistance in horses is associated with increased risk of laminitis. Thus, research efforts to identify causative factors of insulin resistance are important. In other species, increased production of pro-inflammatory cytokines during obesity is known to directly influence decreased insulin sensitivity, thus evidence of this relationship in horses has been evaluated. Although a correlation between plasma concentrations of TNF and obesity are described in horses, the tissue primarily responsible for production of this inflammatory protein has not been identified. Potential tissue sites of increased production of cytokines include several adipose tissue depots, and within adipose tissue there may be cell-specific roles in inflammation. In horses, subcutaneous adipose tissue from the neck has greater basal inflammation than other adipose depots, and stromal-vascular cells have greater response to inflammatory stimulation than mature adipocytes. Potential mechanisms responsible for stimulating cytokine production in adipose tissue include macrophage infiltration, hyperinsulinemia, hypoxia, and LPS. Of these, hyperinsulinemia and LPS stimulation have received the greatest attention. Future work should clearly determine whether inflammation is increased in adipose tissue of obese horses, and further, the role of diet to influence inflammation. This could involve direct investigation of insulin's ability to stimulate inflammatory protein production from different adipose tissue depots, or the ability of diet to influence LPS concentrations. Increasing knowledge of these mechanisms will hopefully enable strategies for preventing and treating insulin resistance. Additionally, with more information we can focus research efforts on a specific adipose depot, which could potentially improve animal welfare, as fewer tissue samples will be required, or provide an incentive to produce a clonal adipocyte cell line.

Improving our understanding of the potential role of inflammation in the relationship between obesity and insulin resistance will hopefully allow for the development of strategies for preventing laminitis in obese horses. These strategies could include diet and pharmacologic agents, in addition to defining a specific marker of increased risk. Specifically, the individual roles of pro-inflammatory cytokines to modulate insulin sensitivity in skeletal muscle and adipose tissue should be identified to determine whether inflammation is indeed active in metabolic diseases similar to that observed in humans. If inflammation is influencing insulin sensitivity in horses, this will provide an incentive for increased resources to be allocated to this research, but if horses are unlike humans in response to cytokines, then research into other mechanisms relating obesity to insulin resistance should be explored.

REFERENCES

1. Jeffcott, L.B.; Field, J.R.; McLean, J.G.; O'Dea, K. Glucose tolerance and insulin sensitivity in ponies and Standardbred horses. *Equine Vet. J.* **1986**, *18*, 97–101.
2. Carter, R.; Treiber, K.H.; Geor, R.J.; Douglass, L.W.; Harris, P.A. Prediction of incipient pasture-associated laminitis from hyperinsulinaemia, hyperleptinaemia and generalised and localised obesity in a cohort of ponies. *Equine Vet. J.***2009**, *41*, 171–178.
3. Treiber, K.; Carter, R.; Gay, L.; Williams, C.; Geor, R. Inflammatory and redox status of ponies with a history of pasture-associated laminitis. *Vet. Immunol. Immunopath.* **2009**, *129*, 216–220.
4. Frank, N. Equine metabolic syndrome. *J. Equine Vet. Sci.* **2009**, *29*, 259–267.
5. Geor, R.J.; Harris, P. Dietary management of obesity and insulin resistance: Countering risk for laminitis. *Vet. Clin. North Am. Equine Prac.* **2009**, *25*, 51–65.
6. Asegaonkar, S.B.; Marathe, A.; Tekade, M.L.; Cherekar, L.; Bavikar, J.; Bardapurkar, J.; Ajay, R. High-sensitivity C-reactive protein: A novel cardiovascular risk predictor in type 2 diabetics with normal lipid profile. *J. Diabet. Complications* **2011**, *25*, 368–370.
7. Pickup, J.C.; Crook, M.A. Is type II diabetes mellitus a disease of the innate immune system. *Diabetologia* **1998**, *41*, 1241–1248.
8. Bonora, E.; Kiechl, S.; Willeit, J.; Oberhollenzer, F.; Egger, G.; Bonadonna, R.C.; Muggeo, M. Metabolic syndrome: Epidemiology and more extensive phenotypic description. Cross sectional data from the Bruneck Study. *Int. J. Obes.***2003**, *27*, 1283–1289.
9. Chan, J.C.N.; Cheung, J.C.K.; Stehouwer, C.D.A.; Emeis, J.J.; Tong, P.C.Y.; Ko, G.T.C.; Yudkin, J.S. The central roles of obesity-associated dyslipidaemia, endothelial activation and cytokines in the Metabolic Syndrome-An analysis by structural equation modelling. *Int. J. Obes.* **2002**, *26*, 994–1008.
10. Schmidt, M.I.; Duncan, B.B.; Sharret, A.R.; Lindberg, G.; Savage, P.J.; Offenbacher, S.; Azambuja, M.I.; Tracey, R.P.; Heiss, G. Markers of inflammation and prediction of diabetes mellitus in adultes (Atherosclerosis risk in communities study): A cohort study. *Lancet* **1999**, *353*, 1649–1652.
11. Cartier, A.; Lemieux, I.; Almeras, N.; Tremblay, A.; Bergeron, J.; Despres, J.-P. Visceral obesity and plasma glucose-insulin homeostasis: Contributions of IL-6 and TNF-α in men. *J. Clin. Endocrinol. Metab.* **2008**, *93*, 1931–1938.
12. Catalán, V.; Gómez-Ambrosi, J.; Ramirez, B.; Rotellar, F.; Pastor, C.; Silva, C.; Rodríguez, A.; Gil, M.; Cienfuegos, J.; Frühbeck, G. Proinflammatory cytokines in obesity: Impact of type 2 diabetes mellitus and gastric bypass. *Obes. Surg.***2007**, *17*, 1464–1474.
13. Pickup, J.C.; Mattock, M.B.; Chusney, G.D.; Burt, D. NIDDM as a disease of the innate immune system: Association of acute-phase reactants and interleukin-6 with metabolic syndrome X. *Diabetologia* **1997**, *40*, 1286–1292.
14. Pickup, J.C. Inflammation and activated innate immunity in the pathogenesis of type 2 diabetes. *Diabet. Care.* **2004**, *27*, 813–823.
15. Vick, M.M.; Adams, A.A.; Murphy, B.A.; Sessions, D.R.; Horohov, D.W.; Cook, R.F.; Shelton, B.J.; Fitzgerald, B.P. Relationships among inflammatory cytokines, obesity, and insulin sensitivity in the horse. *J. Anim. Sci.* **2007**, *85*, 1144–1155.
16. Suagee, J.K.; Burk, A.O.; Quinn, R.W.; Hartsock, T.G.; Douglass, L.W. Effects of diet and weight gain on circulating tumour necrosis factor-a concentrations in Thoroughbred geldings. *J. Anim. Physiol. Anim. Nutr.* **2011**, *95*, 161–170.
17. Holbrook, T.C.; Tipton, T.; McFarlane, D. Neutrophil and cytokine dysregulation in hyperinsulinemic obese horses. *Vet. Immunol. Immunopath.* **2012**, *145*, 283–289.
18. Villaret, A.; Galitzky, J.; Decaunes, P.; Esteve, D.; Marques, M.A.; Sengenes, C.; Chiotasso, P.; Tchkonia, T.; Lafontan, M.; Kirkland, J.L.; Bouloumie, A. Adipose tissue endothelial cells from obese human subjects: Differences among depots in angiogenic, metabolic, and inflammatory gene expression and cellular senescence. *Diabetes* **2010**, *59*, 2755–2763.
19. Burns, T.A.; Geor, R.J.; Mudge, M.C.; McCutcheon, L.J.; Hinchcliff, K.W.; Belknap, J.K. Proinflammatory cytokine and chemokine gene expression profiles in subcutaneous and

visceral adipose tissue depots of insulin-resistant and insulin-sensitive light breed horses. *J. Vet. Intern. Med.* **2010**, *24*, 932–939.

20. Fain, J.N.; Bahouth, S.W.; Madan, A.K. TNF alpha release by the nonfat cells of human adipose tissue. *Int. J. Obes.***2004**, *28*, 616–622.
21. Xu, H.; Barnes, G.T.; Yang, Q.; Tan, G.; Yang, D.; Chou, C.J.; Sole, J.; Nichols, A.; Ross, J.S.; Tartaglia, L.A.; Chen, H. Chronic inflammation in fat plays a crucial role in the development of obesity-related insulin resistance. *J. Clin. Invest.* **2003**, *112*, 1821–1830.
22. Lumeng, C.N.; Deyoung, S.M.; Saltiel, A.R. Macrophages block insulin action in adipocytes by altering expression of signaling and glucose transport proteins. *Am. J. Physiol. Endocrinol. Metab.* **2007**, *292*, E166–E174.
23. Vick, M.M.; Murphy, B.A.; Sessions, D.R.; Reedy, S.E.; Kennedy, E.L.; Horohov, D.W.; Cook, R.F.; Fitzgerald, B.P. Effects of systemic inflammation on insulin sensitivity in horses and inflammatory cytokine expression in adipose tissue. *Am. J. Vet. Res.* **2008**, *69*, 130–139.
24. DeFronzo, R.A. Insulin resistance: A multifaceted syndrome responsible for NIDDM, obesity, hypertension, dyslipidaemia and atheroscleros. *Neth. J. Med.* **1997**, *50*, 191–197. [Google Scholar]
25. Bruunsgaard, H. Physical activity and modulation of systemic low-level inflammation. *J. Leukoc. Biol.* **2005**, *78*, 819–835.
26. Kim, T.H.; Choi, S.E.; Ha, E.S.; Jung, J.G.; Han, S.J.; Kim, H.J.; Kim, D.J.; Kang, Y.; Lee, K.W. IL-6 induction of TLR-4 gene expression via STAT3 has an effect on insulin resistance in human skeletal muscle. *Acta. Diabetol.* **2011**.
27. Saghizadeh, M.; Ong, J.M.; Garvey, W.T.; Henry, R.R.; Kern, P.A. The expression of TNF alpha by human muscle. Relationship to insulin resistance. *J. Clin. Invest.* **1996**, *97*, 1111–1116.
28. Torres, S.H.; de Sanctis, J.B.; de L Briceño, M.; Hernandez, N.; Finol, H.J. Inflammation and nitric oxide production in skeletal muscle of type 2 diabetic patients. *J. Endocrinol.* **2004**, *181*, 419–427.
29. Wellen, K.E.; Hotamisligil, G.S. Inflammation, stress, and diabetes. *J. Clin. Invest.* **2005**, *115*, 1111–1119.
30. Tschop, M.; Thomas, G. Fat fuels insulin resistance through Toll-like receptors. *Nat. Med.* **2006**, *12*, 1359–1361.
31. Boyd, J.H.; Divangahi, M.; Yahiaoui, L.; Gvozdic, D.; Qureshi, S.; Petrof, B.J. Toll-like receptors differentially regulate CC and CXC chemokines in skeletal muscle via NF-kappaB and calcineurin. *Infect. Immun.* **2006**, *74*, 6829–6838.
32. Coletta, D.K.; Balas, B.; Chavez, A.O.; Baig, M.; Abdul-Ghani, M.; Kashyap, S.R.; Folli, F.; Tripathy, D.; Mandarino, L.J.; Cornell, J.E.; DeFronzo, R.A.; Jenkinson, C.P. Effect of acute physiological hyperinsulinemia on gene expression in human skeletal muscle *in vivo*. *Am. J. Phys. Endocrinol. Metab.* **2008**, *294*, E910–E917.
33. Wearn, J.G.; Suagee, J.K.; Crisman, M.V.; Corl, B.A.; Hulver, M.W.; Hodgson, D.R.; Geor, R.J.; McCutcheon, L.J. Effects of the insulin sensitizing drug, pioglitazone, and lipopolysaccharide administration on markers of systemic inflammation and clinical parameters in horses. *Vet. Immunol. Immunopath.* **2012**, *145*, 42–49.
34. Suagee, J.K.; Corl, B.A.; Crisman, M.V.; Hulver, M.W.; McCutcheon, L.J.; Geor, R.J. Effects of acute hyperinsulinemia on inflammatory proteins in horses. *Vet. Immunol. Immunopath.* **2011**, *142*, 141–146.
35. Weisberg, S.P.; McCann, D.; Desai, M.; Rosenbaum, M.; Leibel, R.L.; Ferrante, A.W. Obesity is associated with macrophage accumulation in adipose tissue. *J. Clin. Invest.* **2003**, *112*, 1796–1808.
36. Wellen, K.E.; Hotamisligil, G.S. Obesity-induced inflammatory changes in adipose tissue. *J. Clin.Invest.* **2003**, *112*, 1785–1788.
37. Burns, T.A.; Geor, R.J.; Mudge, M.C.; McCutcheon, L.J.; Belknap, J. Characterization of adipose tissue macrophage infilitration in insulin-resistant and insulin-sensitive light breed horses. In Proceedings of the 28th Annual Forum of the American College of Veterinary Internal Medicine, Anaheim, CA, USA, 9-12 June 2010.

38. Do, M.-S.; Jeong, H.-S.; Choi, B.-H.; Hunter, L.; Langley, S.; Pazmany, L.; Trayhurn, P. Inflammatory gene expression patterns revealed by DNA microarray analysis in TNF-alpha-treated SGBS human adipocytes. *Yonsei Med. J.* **2006**, *47*, 729–736.
39. Kanda, H.; Tateya, S.; Tamori, Y.; Kotani, K.; Hiasa, K.-I.; Kitazawa, R.; Kitazawa, S.; Miyachi, H.; Maeda, S.; Egashira, K.; Kasuga, M. MCP-1 contributes to macrophage infiltration into adipose tissue, insulin resistance, and hepatic steatosis in obesity. *J. Clin. Invest.* **2006**, *116*, 1494–1505.
40. Sartipy, P.; Loskutoff, D.J. Expression profiling identifies genes that continue to respond to insulin in adipocytes made insulin-resistant by treatment with tumor necrosis factor-alpha. *J. Biol. Chem.* **2003**, *278*, 52298–52306.
41. Kim, C.-S.; Park, H.-S.; Kawada, T.; Kim, J.-H.; Lim, D.; Hubbard, N.E.; Kwon, B.-S.; Erickson, K.L.; Yu, R. Circulating levels of MCP-1 and IL-8 are elevated in human obese subjects and associated with obesity-related parameters. *Int. J. Obes.* **2006**, *30*, 1347–1355.
42. Christensen, T.; Richelsen, B.; Bruun, J.M. Monocyte chemoattractant protein-1 is produced in isolated adipocytes, associated with adiposity and reduced after weight loss in morbid obese subjects. *Int. J. Obes.* **2005**, *26*, 146–150.
43. Chen, A.; Mumick, S.; Zhang, C.; Lamb, J.; Dai, H.; Weingarth, D.; Mudgett, J.; Chen, H.; MacNeil, D.J.; Reitman, M.L.; Qian, S. Diet induction of monocyte chemoattractant protein-1 and its impact on obesity. *Obes. Res.* **2005**, *13*, 1311–1320.
44. Tateya, S.; Tamori, Y.; Kawaguchi, T.; Kanda, H.; Kasuga, M. An increase in the circulating concentration of monocyte chemoattractant protein-1 elicits systemic insulin resistance irrespective of adipose tissue inflammation in mice. *Endocrinology* **2010**, *151*, 971–9.
45. Fain, J.N.; Madan, A.K. Regulation of monocyte chemoattractant protein-1 (MCP-1) release by explants of human visceral adipose tissue. *Int. J. Obes.* **2005**, *29*, 1299–1307.
46. Bruun, J.M.; Lihn, A.S.; Pedersen, S.B.; Richelsen, B. Monocyte chemoattractant protein-1 release is higher in visceral than subcutaneous human adipose tissue (AT): Implications of macrophages resident in the AT. *J. Clin. Endocrinol. Metab.* **2005**, *90*, 2282–2289.
47. Gerhardt, C.C.; Romero, I.A.; Cancello, R.; Camoin, L.; Strosberg, A.D. Chemokines control fat accumulation and leptin secretion by cultured human adipocytes. *Mol. Cell. Biol.* **2001**, *175*, 81–92.
48. Vervuert, I.; Voigt, K.; Hollands, T.; Cuddeford, D.; Coenen, M. Effect of feeding increasing quantities of starch on glycaemic and insulinaemic responses in healthy horses. *Vet. J.* **2009**, *182*, 67–72.
49. Ruge, T.; Lockton, J.A.; Renstrom, F.; Lystig, T.; Sukonina, V.; Svensson, M.K.; Eriksson, J.W. Acute hyperinsulinemia raises plasma interleukin-6 in both nondiabetic and type 2 diabetes mellitus subjects, and this effect is inversely associated with body mass index. *Metabolism* **2009**, *58*, 860–866.
50. Siklova-Vitkova, M.P.; Polak, J.; Klimcakova, E.; Vrzalova, J.; Hejnova, J.; Kovacikova, M.; Kovacova, Z.; Bajzova, M.; Rossmeislova, L.; Hnevkovska, Z.; Langin, D.; Stich, V.P. Effect of hyperinsulinemia and very-low-calorie diet on interstitial cytokine levels in subcutaneous adipose tissue of obese women. *Am. J. Physiol. Endocrinol. Metab.* **2009**, *297*, E1154–E1161.
51. Hoffman, R.M.; Boston, R.C.; Stefanovski, D.; Kronfeld, D.S.; Harris, P.A. Obesity and diet affect glucose dynamics and insulin sensitivity in Thoroughbred geldings. *J. Anim. Sci.* **2003**, *81*, 2333–2342.
52. Fasshauer, M.; Klein, J.; Lossner, U.; Paschke, R. Interleukin (IL)-6 mRNA expression is stimulated by insulin, isoproterenol, tumour necrosis factor alpha, growth hormone, and IL-6 in 3T3-L1 adipocytes. *Horm. Metab. Res.* **2003**, *35*, 147–152.
53. Vicennati, V.; Vottero, A.; Friedman, C.; Papanicolaou, D.A. Hormonal regulation of interleukin-6 production in human adipocytes. *Int. J. Obes. Relat. Metab. Disord.* **2002**, *26*, 905–911.
54. Dendorfer, U.; Oettgen, P.; Libermann, T.A. Interleukin-6 gene expression by prostaglandins and cyclic AMP mediated by multiple regulatory elements. *Am. J. Ther.* **1995**, *2*, 660–665.

55. LaPensee, C.R.; Hugo, E.R.; Ben-Jonathan, N. Insulin stimulates interleukin-6 expression and release in LS14 human adipocytes through multiple signaling pathways. *Endocrinology* **2008**, *149*, 5415–22.
56. Kabon, B.; Nagele, A.; Reddy, D.; Eagon, C.; Fleshman, J.W.; Sessler, D.I.; Kurz, A. Obesity decreases perioperative tissue oxygenation. *Anesthesiology* **2004**, *100*, 274–280.
57. Karhausen, J.; Haase, V.H.; Colgan, S.P. Inflammatory hypoxia: Role of hypoxia-inducible factor. *Cell Cycle* **2005**, *4*, 256–258.
58. Lolmede, K.; Durand de Saint Front, V.; Galitzky, J.; Lafontan, M.; Bouloumie, A. Effects of hypoxia on the expression of proangiogenic factors in differentiated 3T3-F442A adipocytes. *Int.J. Obes. Relat. Metab. Disord.* **2003**, *27*, 1187–1195.
59. Thurlby, P.L.; Trayhurn, P. Regional blood flow in genetically obese (ob/ob) mice. The importance of brown adipose tissue to the reduced energy expenditure on non-shivering thermogenesis. *Pflugers Arch.* **1980**, *385*, 193–201.
60. Karpe, F.; Fielding, B.A.; Ilic, V.; Macdonald, I.A.; Summers, L.K.; Frayn, K.N. Impaired postprandial adipose tissue blood flow response is related to aspects of insulin sensitivity. *Diabetes* **2002**, *51*, 2467–2473.
61. Ye, J.; Gao, Z.; Yin, J.; He, Q. Hypoxia is a potential risk factor for chronic inflammation and adiponectin reduction in adipose tissue of ob/ob and dietary obese mice. *Am. J. Physiol. Endocrinol. Metab.* **2007**, *293*, E1118–E1128.
62. Rausch, M.E.; Weisberg, S.; Vardhana, P.; Tortoriello, D.V. Obesity in C57BL//6J mice is characterized by adipose tissue hypoxia and cytotoxic T-cell infiltration. *Int. J. Obes.* **2008**, *32*, 451–463.
63. O'Rourke, R.W.; White, A.E.; Metcalf, M.D.; Olivas, A.S.; Mitra, P.; Larison, W.G.; Cheang, E.C.; Varlamov, O.; Corless, C.L.; Roberts, C.T., Jr.; Marks, D.L. Hypoxia-induced inflammatory cytokine secretion in human adipose tissue stromovascular cells. *Diabetologia* **2011**, *54*, 1480–1490.
64. Semenza, G.L. Life with oxygen. *Science* **2007**, *318*, 62–64.
65. Ivan, M.; Kondo, K.; Yang, H.; Kim, W.; Valiando, J.; Ohh, M.; Salic, A.; Asara, J.M.; Lane, W.S.; Kaelin, W.G., Jr. HIFalpha targeted for VHL-mediated destruction by proline hydroxylation: Implications for O_2 sensing. *Science* **2001**, *292*, 464–468.
66. Jaakkola, P.; Mole, D.R.; Tian, Y.M.; Wilson, M.I.; Gielbert, J.; Gaskell, S.J.; Kriegsheim, A.; Hebestreit, H.F.; Mukherji, M.; Schofield, C.J.; Maxwell, P.H.; Pugh, C.W.; Ratcliffe, P.J. Targeting of HIF-alpha to the von Hippel-Lindau ubiquitylation complex by O_2-regulated prolyl hydroxylation. *Science* **2001**, *292*, 468–472.
67. Semenza, G.L. Targeting HIF-1 for cancer therapy. *Nat. Rev. Cancer* **2003**, *3*, 721–732.
68. Rocha, S. Gene regulation under low oxygen: Holding your breath for transcription. *Trends Biochem. Sci.* **2007**, *32*, 389–397.
69. Wang, B.; Wood, I.S.; Trayhurn, P. Dysregulation of the expression and secretion of inflammation-related adipokines by hypoxia in human adipocytes. *Pflugers Arch.* **2007**, *455*, 479–492.
70. Cancello, R.; Henegar, C.; Viguerie, N.; Taleb, S.; Poitou, C.; Rouault, C.; Coupaye, M.; Pelloux, V.; Hugol, D.; Bouillot, J.-L.; Bouloumie, A.; Barbatelli, G.; Cinti, S.; Svensson, P.A.; Barsh, G.S.; Zucker, J.D.; Basdevant, A.; Langin, D.; Clement, K. Reduction of macrophage infiltration and chemoattractant gene expression changes in white adipose tissue of morbidly obese subjects after surgery-induced weight loss. *Diabetes* **2005**, *54*, 2277–2286.
71. Cummins, E.P.; Taylor, C.T. Hypoxia-responsive transcription factors. *Pflugers Arch.* **2005**, *450*, 363–371.
72. Koong, A.C.; Chen, E.Y.; Giaccia, A.J. Hypoxia causes the activation of nuclear factor kappa B through the phosphorylation of I kappa B alpha on tyrosine residues. *Cancer Res.* **1994**, *54*, 1425–1430.
73. Cummins, E.P.; Berra, E.; Comerford, K.M.; Ginouves, A.; Fitzgerald, K.T.; Seeballuck, F.; Godson, C.; Nielsen, J.E.; Moynagh, P.; Pouyssegur, J.; Taylor, C.T. Prolyl hydroxylase-1 negatively regulates IkappaB kinase-beta, giving insight into hypoxia-induced NFkappaB activity. *Proc. Nat. Acad. Sci. USA* **2006**, *103*, 18154–18159.

74. Belaiba, R.S.; Bonello, S.; Zahringer, C.; Schmidt, S.; Hess, J.; Kietzmann, T.; Gorlach, A. Hypoxia up-regulates hypoxia-inducible factor-1alpha transcription by involving phosphatidylinositol 3-kinase and nuclear factor kappaB in pulmonary artery smooth muscle cells. *Mol. Biol. Cell.* **2007**, *18*, 4691–4697.
75. van Uden, P.; Kenneth, N.S.; Rocha, S. Regulation of hypoxia-inducible factor-1alpha by NF-kappaB. *Biochem. J.* **2008**, *412*, 477–484.
76. Swyers, K.L.; Burk, A.O.; Hartsock, T.G.; Ungerfeld, E.M.; Shelton, J.L. Effects of direct-fed microbial supplementation on digestibility and fermentation end-products in horses fed low- and high-starch concentrates. *J. Anim. Sci.* **2008**, *86*, 2596–2608.
77. Medina, B.; Girard, I.D.; Jacotot, E.; Julliand, V. Effect of a preparation of Saccharomyces cerevisiae on microbial profiles and fermentation patterns in the large intestine of horses fed a high fiber or a high starch diet. *J. Anim. Sci.***2002**, *80*, 2600–2609.
78. Emmanuel, D.G.; Madsen, K.L.; Churchill, T.A.; Dunn, S.M.; Ametaj, B.N. Acidosis and lipopolysaccharide from Escherichia coli B:055 cause hyperpermeability of rumen and colon tissues. *J. Dairy Sci.* **2007**, *90*, 5552–5557.
79. Khafipour, E.; Krause, D.O.; Plaizier, J.C. A grain-based subacute ruminal acidosis challenge causes translocation of lipopolysaccharide and triggers inflammation. *J. Dairy Sci.* **2009**, *92*, 1060–1070.
80. Sprouse, R.F.; Garner, H.E.; Green, E.M. Plasma endotoxin levels in horses subjected to carbohydrate induced laminitis. *Equine Vet. J.* **1987**, *19*, 25–28.
81. Carroll, C.L.; Hazard, G.; Coloe, P.J.; Hooper, P.T. Laminitis and possible enterotoxaemia associated with carbohydrate overload in mares. *Equine Vet. J.* **1987**, *19*, 344–346.
82. Prasse, K.W.; Allen, D.; Moore, J.N.; Duncan, A. Evaluation of coagulation and fibrinolysis during the prodromal stages of carbohydrate-induced acute laminitis in horses. *Am. J. Vet. Res.* **1990**, *51*, 1950–1955.
83. Budak, M.T.; Orsini, J.A.; Pollitt, C.C.; Rubinstein, N.A. Gene expression in the lamellar dermis-epidermis during the developmental phase of carbohydrate overload-induced laminitis in the horse. *Vet. Immunol. Immunopath.* **2009**, *131*, 86–96.
84. Toth, F.; Frank, N.; Chameroy, K.A.; Boston, R.C. Effects of endotoxaemia and carbohydrate overload on glucose and insulin dynamics and the development of laminitis in horses. *Equine Vet. J.* **2009**, *41*, 852–858.
85. Krueger, A.S.; Kinden, D.A.; Garner, H.E.; Sprouse, R.F. Ultrastructural study of the equine cecum during onset of laminitis. *Am. J. Vet. Res.* **1986**, *47*, 1804–1812.
86. Bailey, S.R.; Adair, H.S.; Reinemeyer, C.R.; Morgan, S.J.; Brooks, A.C.; Longhofer, S.L.; Elliott, J. Plasma concentrations of endotoxin and platelet activation in the developmental stage of oligofructose-induced laminitis. *Vet. Immunol. Immunopath.* **2009**, *129*, 167–173.
87. Toth, F.; Frank, N.; Elliott, S.B.; Geor, R.J.; Boston, R.C. Effects of an intravenous endotoxin challenge on glucose and insulin dynamics in horses. *Am. J. Vet. Res.* **2008**, *69*, 82–88.
88. Nieto, J.E.; MacDonald, M.H.; Braim, A.E.P.; Aleman, M. Effect of lipopolysaccharide infusion on gene expression of inflammatory cytokines in normal horses *in vivo*. *Equine Vet. J.* **2009**, *41*, 717–719.
89. Hermann, G.E.; Homes, G.M.; Rogers, R.C. TNF alpha modulation of visceral and spinal sensory processing. *Curr.Opin. Pharma. Des.* **2005**, *11*, 1391–1409.
90. Tracey, K.J.; Wei, H.; Manogue, K.R.; Fong, Y.; Hesse, D.G.; Nguyen, H.T.; Kuo, G.C.; Beutler, B.; Cotran, R.S.; Cerami, A.; Lowry, S.F. Cachectin/tumor necrosis factor induces cachexia, anemia, and inflammation. *J. Exp. Med.***1988**, *167*, 2111–1227.
91. van Miert, A.S.; van Duin, C.T.; Wensing, T. Fever and acute phase response induced in dwarf goats by endotoxin and bovine and human recombinant tumour necrosis factor alpha. *J. Vet. Pharmacol. Ther.* **1992**, *15*, 332–342.
92. Benhariz, M.; Goulet, O.; Salas, J.; Colomb, V.; Ricour, C. Energy cost of fever in children on total parenteral nutrition. *Clin. Nutr.* **1997**, *16*, 251–255.
93. Maratou, E.; Dimitriadis, G.; Kollias, A.; Boutati, E.; Lambadiari, V.; Mitrou, P.; Raptis, S.A. Glucose transporter expression on the plasma membrane of resting and activated white blood cells. *Eur. J. Clin. Invest.* **2007**, *37*, 282–290.

94. Meszaros, K.; Bojta, J.; Bautista, A.P.; Lang, C.H.; Spitzer, J.J. Glucose utilization by Kupffer cells, endothelial cells, and granulocytes in endotoxemic rat liver. *Am. J. Physiol.* **1991**, *260*, G7–G12.
95. Krogh-Madsen, R.; Moller, K.; Dela, F.; Kronborg, G.; Jauffred, S.; Pedersen, B.K. Effect of hyperglycemia and hyperinsulinemia on the response of IL-6, TNF-alpha, and FFAs to low-dose endotoxemia in humans. *Am. J. Physiol. Endocrinol. Metab.* **2004**, *286*, E766–E772.
96. Hotamisligil, G.S. Mechanisms of TNF-alpha induced insulin resistance. *Exp. Clin. Endocrinol. Diabetes* **1999**, *107*, 119–125.
97. Hardardottir, I.; Grunfeld, C.; Feingold, K. Effects of endotoxin and cytokines on lipid metabolism. *Curr. Opin. Lipidol.* **1994**, *5*, 207–215.
98. Das, U.N. Is obesity an inflammatory condition? *Nutrition* **2001**, *17*, 953–966.
99. Engstrom, G.; Hedblad, B.; Stavenow, L.; Lind, P.; Janzon, L.; Lindgarde, F. Inflammation-sensitive plasma proteins are associated with future weight gain. *Diabetes* **2003**, *52*, 2097–2101.
100. Festa, A.; D'Agostino, R.; Howard, G.; Mykkanen, L.; Tracey, R.P.; Haffner, S.M. Chronic subclinical inflammation as part of the insulin resistance syndrome: The insulin resistance atherosclerosis study (IRAS). *Circulation* **2000**, *102*, 42–47.
101. Kershaw, E.E.; Flier, J.S. Adipose tissue as an endocrine gland. *J. Clin. Endocrinol. Metab.* **2004**, *89*, 2548–2556.
102. Trayhurn, P.; Wood, I.S. Adipokines: Inflammation and the pleitropic role of white adipose tissue. *Br.J. Nutr.* **2004**, *92*, 347–355.
103. Vettor, R.; Milan, G.; Rossato, M.; Federspil, G. Review article: Adipocytokines and insulin resistance. *Aliment. Pharmacol. Ther.* **2005**, *22*, 3–10.
104. Yudkin, J.S.; Stehouwer, C.D.A.; Emeis, J.J.; Coppack, S.W. C-reactive protein in healthy subjects: Associations with obesity, insulin resistance, and endothelial dysfunction: A potential role for cytokines originating from adipose tissue? *Arterioscle. Thromb. Vasc. Biol.* **1999**, *19*, 972–978.
105. Stephens, J.M.; Lee, J.; Pilch, P.F. Tumor necrosis factor-a-induced insulin resistance in 3T3-L1 adipocytes is accompanied by a loss of insulin receptor substrate-1 and GLUT4 expression without a loss of insulin receptor-mediated signal transduction. *J. Biol. Chem.* **1997**, *272*, 971–976.
106. Hauner, H.; Petruschke, T.H.; Russ, M.; Rohrig, K.; Eckel, J. Effects of tumor necrosis factor alpha (TNFa) on glucose transport and lipid metabolism of newly-differentiated human fat cells in cell culture. *Diabetologia* **1995**, *38*, 764–771.
107. Ishizuka, K.; Usui, I.; Kanatani, Y.; Bukhari, A.; He, J.; Fujisaka, S.; Yamazaki, Y.; Suzuki, H.; Hiratani, K.; Ishiki, M.; Iwata, M.; Urakaze, M.; Haruta, T.; Kobayashi, M. Chronic tumor necrosis factor-alpha treatment causes insulin resistance via insulin receptor substrate-1 serine phosphorylation and suppressor of cytokine signaling-3 induction in 3T3-L1 adipocytes. *Endocrinology* **2007**, *148*, 2994–3003.
108. Bouzakri, K.; Zierath, J.R. MAP4K4 gene silencing in human skeletal muscle prevents tumor necrosis factor-alpha-induced insulin resistance. *J. Biol. Chem.* **2007**, *282*, 7783–7789.
109. Jager, J.; Gremeaux, T.; Cormont, M.; Le Marchand-Brustel, Y.; Tanti, J.-F. Interleukin-1b-induced insulin resistance in adipocytes through down-regulation of insulin receptor substrate-1 expression. *Endocrinology* **2007**, *148*, 241–251.
110. Lagathu, C.; Yvan-Charvet, L.; Bastard, J.P.; Maachi, M.; Quignard-Boulange, A.; Capeau, J.; Caron, M. Long-term treatment with interleukin-1beta induces insulin resistance in murine and human adipocytes. *Diabetologia* **2006**, *49*, 2162–2173.
111. Klover, P.J.; Zimmers, T.A.; Koniaris, L.G.; Mooney, R.A. Chronic exposure to interleukin-6 causes hepatic insulin resistance in mice. *Diabetes* **2003**, *52*, 2784–2789.
112. Lagathu, C.; Bastard, J.P.; Auclair, M.; Maachi, M.; Capeau, J.; Caron, M. Chronic interleukin-6 (IL-6) treatment increased IL-6 secretion and induced insulin resistance in adipocyte: Prevention by rosiglitazone. *Biochem. Biophys. Res. Commun.* **2003**, *311*, 372–379.

113. Rotter, V.; Nagaev, I.; Smith, U. Interleukin-6 (IL-6) induces insulin resistance in 3T3-L1 adipocytes and is, like IL-8 and tumor necrosis factor-alpha, overexpressed in human fat cells from insulin-resistant subjects. *J. Biol. Chem.* **2003**, *278*, 45777–84.
114. Bastard, J.P.; Maachi, M.; Van Nhieu, J.T.; Jardel, C.; Bruckert, E.; Grimaldi, A.; Robert, J.J.; Capeau, J.; Hainque, B. Adipose tissue IL-6 content correlates with resistance to insulin activation of glucose uptake both *in vivo* and *in vitro*. *J. Clin. Endocrinol. Metab.* **2002**, *87*, 2084–2089.
115. Pedersen, B.K.; Akerstrom, T.C.; Nielsen, A.R.; Fischer, C.P. Role of myokines in exercise and metabolism. *J. Appl. Physiol.* **2007**, *103*, 1093–1098.
116. Plomgaard, P.; Bouzakri, K.; Krogh-Madsen, R.; Mittendorfer, B.; Zierath, J.R.; Pedersen, B.K. Tumor necrosis factor-α induces skeletal muscle insulin resistance in healthy human subjects via inhibition of Akt substrate 160 phosphorylation. *Diabetes* **2005**, *54*, 2939–2945.
117. Begum, N.; Ragolia, L. Effect of tumor necrosis factor-alpha on insulin action in cultured rat skeletal muscle cells. *Endocrinology* **1996**, *137*, 2441–2446.
118. de Alvaro, C.; Teruel, T.; Hernandez, R.; Lorenzo, M. Tumor necrosis factor alpha produces insulin resistance in skeletal muscle by activation of inhibitor kappaB kinase in a p38 MAPK-dependent manner. *J. Biol. Chem.* **2004**, *279*, 17070–17078.
119. Rosenzweig, T.; Braiman, L.; Bak, A.; Alt, A.; Kuroki, T.; Sampson, S.R. Differential effects of tumor necrosis factor-alpha on protein kinase C isoforms alpha and delta mediate inhibition of insulin receptor signaling. *Diabetes* **2002**, *51*, 1921–1930
120. Grzelkowska-Kowalczyk, K.; Wieteska-Skrzeczynska, W. Treatment with TNF-α and IFNγ alters activation of SER/THR protein kinases and the metabolic response to IGF-1 in mouse C2C12 myogenic cells. *Cell Mol. Biol. Lett.***2010**, *15*, 13–31.
121. Kim, H.J.; Higashimori, T.; Park, S.Y.; Choi, H.; Dong, J.; Kim, Y.J.; Noh, H.L.; Cho, Y.R.; Cline, G.; Kim, Y.B.; Kim, J.K. Differential effects of interleukin-6 and -10 on skeletal muscle and liver insulin action *in vivo*. *Diabetes* **2004**, *53*, 1060–1067.
122. Seyoum, B.; Fite, A.; Abou-Samra, A.B. Effects of 3T3 adipocytes on interleukin-6 expression and insulin signaling in L6 skeletal muscle cells. *Biochem. Biophys. Res. Commun.* **2011**, *410*, 13–18.

CHAPTER 14

Effect of feeding Actigen™ to sows during gestation and lactation and on piglet performance

J. Taylor-Pickard[1*], T. McArdle1 and S. Icely[2]

[1]*Alltech Biotechnology Centre, Sarney, Summer Hill Road, Dunboyne, County Meath, Ireland*

[2]*Harper Adams University, Edgmond, Shropshire, UK*

ABSTRACT

Actigen™ is a second generation, unique bioactive fraction derived from the outer cell wall of a specific strain of yeast, *Saccharomyces cerevisiae*, where it acts in the gut to bind pathogenic bacteria, preventing disease and competition for nutrients. In the current trial, 260 sows were used and fed either an unsupplemented control diet or one containing Actigen™ at 0.08% of the formulation during gestation, farrowing and lactation, to assess its impact on sow and piglet performance. Results showed significant increases in feed intake for sows fed Actigen™ pre-farrowing, but this was reversed during lactation, leading to a 7 kg saving in feed intake per animal. At birth, piglets weighed 42 g more in litters from Actigen™-fed sows compared to the control group. Weaning weight of the total litter from Actigen™ sows was significantly higher ($P < 0.05$) by 3.3 kg, although individual piglet weights were the same. Sows showed no significant loss in body weight during the trial, indicating that the improved piglet production was related to improved nutrient availability and perhaps Ig status of piglets due to the addition of Actigen™ in the diet.

Keywords: Actigen, yeast, sows, piglets, performance

1. INTRODUCTION

Actigen™ (ACT) is a second generation, unique bioactive fraction derived from the outer cell wall of a specific strain of yeast, *Saccharomyces cerevisiae*, formulated to be used in animal diets. This yeast-derived product has been developed for more than twenty years, with a large body of research demonstrating its mode of action and efficacy in many species, especially where antibiotics have been removed from feeds. The active component attaches to the fibriae present on the surface of pathogenic bacteria, preventing their attachment and colonisation in the gut (Spring *et al.*, 2000). Its ability to interact with the gut wall and associated immune moieties has been proven to promote immune-competence in young animals, especially when suckling and at weaning (Brennan *et al.*, 2012). In sows, it makes more nutrients available, due to the removal of competition for nutrients by gut microflora, leading to better maintenance of body condition, more piglets born and improved return to service intervals (Miguel *et al.*, 2002; Le Dividich *et al.*, 2009; Lazarevic *et al.*, 2010).

Commercial trials with gestating and lactating sows have proven that ACT can increase immunoglobulin (Ig) expression in milk (Spring *et al.*, 2003) thereby promoting immune protection in young piglets. The prevention of competition for nutrients in sows due to better gut health, as well as immune status, body composition and feeding regime can have a major influence not only on Ig status, but also on colostrum and milk quality, allowing piglets better availability of nutrients from the start of suckling (Fahmy, 1972; Klobasa *et al.*, 1987; Goransson, 1990; Zou *et al.*, 1992; Darragh and Moughan, 1998). More efficient digestion with less competition for nutrients is important to maintaining sow condition, allowing her to return to service in a timely manner (Klaver *et al.*, 1981). These changes lead to fewer digestive disorders and diarrhoea at weaning, and better productive performance through to slaughter weight in growing and finishing pigs. In recent trials conducted in 2014 in a commercial pig breeding herd monitored for a 12 month period, ACT supplementation in sow diets prior to parturition gave improved milk Ig levels by 59% and piglet birth weight was increased by 0.19 kg relative to unsupplemented control sows and their litters (McArdle, unpublished).

When piglets undergo weaning, their own acquired immunity needs to be robust enough to prevent digestive problems that can reduce growth performance and welfare. Changing from a milk-based to a cereal-based diet resulted in changes in the gut and increased enzyme secretion which can impact digestion and uptake of nutrients. Ensuring the successful transfer of piglets from milk to meal is essential for productive performance, as each extra 100 g in body weight at weaning is considered equivalent to one day less to reach final slaughter weight, saving time, labour and feed costs (Cooper *et al.*, 2001). Heavier weaners are considered more efficient at converting feed into gain, and have leaner carcasses at slaughter. Over a twelve month period where a commercial pig breeding herd was monitored, 0.61 more piglets were born per litter, with a 0.5 kg higher average weaning weight at 1.5 days earlier from ACT supplemented sows (Alltech, personal communication). After weaning, time to service was reduced by 0.6 days. Such improvements could have major economic implications for pig producers. Therefore, the objective of the following trial was to evaluate the effect of supplementing ACT in feed throughout the gestation and lactation on both sow and piglet performance.

2. MATERIALS AND METHODS

Two hundred and sixty sows and their litters were used in the trial. The sows were crossbred, using JSR Genepacker 90 (LWxLR) and PIC Camborough (LWxLR) xWhite Duroc), with a JSR 900 sire being used for breeding the trial piglets. All pigs were sourced from the Harper Adams University sow herd. PMWS was present in the herd but with low mortality (<3% post-weaning), and the herd was PRRS and enzootic pneumonia positive, and vaccinated accordingly for both.

Sows were allocated to one of two dietary treatments – either a control (unsupplemented) diet or one containing ACT (Alltech Inc. KY) at 0.08% of the total diet formulation. Sows were allocated to diets on the basis of genotype and equalised for parity between treatments. Pens containing eight sows each were allocated to the dietary treatments and these animals remained in the same group throughout gestation. Animals that returned to oestrus and were re-bred stayed within their group but were housed separately when pen mates moved to the farrowing room, which occurred seven to eight days prior to farrowing. Farrowing pens were fully slatted crate systems arranged in two rooms with sows allocated evenly between rooms.

All sows were fed to appetite from previous litter weaning to mating and then restricted to 2.4 kg per day supplied via volumetric dispensers during gestation. Sows received the lactation ration when they entered the farrowing house and were fed 2.5 kg per day until farrowing, whereby the amount rose on the standard Stotfold scale (a standardised feeding strategy for sows) to 11 kg per day depending on litter size. Sows were weighed on entry and exit of the farrowing house. Milk replacer was made available to piglets from birth, and prestarter feed from day 14. Milk replacer was formulated containing whey protein, palm and coconut oil and premix. The prestarter diet main ingredients comprised whey protein, full fat soya, oat flakes, palm oil and wheat with premix to balance. Piglets received 1.3 litres of milk replacer each and were offered 0.85 kg of prestarter diets.

Table 1. Diet specifications and nutrient analysis

Raw Material	Dry sow diet (%*)	Lactating sow diet (%*)
Barley (62 kg/hl)	10.00	11.90
Wheat	36.90	37.90
Wheatfeed	46.60	23.20
Maize Germ	0.00	8.30
Soya HiPro extruded	0.00	7.20
Peas	0.00	5.00
Fat Hispec (vegetable base)	1.00	1.00
Molasses	2.50	2.50
Mineral and mineral premix	3.00	3.00
Total	100.00	100.00
Nutrient Analysis		
Oil B	4.58	4.60
Protein	12.04	14.28
Fibre	5.08	4.25
Ash	4.61	4.98
Salt	0.45	0.58
Calcium	0.64	0.73
Phosphorus	0.49	0.50
Energy (DE MJ/kg)	12.57	13.16
Lysine	0.46	0.81
Methionine	0.19	0.25
Vitamin A (IU/kg)	100000.00	100000.00
Vitamin D3 (IU/kg)	1875.00	1875.00
Vitamin E (IU/kg)	50.00	100.00

NB Treatment diets contained 0.08% ACT.
*all as % unless otherwise stated in first column.

Diets were formulated to meet NRC (2012) requirements and based on commercial (UK) gestation and lactation rations (Table 1). Diets were manufactured by a UK national feed compounder as 3 mm pellets. Milk replacer and piglet feed composition and analyses are shown in Table 2.

Table 2. Analysis of milk replacer and prestarter diets for piglets

Nutrient	Milk replacer	Pre-starter diet
Crude protein %	22	20
Crude fibre %	0	1.9
Crude fat %	14	13
Ash %	7.5	8
Lysine %	2	1.5
Calcium %	0.8	0.6
Sodium %	0.5	0.5
Phosphorus %	0.7	0.6
Vitamin A IU/kg	25000	12500
Vitamin D3 IU/kg	6000	2000
Vitamin E IU/kg	250	250

All piglets were tagged and individually weighed at birth and at approximately three days of age and at weaning. Any cross fostering was done within the first 24 hours within treatments and recorded for each piglet. Feed intake was measured on a daily basis and averaged per sow when group housed, or expressed on an individual basis in the farrowing house. During lactation, weekly feed intakes were calculated at seven day intervals. The actual duration of gestation was calculated for each sow, and no farrowing was induced. The weaning to first service and effective service intervals were recorded post-weaning. Total piglets born, numbers alive, stillborn, mummified, fostered, joint ill incidence and all health treatments and mortality were recorded. Pigs were observed daily for signs of ill health. Sick animals were treated as necessary and recorded by ear tag including the reason for treatment, medication received, date and duration of treatment. For welfare reasons, sick piglets were removed and the weight recorded. Mortality was recorded on an individual piglet basis.

The trial was run according to Harper Adams University standards of ethical treatment of animals. Data was analysed using the GLM procedure of UNISTAT release 5.5 (London, UK), with confidence limits sets at 95% ($P < 0.05$).

3. RESULTS

At the start of the trial, there were no significant differences between the control group or ACT-fed groups of sows. From entry into crates until farrowing, feed intake was higher for the sows fed ACT (P = 0.0014), however intake was significantly lower for the sows fed ACT during the first two weeks post-farrowing (data not shown), leading to an overall decrease in total feed consumed for the ACT group (P = 0.0128; Table 3).

Table 3. Performance of sows fed control or ACT-supplemented feeds during gestation and lactation

Parameter	Control	ACT	P value
Dry sow start weight (kg)	207.799	205.131	0.6639
Sow wt entry farrowing (kg)	257.341	256.947	0.9442
Sow wt at weaning (kg)	228.377	225.404	0.5989
Feed intake duration of gestation (kg/sow):	288.31	287.27	0.5202
Feed eaten entry-farrowing (kg)[1]	18.16^a	19.68^b	0.0014
Feed intake wk 1 after farrowing (kg)	26.43	26.19	0.4956
Total feed eaten post farrowing (kg)[2]	197.83^b	190.55^a	0.0128
Wean to effective service interval (days)	5.6	5.7	0.7343

[1]Sows house 7-8 days pre farrowing. Means with different superscripts vary significantly (P < 0.05). [2]Although offered on the Stotfold scale for intake, sows were allowed to eat to their biological requirement.

Piglets from sows fed ACT were significantly heavier at birth (P = 0.0009) compared to those from the control group. Sows fed ACT had more piglets at day three and weaning (Table 4), with the ACT sow litters being younger and more numerous (P < 0.05) compared to the control. Litter weight at weaning was significantly higher (P = 0.0395) and the piglets were 0.6 days younger in litters from sows fed ACT versus the control group.

4. DISCUSSION

Feeding ACT to sows during gestation and lactation had various benefits, including reduced feed costs due to decreased feed intake post-farrowing. The piglets from the ACT supplemented group were heavier at birth and weaning, more numerous and attained weaning at a younger age compared to the control group. As the ACT supplemented sows consumed less feed during lactation, this suggested that milk production and quality was enhanced by supplementary ACT, despite intakes being significantly lower, and hence sows were using feed more efficiently without loss in body weight. This is in agreement with the published work of Miguel *et al.* (2002) and Lazarevic *et al.* (2010) who reported that ACT supplements made more nutrients available to the sow, by promoting correct gut microflora, promoting body weight and reproductive performance. Although not measured in this trial, it may be assumed (based on previous data) that the sows fed ACT would have had higher levels of Ig in their colostrum and increased milk nutrient profiles, both of which would have improved piglet health, viability and growth. Maintaining sow body weight, achieving heavier piglets at weaning, requiring less time to attain weaning and reduced wean to oestrus intervals are all important to the economics of pig production. Time, labour and feed costs are saved by weaning piglets earlier. In addition, heavier piglets at weaning have been shown to need less time to reach target slaughter weights - which has been calculated at one day saved for each 100 g higher weaning weight (Cooper *et al.*, 2001). In the case of this trial, nearly half a day was saved.

Table 4. Effects of ACT fed to sows on piglet litters

Parameter	Control	ACT	P value
Total piglets born	14.567	14.102	0.3051
No piglets born alive	13.693	13.448	0.5950
No piglets stillborn	0.947	0.712	0.1792
No piglets mummified	0.439	0.408	0.7781
Piglet birth weight (kg)	1.509[a]	1.551[b]	0.0009
Total litter weight born (kg)	20.976	20.996	0.9755
No. piglets day 3 after birth	12.123[a]	12.464[b]	0.0339
Litter weight day 3 (kg)	23.515	23.660	0.7967
Litter age weaning	26.614[b]	26.024[a]	0.0028
No piglets weaned	11.658[a]	12.080[b]	0.0081
Litter weight weaning (kg)	95.9[a]	99.2[b]	0.0395
Age weaning per piglet (days)	26.6[b]	26.1[a]	<0.0001
Weaning wt per piglet (kg)	8.2	8.2	0.6254

Means with different superscripts vary significantly ($P < 0.05$).

5. CONCLUSIONS

The trial demonstrated various economic and performance benefits for sows and their piglets when gestation and lactation diets were supplemented with ACT. Such improvements in performance and reductions in feed costs demonstrated a net improvement of £22 per sow (at current pricing on trial date) per year, which amounted to 3.8:1 return on investment in the breeding herd. This was attributed to the combination of performance factors that were enhanced by feeding ACT. Economics of pig production, and benefits in early piglet growth are typically carried over throughout the growing period until slaughter weight, and it would be expected that pigs from dams fed ACT would have higher subsequent growth rates and higher carcass yields at slaughter.

ACKNOWLEDGEMENTS

The authors would like to thank Alltech Inc. for their financial support of this project.

DECLARATION OF INTEREST

Dr. Jules Taylor-Pickard and Mr. Terry McArdle are employees of Alltech Biotechnology Centre, Ireland.

REFERENCES

1. Brennan K.M., Graugnard D.E., Paul M. and Xiao R. (2012) Effect of maternal Actigen supplementation during gestation and lactation on piglet gut development and gene expression. Journal of Animal Science, 90: 390, 2012
2. Cooper D.R., Patience J.F., Zijlstra R.T. and Rademacher M. (2001) Effect of nutrient intake in lactation on sow performance: Determining the threonine requirementof the high producing lactating sow. Journal of Animal Science, 79: 2378–2387
3. Darragh A.J. and Moughan P.J. (1998) The Composition of Colostrum and Milk. In: The Lactating Sow (Eds: Verstegen Moughan and Schrama) Wageningen Press, Wageningen, Netherlands pp. 3–21
4. Fahmy M.H. (1972) Comparative study of colostrum and milk composition of seven breeds of swine. Canadian Journal of Animal Science, 52: 621–627
5. Göransson L. (1990) The effect of late pregnancy feed allowance on the composition of the sow's colostrum and milk (sic). Acta Veterinaria Scandinavia, 31: 109–115
6. Klaver J., van Kempen G.J.M., de Lange P.G.B., Verstegen M.W.A. and Boer H. (1981) Milk composition and daily yield of different milk components as affected by sow condition and lactation/feeding regimen. Journal of Animal Science, 52: 1091–1097

7. Klobasa F., Werhan E. and Butler J.E. (1987) Composition of sow milk during lactation. Journal of Animal Science, 64: 1458–1466
8. Lazarevic M., Spring P., Shabanovic M., Tokic V. and Tucker L.A. (2010) Effect of gut active carbohydrates on plasma IgG concentrations in piglets and calves. Animal, 4: 938–943.
9. Le Dividich J., Martel-Kennes Y. and Coupel A. (2009) Bio-Mos in diets for sows: effects on piglet performance. Journees Recherche Porcine, 41: 249–250
10. Miguel J.C., Rodriguez-Zas S.L. and Pettigrew J.E. (2002) Practical responses to Bio-Mos in nursery pigs: a meta-analyses. In: Nutritional biotechnology in the feed and food industry. (Eds: Lyons and Jacques) Nottingham University Press, UK.
11. National Research Council (2012) Nutrient Requirements of Swine.
12. Spring P., Geliot P. and Newman K.E. (2003) Effect of dietary mannan oligosaccharide in sow diets on performance of sows and nursing piglets. Proceedings of the 9th Symposium on Vitamins and Additives in Nutrition of Man and Animals. Jena/Thurinigia, Germany.
13. Spring P., Wenk C., Dawson K.A. and Newman K.E. (2000) The effects of dietary mannanoligosaccharides on cecal parameters and the concentrations of enteric bacteria in the ceca of salmonellachallenged broiler chicks. Poultry Science, 79: 205–211.
14. Zou S., McLaren D.G. and HurleyW.L. (1992) Pig colostrum and milk composition: comparisons between Chinese Meishan and US breeds. Livestock Production Science, 30: 115–127.

CHAPTER 15

Fatty Acid Profile and Conjugated Linoleic Acid Content of Milk from Confined Holstein Cows During the Summer and Winter Seasons

Alfonso Martínez-Borraz[1], Silvia Y. Moya-Camarena[1], Humberto González-Rios[1], Jesús Hernández[1] and Araceli Pinelli-Saavedra[1]

[1]*Centro de Investigación en Alimentación y Desarrollo, A.C., Hermosillo, Sonora, México*

1. INTRODUCTION

In recent decades, research in dairy cattle has been focused on the evaluation of factors that may cause changes in the lipid composition of milk, due to the fact that unsaturated fatty acids, conjugated linoleic acids (CLA) and high monounsaturated fatty acids/saturated fatty acids (MUFA/SFA) and polyunsaturated fatty acids/saturated fatty acids (PUFA/MUFA) ratios in milk have shown beneficial properties, including antiatherogenic and anticarcinogenic effects in humans. The principal factors that determine these effects are the breed (Lawless et al., 1999; Wood et al., 1980) and feeding regime (Cooper et al., 2004) in addition to less-studied factors, such as the parity, days in milk (DIM) and extreme temperatures. CLA are a mixture of linoleic acid isomers that contain conjugated double bonds. Studies in experimental animals have demonstrated that CLA has properties that may be beneficial for humans, providing anticarcinogenic, antidiabetic, antiobesity, antiatherogenic and immune stimulatory effects (Huth et al., 2006; Nirvair et al., 2007; Pariza et al., 2001; Parodi., 1999). Milk and dairy products are the primary sources of CLA, and approximately 75 to 90% of the total CLA content in milk fat is represented by *cis*-9, *trans*-11-CLA (Chin et al., 1992; Kay et al., 2004). CLA (9-*cis,* 11-*trans*-CLA) in milk fat is produced in the mammary glands via an endogenous synthesis pathway in which Δ^9-desaturase converts vaccenic acid (*trans*-11 C18:1) to CLA (Bauman et al., 2001; Kay et al., 2005). Under conditions fostered by a certain combination of feed production system, additives, diet, breed, stage of lactation, and season, vaccenic acid is increased in the rumen, which results in a concomitant increase in the 9-*cis*, 11-*trans*-CLA content in milk fat (Griinari et al., 2000; Lock & Bauman, 2004). However, the factors mentioned above have been studied under controlled conditions. In contrast, our study focused on the evaluation of some of these factors on the fatty acid profile and 9-*cis*, 11-*trans*-CLA content in milk under commercial conditions where extreme temperatures occur. To this end, we evaluated the fatty acid profile and 9-*cis*, 11-*trans*-CLA content in Holstein cow milk during the winter (14 °C) and summer (40 °C) on a commercial dairy farm in Northwestern Mexico (Sonora) by studying the effects of feed composition, parity, stage of lactation (DIM), and milk yield.

2. MATERIAL AND METHODS

2.1. Location and sampling

This study was carried out on a commercial dairy farm in Sonora, México, which was located at 29.07 ° longitude and 110.90 ° latitude. A total of 240 Holstein dairy cows were included in the study, and the parity of the cows ranged from 1 to 6. The cows were exposed to all of the environmental elements that can affect the behaviour and milk yield of dairy cattle, including solar radiation, rain, and wind, as shown in Figure 1. The cows were divided into parity groups, including the following: primiparous (P, 1 parity, n = 84); earlier multiparous (EM, from 2 to 3 parities, n = 96); and late multiparous (LM, from 4 to 6 parities; n = 60). The milk was sampled in proportion to the parity of the cows; exclusion criteria included the presence of mastitis and more than 350 DIM.

Figure 1. Physical and environmental conditions of the dairy farm.

The diets were formulated using the Cornell-Penn-Miner Dairy model (Moate et al., 2004), and the ingredients and chemical compositions of summer and winter feed are shown in Table 1. Milking was carried out twice per day (0400 h and 1600 h), and the two milk samples (50 mL each) were combined. Milk samples were collected during the summer (from June to August 2006) and the winter (December 2006 to February 2007). The milk yield was measured using a Waikato MKV milk meter (Waikato MKV, Milking Systems, NZ) and evaluated on the last day of each month, which was standard practice on this dairy farm. The samples were transported on ice to the laboratory and stored at -20 °C for later analysis. Feed samples were collected on each sampling date. The DIM were also recorded.

The meteorological station "El Perico", which was located 5 km from the farm, recorded the daily weather data, and this information was used to calculate the temperature-humidity index (THI) for the winter and summer seasons. The THI was calculated as follows: THI = td – (0.55 – 0.55RH) × (td – 58), where td = the dry bulb thermometer in °F and RH = the relative humidity expressed as a decimal (NOAA, 1976). A THI < 72 was taken as an indicator of zero stress for the dairy cows. THI values from 72 to 79 indicated that the dairy cows were under mild stress, whereas THI values from 79 to 88 indicated that cows were under high stress (Armstrong, 1994).

2.2. Feed analysis

2.2.1. Feed chemical composition

The chemical composition of the diets was analysed by the AgroLab México laboratory (Gómez Palacio, Dgo). The components of the feeds that were evaluated in winter and summer were the crude protein, neutral detergent fibre, acid detergent fibre, ether extract, and net absorbable energy.

2.2.2. Feed fatty acid analysis

Total lipids were extracted with a chloroform and methanol mixture (2:1, vol:vol), as described by Folch et al., (1957). Fatty acid methyl esters (FAMEs) were prepared by the standard method (AOAC, 1995). FAMEs were identified and quantified by gas chromatography using a Varian Star 3400 CX Gas Chromatograph (VARIAN Inc, Walnut Creek, CA) with a DB-23 (J & W Scientific, Folsom CA) capillary column (30 m x 0.25 mm). Helium was used as the carrier gas with a flow rate of 1 mL/min; the airflow was 300 mL/min, and the hydrogen flow was 30 mL/min. The column temperature was initially set at 50 °C and held for 1 min. It was then increased at 10 °C/min to 166 °C, and then at 1 °C/min to 174 °C and held for 30 s; next, it was increased at 2 °C/min to 194 °C and held for 30s. As a final step, the column temperature was increased at a rate of 3.5 °C/min to 215 °C and held for 5 min. The total running time was 42.6 min. Identification of the fatty acid profile was performed with an external standard FAME mix (C4-C24, Sigma, USA).

2.3. Milk quality

The milk composition (fat, protein, lactose, and total solids) was determined by near infrared spectrometry (Milkoscan Minor FT120).

2.3.1. Fatty acid analysis of the milk

2.3.1.1. Extraction

Milk fat extraction was carried out according Luna et al., (2005). In brief, 30 mL of milk was centrifuged at 17,800 × g for 30 min at 8 °C (Beckman Coulter centrifuge, Mod. Allegra 64R). Approximately 350 mg of fat was subsequently removed for lipid extraction with 18 mL of a hexane:isopropanol mixture (3:2 v:v) per g of fat; next, 12 mL of sodium sulphate per g of fat was added as described by Hara & Radin (1978).

2.3.1.2. Methylation

The milk fatty acids were transesterified with methoxide sodium according the method of Christie (1982) as modified by Chouinard et al., (1999). Briefly, the fatty acids were mixed with 2 mL of hexane per 40 mg of fatty acid, and 40 μL of methyl acetate was added. We then vortexed the mixture and added 40 μL of methylation reagent (1.75 mL of methanol mixed with 0.4 mL of 5.4 M sodium methoxide). The mixture was again vortexed and allowed to react for 10 min. We then added 60 μL of termination reagent (1 g of oxalic acid in 30 mL of diethyl ether). Next, the sample was centrifuged for 5 min at 2400 × g at 5 °C, and the hexane layer was removed and transferred to a new tube for evaporation with nitrogen. When the resulting methylated fatty acids had dried, 50 mg was transferred to an amber vial, and 200 μL of hexane was added. At that point, the sample was ready for quantification of fatty acids and 9-*cis*, 11-*trans*-CLA.

2.3.1.3. Identification and quantification of FAMEs by gas chromatography

The FAMEs were identified and quantified by gas chromatography using a Varian Star 3400 CX Gas Chromatograph (VARIAN Inc., Walnut Creek, CA), with a DB-23 (J & W Scientific, Folsom, CA) capillary column (30 m x 0.25 mm). The procedure was identical to that described above for the *FeedAnalysis of Fatty Acids* with the same total running time of 42.6 min. Identification of the fatty acid profile was performed with an external standard fatty acid methyl ester mix (C4-C24, Sigma, USA). Conjugated linoleic acid isomers were identified using 98% pure standards (Matreya Inc., PA).

2.3.2. Desaturase index

The desaturase index was calculated as reported by Kelsey et al. (2003), using four pairs of fatty acids where each pair represented one product and one substrate of Δ^9-desaturase. The fatty acid pairs (product and substrate) were as follows: *cis*-9 14:1 and 14:0; *cis*-9 16:1 and 16:0; *cis*-9 18:1 and 18:0; and *cis*-9 *trans*-11 CLA and *trans*-11 18:1. Kelsey et al. (2003) have defined the desaturase index as follows: [product of Δ^9-desaturase]/[product of Δ^9-desaturase + substrate of Δ^9-desaturase]. For example, the desaturase index for *cis*-9 16:1 would be (*cis*-9 16:1)/(*cis*-9 16:1 + 16:0).

2.4. Statistical analysis

The PROC MIXED procedure provided in the SAS software (SAS Inst. Inc., Cary, NC, 2001) was used to adjust a model for analysing the fatty acid data. The model included the fixed effects of season and month within the season and parity; the random effects included milk yield and DIM (with linear and quadratic effects) and the residual term of animal per parity. The data were considered significant when $P < 0.05$. The response variable was reported as the least square mean ± SEM. The adjusted model was evaluated as follows:

$$Y_{ijklmn} = \mu + A_i + B_j(A_i) + C_k + D_l + E_m + F_n + \varepsilon_{ijklmn},$$

where Y_{ijklm} is the response variable (fatty acid content), A_i is the season, B_j is the month in the season, C_k is the parity, D_l is the days in milk, E_m is the quadratic of days in milk, F_n is the milk yield, and ε_{ijklmn} is the residual error term.

When significant effects of the factors studied were found, mean comparisons of the response variables were performed by the LSMEANS procedure of SAS.

3. RESULTS AND DISCUSSION

3.1. Temperature-humidity index

In desert climates, the ambient temperatures are extreme and vary widely between winter and summer. To our knowledge, the previous studies that have evaluated variations in the milk 9-*cis*, 11-*trans*-CLA content in different seasons had been performed in countries with moderate ambient temperatures. Therefore, our objective was to evaluate the 9-*cis*, 11-*trans*-CLA content in Holstein cow milk in the winter and summer in a hot environment in the Northwestern region of México. We used the THI as an indicator of the heat stress of the dairy cattle in both seasons (Figure 2). As anticipated, during the winter, with an average maximum temperature of 24 °C and 20% RH, the dairy cows were not under heat stress according to the maximum THI measured (average 57 ± 8.7, range from 33 to 74). In contrast, with an average maximum temperature of 43 °C and 50% RH, the dairy cows were under heat stress during the summer (average THI maximum = 80 ± 5.1, range from 63 to 88).

3.2. Feed composition

According to the components and chemical analysis of the feed (Table 1), the summer and winter diets were similar in caloric, nitrogen, and fibre content. Nevertheless, changes in the ingredients were made according to the standard practice on the dairy farm. These changes included the use of whole cottonseeds, maize silage and less alfalfa content in the feed during the winter. The final forage-to-concentrate ratios of the diets were 58:42 and 59:41 for winter and summer, respectively. Accordingly, the fatty acid composition of the diets showed variations between the two seasons (Table 1). Because these compounds are precursors of CLA in both ruminal and mammary gland biosynthesis, the discussion is focused on the concentrations of *cis*-9 18:1, C18:2 and C18:3 in the feed. Despite a lower 18:2 concentration in the summer diet (37.21% of FAMEs) compared to the winter diet (45.45% of FAMEs), the 9-*cis*, 11-*trans*-CLA content in the milk during the summer was higher (9.36 mg/g fatty acids) compared to that observed during the winter (7.03 mg/g fatty acids). However, the concentration of the other CLA precursors *cis*-9 18:1 and C18:3 were higher ($P < 0.05$) during the summer compared to the winter diets. This finding may explain, in part, the increased content of CLAs in the milk during the summer. Although the *cis*-9 18:1 isomer is not a well-recognised CLA precursor, Mosley et al. (2002) have demonstrated the ability of mixed ruminal microbes to convert *cis*-9 18:1 to several *trans* positional isomers, including *trans*-11 18:1, which is a precursor of CLA (Griinari et al., 2000; Lock & Bauman, 2004).

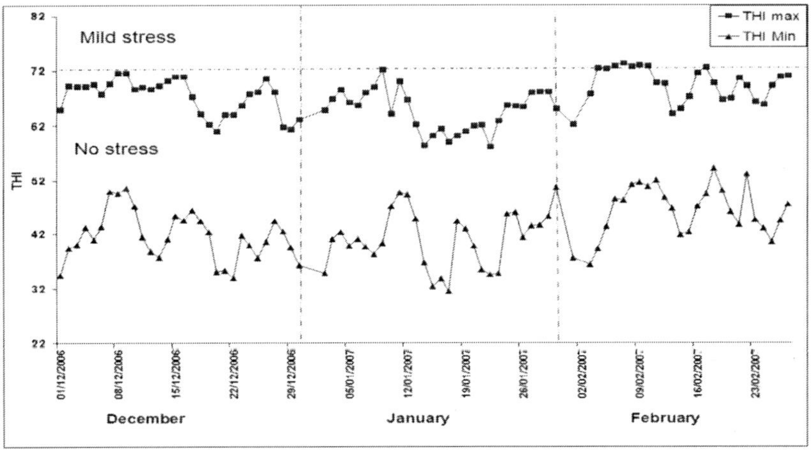

A

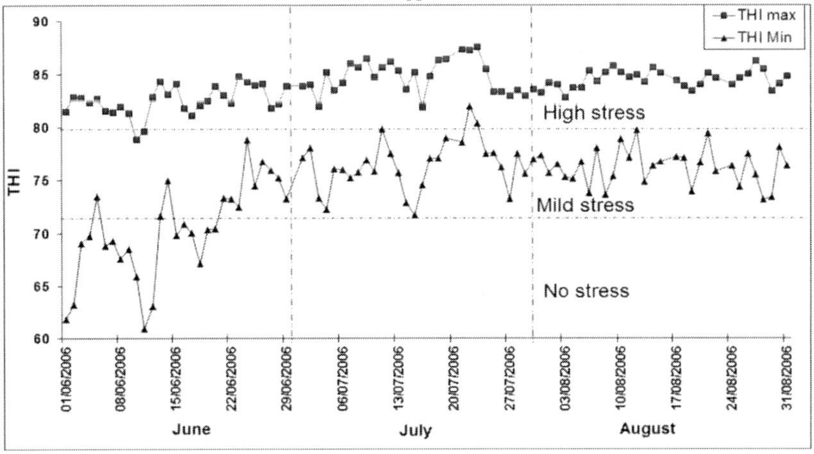

B

Figure 2. Maximal and minimal THI values during the winter (A) and summer (B) and the relation of THI to heat stress in dairy cows.

These data suggest that biosynthesis in the mammary gland may explain the increased milk 9-*cis*, 11-*trans*-CLA content observed during the summer (Soyeurt et al., 2008).

Table 1. Ingredients and chemical composition of the feed in the winter and summer.

	Winter	Summer
Ingredients[1]		
Concentrate	42.21	41.13
Maize ground	21.09	25.49
Whole Cottonseed	6.48	0
Soya 47% Protein	3.53	5.59
Soybean 70	4.27	2.91
Vitamins and minerals	2.28	2.08
Fat mixture[2]	1.42	1.85
Cane Molasses	3.15	3.07
Alfalfa	30.19	58.87
Maize silage	27.60	0
F:C[3]	58:42	59:41
Chemical analysis[1]		
CP	15.7	15.4
NDF	24.2	25.2
ADF	33.6	34.4
EE	5.56	4.86
ENl	1.55	1.54
Fatty acids of the diet[4]		
C12:0	0.00^{b}	0.78^{a}
C14:0	0.00^{b}	0.77^{a}
C16:0	20.54^{a}	20.04^{a}
C18:0	6.71^{a}	5.70^{a}
C18:1 c9	21.56^{b}	26.15^{a}
C18:2	45.45^{a}	37.21^{b}
C18:3	5.74^{b}	9.35^{a}

[1]Values were determined using dry matter (% of DM); [2]Tallow and soybean oil; [3]Forage:concentrate ratio; [4]Fatty acid values were calculated as the percentage of methylated fatty acids; Means within a row with different superscripts were significantly different.

3.3. Milk yield and composition

The different climate conditions also had effects on the milk yield and composition (fat, protein, lactose, and total solids) as shown in Table 2. The milk yield was significantly lower ($P < 0.05$) in the summer (15.8 ± 0.5

kg/d) compared to the winter (18.9 ± 0.4 kg/d). This reduced yield may have been due, at least in part, to the reduced energy intake and the heat stress observed during the summer season, similar to the results reported by others (Fuquay, 1981; Granzin, 2006, Rhone et al., 2008; West, 2003). In addition, the reduction in milk yield agreed with the findings of West (2003), who has reported a reduction in the milk yield of 0.2 kg per unit increase in THI when the THI was above 72. Accordingly, in our study, the milk yield was 3.1 kg/d lower in the summer than in the winter; this reduction corresponded to the 16 THI units above 72 that were recorded during the summer (average 80 ± 5.1, range from 63 to 88). The percentages of fat, protein, lactose, and total solids in the milk were also lower during the summer season. This result may also have been attributable to the reduced feed intake induced by heat stress (Collier et al., 2006).

Table 2. Cow milk yield and composition during the summer and winter.

	Winter	Summer
Milk yield, kg/d	$18.9^a \pm 0.40$	$15.8^b \pm 0.50$
Milk fat, %	$2.69^a \pm 0.10$	$1.91^b \pm 0.06$
Milk protein, %	$3.45^a \pm 0.03$	$3.34^b \pm 0.03$
Lactose, %	$4.41^a \pm 0.02$	$4.30^b \pm 0.02$
Total solids, %	$11.5^a \pm 0.11$	$10.4^b \pm 0.09$

[a,b] Means within a row with different superscripts were significantly different ($P < 0.05$)

3.4. Fatty acid profile in milk

Table 3 shows the significance, determined by the analysis of variance, of the effects of the season, month in the season, parity, DIM, and milk yield on the content of CLA, individual fatty acids, SFA, MUFAs, PUFAs, and the desaturase index. Most of the individual fatty acids in milk and the sum of the saturated fatty acids, the sum of the monounsaturated fatty acids and the sum of the polyunsaturated fatty acids, the MUFA/SFA and PUFA/SFA ratios and desaturase index were affected (P<0.05) by the season. In contrast, only the C14:0, C16:0, and unsaturated fatty acids, such as C16:1, C18:1 t11, C18:2 and 9-*cis*, 11-*trans*-CLA, and the sum of PUFA and both ratios MUFA/SFA and PUFA/SFA were affected by parity. Only the season, month in the season, parity, and DIM had significant effects (P < 0.001) on the milk CLA content. In addition, the season, month in the season, and milk yield had significant effects (P<0.05) on the desaturase activity, as evaluated indirectly by the desaturase index for *cis*-9 14:1.

Table 4 shows the fatty acid profiles of the milk during both seasons. A low content of saturated fatty acids (C14:0, C16:0, and C18:0) and a low sum of saturated fatty acids were observed during the summer. In contrast, the sum of monounsaturated fatty acids, the sum of polyunsaturated fatty acids, 9-*cis*, 11-*trans*-CLA, and the MUFA/SFA and PUFA/SFA ratios were higher in the summer than in the winter. The milk fat samples obtained during the summer had higher contents of 9-*cis*, 11-*trans*-CLA ($P < 0.001$) than those obtained during the winter (9.36 ± 0.28 mg/g fatty acids vs. 7.03 ± 0.29 mg/g fatty acids, respectively). As mentioned above, this difference in the 9-*cis*, 11-*trans*-CLA content was attributable to the summer feed content of 18:2 and mammary gland precursors of CLA (*cis*-9 18:1 and 18:3).

These results show that the attributes of the milk were modified by the season, but this modification of the fatty profile and 9-*cis*, 11-*trans*-CLA content was mainly attributable to the diet because the summer diet (28% more of alfalfa) was richer in PUFA and CLA precursors than the winter diet, promoting an increase in the *Butirivibrio fibrisolvens* isomerase activity and resulting in an increase in 9-*cis*, 11-*trans*-CLA production and MUFAs and PUFAs during the summer.

Table 3. Significance of factors of the adjusted model on individual milk fatty acids.

Fatty acid	Model adjusted term					
	Season	Month (season)	Parity	DIM	DIM[2, 1]	Milk yield
C4:0	*	**	NS	NS	NS	NS
C6:0	**	**	NS	NS	NS	NS
C8:0	**	**	*	*	*	NS
C10:0	**	**	**	*	*	NS
C12:0	*	**	*	NS	NS	NS
C14:0	*	**	*	NS	NS	NS
C14:1	**	NS	NS	NS	NS	*
C15:0	NS	**	*	NS	NS	NS
C16:0	**	*	*	NS	NS	NS
C16:1	NS	0.064	*	*	*	NS
C17:0	NS	**	NS	NS	NS	NS
C18:0	*	*	NS	NS	NS	NS
C18:1, t11	*	**	*	NS	NS	NS
C18:1, c9	**	**	NS	NS	NS	NS
C18:2	**	**	*	NS	NS	*
C18:3	**	**	NS	NS	NS	NS

CLA[2]	**	**	*	**0.051**	**0.070**	NS
C20:0	*	**	NS	NS	NS	NS
Unknown	NS	NS	*	NS	NS	NS
∑ SFA[3]	**	**	NS	NS	NS	NS
∑ MUFA[4]	**	**	NS	NS	NS	NS
∑ PUFA[5]	**	**	*	NS	NS	*
MUFA/SFA ratio	**	**	**	NS	NS	NS
PUFA/SFA ratio	**	**	*	NS	NS	NS
Desaturase index						
cis-9 14:1	**	*	NS	NS	NS	*
cis-9 16:1	*	*	NS	*	*	NS
cis-9 18:1	**	**	NS	*	*	NS
cis-9, *trans*-11 CLA	**	**	NS	*	*	NS

[1]Quadratic term of days in milk; [2]9-*cis*, 11-*trans*-CLA; [3]Sum of saturated fatty acids, [4]Sum of monounsaturated fatty acids, [5]Sum of polyunsaturated fatty acids
*$P < 0.05$; **$P < 0.001$; NS = Not significant

Table 4. Cow milk fat composition (mg/g fatty acids) by season.

Fatty acid	Winter		Summer		P
	Mean ± SE	CI (95%)	Mean ± SE	CI (95%)	
C4:0	1.80 ± 0.18	1.44 – 2.14	0.90 ± 0.12	0.67 – 1.13	*
C6:0	9.69 ± 0.25	9.19 – 10.18	6.45 ± 0.30	5.86 – 7.03	**
C8:0	10.36 ± 0.14	10.07 – 10.64	8.28 ± 0.18	7.90 – 8.64	**
C10:0	26.10 ± 0.40	25.30 – 26.90	21.21 ± 0.41	20.40 – 22.02	**
C12:0	31.70 ± 0.48	30.74 – 32.65	27.74 ± 0.45	26.85 – 28.63	*
C14:0	109.05 ± 0.97	107.13 – 110.96	100.53 ±1.01	98.51 – 102.54	*
C14:1	8.71 ± 0.21	8.29 – 9.13	11.83 ± 0.28	11.27 – 12.39	**
C15:0	11.53 ± 0.13	11.28 – 11.78	11.69 ± 0.19	11.30 – 12.08	NS
C16:0	332.31 ± 2.37	327.62 – 337.00	299.90 ± 2.73	294.54 – 305.37	**
C16:1	14.04 ± 0.30	13.45 – 14.63	15.97 ± 0.38	15.22 – 16.72	NS
C17:0	6.09 ± 0.10	5.88 – 6.30	5.54 ± 0.11	5.32 – 5.75	NS
C18:0	125.18 ± 1.63	121.90 – 128.42	105.86 ± 1.68	102.50 – 109.21	*
C18:1, t11	9.64 ± 0.23	9.17 – 10.11	8.18 ± 0.28	7.62 – 8.73	*

C18:1, *cis*-9	237.75 ± 2.50	232.80 – 242.70	288.23 ± 2.77	282.70 – 293.70	**
C18:2[1]	22.79 ± 0.29	22.21 – 23.37	39.59 ± 0.84	37.92 – 41.26	**
C18:3[2]	2.82 ± 0.08	2.65 – 2.98	4.37 ± 0.09	4.18 – 4.56	**
CLA[3]	**7.03 ± 0.29**	**6.46 – 7.59**	**9.36 ± 0.28**	**8.81 – 9.91**	**
C20:0	1.25 ± 0.07	1.10 – 1.39	0.68 ± 0.08	0.51 – 0.84	*
Others	32.62 ± 0.30	32.01 – 33.22	33.12 ± 0.63	31.86 – 34.38	NS
Σ SFA[4]	665 ± 28.8	659.8 – 670.3	588.8 ± 41.4	581.3 – 596.3	**
Σ MUFA[5]	270.1± 25.3	265.6 – 274.7	324.2 ± 31.2	318.5 – 329.9	**
Σ PUFA[6]	32.3 ± 4.0	31.5 – 33.0	53.8 ± 11.2	51.7 – 55.8	**
MUFA/SFA ratio	0.408±0.007	0.398- 0.419	0.557±0.007	0.539-0.574	**
PUFA/SFA ratio	0.048±0.001	0.047-0.05	0.092±0.001	0.088-0.097	**

[1]Linoleic acid; [2]Linolenic acid; [3]9-*cis*, 11-*trans*-CLA; [4]Sum of saturated fatty acids; [5]Sum of monounsaturated fatty acids; [6]Sum of polyunsaturated fatty acids; *P < 0.05; **P < 0.001; NS = Not significant; N = 240 cows (120 in winter and 120 in summer); ± Standard error

The importance of manipulating the fatty acid composition in the milk, thus favouring a high content of unsaturated fatty that includes 9-*cis*, 11-*trans*-CLA, is due to the fact that their consumption can help to prevent chronic diseases in humans. A decrease in the SFA levels and/or a concomitant increase in the MUFA and PUFA contents of ruminant milk may confer benefits for human health and may provide a basis for marketing claims. The PUFA/SFA ratio in humans is an important risk factor for cardiovascular diseases (Sacks & Katana, 2002; Simopoulos, 1999), and thus the proportions of total unsaturated and PUFAs in regard to the total SFAs are relevant from a human health perspective. Indeed, the PUFA/SFA ratio has been used to calculate the risk factor of foods. According to Wood et al., (2008), the recommended value is at least 0.4, although, others have suggested that the minimum PUFA/SFA ratio is 0.12 (Hoffman, Muller, & Cloete, 2003). However, in our study, the PUFA/SFA ratios for the milk in both seasons (0.048 for winter and 0.092 for summer season) was lower than 0.12. It is important to note that there was no manipulation on the original formulation of the diets to modify the PUFA/SFA ratios obtained.

In contrast, it is well known that most CLA are produced in the mammary gland via Δ^9 desaturase. As shown in Table 5, the activity of this enzyme was higher in summer than in winter for the four substrates (C14:0, C16:0, C18:1, and *trans*-C18:1) of the desaturated products, C14:1, C16:1, *cis* 9-C18:1, and CLA, respectively. Because the C14:0 content in milk fat is produced via de novo synthesis in the mammary gland, desaturation is the only source of C14:1; therefore, the ratio of C14:1 to C14:0 is the best indicator of Δ^9desaturase activity (Lock & Garnsworthy, 2003). Our results showed that the desaturase index was lower for C14:1 and C16:1 compared to 18:1 and CLA. Kelsey et al. (2003) have reported a similar pattern in the desaturase index (0.06, 0.04, and 0.67 for C14:1, C16:1, and 18:1, respectively).

Table 5. Estimated Δ9-desaturase activity in milk fatty acids during the winter and summer.

Desaturase index[1]	Winter		Summer		P
	Mean ± SE	CI (95%)	Mean ± SE[2]	CI (95%)	
cis-9 14:1	0.074 ± 0.001	0.071 – 0.077	0.104 ± 0.002	0.100 – 0.109	**
cis-9 16:1	0.040 ± 0.001	0.038 – 0.042	0.05 ± 0.001	0.048 – 0.052	**
cis-9 18:1	0.65 ± 0.002	0.65 – 0.66	0.73 ± 0.003	0.72 – 0.73	**
c-9, t-11 CLA	0.41 ± 0.007	0.39 – 0.42	0.55 ± 0.008	0.53 – 0.57	**

[1]Calculated as the Δ^9-desaturase product divided by the sum of the Δ^9-desaturase product and substrate. For example, the desaturase index for *cis* 9 16:1 would be (*cis*-9 16:1)/(*cis*-9 16:1 + 16:0) . **P < 0.001; [2]Standard error

Figure 3 shows the frequency distributions of the 9-*cis*, 11-*trans*-CLA content in milk obtained during the winter and summer. In the winter, the CLA content ranged from 3.49 to 10.45 mg/g fatty acids (with a mean value of 6.65 ± 1.4 mg/g fatty acids), with a peak frequency (59/120) within the range of 5.0 to 6.99 mg/g fatty acids. In the summer, the CLA content ranged from 4.03 to 17.86 mg/g fatty acids (with a mean value of 9.83 ± 2.58 mg/g fatty acids), with a peak frequency (40/120) within the range of 9.0 to 10.99 mg/g fatty acids. Additionally, in the summer, 34 out of 120 cows had CLA values above 10.99 mg/g fatty acids. There was a broad variation in the content of 9-*cis*, 11-*trans* CLA in milk within a season and among animals despite the same diet and environmental conditions. Similar results have been reported by other authors (Bell et al., 2006; Lawless et al., 1999; Stanton et al., 1997; Staszak, 2005;).

In the present study, parity affected the CLA content in milk fat as shown in Figure 4. For this analysis, dairy cows were grouped by parity status by season. In the summer, there were no differences in the 9-*cis*, 11-*trans*-CLA content between parity levels, and the range of CLA was from 9 to 10 mg/g fatty acids. In contrast, we found differences between the parity groups and the 9-*cis*, 11-*trans*-CLA content in milk during the winter where primiparous cows had a higher CLA content (6.96 ± 0.2 mg/g fatty acids, P < 0.05) compared to the milk from LM cows (6.17 ± 0.2 mg/g fatty acids), but the difference was not significantly greater than with the milk from EM cows (6.69 ± 0.25 mg/g fatty acids, P > 0.05). To date, studies have been scarce on the effect of parity on the CLA content in milk fat. Peng et al.

(2008) have reported that the milk from primiparous yak cows had a lower CLA content than multiparous yak cows. These authors have also suggested that the differences found may be related to the stage of growth of the mammary glands, which are one of the sites of the synthesis of fatty acids in milk, or that the differences may be related to forage, as there were differences on forage intake in the Yaks of different parities. In contrast, Kelsey et al., (2003) have studied the effect of parity and did not find differences in the milk CLA content between primiparous and multiparous dairy cows. This discrepancy with respect to our results for the winter season led us to hypothesise that the differences in CLA may be due, at least in part, to the hierarchy of the cows within the paddock area. Multiparous cows are generally older, bigger, and heavier than primiparous cows, therefore, they are the first to attain feed (Phillips & Rind, 2002). This hierarchy may have resulted in the consumption by the multiparous cows of high amounts of concentrate, which provoked a decrease in the ruminal pH, which, in turn, induced an adverse environment for the *Butirivibrio fibrisolvens* bacteria in the rumen that are responsible for CLA production.

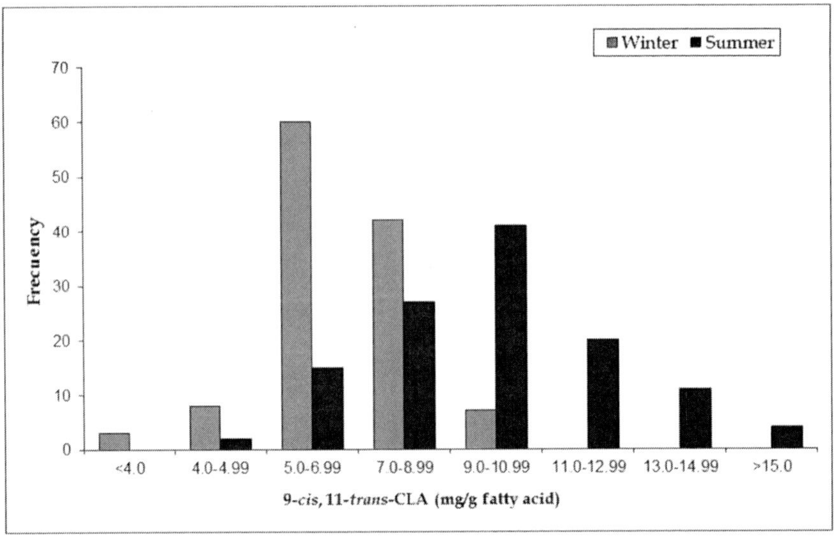

FIGURE 3. Frequency distribution for the 9-*cis*, 11-*trans*-CLA content in cow's milk collected during the winter and summer seasons.

The relationships between the milk CLA content and DIM, milk fat, and milk yield are shown in Figure 5 (Panels A, B, and C, respectively). The DIM had an effect on the milk CLAs with an $R^2 = 0.338$, $P < 0.05$. In our study, the DIM explained a third (33.8%) of the variation in the CLA content. This result contrasted with previous reports that have shown that this factor explained less than 10% of the variation (Lock et al., 2005; Kelsey et al., 2003). The milk fat and milk yield did not have effects on the milk CLA content with $R^2 = 0.063$ and $R^2 = 0.017$, respectively; this result is in agreement with Kelsey et al., (2003).

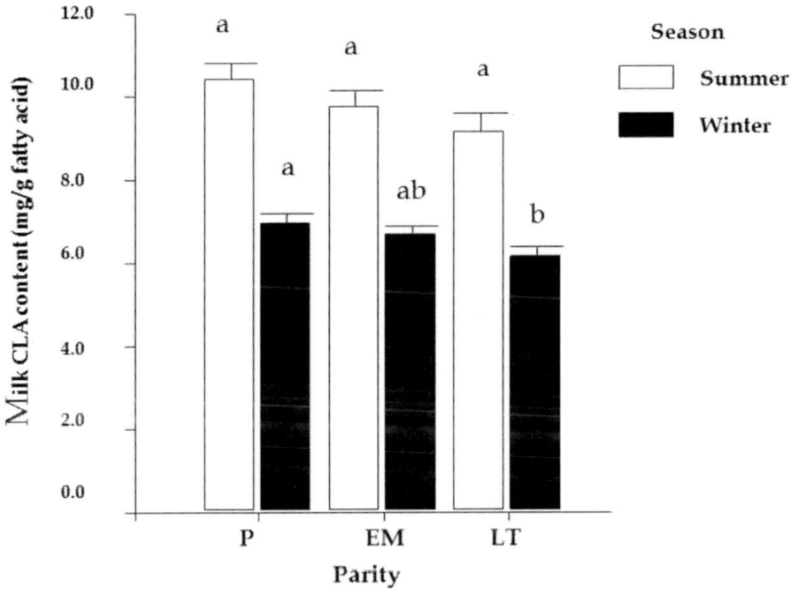

Figure 4. Milk CLA (9-*cis*, 11-*trans*) relative to parity by season. Bars with different superscripts (a, b, and ab) within the season indicate significant differences ($P < 0.05$). Primiparous (P): one parity; early multiparous (EM): dairy cows with two and three parities; late multiparous (LM): dairy cows with four to six parities.

4. CONCLUSIONS

We found that the practical management of the adjustment of the ingredients in the diet to maintain energy intake and avoid decreased milk yield during the summer (when temperatures may exceed 40 °C) was beneficial because this adjustment of the diet allowed an improvement in the unsaturated fatty acid profile and CLA content in the milk, despite the animals' experiencing heat stress. The alterations in the fatty acid profile and CLA content in milk suggest that this milk could be healthier for the consumer in the summer, due to the increased MUFA/SFA and PUFA/SFA ratios.

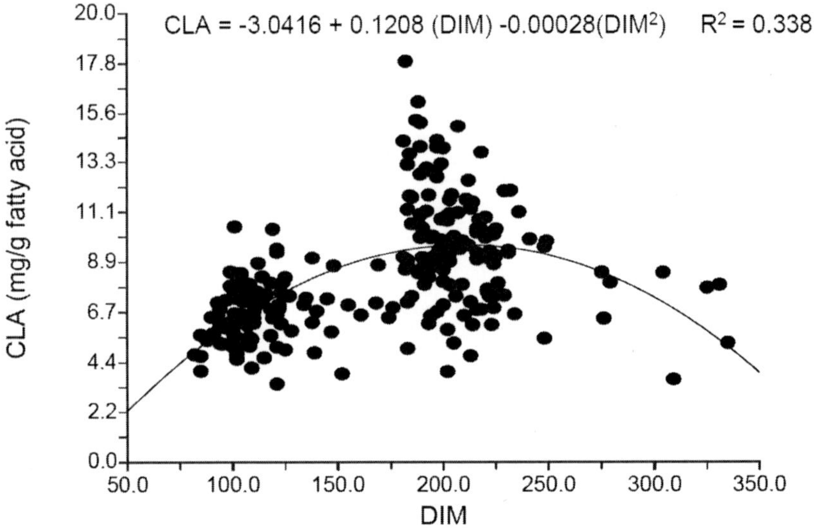

A

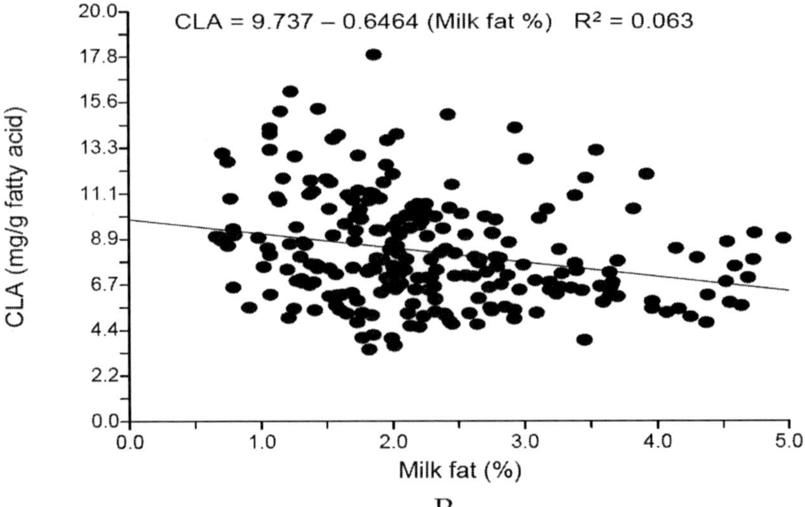

B

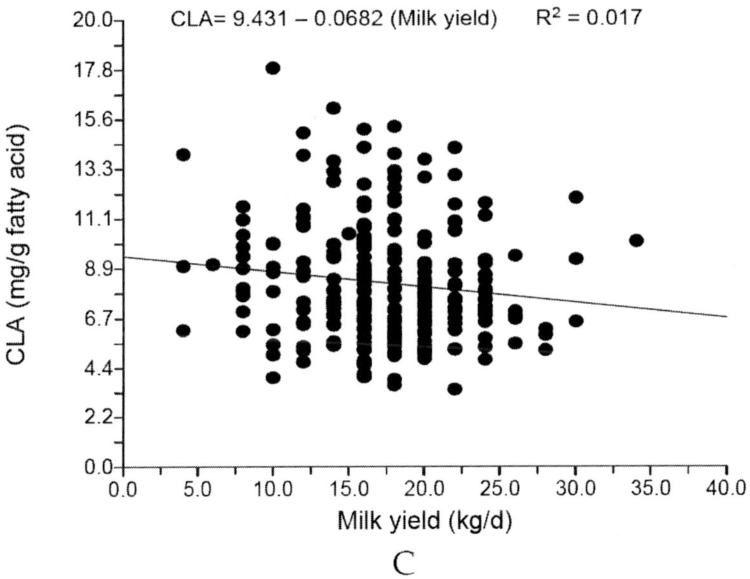

Figure 5. Relationships between the milk CLA content and the (A) DIM, (B) milk fat, and (C) milk yield. Data points represent milk samples collected during both the winter and summer seasons from 240 dairy cows

ACKNOWLEDGEMENT

The authors thank Dr. Arturo Madrid Lopez, Nutritionist, for his advice and assistance. We also thank Ing. Roy Swanson, the manager of the dairy farm where the present study was conducted. We also thank Amparo Nieblas Almada, Ingrid Rebeca Esquerra Brauer and Francisco A. Vazquez Ortiz for their excellent technical assistance.

REFERENCES

1. Armstrong, DV. (1994). Heat stress interaction with shade and cooling. *J. Dairy Sci.* 77:2044-2050.
2. Bauman, DE., Corl, BA., Baumgard, HH., & Griinari, JM. (2001). Conjugated linoleic acid (CLA) and the dairy cow. in Garnstworthy PC. & Wiseman, J. (ed.). *Recent Advances in Animal Nutrition.* Nottingham University Press, ISBN 1897676085 Nottingham, UK. Pp 221-250
3. Bell, JA., JM Griinari, & Kennelly, JJ. (2006). Effect of safflower oil, flaxseed oil, monensin, and vitamin E on concentration of conjugated linoleic acid in bovine milk fat. *J. Dairy Sci*, 89:733-748

4. Chin, SF., Liu, W., Storkson, JM., Ha, YL. & Pariza, M.W. (1992). Dietary sources of conjugated dienoic isomers of linoleic acid, a newly recognized class of anticarcinogens. *J. Food Compos. Anal*, 5:185-197
5. Chouinard, PY.; Corneau, L.; Saebo, A.; Bauman, DE. (1999). Milk yield and composition during abomasal infusion of conjugated linoleic acids in dairy cows. *J. Dairy Sci*, 82:2737–2745
6. Christie, WW. (1982). A simple procedure for rapid transmethylation of glycerolipids and cholesterol esters. *J. Lipid Res*, 23:1072–1075
7. Collier, RJ., Dahl, GE., & VanBaale MJ. (2006). Major advances associated with environmental effects on dairy cattle. *J. Dairy Sci*, 89:1244-1253
8. Cooper, S. L., Sinclair, L. A., Wilkinson, R. G., Hallett, K. G., Enser, M., & Wood, J. D. (2004). Manipulation of the n-3 polyunsaturated fatty acid content of muscle and adipose tissue in lambs. *J. Anim. Sci*, 82: 1461-1470
9. Cunniff, P.(16th Edition).(1995). *Official Methods of Analysis of AOAC International*, Association of Official Analytical Chemists, ISBN 0935584544, Arlington, VA.
10. Folch, J., Lees, M. & Sloane-Stanley, GH. (1957). A simple method for the isolation and purification of total lipides from animal tissues. *J. Biol. Chem*, 226:497-509
11. Fuquay, JW. (1981). Heat stress as it affects animal production. *J. Anim. Sci*, 52: 164-174
12. Granzin, BC. (2006). Cooling and forage supplementation of grass-fec Holstein cows during hot conditions. *Trop. Anim. Health Prod.* 38: 141-150
13. Griinari, JM., Corl, BA., Lacy, SH., Chouinard, PY., Nurmela, KVV. & Bauman, DE. (2000). Conjugated linoleic acid is synthesized endogenously in lactating dairy cows by Δ9-Desaturase. *J. Dairy Sci*, 130: 2285-2291
14. Hara, A. & Radin, NS. (1978). Lipid extraction of tissues with a low-toxicity solvent. *Anal. Biochem*, 90:420–426
15. Hoffman, L.C.; Muller, M., & Cloete, S. W. (2003). Comparison of six crossbreed lamb types: sensory, physical and nutritional meat quality characteristics. *Meat Si,* 65: 1265-1274.
16. Huth, PJ., DiRienzo, DB., & Miller, GD. (2006). Major scientific advances with dairy foods in nutrition and health. *J. Dairy Sci*, 89:1207-1221.
17. Kay, JK., Mackle, TR., Alduist, MJ., Thomson, NA. & Bauman, DE. (2004). Endogennos synthesis of cis-9, trans-11 conjugated linoleic acid in dairy cows fed fresh pasture. *J. Dairy Sci*, 87:369-378
18. Kay, JK., Weber, WJ., Moore, CE., Bauman, DE., Hansen, LB., Chester-Jones, H., Crooker, BA. & Baumgard, LH. (2005). Effects of week of lactation and genetic selection for milk yield on milk fatty acid composition in Holstein cows. *J. Dairy Sci*, 88:3886- 3893
19. Kelsey, JA., Corl, BA., Collier, RJ. & Bauman, DE. (2003). The effect of breed, parity, and stage of lactation on conjugated linoleic acid (CLA) on milk fat dairy cows. *J. Dairy Sci,* 86: 2588-2587
20. Lawless, F., Stanton, C., Escop, PL., Devery, R., Dillon, P. & Murphy, JJ. (1999). Influence of breed on bovine milk cis-9, trans-11 conjugated linoleic acid. *Livest. Prod. Sci*, 62:43- 49
21. Lock, AL. & Bauman, DE. (2004). Modifying milk fat composition of dairy cows to enhance fatty acids beneficial to human health. *Lipids*, 39: 1197-1206
22. Lock, AL., Bauman, DE. & Garnsworthy, PC. (2005). Effect of production variables on the cis-9, trans-11 conjugated linoleic acid content of cows' milk. *J. Dairy Sci*, 88:2714–2717.
23. Lock, AL. & Garnsworthy, PC. (2003). Seasonal variation in milk conjugated linoleic acid and Δ9-desaturase activity in dairy cows. *Livest. Prod. Sci*, 79:47-59.
24. Luna, P., Juárez, M. & de la Fuente, MA. (2005). Validation of a rapid milk fat separation method to determine the fatty acid profile by gas chromatography. *J. Dairy Sci,* 88:3377–3381
25. Moate, PJ., Chalupa, W., Jenkins, TC. & Boston, RC. (2004). A model to describe ruminal metabolism and intestinal absorption on LCFA in dairy cows. *Anim. Feed Sci. Technol,* 112: 79-105.
26. Mosley, SA., Mosley, EE., Hatch, B., Szasz, JI., Corato, A., Zacharias, N., Howes, D. & McGuire, MA. (2007). Effect of varying levels of fatty acids from palm oil on feed intake and milk production in Holstein cows. *J. Dairy Sci*, 90:987-993.

27. National Oceanic and Atmospheric Administration (NOAA). (1976). Livestock hot weather stress. US Dept. Commerce, Natl. Weather Serv. Central Reg., Reg. Operations Manual Lett C-31–76. US Govt. Printing Office, Washington, DC.
28. Nirvair, SK., Neil, EH., & Kent LE. (2007). Conjugated linoleic acid isomers and cancer. *J. Nutr,* 137:2599-2607.
29. Pariza, MW., Yeonhwa P., & Mark EC. (2001). The biologically active isomers of conjugated linoleic acid. *Progress in lipid research,* 40:283-298.
30. Parodi, PW. (1999). Conjugated linoleic acid and other anticarcinogenic agents of bovine milk fat. *J. Dairy Sci,* 82:1339-1349.
31. Peng, YS., Brown, MA., PAS, Wu, JP., Wei, LX., Wu, JL. & Sanbei, DZ. (2008) Fatty acid profile in milk fat from Qinghai Plateau Yak at different altitudes and parities. *Professional Animal Scientist,* 24 :479–487
32. Phillips, CJC. & Rind, MI. (2002). The Effects of social dominance on the production and behavior of grazing dairy cows offered forage supplements. *J. Dairy Sc,* 85:51–59
33. Rhone, JA., Koonawootrittriron, S. & Elzo, MA. (2008). Factors affecting milk yield, milk fat, bacterial score, and bulk tank somatic cell count of dairy farms in the central region of Thailand. *Trop. Anim. Health Prod,* 40: 147-153
34. Sacks, A. M. & Katana, M. (2002). Randomized clinical trials on the effects of dietary fat and carbohydrate on plasma lipoproteins and cardiovascular disease. *Am J Med,* 113: 13-24.
35. SAS. (2001). SAS/STAT User's Guide (release 8.02) SAS Inst. Inc. Cary, North Caroline, U.S.A.
36. Simopoulos, A. P. (1999). Esential fatty acids in health and chronic disease. *Am J Clin Nutr,* 70: 560-569.
37. Soyeurt, H., Dehareng, F., Mayeres, P., Bertozzi, C., & Gengler, N. (2008). Variation of Δ^9-desaturase activity in dairy cattle. *J. Dairy Sci,* 91:3211-3224.
38. Staszak, E. (2005). Conjugated linoleic acid (CLA) content in milk from cows fed different diets. *Folia Biol* (Kraków), 53:103-106
39. Stanton, C., Lawless, F., Kjellmer, G., Harrington, D., Devery, R., Connolly, JF. & Murphy, J. (1997). Dietary influences on bovine milk cis-9, trans 11-conjugated linoleic acid content. *J. Food Sci,* 62:1083-1086
40. West, JW. (2003). Effects of heat-stress on production in dairy cattle. *J. Dairy Sci,* 86:2131–2144
41. Wood, J. D., Macfie, H. J., Pomeroy, R. W., & Twinn, D. J. (1980). Carcass composition of four sheep breeds. The importance of type of breed and stage of maturity. *Anim Prod,* 30: 135-152
42. Wood, J. D., Enser, M., & Fisher, A. V. (2008). Fat deposition, fatty acid composition and meat quality: a review. *Meat Sci,* 78: 343-358

CHAPTER 16

Intermittent Suckling in Combination with an Older Weaning Age Improves Growth, Feed Intake and Aspects of Gastrointestinal Tract Carbohydrate Absorption in Pigs after Weaning †

Diana L. Turpin [1,*], Pieter Langendijk [2,‡], Tai-Yuan Chen [2] and John R. Pluske [1]

[1]*School of Veterinary and Life Sciences, Murdoch University, Murdoch, WA 6150, Australia*

[2]*South Australian Research and Development Institute, Roseworthy Campus, JS Davis Building, Roseworthy, SA 5371, Australia*

SIMPLE SUMMARY

The weaning of piglets involves abrupt changes that often cause reduced production and predisposition to gastrointestinal tract dysfunction. Increasing the weaning age or using gradual weaning are techniques that can improve post-weaning performance. In this study, it is shown that an older weaning age (day 35) in combination with intermittent suckling (daily separation of piglets from the sow for 8 h in the week before weaning) improved growth rate, feed intake and aspects of gastrointestinal tract carbohydrate absorption after weaning. However, intermittent suckling and weaning at 28 days did not seem to improve post-weaning performance in this study.

ABSTRACT

This study tested the hypothesis that intermittent suckling (IS) with or without an older weaning age would improve post-weaning gastrointestinal tract (GIT) carbohydrate absorptive capacity in pigs while reducing post-weaning stress and aspects of the inflammatory response. Three weaning regimes using primiparous sows were compared: (1) conventional weaning (CW28) (n = 22), where piglets were weaned conventionally at day 28; (2) IS28 (n = 21), where IS started at day 21 until weaning at day 28; and (3) IS35 (n = 21), where IS started at day 28 until weaning at day 35. Sugar absorption tests (10% mannitol or 10% galactose) were used to measure GIT absorptive capacity. All measured parameters were compared in relation to weaning across treatments (i.e., different physiological ages were compared). The IS35 pigs grew fastest in the 12 days after weaning ($p < 0.01$) and had the highest solid feed intake before and after weaning ($p < 0.05$). Irrespective of treatment, pre-weaning mannitol levels were higher than post-weaning levels ($p < 0.01$), whereas post-weaning galactose levels were highest in IS35 pigs ($p < 0.01$). Cytokine data did not show any treatment effects. In conclusion, these data suggest that IS in combination with an older weaning age (day 35) improved post-weaning adaptation as evidenced by improvements in performance measures and galactose absorption. However, IS28 did not improve post-weaning performance.

Keywords: pig; intermittent suckling; weaning; intestine

1. INTRODUCTION

Weaning under commercial conditions at three to four weeks of age is a stressful and abrupt process, with pigs typically having difficulty adapting to the sudden dietary, environmental and social changes involved. Consequently, their level of feed intake in the immediate post-weaning period is usually insufficient to meet maintenance energy requirements [1]. Inadequate feed intake also contributes to intestinal inflammation [2], thereby compromising gastrointestinal tract (GIT) structure and function, predisposing piglets to enteric disease such as post-weaning collibacillosis [3]. Furthermore, stress associated with the process of weaning compromises GIT barrier function, mediated in part by the upregulation of receptors for corticotrophin releasing hormone (CRH) [4]. Collectively, these processes cause a post-weaning 'growth check' and contribute to poorer production efficiency and increased mortality and (or) morbidity. This effect appears to be more marked in progeny from primiparous sows as evidenced by reduced growth rates, a higher acute phase protein response and higher mortality rates than sow progeny during the nursery period (day 28 to 60) [5].

Increasing weaning age (five to seven weeks) improves GIT structure [6], increases post-weaning feed intake [7] and reduces stress [7]. However, an extended lactation causes fewer litters per sow per year and may impact production in the breeding herd [8]. Intermittent suckling (IS) is a process whereby piglets are separated from the sow for a specified period of time each day in the latter stage of lactation, and has the potential to uncouple sow reproductive performance from weaning age. A reduction in suckling stimulus increases the percentage of sows ovulating [9], which, in turn, may permit mating during lactation [10,11]. By mimicking the increasing amount of time the sow would naturally spend away from her piglets, IS also has the potential to increase solid feed intake in piglets before weaning, resulting in improved dry feed intake and growth after weaning [9,12]. Additionally, it has been suggested that short periods of separation from the sow with [13] or without [14] fasting before weaning could induce some essential adaptive changes (structural and functional) in the GIT, resulting in quicker maturation, partly protecting piglets from some of the adverse effects of weaning. A study by Berkeveld et al. [15], in which piglets were subjected to IS for seven days before weaning, reported favourable post-weaning outcomes with respect to small intestinal (SI) morphology, which were improved even more when IS was combined with an extended lactation (day 33 versus day 26). However, pre-weaning GIT adaptive changes were not examined in the study by Berkeveld et al. [15], and neither were stress and inflammatory responses.

The hypothesis tested in this study was that piglets subjected to IS would have improved post-weaning performance, GIT absorptive capacity of selected carbohydrates and a reduced inflammatory and stress response, with minimal effect on pre-weaning performance compared to piglets from a conventional weaning (CW) regimen. It was anticipated that extending lactation by one more week in IS piglets (weaning at day 35 versus day 28) would improve post-weaning performance and GIT absorptive capacity to a greater extent. Progeny from primiparous sows was used, as there was limited data on their responses to IS regimens.

2. MATERIALS AND METHODS

The experimental design and procedures were approved by the Animal Ethics Committees of Primary Industries and Resources South Australia (07/12) and Murdoch University (N2553/13).

2.1. Animals, Housing and Diet

The experiment was a completely randomised block design conducted in six replicates between April and October 2013 at the Pig and Poultry Production Institute (PPPI) (Roseworthy Campus, University of Adelaide, Roseworthy, Australia). Sixty-four primiparous sows (Large White and Large White Landrace terminal line) and their litters were used in six batches consisting of 9, 11, 11, 10, 11 and 12 sows. Sows were mated at the experimental farm and group-housed during gestation. A week before farrowing, sows were moved into individual farrowing crates (2.1 m × 0.6 m) within farrowing pens (2.1 m × 1.92 m). Once born, litters were provided with an infrared lamp until day 10, and floor heating until weaning. Artificial lighting was provided between 08:00 and 17:00. Litter size varied at birth from 3 to 15 live piglets and was standardised within three days of farrowing by cross fostering within each batch, resulting in an average of 11 piglets per sow. Within 24 h of farrowing, piglets received a 1 mL intramuscular (IM) iron injection (Feron 200 + B12, 200 mg/mL iron dextran and 40 μg/mL cyanocobalamin; Bayer Healthcare, Pymble, NSW 2073, Australia), 0.3 mL IM of Ceftiofur (50 mg/mL; Pfizer, West Ryde, NSW 2114, Australia), and their eyeteeth were clipped and their tails cut and cauterised. At three days after farrowing, all piglets received an oral 1 mL dose of Toltrazuril (50 g/L; Bayer Healthcare, Pymble, NSW 2073, Australia), a repeat 0.3 mL IM of Ceftiofur (50 mg/mL), and 2 mL IM of a *Mycoplasma hypopneumoniae* vaccination (Pfizer, West Ryde, NSW 2114, Australia).

All litters had unlimited access to drinking water from one nipple drinker per farrowing or weaner pen. All piglets were offered creep feed (15.7 MJ digestible energy (DE)/kg; CP, 220 g/kg; 0.88 g available Lysine (av. Lys)/MJ DE; Lienerts Australia 700 Creep Pellet, Roseworthy, South Australia) ad libitum from a rotary feeder with hopper (one feeder per farrowing crate) from 10 days of age before weaning and from a 3-hole weaner hopper (one feeder per weaner pen) after weaning. Sows were fed to appetite with a standard lactation diet (14.1 MJ DE/kg; CP, 160.9 g/kg; 1.05 g av. Lys/MJ DE) during the 28 or 35 days lactation period. Sows had ad libitum access to water through one nipple drinker per farrowing pen.

2.2. Treatments

Within each batch, sows were allocated to one of three weaning regimens: (1) conventional weaning (CW28) (n = 22); (2) IS with a conventional weaning age (IS28) (n = 21); and (3) IS with an older weaning age (IS35) (n = 21). The number of litters per treatment per batch ranged from three to four litters. Litters from each of the three treatment groups within a batch were housed in the same farrowing room. Within each weaning regime, the beginning of the experimental procedure (day 0) was designated as the day on which most of the litters were born within that treatment.

The CW28 litters had continuous access to the sow during the 4-week lactation period (day 0 to day 28). In the IS treatments, sows were separated from the litter from 08:00 to 16:00 h daily in the week before weaning (day 21 to day 28 and day 28 to day 35 for IS28 and IS35, respectively). During separation, sows were exposed to a boar for 20 min twice daily in a detection mating area (as part of another experiment measuring ovulation rates; Chen et al. [16]) and then housed individually in a separate building. Within each litter, at day 14, five piglets closest to the average body weight of that litter were selected and individually ear-tagged. Of the piglets used, 52% were female and 48% were males. At weaning, each litter was removed from the farrowing pen. The five individually tagged piglets within each litter remained together and were moved to pens (2.09 m × 1.35 m, consisting of 1.26 m^2 slatted flooring and 1.56 m^2 solid flooring), while the rest of the litter was returned to the herd. After day 46, the five individually tagged piglets from each litter were also returned to the herd.

2.3. Measurements

Piglets were individually weighed at days 3, 14 (batches 4, 5 and 6 only), 20, 23 (CW28 and IS28 only), 27, 30, 34, 37 (IS35 only), 41, 46 and at 90 days of age. The weights of the individually tagged piglets were recorded to ensure that the performance of the litter before weaning corresponded to the subset of piglets that were followed after weaning. Starting at day 20 (CW28 and IS28) and day 27 of lactation (IS35), creep feed residues were measured on the same days as piglet weights and average daily feed intake (ADFI) was calculated. No food wastage was observed due to the design of the rotary feeder; therefore, disappeared creep feed was considered eaten.

A subsample of piglets from sows in three of the batches (a total of nine piglets per treatment) were subjected to sugar absorption tests (SAT) as markers of intestinal absorption based on methodologies described by Thymann et al. [17], Cox et al. [18] and Berkeveld et al. [19]. Briefly, two piglets were selected per litter and sugar absorption across the GIT using either 10% mannitol (≥98%; Sigma Aldrich, St Louis, MO, USA) (5 mL/kg body weight (BW)) or 10% galactose (≥99%; Sigma Aldrich, St Louis, MO, USA) (5 mL/kg BW) was assessed. The mannitol SAT was a longitudinal study performed 3 days before (4 days after the start of IS for IS28 and IS35 piglets) and 4 days after weaning. A different piglet was used for the galactose SAT, which was performed 4 days after weaning. Piglets in both tests were fasted for 3 h by separation from the sow and removal of creep feed (pre-weaning SAT) or by removal of solid feed (post-weaning SAT), but had free access to water. Piglets from the CW28 group, selected for fasting during the pre-weaning phase, were housed with three other piglets from their litter in pens in a separate room to the experiment litters. To keep conditions consistent, selected piglets from the IS28 and IS35 groups were also housed in pens in a separate room with three other piglets from their litter during the fasting period. An oral dose containing the sugar solution was then administered via a nasogastric tube. Twenty minutes after administration, a blood sample was taken by venipuncture of the jugular vein. The sample was collected in a lithium-heparin coated and an ethylenediaminetetraacetic acid (EDTA)-coated tube and immediately chilled on ice. The blood was centrifuged for 20 min at 2800× *g* at 4 °C and aliquots of plasma were stored at −80 °C.

2.4. Determination of Plasma Mannitol and Galactose Concentrations, Cytokine Concentrations, and Cortisol and CRH Concentrations

Plasma mannitol and galactose concentrations were determined using commercial kits (Abcam, ab 155890 D-Mannitol colorimetric assay kit; ab83382 Galactose assay kit, Waterloo, NSW, Australia) in accordance with the manufacturers' instructions. Plasma from piglets dosed orally with mannitol was analysed for interferon (IFN)-γ, interleukin (IL)-1α, IL-1β, IL-6, IL-10 and tumor necrosis factor (TNF)-α using a multiplex cytokine assay system according to the manufacturers' instructions (product code MPPCYTMAG23 K06; Millipore; Billercia, MA, USA). Samples and standards were read on a multiplex magnetic bead reader (Bio-plex, Bio-Rad 200, Hercules, CA, USA) and results were analysed using Bio-plex manager's software (version 5, Bio-Rad, Hercules, CA, USA).

Plasma from the piglets dosed orally with mannitol was also analysed for cortisol and CRH. Plasma cortisol and CRH were determined using commercially available ELISA kits (Enzo Life Sciences, Cortisol ELISA kit, AD-901-071, Farmingdale, NY, USA and BioSource Porcine Corticotrophin Releasing Hormone ELISA kit, MBS267253, San Diego, CA, USA) in accordance with the manufacturers' instructions with the exception of the optical density for cortisol, which was read at 415 nm instead of the recommended 405 nm. The intra-assay coefficient of variation (CV) for cortisol was 10.5% (low), 6.6% (medium) and 7.3% (high), and ≤8% for CRH. Samples for cortisol and CRH were diluted to 1:25 and 1:40, respectively.

2.5. Statistical Analysis and Presentation of Data

Data are presented as means with overall standard error mean (SEM). Performance data were analysed using the generalized linear model (GLM) procedures of SAS (SAS Inst. Inc., Cary, NC, USA) with the litter as the experimental unit. Data for BW, average daily gain (ADG) and ADFI are expressed on a per piglet basis. Body weights were averaged per litter for each day of measurement. Day 3 weights influenced the values for ADG for the period before treatment intervention, eight and five days before weaning, as well as the weaning weights (taken on day 27 for CW28 and IS28 and day 34 for IS35) and the weights on the last day of the experiment (day 46) ($p < 0.01$). Therefore, effects on ADG were always corrected for BW at day 3 for the above-mentioned time points. Results from the GLM for litter ADG between 5 days and 1 day before weaning yielded the same results as that of the subsample of piglets selected for weaning. Data on BW, ADG and ADFI were normally distributed and analysed using the following model:

$$Y_{ij} = \mu + TRT_i + bw3 + e_j,$$

where μ is the overall mean, TRT_i is the treatment, bw3 is the body weight at 3 days of age and e_j is the residual error.

The growth check experienced over the weaning period was calculated according to a previous report [20] ($ADG_{\text{2 to 6 days after weaning}} - ADG_{\text{5 to 1 days before weaning}}$).

Plasma mannitol, cortisol, CRH and cytokine concentrations were analysed on a per pig basis (one piglet per litter) using the GLM procedures of SPSS (IBM Corp, Version 21, Armonk, NY, USA) using the following model:

$$Y_{ij} = \mu + TRT_i + e_j,$$

where μ is the overall mean, TRT_i is the treatment, and e_j is the residual error. Whenever there was an overall treatment effect, differences between two treatments were tested using post hoc Tukey test, maintaining overall Type II error, and litter as the experimental unit. A GLM was also used to compare differences between days within treatment.

Data for galactose and IL-10 were negatively skewed, requiring transformation to force normality using a $\log_{10}$transformation before analysis. The mean values and confidence intervals were then back-transformed and expressed as least square means with 95% confidence intervals. All correlations were investigated using the Pearson's correlation coefficient. Preliminary analyses were performed to ensure no violation of the assumptions of normality and linearity. Interpretation of batch variation was not reliable due to the small numbers of litters per treatment per batch and therefore batch was not included in any of the statistical models. Statistical significance was accepted at $p \leq 0.05$ and a trend was considered at $p \leq 0.1$ and $p > 0.05$.

3. RESULTS

3.1. Production Measurements

There was no difference in piglet BW between the three groups from 3 to 46 days of age (Table 1). At 90 days of age there was also no difference in average BW between the groups, resulting in a similar ADG from birth to this time (an average of 0.6 kg per day for all treatment groups, data not shown). Before any treatment intervention, there was a trend for CW28 to have a lower ADG than the IS35 group (Figure 1; $p < 0.1$). However, between five and eight days before weaning, after the start of IS, there was no treatment effect (Figure 1; $p > 0.05$) for ADG. Compared to CW28, IS reduced the growth rate prior to weaning in the IS28 group (Figure 1; $p < 0.01$) but not in the IS35 group. Weaning was associated with a growth check in the CW28 group (−34 g/day); however, in both IS groups, there was no such growth check (105 g/day and 160 g/day, for IS28 and IS35, respectively). Between one day before weaning to two days after weaning, two days after weaning to six days after weaning, and 6 days after weaning to 12 days after weaning, the IS35 group maintained a higher ADG ($p < 0.01$) than the IS28 and CW28 groups (Figure 1).

3.2. Piglet Feed Intake

Two days after the start of IS (between five and eight days before weaning), the IS35 treatment group were eating more than twice the amount of the CW28 and IS28 groups (Table 2; $p < 0.05$). While the changes in feed intake from one and five days before weaning remained negligible for CW28 and IS28, the feed intake of IS35 piglets continued to increase, remaining higher than CW28 and IS28 (Table 2; $p < 0.001$). In the peri-weaning period (one day before weaning to two days after weaning), the CW28 and IS28 groups more than doubled their feed intake compared with pre-weaning values. There was no treatment effect ($p > 0.05$) between CW28 and IS28 groups during the post-weaning period; however, the IS35 group continued to have the highest feed intake ($p < 0.001$) in the 12 day after weaning (Table 2).

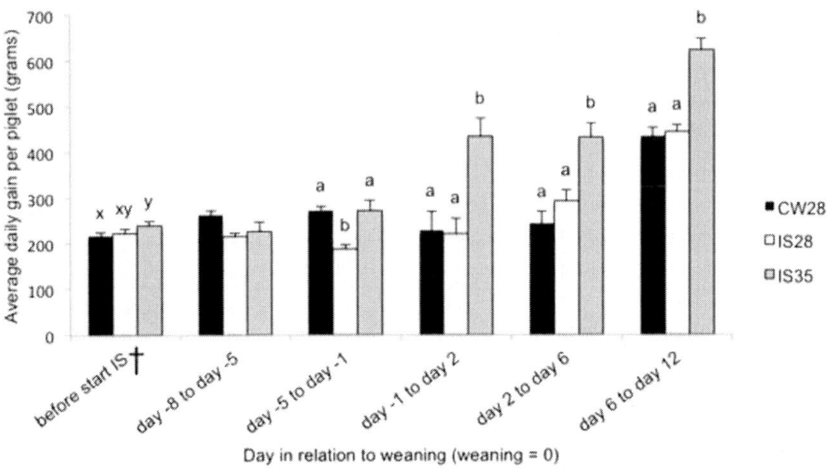

Figure 1. The average daily gain (ADG) of piglets from birth (day three) to 12 days after weaning in conventional weaning and intermittent suckling with or without an older weaning age. The figure depicts the results of three different weaning regimes (CW28 = conventional weaning (n = 22); IS28 = intermittent suckling starting at day 28 (n = 21); and IS35 = intermittent suckling starting at day 28 (n = 21). [ab] Values on a specific day not having the same superscript are significantly different. [xy] values on a specific day not having the same superscript are trends. † Before the start of the intermittent suckling (IS) treatment intervention for IS28 and IS35.

When data from all treatment groups were combined, ADFI measured between five and eight days before weaning was highly correlated to ADFI between one to five days before weaning (r = 0.77, $p < 0.001$) and feed intake from one day before weaning to two days after weaning ($r = 0.72, p < 0.001$). However, the strongest correlation was seen between ADFI from one to five days before weaning and one day before weaning to two days after weaning (r = 0.80, $p < 0.001$). A weaker positive correlation also existed between growth rate at one day before weaning (between one to five days before weaning) and feed intake at two days after weaning (between one day before weaning and two days after weaning) ($r = 0.55, p < 0.001$).

Table 1. Effect of intermittent suckling with and without older weaning age on the body weight of piglets between 3 and 90 days of age.

	Treatment [1]			SEM	*p* Value
	CW28	IS28	IS35		
BW, kg					
Day [2]					
3	1.8	1.6	1.6	0.08	0.27
14 [3]	3.7	3.7	3.7	0.24	0.97
20	5.5	5.5	5.0	0.23	0.30
27	7.4	6.8	6.9	0.23	0.19
30	8.1 *	7.5 *	7.6	0.24	0.20
34	9.0	8.7	8.7	0.27	0.55
37	-	-	10.0 *	0.36	-
	Treatment [1]			**SEM**	***p* Value**
	CW28	**IS28**	**IS35**		
41	12.1	11.8	11.7	0.35	0.72
46	14.8	14.5	14.5	0.43	0.83
90	75.2	77.1	74.7	1.51	0.50

Table 2. Effect of intermittent suckling with and without an older weaning age on the average daily feed intake of piglets before and after weaning.

	Treatment [1]			SEM	*p* Value
	CW28	IS28	IS35		
ADFI (g)					
Days before weaning					
−8 to day −5 [2]	35 [a]	30 [a]	75 [b]	8.3	<0.05
−5 to day −1	43 [a]	33 [a]	114 [b]	9.1	<0.001
Days after weaning					
−1 [3] to day 2	111 [a]	106 [a]	255 [b]	14.9	<0.001
2 to day 6	274 [a]	290 [a]	435 [b]	23.3	<0.001
6 to day 12	459 [a]	511 [a]	670 [b]	25.6	<0.001

[1] CW28 = Conventional weaning (n = 22); IS28 = Intermittent suckling starting at day 21 (n = 21); IS35 = Intermittent suckling starting at day 28 (n = 21). [2] Day in relation to weaning; the minus symbol is the day of measurement prior to weaning (e.g., −8 is eight days before weaning).[3] Creep feed intake was measured the day before weaning. [a,b] Values within a row not having the same superscript are significantly different. ADFI: average daily feed intake.

3.3. Sugar Absorption Tests

Mannitol concentrations did not differ ($p > 0.05$) between treatments at three days before or four days after weaning (Table 3). When data from the IS groups were combined and compared to CW28, there was a tendency for the CW28 group to have higher mannitol concentrations than IS at three days before weaning (1250 ± 220.8 nmol/mL versus 811 ± 156.1, respectively; $p < 0.1$). The CW28 piglets had the biggest decrease in mannitol concentration when comparing pre-weaning to post-weaning values within treatment ($p = 0.01$). The IS28 group showed a trending decrease in mannitol concentrations from before weaning to after weaning ($p = 0.1$), and there was no difference in IS35 piglets for pre-weaning versus post-weaning mannitol levels. When data from all three treatment groups were combined, values for mannitol concentrations before weaning were higher than values after weaning (957 ± 131.4 nmol/mL versus 431 ± 41.5 nmol/mL, respectively; $p < 0.01$). Four days after weaning, the IS35 group had the highest galactose concentrations (Table 3; $p < 0.01$).

Table 3. Effect of intermittent suckling with and without an older weaning age on plasma mannitol and galactose concentrations (nmol/mL) of piglets before and after weaning.

		Treatment [1]			SEM	p Value
		CW28	IS28	IS35		
	Day [2]					
Mannitol (nmol/mL)	−3	1250	874	747	224.6	0.28
	4	471	411	417	79.7	0.83
Galactose [3] (nmol/mL)	4	70 [a] (34.3–143.9)	133 [ax] (64.9–272.3)	316 [by] (135.2–751.6)		0.03

[1] CW28 = Conventional weaning (n = 9); IS28 = Intermittent suckling starting at d 21 (n = 9); IS35 = Intermittent suckling starting at d 28 (n = 9). [2] Day in relation to weaning, the minus symbol is the day of measurement prior to weaning (e.g., −3 is three days before weaning). [3] Data were logarithmically transformed before being subjected to the generalized linear model (GLM). Values were then back-transformed and expressed as least square means with 95% confidence intervals (in parenthesis). [a,b] Values within a row not having the same superscript are significantly different. [x,y] Values within a row with different superscripts are trends.

3.4. Cytokine Concentrations

A high proportion of the samples for IFN-γ, IL-α, IL-1β and IL-6 fell below the detection limit of the respective standard curve, meaning that the remaining sample size was insufficient for statistical analysis. However, a sufficient number of samples read in the range of the standard curve for IL-10 and TNF-α. Weaning had an effect on IL-10 concentrations within the CW28 and IS28 treatment groups, with higher post-weaning values compared with pre-weaning values (137 (66.4–282.5) versus 335 (186.6–599.8) pg/mL for CW28, $p < 0.05$; and 153 (71.9–326.6) versus 258 (147.9–450.8) pg/mL for IS28, $p < 0.05$). Although the post-weaning IL-10 concentration for IS35 piglets was numerically higher than the pre-weaning value (126 (54.5–290.4) versus 288 (160.7–516.4) pg/mL), it was not statistically significant. Trends within the IS35 treatment group between pre-weaning and post-weaning values were evident with TNF-α concentrations increasing after weaning (25 ± 10.4 pg/mL versus 44 ± 29.1 pg/mL, $p < 0.1$).

3.5. Cortisol and CRH Concentrations

The CW28 group had the highest ($p < 0.05$) cortisol concentration before weaning compared with IS35, and tended ($p < 0.1$) to have a higher cortisol concentration than IS28 (Table 4). This effect disappeared by four days after weaning. The IS35 group had the highest plasma CRH concentration three days before weaning compared with the other treatment groups ($p < 0.05$), but there was no treatment effect for CRH at four days after weaning. Overall, when data from all treatments were combined, the pre-weaning cortisol values were higher than the post-weaning values (118 versus 75 ng/mL; $p < 0.05$). There was no significant difference between pre-weaning and post-weaning values for CRH (Table 4). When comparing pre-weaning and post-weaning values within treatment, the CW28 and IS28 groups both had a trending decrease in cortisol from before to after weaning ($p = 0.10$ and $p < 0.1$, respectively), and there was no change for the IS35 group ($p > 0.05$). There was no change ($p > 0.05$) in CRH concentration within any of the treatment groups over the weaning period.

4. DISCUSSION

The aim of this study was to determine if IS with or without an older weaning age provides an adaptation advantage for the GIT during the pre-weaning period, possibly through a reduction in the stress and inflammatory responses, resulting in improved post-weaning piglet performance and GIT function. Results from the current study showed that IS for one week before weaning at day 28 made no difference to the absorption of mannitol and galactose across the GIT compared with a conventional weaning regime. There was also no difference in the cytokine response, and, in contrast to results from previous studies [9,12,15], post-weaning growth and solid feed intake was not increased by IS. Intermittent suckling for one week in combination with an older weaning age (day 35), however, improved feed intake in the 8 days before and the 12 days after weaning, improved growth rate within 12 days of weaning, and improved galactose absorption at 4 days after weaning. There was no effect on the cytokine response. A CW group with an older weaning (day 35) could not be included in this study, and, therefore, comparing measured parameters of IS35 piglets with that of CW28 and IS28 piglets implicates the effects of both IS and an older weaning age.

Table 4. Effect of intermittent suckling with and without an older weaning age on cortisol and corticotrophin releasing hormone (CRH) concentrations (ng/mL) of piglets before and after weaning.

		Treatment [1]			SEM	*p* Value
		CW28	IS28	IS35		
	Day [2]					
Cortisol (ng/mL)	−3	160 ^{ax}	110 ^{ay}	83 ^{b}	19.5	0.03
	4	96	60	72	23.5	0.55
CRH (ng/mL)	−3	10 ^{a}	9 ^{a}	13 ^{b}	1.1	0.03
	4	10	10	12	1.0	0.21

[1] CW28 = Conventional weaning (*n* = 9); IS28 = Intermittent suckling starting at day 21 (*n* = 9); IS35 = Intermittent suckling starting at day 28 (*n* = 9). [2] Day in relation to weaning, the minus symbol is the day of measurement prior to weaning (e.g., −3 is three days before weaning). [a,b] Values within a row not having the same superscript are significantly different. [x,y] Values within a row with different superscripts are trends.

During lactation, piglets subjected to IS between day 21 and 28 of age experienced a reduction in growth after the onset of IS which lasted until one day before weaning. This pattern is consistent with results from other studies [9,12,15], with Berkeveld et al. [15] suggesting that creep feed intake was insufficient to compensate for the loss of nutrient and energy intake from the reduction in milk consumption due to the separation. Despite a reduction in growth during lactation in the same IS studies, a 12 h/day separation for 11 days before weaning still stimulated solid feed intake before and after weaning [9,12], whereas a 10 h/day separation for 7 days did not stimulate solid feed intake before weaning, but solid feed intake was higher than conventionally weaned pigs for 2 days after weaning [15]. In the current study, an 8 h/day separation for 7 days before weaning neither stimulated solid feed intake before or after weaning compared with conventionally weaned pigs. The difference in results, especially with regard to post-weaning performance, could be due to the fact that primiparous progeny were used, although a study by Turpin et al. [21] showed primiparous piglets separated for 8 h per day 7 days before weaning ate more creep feed before weaning and had a tendency to grow faster than conventionally weaned pigs after weaning. It is also possible that CW28 piglets in the current study may have been exposed to less severe stressors at weaning than under commercial conditions allowing them to perform better than expected. The current study was performed at a research facility where piglets were not mixed or transported at weaning. The housing of piglets with littermates after weaning is associated with less aggression [22] and reduced elevations in cortisol [23]. In this regard, the effects of weaning on CW28 piglet performance would likely be exacerbated under commercial conditions.

In studies by Berkeveld et al. [20] and Berkeveld et al. [15], an IS period lasting two weeks or more caused a marked reduction of the post-weaning growth check. Since ADG measured in the first experimental period after weaning was between one day before weaning to two days after weaning and not from the day of weaning in the current study, it is difficult to draw firm conclusions regarding the growth check since it is not uncommon for conventionally weaned piglets to grow more than 200 g/day just before weaning with zero, or indeed negative growth, immediately after weaning [24]. However, CW28 pigs did experience a reduction in growth during the week after weaning compared with the week before weaning, whereas IS28 pigs experienced a reduction in growth during the week of IS with no growth check over the weaning period. In this regard, it seems that IS is associated with similar problems to abrupt weaning, shifting part of the problems associated with weaning into the lactation period [15]. In saying this, however, given IS piglets did not experience a growth check over the weaning period, one could assume that IS piglets were more familiar with the weaning process and were perhaps less stressed, although the neuroendocrine data from this experiment does not support this assumption, possibly due to sampling time and additional procedures. When IS is combined with an extended lactation, however, recovery is much quicker as evidenced by the IS35 group's ADG results, but caution must be used to attribute the beneficial effects of the longer IS treatment, since an older weaning age per se has also been shown to improve post-weaning performance [7]. In the current study, due to the lack of a non-IS group weaned at 35 days, the effect of age and IS in the IS35 group performance cannot be separated.

An important aspect of a modified weaning regime, such as IS, is its potential impact on the GIT, since weaning causes marked changes in GIT structure and function, as evidenced by altered nutrient and electrolyte transport, villous atrophy, increased mucosal permeability and inflammation [3]. The SI absorptive capacity of two sugars, mannitol and galactose, was used as an indicator of overall GIT function in the current study. The observed pattern of mannitol absorption in the CW28 piglets corresponded to previously reported values [19]. Although some mannitol passes through the epithelial layer via the paracellular route, it is mostly absorbed via the transcellular route via water-filled pores in the enterocyte membrane, making mannitol a surrogate marker of functional mucosal mass in the SI [18]. In the pre-weaning period, the trend for the IS groups (when combined) to have lower mannitol concentrations than the CW28 group suggests that IS reduces intestinal surface area before weaning. Since solid feed intake at the time of the pre-weaning SAT was similar across treatments and no inflammatory response was evident in the measured cytokines, it is possible that the reduction in intestinal surface area may be a normal physiological adaptive response made more evident by the gradual nature of the IS weaning technique [25]. This is also highlighted by the nearly three-fold decrease in mannitol in the CW28 group over the peri-weaning period compared to an approximate two-fold decrease in the IS groups, suggesting more sudden changes to GIT morphology in CW28 piglets. The lack of treatment effect after weaning is consistent with results from a study by Berkeveld et al. [15], in which no differences in villous height were observed between piglets weaned at day 26 with or without a week of IS. However, in the same study, when lactation was extended to day 33, villous height was higher eight days after weaning, suggesting that improvements in SI morphology seemed to be an effect of age and possibly secondary to a higher feed intake. However, the current study found no difference in IS35 mannitol concentrations compared to the other treatment groups after weaning, which may be a result of timing of the measurement (four versus eight days after weaning in the Berkeveld et al. [15] study) and/or a result of the limitation of mannitol as a measure of intestinal surface area.

In contrast, the post-weaning absorption of galactose was improved in the IS35 group compared to the other treatment groups. The absorption of galactose is similar to glucose via a sodium-glucose linked transporter (SGLT1), which relies on the function of the Na^+-K^+ ATP pump to create a Na^+ gradient across the cell membrane to function [26]. The change from a milk-based feed to a solid feed increases Na^+-K^+ ATP pump activity [27], therefore enhancing glucose and galactose absorption. In the current study, the greater solid feed intake in the IS35 group four days after weaning may have had the same effect on Na^+-K^+ ATP pump activity and resulted in a higher absorption of galactose.

Cytokines are essential for the activation of neutrophils, macrophages, T cells, B cells and dendritic cells to protect the integrity of the epithelium, and also play a role in body defense [28]. The predominant effect of IL-10 is to reduce inflammation; therefore, the upregulation of IL-10 after weaning is consistent with the inflammatory response seen around weaning [29]; however, there are no weaning studies to our knowledge that have examined the IL-10 response. The values for TNF-α over the weaning period in the IS35 group are much lower than other studies where weaned pigs were subjected to an immune stressor (4000 pg/mL for the stressed pigs) and resemble values reported in controls in the same experiment (50 pg/mL) [30]. Therefore, the results in the current study are not likely to be associated with an inflammatory response. The lack of response in the other cytokines measured may have been the result of low sample numbers, the low stress research environment (mentioned previously) or the timing of when the sample was taken (four days post-weaning). Pié et al. [31] reported increases in cytokine gene expression along the GIT one to two days after weaning. The late sampling may also explain the post-weaning increase in an anti-inflammatory cytokine (IL-10) rather than pro-inflammatory cytokines.

In the current study, blood samples to measure cytokines, cortisol and CRH were taken from piglets that had fasted for 3 h and underwent nasogastric intubation to receive an oral dose of mannitol or galactose. Although these procedures did not appear to impact the cytokine response, when interpreting the cortisol data, one needs to consider that CW28 piglets had to be separated from their dams during the pre-weaning fasting period. Psychosocial stress in the form of maternal separation during early postnatal life has been shown to induce deleterious changes in neuroendocrine functions and behaviour in many different species [32,33,34]. The higher levels of cortisol seen in the CW28 piglets before weaning during their first experience of maternal separation compared with piglets that had been subjected to maternal separation for three days (IS28 and IS35 groups) were consistent with results from a previous study from the same research group, where IS piglets experienced a transient increase in cortisol on the first day of separation but were attenuated by the last day of separation [21]. Furthermore, the plasma cortisol values in the current study were generally much higher than one would expect from weaning associated stress [29,35]. Plasma cortisol concentrations ranging from 60–108 ng/mL have previously been associated with procedures such as electrical stimulation, negative handling and the use of a snout rope [36]. Since previous weaning studies only involved venipuncture of the piglets at weaning, it is most likely that the nasogastric intubation strongly influenced the cortisol values in the current study. In saying this, however, all treatment groups were treated the same way before and after weaning, making comparison between the groups and time points possible. Overall, pre-weaning cortisol values across all treatments were higher than the post-weaning values. This was unexpected given reported results from most other experiments measuring plasma or urinary cortisol concentrations over the weaning period showed an increase in cortisol after weaning [4,37,38,39]. Alternatively, there are some studies that failed to find any increase in cortisol concentrations associated with weaning despite changes in other measures including growth, behavior and hormones [40,41]. It is possible that the choice of sample time meant that the acute post-weaning rise in cortisol was missed or intubation with a nasogastric tube the first time (before weaning) could have been more stressful than the second time (after weaning).

The number of studies examining plasma CRH in the peri-weaning period are limited, however, results generally have shown that CRH follows the same pattern as cortisol [4,42] but with changes in CRH lasting for a shorter period of time after weaning. This is presumably due to the negative feedback response exerted when glucocorticoid concentrations are high [43]. The reason why the IS35 group had higher pre-weaning plasma CRH concentrations compared with the other treatment groups is not known. Once again, the process of nasogastric intubation most likely influenced the CRH values, resulting in much higher concentrations than one would expect for weaning-associated stress [4,29,42]. However, this difference could also be explained by the effect of sedation with piglets in the current study being fully conscious during venipuncture, whereas piglets in the previously mentioned weaning studies were sedated. It has previously been mentioned that CRH may be a more sensitive stress indicator of GIT barrier dysfunction [4]; however, the authors are unable to comment on the effect that high CRH values may have had on the intestinal mucosa of piglets in the current study, since GIT permeability was not measured. Overall, the lack of treatment effect in the CRH values between CW28 and IS28 before weaning suggests that short periods of maternal separation (8 h/day) do not affect the piglet stress response before or after weaning.

5. CONCLUSIONS

Overall, the combination of IS with an older weaning age (day 35) increased solid feed intake before and after weaning resulting in higher growth rates 12 days after weaning and an improvement in GIT absorptive function (as assessed by galactose absorption). However, and due to the lack of non-IS group weaned at 35 days, the effect of age and IS in the IS35 group on performance cannot be separated. Differences in ADG had disappeared by the time pigs reached 90 days of age. Unfortunately, since cytokine and neuroendocrine data were potentially influenced by procedures, such as the oral dosing of sugars, it is difficult to draw firm and unifying conclusions about the effect IS has on the neuroendocrine and inflammatory responses. Nevertheless, the minimal differences between the treatments suggest that, apart from an initial peak in cortisol, short periods of maternal separation (8 h/day) do not affect the piglet stress response before or after weaning, but neither do they add any adaptation advantage to the process of weaning. In conclusion, IS involving an 8 h daily separation the week before weaning does not appear to prevent weaning-associated changes, but rather advances them in an attenuated way. This provided a more gradual transition into weaning and was seemingly more pronounced when piglets were weaned at a later age (i.e., 35 days).

ACKNOWLEDGMENTS

Appreciation is extended to the Cooperative Research Centre for High Integrity Australian Pork for funding and provision of a postgraduate scholarship to the senior author. The technical assistance of Emmy Bouwman and Josephine Mansfield is also gratefully acknowledged.

REFERENCES

1. Leibbrandt, V.D.; Ewan, R.C.; Speer, V.C.; Zimmerman, D.R. Effect of weaning and age at weaning on baby pig performance. *J. Anim. Sci.* 1975, *40*, 1077–1080.
2. McCracken, B.A.; Spurlock, M.E.; Roos, M.A.; Zuckermann, F.A.; Gaskins, H.R. Weaning anorexia may contribute to local inflammation in the piglet small intestine. *J. Nutr.* 1999, *129*, 613–619.
3. Pluske, J.R.; Hampson, D.J.; Williams, I.H. Factors influencing the structure and function of the small intestine in the weaned pig: A review. *Livest. Product. Sci.* 1997, *51*, 215–236.
4. Moeser, A.J.; Vander Klok, C.; Ryan, K.A.; Wooten, J.G.; Little, D.; Cook, V.L.; Blikslager, A.T. Stress signaling pathways activated by weaning mediate intestinal dysfunction in the pig. *Am. J. Physiol.-Gastrointest. Liver Physiol.*2007, *292*, G173–G181.
5. Pineiro, C.; Pineiro, M.; Morales, J.; Manteca, X.; Mateos, G. *Comparison of Productivity of Primiparous with Progeny of Multiparous Sows, XI Jornadas Sobre Producción Animal, Zaragoza, Spain, 11–12 Mayo, 2005 (Volúmenes I & II). 2005*; Gobierno de Aragón, Servicio de Investigación Agroalimentaria: Zaragoza, Spain; pp. 261–263.
6. Cera, K.; Mahan, D.; Cross, R.; Reinhart, G.; Whitmoyer, R. Effect of age, weaning and postweaning diet on small intestinal growth and jejunal morphology in young swine. *J. Anim. Sci.* 1988, *66*, 574–584.
7. Van der Meulen, J.; Koopmans, S.; Dekker, R.; Hoogendoorn, A. Increasing weaning age of piglets from 4 to 7 weeks reduces stress, increases post-weaning feed intake but does not improve intestinal functionality. *Animal* 2010, *4*, 1653–1661.
8. Mabry, J.W.; Culbertson, M.S.; Reeves, D. Effects of lactation length on weaning-to-first-service interval, first-service farrowing rate, and subsequent litter size. *Swine Health Product.* 1996, *4*, 185–188.
9. Kuller, W.; Soede, N.; van Beers-Schreurs, H.; Langendijk, P.; Taverne, M.; Verheijden, J.; Kemp, B. Intermittent suckling: Effects on piglet and sow performance before and after weaning. *J. Anim. Sci.* 2004, *82*, 405–413.
10. Gerritsen, R.; Soede, N.; Langendijk, P.; Taverne, M.; Kemp, B. Early embryo survival and development in sows with lactational ovulation. *Reprod. Domest. Anim.* 2008, *43*, 59–65.
11. Downing, J.; Brock, D.; Smits, R.; Giles, L. *Induction of Oestrus in Lactation Results in Normal Mating and Farrowing Performance, Manipulating Pig Production XII (eds RJ van Barneveld), 2009*; van Barneveld, R., Ed.; Australasian Pig Science Association (Inc.): Werribee, Australia, 2009; p. 144.

12. Kuller, W.; Soede, N.; van Beers-Schreurs, H.; Langendijk, P.; Taverne, M.; Kemp, B.; Verheijden, J. Effects of intermittent suckling and creep feed intake on pig performance from birth to slaughter. *J. Anim. Sci.* 2007, *85*, 1295–1301.
13. Thymann, T.; Gudbergsen, C.; Bresson, S.; Kristensen, N.B.; Hansen, C.F. Intermittent suckling improves post-weaning feed uptake but does not change functional gut characteristics of piglets. *Livest. Sci.* 2007, *108*, 132–136.
14. Nabuurs, M.; Hoogendoorn, A.; Van Zijderveld-Van Bemmel, A. Effect of supplementary feeding during the sucking period on net absorption from the small intestine of weaned pigs. *Res. Vet. Sci.* 1996, *61*, 72–77.
15. Berkeveld, M.; Langendijk, P.; Soede, N.M.; Kemp, B.; Taverne, M.A.; Verheijden, J.H.; Kuijken, N.; Koets, A.P. Improving adaptation to weaning: Effect of intermittent suckling regimens on piglet feed intake, growth, and gut characteristics. *J. Anim. Sci.* 2009, *87*, 3156–3166.
16. Chen, T.-Y.; Knight, A.L.; Bouwman, E.G.; Turpin, D.L.; Langendijk, P. Oestrus Induce in Primiparous Sows by Intermittent Suckling Shows a Bimodal Distribution. In Proceedings of the Fourtheenth Biennial Conference of the Australasian Pig Science Association, Melbourne, Australia, 24–27 November 2013; Pluske, J.R., Pluske, J.M., Eds.; Australasian Pig Science Association (Inc.): Melbourne, Australia, 2013.
17. Thymann, T.; Burrin, D.G.; Tappenden, K.A.; Bjornvad, C.R.; Jensen, S.K.; Sangild, P.T. Formula-feeding reduces lactose digestive capacity in neonatal pigs. *Br. J. Nutr.* 2006, *95*, 1075–1081.
18. Cox, M.; Lewis, K.; Cooper, B. Measurement of small intestinal permeability markers, lactulose, and mannitol in serum (results in celiac disease). *Dig. Dis. Sci.* 1999, *44*, 402–406.
19. Berkeveld, M.; Langendijk, P.; Verheijden, J.; Taverne, M.; van Nes, A.; van Haard, P.; Koets, A. Citrulline and intestinal fatty acid-binding protein: Longitudinal markers of postweaning small intestinal function in pigs? *J. Anim. Sci.* 2008, *86*, 3440–3449.
20. Berkeveld, M.; Langendijk, P.; van Beers-Schreurs, H.M.; Koets, A.P.; Taverne, M.A.; Verheijden, J.H. Postweaning growth check in pigs is markedly reduced by intermittent suckling and extended lactation. *J. Anim. Sci.* 2007, *85*, 258–266.
21. Turpin, D.L.; Langendijk, P.; Chen, T.-Y.; Lines, D.; Pluske, J.R. Intermittent suckling causes a transient increase in cortisol that does not appear to compromise selected measures of piglet welfare and stress. *Animals* 2016, 24.
22. Friend, T.; Knabe, D.; Tanskley, T. Behavior and performance of pigs grouped by three different methods at weaning. *J. Anim. Sci.* 1983, *57*, 1406–1411.
23. Blecha, F.; Pollmann, D.; Nichols, D. Immunologic reactions of pigs regrouped at or near weaning. *Am. J. Vet. Res.*1985, *46*, 1934–1937.
24. Weary, D.M.; Pajor, E.A.; Bonenfant, M.; Fraser, D.; Kramer, D.L. Alternative housing for sows and litters.: Part 4. Effects of sow-controlled housing combined with a communal piglet area on pre-and post-weaning behaviour and performance. *Appl. Anim. Behav. Sci.* 2002, *76*, 279–290.
25. Wijtten, P.J.; Meulen, J.V.D.; Verstegen, M.W. Intestinal barrier function and absorption in pigs after weaning: A review. *Br. J. Nutr.* 2011, *105*, 967–981.
26. Wright, E.M.I. Glucose galactose malabsorption. *Am. J. Physiol.-Gastrointest. Liver Physiol.* 1998, *275*, G879–G882.
27. Boudry, G.; Lallès, J.P.; Malbert, C.H.; Bobillier, E.; Sève, B. Diet-related adaptation of the small intestine at weaning in pigs is functional rather than structural. *J. Pediatr. Gastroenterol. Nutr.* 2002, *34*, 180–187.
28. Pitman, R.S.; Blumberg, R.S. First line of defense: The role of the intestinal epithelium as an active component of the mucosal immune system. *J. Gastroenterol.* 2000, *35*, 805–814.
29. Moeser, A.J.; Ryan, K.A.; Nighot, P.K.; Blikslager, A.T. Gastrointestinal dysfunction induced by early weaning is attenuated by delayed weaning and mast cell blockade in pigs. *Am. J. Physiol.-Gastrointest. Liver Physiol.* 2007, *293*, G413–G421.
30. Wright, K.; Balaji, R.; Hill, C.; Dritz, S.; Knoppel, E.; Minton, J. Integrated adrenal, somatotropic, and immune responses of growing pigs to treatment with lipopolysaccharide. *J. Anim. Sci.* 2000, *78*, 1892–1899.

31. Pié, S.; Lalles, J.; Blazy, F.; Laffitte, J.; Sève, B.; Oswald, I. Weaning is associated with an upregulation of expression of inflammatory cytokines in the intestine of piglets. *J. Nutr.* 2004, *134*, 641–647.
32. Kaufman, C.I.; Rosenblum, L.A. The reaction to separation in infant monkeys: Anaclitic depression and conservation-withdrawal. *Psychosom. Med.* 1967, *29*, 648–675.
33. Coe, C.L.; Wiener, S.G.; Rosenberg, L.T.; Levine, S. Physiological consequences of maternal separation and loss in the squirrel monkey. In *Handbook of Squirrel Monkey Research*; Rosenblum, L.A., Coe, C.L., Eds.; Springer Science + Business Media: New York, NY, USA, 1985; pp. 127–148.
34. Kanitz, E.; Tuchscherer, M.; Puppe, B.; Tuchscherer, A.; Stabenow, B. Consequences of repeated early isolation in domestic piglets (sus scrofa) on their behavioural, neuroendocrine, and immunological responses. *Brain Behav. Immun.* 2004, *18*, 35–45.
35. Turpin, D.; Langendijk, P.; Chen, T.; Lines, D.; Pluske, J. Intermittent suckling causes a stress response in piglets that is attenuated over time. *Anim. Product. Sci.* 2015, *55*, 1480–1480.
36. Turner, A.; Hemsworth, P.; Tilbrook, A. Susceptibility of reproduction in female pigs to impairment by stress or elevation of cortisol. *Domest. Anim. Endocrinol.* 2005, *29*, 398–410.
37. Hay, M.; Orgeur, P.; Lévy, F.; Le Dividich, J.; Concordet, D.; Nowak, R.; Schaal, B.; Mormède, P. Neuroendocrine consequences of very early weaning in swine. *Physiol. Behav.* 2001, *72*, 263–269.
38. Worsaae, H.; Schmidt, M. Plasma cortisol and behaviour in early weaned piglets. *Acta Vet. Scand.* 1980, *21*, 640–657.
39. Kanitz, E.; Tuchscherer, M.; Tuchscherer, A.; Stabenow, B.; Manteuffel, G. Neuroendocrine and immune responses to acute endotoxemia in suckling and weaned piglets. *Neonatology* 2002, *81*, 203–209.
40. Carroll, J.A.; Veum, T.L.; Matteri, R.L. Endocrine responses to weaning and changes in post-weaning diet in the young pig. *Domest. Anim. Endocrinol.* 1998, *15*, 183–194.
41. Colson, V.; Orgeur, P.; Foury, A.; Mormède, P. Consequences of weaning piglets at 21 and 28 days on growth, behaviour and hormonal responses. *Appl. Anim. Behav. Sci.* 2006, *98*, 70–88.
42. Smith, F.; Clark, J.E.; Overman, B.L.; Tozel, C.C.; Huang, J.H.; Rivier, J.E.; Blisklager, A.T.; Moeser, A.J. Early weaning stress impairs development of mucosal barrier function in the porcine intestine. *Am. J. Physiol.-Gastrointest. Liver Physiol.* 2010, *298*, G352–G363.
43. Munck, A.; Náray-Fejes-Tóth, A. The ups and downs of glucocorticoid physiology permissive and suppressive effects revisited. *Mol. Cell. Endocrinol.* 1992, *90*, C1–C4.

CHAPTER 17

Chemical Composition and Nutritive Value of Agro-Industrial By-Products in Ruminant Nutrition

Eyob Haile[1], Francis K. Njonge[2], Goitom Asgedom[1], Mathew Gicheha[2*]

[1]*Department of Animal Sciences, Hamelmalo Agricultural College, Keren, Eritrea*

[2]*Department of Animal Sciences, Jomo Kenyatta University of Agriculture and Technology, Nairobi, Kenya*

ABSTRACT

This study was carried out to determine the chemical composition and in situ degradability of agro-industrial by-products found in Eritrea. Three categories of by- products were evaluated and were the milling industry (wheat bran; WB, short; WS, and middling; WM), brewery (brewers' dry grain; BDG, hops; BDH, and yeast; BDY) and sesame cakes (sesame cake machine extracted; SCM and manually extracted; SCT). The dry matter (DM) varied between 88.46% in BDY to 92.39% in SCT. The lowest ($P < 0.05$) crude protein (CP) content was recorded in WM at 10.11% while the highest was from the BDY at 48.20%. The metabolisable energy (ME) value of the agro-industrial by-products ranged from 8.72 to 11.18 MJ per kg DM with the BDH recording the lowest value ($P < 0.05$). The sesame cakes (SCM and SCT) recorded higher values of 11.17 and 11.18 MJ per kg DM respectively. The SCT recorded the highest ash content at 10.93% followed by BDY at 10.16% with the least being obtained from WM at 2.48%. The ether extract and acid detergent lignin contents were generally low in all cases for all the by-products. Generally, the results indicated that there was no clear pattern in terms of nutrients content amongst the by-products. The in situ DM, organic matter (OM) and CP degradability differed amongst and within the sesame cakes, milling and brewery by-products. The wide variation in chemical composition, DM, OM, CP degradability, and ME obtained from this study offer farmers huge flexibility in formulating rations according to the productive performance of target animals.

Keywords: Chemical Composition, Nutritional Value, Agro-Industrial By-Products

1. INTRODUCTION

There is potential for increased livestock production in Eritrea through use of agro- industrial by-products as a supplement in ruminant diets. The bulk of livestock feed (estimated to be about 90%) comes from grazing on pastures and stubble. According to [1] , grazed pastures in Eritrea are declining partly as a result of overgrazing aimed at producing more protein to satisfy the increasing demand resulting from an increase in human population. This coupled with climate change and increased climatic variability has resulted in reduction in nutrients available for grazing stock thus compromising productivity. Besides the increase in the protein demand, competition for cereal grains between animal and human has been on the rise leaving very little for use in animal feeding. This implies the need to have alternative source of animal feed and/or sustainable supplementation. Cereal and legume crop residues and agro-industrial by-prod- ucts have played a major role in supplying supplemental nutrients to the grazed stock in Eritrea [1]. Studies by [2] evaluated the chemical characteristics and nutritive value of the cereal and legume residues respectively and found out that they contained nutrients that can supply up to 10% of the total nutrients required by grazing livestock. Furthermore, chemical composition and nutritive value data available for most of crop residues and agro-industrial by-products available in the tropics indicate that they can satisfy significant proportion of the nutritional requirement of ruminant animals [2] - [8].

Despite their potential use in supplementing ruminant nutrition, crop residues have been shown to have high fibre content besides being low in metabolized energy (ME) and crude protein [9] [10] [11] that result in low degradability in the rumen. Various approaches that have potential to increase digestibility and intake have been proposed. Besides the crop residues, Eritrea's agro-industries produce substantial amounts of by-products [2] with potential to be used in supplementing the grazing animals thus further availing the much needed nutrients to increase livestock production. However, the chemical composition and nutritive value data for the by-products are scanty. This implies that there is a need to determine the chemical and nutritional characteristics of the by-products so as to optimally integrate them in the animal feeding strategies utilised in Eritrea. The aim of this study was to determine the chemical composition and nutritive value of agro-industrial by-products in ruminant nutrition in Eritrea.

2. MATERIAL AND METHODS

2.1. Sample Collection and Chemical Analysis

A total of 8 different samples including wheat bran (WB), short (WS) and middling (WM); brewers' dry grain (BDG), hops (BDH) and yeast (BDY); and sesame cakes (SCM and SCT) were collected from milling, beer and oil industries respectively. Dry samples were collected from milling and oil industries while those from the beer industries were wet. The wet samples were sun dried before grinding. Samples for each feedstuff were pooled and milled for the chemical and in situ (nylon bag technique) analysis.

2.2. Chemical Analysis

The dry matter (DM) and organic matter (OM) were determined according to the standard methods [12] . The ash content was determined by ashing samples in a muffle furnace at 550°C for 6 h while the nitrogen (N) content was determined using Kjeldahl method [12] . The crude protein (CP) was calculated as:

$$CP = N * 6.25$$

The crude fiber (*CF*) and ether extract (*EE*) were determined by the methods described in the [12] . The nitrogen free extract (*NFE*) was determined as:

$$NFE = \left[100 - \left(CP + EE + CF + Ash\right)\right]$$

The cell wall components (NDF, ADF and ADL) were determined according to [13] .

2.3. In Situ Degradation Procedures

The nylon bag procedure described by [14] was used in determining the nutritive value of the crop residues considered in this study. In all, a 5 g of dried sample of the crop residues were milled through a 3 mm screen. The sample was then weighed in nylon bags (16 × 8 cm, pore size 45 to 60 um) which were then incubated in the rumen of two cattle fitted with rumen cannula. The research adhered to the guidelines proposed in the *Guide for the Care and Use of Agricultural Animals in Agricultural Research and Teaching* [15].

The bags were withdrawn at 4, 8, 16, 24, 48, 72 and 96 h intervals following insertion. They were subsequently rinsed with cold water until it became clear. This was followed by drying of the bags and samples at 60°C for 48 h. The soluble fraction (0 h) value was obtained by soaking two bags of the sample in warm water (38°C) bath for 1 h which was then followed by washing in cold water for 15 min in a washing machine. The samples were then dried for 48 h at 60°C. The rumen degradation kinetics of DM and OM was calculated using the exponential equation by [14] as:

$$P = a + b\left(1 - e^{-ct}\right)$$

where p is the percentage degradability for response variable at time *t* which is the time relative to incubation (hours), *a* represents the highly soluble and readily degradable fraction (%), *b* the insoluble and slowly degradable fraction (%), *c* is the rate constant for degradation (h^{-1}) and *e* is the natural logarithm base (2.7182). The effective degradability (*ED*) of the DM and OM of each sample was determined using the equation proposed by [16] :

$$ED = a + \frac{(b * c)}{(c + k)}$$

where parameters *a*, *b* and *c* are as previously defined while *k* is the rate constant of passage (h^{-1}) which was assumed to be 0.02, 0.05 and 0.08 per hour [16] . The metabolisable Energy (*ME*) content was estimated using equation described by [16] as:

$$ME(\text{MJ/kg } DM) = 2.27563 + 0.1073DMD$$

where *DMD* is rumen dry matter degradability at 48 h of incubation.

2.4. Statistical Analysis

Data on chemical composition and degradation characteristics were subjected to analysis of variance while the least significant differences (LSD) test was used in all cases to compare the samples means. Differences were accepted when $P \leq 0.05$.

3. RESULTS

3.1. Chemical Composition of Crop Residues

The chemical composition of the agro-industrial by-products are presented in Table 1. The DM varied between 88.39% for WB to 92.39% for the SCT. The WM recorded the highest OM content at 97.52% while the lowest recorded was from SCT at 89.07%. The SCT recorded the highest ash content at 10.93% followed by BDY at 10.16% with the least being obtained from WM at 2.48%. The CP was highest in SCT at 40.41% and lowest in WM at 10.11% while the SCM and SCT contained the highest ME at 11.17 and 11.18 MJ per kg DM respectively. The EE and ADL contents were generally low in all cases for all the by-products. Generally, the results indicated that there was no clear pattern in terms of nutrients content amongst the by-products.

The CP content of the agro-industrial by-products ranged from 10.11% to 48.20%. The lowest ($P < 0.05$) was recorded in WM while the highest ($P < 0.05$) was obtained from the BDY. The WS had higher ($P < 0.05$) CP amongst all the milling by-products with the BDG recording lower ($P < 0.05$) CP when compared to the other brewers' by-products. The CF content for different by-products ranged from 1.14% in BDY to 22.58% in BDG. The NDF, ADF, and ADL ranged between 3.30% in BDY to 74.43 in BDG, 1.55% in BDY to 21.18% in BDG and 0.9% in BDY to 5.23% in BDH respectively. Generally, the BDY had lower ($P < 0.05$) CF, NDF and ADF content. Conversely, the BDG recorded higher values ($P < 0.05$) for the parameters when compared to all other agro-industrial by-products. The WM resulted in lower ($P < 0.05$) CF, NDF and ADF than the other milling by-products.The metabolizable energy (ME) value of the agro- industrial by-products ranged from 8.72 to 11.18 MJ per kg DM with the BDH recording the lowest value ($P < 0.05$) of 8.72 MJ per kg DM. The sesame cakes (SCM and SCT) recorded higher values ($P < 0.05$) of 11.17 and 11.18 MJ per kg DM respectively.

Table 1. Chemical composition of agro-industrial by-products.

	Agro-Industrial By-Products[1]							
Nutrients[2]	WB	WS	WM	BDG	BDH	BDY	SCM	SCT
DM	88.39^{a}	90.42^{b}	88.9^{a}	90.17^{b}	91.35^{b}	88.46^{a}	92.00^{c}	92.39^{c}
Ash	5.32^{b}	4.45^{c}	2.48^{a}	5.32^{b}	2.53^{a}	10.16^{d}	9.82^{d}	10.93^{f}
OM	94.68^{c}	95.55^{a}	97.52^{b}	94.68^{c}	97.47^{b}	89.84^{e}	90.18^{e}	89.07^{d}
CP	11.05^{h}	12.27^{g}	10.11^{f}	19.96^{e}	21.63^{d}	48.20^{c}	37.45^{b}	40.41^{a}
EE	2.98^{ab}	4.03^{b}	2.25^{a}	5.92^{d}	2.52^{ab}	0.47^{e}	16.84^{c}	15.43^{c}
NFE	67.96^{d}	66.46^{d}	79.2^{f}	46.22^{e}	60.12^{c}	40.02^{b}	27.86^{a}	26.17^{a}
CF	12.7^{e}	12.79^{e}	5.96^{a}	22.58^{b}	13.2^{e}	1.14^{c}	8.04^{d}	7.05^{d}
NDF	61.95^{c}	46.91^{d}	20.93^{b}	74.43^{e}	22.63^{b}	3.3^{f}	18.54^{a}	18.98^{a}
ADF	19.73^{c}	13.78^{a}	5.88^{d}	21.18^{f}	13.51^{a}	1.55^{e}	9.18^{b}	9.02^{b}
ADL	3.96^{ac}	3.66^{ac}	2.51^{ac}	4.07^{ac}	5.23^{a}	0.9^{c}	1.5^{c}	1.29^{c}
ME (MJ/kg DM)	10.32^{b}	10.52^{b}	9.89^{a}	8.85^{c}	8.72^{c}	ND	11.17^{d}	11.18^{d}

Means within the same row with different superscript are significantly different ($P < 0.05$). [1]See text for the description of the Agro-Industrial By-Products. [2]See text for the description of the nutrients; the values for the Ash, OM, CP, EE, NFE, CF, NDF, ADL are presented as per cent of the DM content. ND = Not Determined.

The in situ DMD, OMD and crude protein (CP) degradability of agro-industrial by- products are presented in Table 2. The CPD data is presented for the 72 hours beyond which the entire sample had been degraded in most of the by-products. The sesame cakes (SCM and SCT) had the highest ($P < 0.05$) DMD at all incubation times while the brewer's yeast by-products had the least (BDG and BDH). The WM had lower values ($P < 0.05$) compared to the other milling industry by-products except at 0 h where WB recorded the lowest value of 32.57%.

The in situ OMD differed amongst and within the sesame cakes, milling and brewery by-products. However, a pattern similar to the one in the DMD was obtained with the cakes having the highest OMD content followed by the milling by-products with least content being obtained from the brewery by-products.

The CPD pattern differed from that of the DMD and OMD except for the sesame cakes which remained higher ($P < 0.05$) than the brewery and milling by-products. The highest CPD was obtained from SCT at 94.61% in the 96th hour of incubation while the lowest was from WM at 22.05% before start of the incubation in the rumen. Generally, an increase in the incubation time led to increase in the CPD for all by-products.

Table 2. In situ DM, OM and CP degradability of agro-industrial by-products (% DMD, OMD and CPD).

AIBP[1]	Time of Incubation in the Rumen (Hrs.)								SEM
	0	4	8	16	24	48	72	96	
DMD									
WB	32.57^{ab}	48.33^{c}	53.72^{d}	61.57^{e}	66.93^{f}	74.98^{g}	77.56^{gi}	77.67^{gi}	0.169
WS	36.63^{a}	47.72^{c}	53.23^{d}	60.62^{e}	66.16^{f}	76.84^{g}	79.03^{gi}	79.03^{gi}	0.140
WM	34.34^{ab}	42.21^{d}	46.15^{c}	50.91^{f}	57.91^{e}	70.95^{k}	73.97^{ik}	73.98^{ik}	0.208
BDG	21.77^{c}	34.62^{e}	39.11^{f}	44.77^{d}	49.27^{g}	61.31^{h}	64.47^{h}	65.05^{h}	0.293
BDH	29.17^{b}	32.87^{a}	37.43^{f}	43.14^{d}	47.72^{g}	60.04^{h}	61.56^{h}	61.68^{h}	0.057
SCM	63.88^{d}	65.35^{b}	67.45^{a}	71.15^{c}	74.90^{h}	82.87^{j}	85.30^{j}	85.34^{j}	0.178
SCT	62.01^{d}	64.39^{e}	67.50^{a}	71.53^{c}	74.91^{h}	82.98^{j}	85.36^{j}	85.36^{j}	0.081
SEM	4.200	3.334	3.140	3.001	2.882	2.403	2.421	2.378	
OMD									
WB	32.59^{a}	47.10^{e}	52.40^{b}	59.75^{d}	65.19^{c}	73.01^{f}	76.27^{k}	76.27^{k}	0.085
WS	33.94^{a}	45.56^{e}	52.46^{b}	60.47^{d}	65.44^{c}	75.78^{g}	78.15^{j}	78.15^{j}	0.040
WM	33.87^{a}	40.57^{d}	45.57^{c}	49.40^{b}	56.23^{e}	69.67^{h}	73.11^{i}	73.12^{i}	0.023
BDG	19.23^{b}	33.68^{c}	38.00^{d}	44.22^{a}	48.78^{f}	60.38^{e}	63.38^{h}	63.98^{h}	0.091
BDH	27.63^{c}	31.43^{b}	36.18^{a}	42.06^{e}	45.84^{d}	58.87^{i}	60.72^{g}	60.89^{g}	0.036
SCM	64.03^{d}	65.71^{a}	67.88^{e}	71.53^{c}	75.44^{b}	83.18^{j}	85.78^{f}	85.78^{f}	0.016
SCT	63.51^{d}	65.89^{a}	68.43^{e}	71.34^{c}	75.10^{b}	83.37^{j}	85.86^{f}	85.86^{f}	0.089
SEM	4.498	3.586	3.330	3.112	3.044	2.530	2.542	2.502	
CPD									
WB	28.22^{e}	50.26^{b}	67.36^{fj}	74.14^{dgj}	77.58^{g}	79.79^{ghi}	79.96^{gh}	-	0.348
WS	35.22^{c}	54.98^{a}	69.82^{f}	76.70^{d}	78.50^{g}	80.96^{hi}	81.08^{hi}	-	0.216
WM	22.05^{d}	45.62^{c}	56.11^{e}	66.97^{b}	73.03^{g}	77.10^{h}	77.50^{h}		0.179
BDG	27.19^{e}	36.64^{d}	48.39^{c}	67.68^{f}	77.52^{g}	83.78^{i}	84.18^{i}	-	0.114
BDH	30.13^{a}	45.03^{c}	56.02^{e}	69.63^{f}	77.53^{g}	83.76^{i}	84.15^{i}	-	0.188
SCM	77.80^{b}	82.57^{e}	90.75^{ad}	91.81^{ac}	92.78^{cf}	93.82^{f}	93.82^{f}	-	0.142
SCT	78.40^{b}	82.34^{e}	90.52^{ad}	91.87^{ac}	92.82^{cf}	94.28^{f}	94.61^{f}	-	0.088
SEM	6.287	4.727	4.321	2.968	2.067	1.737	1.719	-	

Means within the same row with different superscript are significantly different ($P < 0.05$) while those within the same column with different superscript are significantly different ($P < 0.05$). [1]See text for the description of the crop residues.

The DM, OM and CP disappearance from the nylon bags incubated in the rumen increased with increase in time. The rate of increase however reduced with increase in the incubation time. In all cases, there was minimal change in DMD, OMD and CPD between hours 72 and 96. Although, the level of degradability was different at the various incubation times both between and within the feedstuff, it was revealed that there was no significant difference ($P < 0.05$) within the feedstuff and between the by-products from milling, or brewers' industry after 48 h.

3.2. Degradability Characteristics

The percentage DM, OM and CP rapidly soluble fraction (a), potentially degradable fraction (b), rate of degradation of b fraction (c) and the effective degradability (ED) of different by-products are presented in Table 3. The respective values of the DM, OM and CP a, b, c, ($a + b$) and ED parameters varied amongst by-products. The highest values of DM, OM and CP a were obtained from SCT at 61.85%, SCM at 63.86% and 78.40% from SCT respectively. The respective b values were 44.73% in WB, 46.65% in WS and 57.08% in BDG. The DM ($a + b$) varied between 62.01% in BDH and 85.43% in SCT. Respective values for the OM ($a + b$) were 61.25% in BDH and 85.94% in SCT while they were between 77.58% in WM and 94.61% in SCT for the CP. The rate of the degradation of b fraction varied minimally between the DM whose range was between 0.040 in WM and 0.071 in SCM and SCT as the two by-products resulted in a similar value. The rate in OM was similar to that in DM. The CP c values were higher than the corresponding DM and OM c values and ranged between 0.077 for BDG and 0.186 for SCM and SCT which had a common value. In general, the value of c was highest in SCM and SCT in all cases. The SEM was generally high for the percentage DM, OM and CP degradability characteristics and the effective degradability values of different agro-industrial by-products. This could result from variation in the processing procedure rather than the differences in the by-products.

The values for DMED at 0.02, 0.05 and 0.08 hourly rate of passage ranged between 51.90% to 80.61%, 43.16% to 76.45% and 38.18% to 73.95% respectively. Similarly, respective rates in the OM were 50.50% to 80.98%, 40.77% to 76.76% and 35.65% to 74.23% while in the CP the rates ranged between 69.58% to 93.04%, 61.13% to 91.18% and 55.23% to 89.74% respectively. The sesame cakes (SCT and SCM) had higher ($P < 0.05$) DM, OM and CP ED than all the by-products at all the passage rates.

The DM, OM and CP ED varied among by-products and passage rates. However, there was no significant difference for the DM ED amongst the milling (WB, WS and WM), the brewery (BDG and BDH) and sesame cakes (SCM and SCT) by-products. The OM ED for the WS within the milling by-products was always higher ($P < 0.05$) as did the BDH and SCM between the brewery and sesame cake by-products respectively.

Table 3. The percentage DM, OM and CP degradability characteristics and the effective degradability values of different agro-industrial by-products.

	Agro-industrial by-products[1]							SEM
	WB	WS	WM	BDG	BDH	SCM	SCT	
DM								
a	30.61^{ab}	34.86^{a}	31.89^{ab}	20.97^{c}	28.12^{b}	63.71^{d}	61.85^{d}	4.342
b	47.64^{c}	44.73^{cd}	42.86^{d}	44.28^{cd}	33.89^{a}	21.70^{b}	23.58^{b}	2.746
(*a* + *b*)	78.25^{f}	79.59^{e}	74.75^{d}	65.25^{b}	62.01^{c}	85.41^{a}	85.43^{a}	2.363
c per h	0.042^{a}	0.044^{ab}	0.040^{a}	0.049^{b}	0.047^{b}	0.071^{c}	0.071^{c}	0.004
ED (%)								
0.02	63.13^{b}	65.69^{b}	60.65^{b}	52.48^{c}	51.90^{c}	80.61^{a}	80.22^{a}	3.218
0.05	53.16^{a}	56.37^{a}	51.83^{a}	43.16^{b}	44.84^{b}	76.45^{c}	75.70^{c}	3.603
0.08	48.10^{c}	51.58^{c}	47.44^{c}	38.18^{a}	41.12^{a}	73.95^{b}	72.98^{b}	3.785
OM								
a	30.41^{a}	32.07^{a}	31.51^{a}	18.26^{b}	26.50^{c}	63.86^{d}	63.34^{d}	4.651
b	46.51^{b}	46.65^{b}	42.36^{c}	45.94^{b}	34.75^{d}	21.99^{a}	22.60^{a}	2.832
(*a* + *b*)	76.92^{a}	78.72^{b}	73.87^{c}	64.20^{d}	61.25^{e}	85.85^{f}	85.94^{f}	2.484
c per h	0.041^{c}	0.043^{cd}	0.040^{c}	0.046^{b}	0.045^{bd}	0.071^{a}	0.071^{a}	0.004
ED (%)								
0.02	61.90^{c}	64.03^{d}	59.97^{e}	50.50^{a}	50.71^{a}	80.98^{b}	80.93^{b}	3.238
0.05	52.21^{c}	54.28^{a}	51.24^{b}	40.77^{d}	43.45^{e}	76.76^{f}	76.60^{f}	3.755
0.08	47.34^{b}	49.30^{c}	46.89^{b}	35.65^{a}	39.67^{d}	74.23^{e}	73.99^{e}	3.993
CP								
a	28.22^{a}	35.22^{b}	22.05^{c}	27.12^{a}	30.09^{d}	77.80^{e}	78.40^{e}	6.289
b	51.74^{b}	45.85^{a}	55.53^{c}	57.08^{d}	54.07^{e}	16.03^{f}	16.22^{f}	4.693
(*a* + *b*)	79.96	81.08	77.58	84.20	84.16	93.82	94.61	1.714
c per h	0.175^{a}	0.176^{a}	0.119^{b}	0.077^{c}	0.082^{c}	0.186^{d}	0.186^{d}	0.013
ED (%)								
0.02	74.65^{a}	76.39^{b}	69.58^{c}	72.49^{d}	73.51^{e}	92.27^{f}	93.04^{g}	2.481
0.05	68.46^{g}	70.91^{f}	61.13^{e}	61.82^{d}	63.62^{c}	90.43^{b}	91.18^{a}	3.334
0.08	63.72^{a}	66.72^{b}	55.23^{c}	55.23^{c}	57.40^{d}	89.01^{e}	89.74^{f}	3.884

Means within the same row with different superscript are significantly different ($P < 0.05$). [1]See text for the description of the agro-industrial by-products. [2]Not significant.

4. DISCUSSION

This study has provided vital information on the chemical composition as well as the nutritive value of different agro-industrial by-products produced in Eritrea. As expected, the composition and nutritional values varied within (by-products from a common crop) and between by-products. This implies that they would supply different quantity and quality nutrients when used in livestock feeding [17] , while studying the chemical composition, DM and in vitro gas production of different ruminant feedstuff noted that the difference in chemical composition as well as the feeding quality between alternative feedstuffs could be explained by the variation in the composition of the original material, the method used in processing (physical or chemical) and type of components extracted.

The values for different parameters used in evaluating the ruminant feedstuffs obtained in this study were compared to findings from other studies with care being taken to consider those obtained from study areas and production systems similar to the conditions found in Eritrea to avoid bias. It was found that the CP and EE for WB reported by [18] were higher than those obtained in the present study while the NDF and ADF were lower. Similarly, [19] reported higher values for the CP, EE and NDF in WM than those obtained from this study. However, the CP for BDG in the present study was similar to that reported by [20] as did the CP content in SCM which matched that reported by [21] and [22].

The DMD, which is an indicator of the level of cell walls in a feedstuff, varied within and between by-products. This implies that the values of DM, OM and CP *a*, *b* and (*a* + *b*) would be dependent on the content and the composition of the cell walls of particular feedstuff. [23] reported that there are strong correlations among the *b* fraction and the NDF, ADF and ADL contents of test feedstuff. The study further demonstrated that the content and the composition of the cell walls are the best variables to describe the variation of DM (*a* + *b*) of the feedstuff. It is [24] who determined the relationship between the degradability parameters *a*, *b* and *c* and the chemical composition of 60 test feeds and noted that the slowly fermentable structural carbohydrates in feedstuff are the determinant of the degradation characteristics in the rumen. Generally, SCM and SCT had high levels of DMD, OMD and CPD at all incubation times. This can be attributed to the availability of rumen degradability nitrogen thus optimizing on the microbial activity which would then enhance the sesame cake by-products to be degraded to their potential. Despite high NDF in WB, its corresponding DMD and CPD were higher than in the WM which can be explained by high levels of soluble containing fraction of DM and CP as well as the high hemicellulose in NDF content that tend to have high degradability.

The values for the degradability parameters *a*, *b* and *c* obtained from this study match most of those reported in other studies. For instance, the values of the DM *a* and *c* fractions for WB were similar to those obtained by [20] while the DM *c* fraction in BDG reported by [19] is similar to that of the present study. [22] also presented CP *c* fraction similar to the one obtained in this study. This is possible since the industrial processing would be expected to follow the same principles and probably use similar machinery in the extraction resulting in by-products that only differ as a result of other factors rather than the extraction procedures. Besides, it is common to find commonality in different crop production conditions from different countries implying that the differences might not be such huge. It is however possible to find small differences in similar by-products in different studies such as the one by [20] who reported a slightly higher DM *c* fraction in BDG this would be explained by the use of test feed that was low in NDF. Similarly, [21] reported slightly lower DM a c and higher *b* fractions in SCT and SCM than those obtained in the current study. The small differences were expected and could be explained by factors such as variation in laboratory equipment and techniques as well as differences in the milling screen sizes among others.

5. CONCLUSION

This study has shown that different agro-industrial by-products differ in terms of chemical composition as well as in nutritional value. Generally, the wide variation in chemical composition, DMD, OMD, CPD, and ME offer farmers huge flexibility in formulating rations according to the productive performance of target animals. By determining the chemical composition and the nutritional value of different by-products, the findings have demonstrated the opportunity available in utilizing them in ruminants feeding. It is evident that legume by-products are superior in terms of nutritional value when compared to the cereal based ones. However, it is important to evaluate the availability of the nutrients to the target animal by carrying out feeding trials and determining the performance (such as growth rate) of animals fed on diet incorporating different by- products.

ACKNOWLEDGEMENTS

The authors would like to recognize Japan International Corporation Agency (JICA) for provision of financial support for this study and Hamelmalo Agricultural College (HAC) of Eritrea for granting the first author permission to participate in the research.

REFERENCES

1. FAO (2006) Food and Agricultural Organization of the United Nation. Suttie, J.M. and Reynolds, S.G., Eds.
2. Haile, E., Njonge, F.K., Goitom, A. and Gicheha, M.G. (2016) Production and Potential Utilization of Crop Residues and Agro-Industrial By-Products in Ruminant Nutrition in Eritrea. RUFORUM Biennial Conference: Linking Universities with Private Sector, Governments and Other Stakeholders in Support of Agricultural Development in Africa, Cape Town, 17-21 October 2016 (in Press).
3. Chimwano, A.M. (1990) Zambia's Farm Livestock Feedstuffs: Their Nutrient Content and Nutritive Value. University of Zambia, Lusaka, Zambia, 12 p.
4. Aregheore, E.M. (2000) Crop Residues and Agro-Industrial By-Products in Four Pacific Island Countries: Availability, Utilization and Potential Value in Ruminant Nutrition. Asian-Australian Journal of Animal Sciences, 13, 266-269.
5. Abrar, A.S., Khan, M.A., Mahr-Un-Nisa and Sarwar, M. (2004) Agro-Industrial By-Products as a Potential Source of Livestock Feed: A Review. International Journal of Agriculture and Biology, 2, 307-310.
6. Afshar, M.A. and Nader, M.S. (2008) Nutritive Value of Some Agro-Industrial By-Products for Ruminants: A Review. World Journal of Zoology, 3, 40-46.
7. Naser, M.S., Bayaz, A.Z., Ramin, S., Alireza, A., Abolfaze, A.G. and Mehdi, M. (2011) Determining Nutritive Value of Soybean Straw for Ruminant Using Nylon Bag Technique. Pakistan Journal of Nutrition, 10, 838-841.
8. Mugerwa, S., Kabirizi, J., Zziwa, E. and Lukwago, G. (2012) Utilization of Crop Residues and Agro-Industrial By-Products in Livestock Feed and Feeding Systems in Uganda. International Journal of Biosciences, 2, 82-89.
9. Devendra, C. (1991) Technologies Currently Used for the Improvement of Straw Utilization in Ruminant Feeding Systems in Asia. In: Romey, D.K., Orskov, E.R. and Gill, M., Eds., Proceedings of the NRI/MARDI Workshop on the Utilization of Straw in Ruminant Production Systems, 7-11 October 1991, Natural Resources Institute and Malaysian Agricultural Research and Development Institute, Kuala Lumpur, 1-19.
10. Preston, T.R. (1995) Tropical Animal Feeding. A Manual for Research Workers. FAO Animal Production and Health, Rome, Paper No. 126, 305.
11. Tingshuang, G., Sánchez, M.D. and Yu, G.P. (2002) Composition, Nutritive Value and Upgrading of Crop Residues. Animal Production Based on Crop Residues—Chinese Experiences. FAO Animal Production and Health Paper 149.
12. AOAC International (2005) Official Methods of Analysis of AOAC International. AOAC International, Rockville.
13. Van Soest, P.J., Robertson, J.D. and Lewis, B.A. (1991) Methods for Dietary Fibre, Neutral Detergent Fibre and Non-Starch Polysaccharides in Relation to Animal's Nutrition. Journal of Dairy Science, 74, 3583-3597. https://doi.org/10.3168/jds.S0022-0302(91)78551-2
14. Orskov, E.R., Hovell, F.D.D. and Mould, F. (1980) The Use of the Nylon Bag Technique for the Evaluation of Feedstuffs. Tropical Animal Production, 5, 195-213.
15. Federation of Animal Science Societies (1999) Guide for the Care and Use of Agricultural Animals in Agricultural Research and Teaching. Committees to Revise the Guide for the Care and Use of Agricultural Animals in Agricultural Research and Teaching, Savoy.
16. Bhargava, P.K. and Orskov, E.R. (1987) Manual for the Use of Nylon Bag Technique in the Evaluation of Feedstuffs. Rowett Research Institute, Aberdeen.
17. Getachew, G., Robinson, P.H., de Peter, E.J. and Taylor, S.J. (2004) Relationship between Chemical Composition, Dry Matter Degradability and in Vitro Gas Production of Several Ruminants Feed. Animal Feed Science Technology, 111, 57-71. https://doi.org/10.1016/S0377-8401(03)00217-7
18. Paya, H., Tagfizadeh, A., Janamohanadi, H. and Moghadam, G.M. (2008) Ruminal Dry Matter and Crude Protein Degradability of Some Tropical (Iranian) Feed Used in

Ruminant Diet Estimated Using in Situ and in Vitro Techniques. Research Journal of Biological Sciences, 3, 720-725.

19. Batajoo, K.K. and Shaver, D.R. (1998) In Situ Dry Matter, Crude Protein, and Starch Degradability of Selected Grains and By-Product Feeds. Animal Feed Science Technology, 71, 165-176. https://doi.org/10.1016/S0377-8401(97)00132-6
20. Paya, H., Taghizadeh, A., Lashkari, S. and Shirmohammadi, S. (2012) Evaluation of Rumen Fermentation Kinetics of Some By-Products Using in Situ and in Vitro Gas Production Technique. Slovak Journal of Animal Sciences, 45, 127-133.
21. Mahala, A.G. and Gomaa, A.S. (2007) Effect of Heat Treatment on Sesame Cake Protein Degradation. Research Journal of Animal and Veterinary Sciences, 2, 39-42.
22. Marghazani, I.B., Jabbar, M.A., Pasha, T.N. and Abdullah, M. (2013) Ruminal Degradability Characteristics in Vegetable Protein Sources of Pakistan. The Journal of Animal and Plant Sciences, 23, 1578-1582.
23. Bruno-Soaresa, A.M., Abreua, J.M., Guedesb, C.V. and Dias-da-Silvab, A.A. (2000) Chemical Composition, DM and NDF Degradation Kinetics in Rumen of Seven Legume Straws. Animal Feed Science and Technology, 83, 75-80. https://doi.org/10.1016/S0377-8401(99)00113-3
24. Woods, V.B, O'Mara, F.P. and Moloney, A.P. (2003) The Nutritive Value of Concentrate Feedstuff for Ruminant Animals. Part I: In Situ Ruminal Degradability of Dry Matter and Organic Matter. Animal Feed Science Technology, 110, 111-130. https://doi.org/10.1016/S0377-8401(03)00220-7

CHAPTER 18

The Effects of Fiber Inclusion on Pet Food Sensory Characteristics and Palatability

Kadri Koppel[1,*], Mariana Monti[2], Michael Gibson[3], Sajid Alavi , Brizio Di Donfrancesco[1] and Aulus Cavalieri Carciofi[2]

[1]*Sensory Analysis Center, Department of Human Nutrition, Kansas State University, 1310 Research Park Drive, Manhattan, KS 66502, USA*

[2]*Department of Veterinary Clinic and Surgery, College of Agricultural and Veterinarian Sciences, São Paulo State University (UNESP), Via de Acesso Prof. Paulo Donato Castellane, s/n, Jaboticabal, SP 14.884-900, Brazil*

[3]*Department of Grain Science and Industry, Kansas State University, Manhattan, KS 66506, USA*

AUTHOR CONTRIBUTIONS

Kadri Koppel was responsible for planning the study, conducting sensory analysis, writing and editing the manuscript, and analyzing the sensory data. Mariana Monti was responsible for planning the study, manufacturing the diets, reviewing the manuscript, and analyzing the diets composition. Michael Gibson was responsible for manufacturing the diets, planning the study, and reviewing the manuscript. Sajid Alavi was responsible for planning the study and composing and reviewing the manuscript. Brizio Di Donfrancesco was responsible for conducting the sensory analysis and reviewing the manuscript. Aulus Carciofi was responsible for planning the study, manufacturing the diets, organizing palatability trials, and composing and reviewing the manuscript.

SIMPLE SUMMARY

The results from this research indicate that fibers have an effect on extruded pet food texture and palatability. These results may help pet food companies select ingredients for successful product formulations.

ABSTRACT

The objectives of this study were to determine (a) the influence of fiber on the sensory characteristics of dry dog foods; (b) differences of coated and uncoated kibbles for aroma and flavor characteristics; (c) palatability of these dry dog foods; and (d) potential associations between palatability and sensory attributes. A total of eight fiber treatments were manufactured: a control (no fiber addition), guava fiber (3%, 6%, and 12%), sugar cane fiber (9%; large and small particle size), and wheat bran fiber (32%; large and small particle size). The results indicated significant effects of fibers on both flavor and texture properties of the samples. Bitter taste and iron and stale aftertaste were examples of flavor attributes that differed with treatment, with highest intensity observed for 12% guava fiber and small particle size sugar cane fiber treatments. Fracturability and initial crispness attributes were lowest for the sugar cane fiber treatments. Flavor of all treatments changed after coating with a palatant, increasing in toasted, brothy, and grainy attributes. The coating also had a masking effect on aroma attributes such as stale, flavor attributes such as iron and bitter taste, and appearance attributes such as porosity. Palatability testing results indicated that the control treatment was preferred over the sugar cane or the wheat bran treatment. The treatment with large sugarcane fiber particles was preferred over the treatment with small particles, while both of the wheat bran treatments were eaten at a similar level. Descriptive sensory analysis data, especially textural attributes, were useful in pinpointing the underlying characteristics and were considered to be reasons that may influence palatability of dog foods manufactured with inclusion of different fibers.

Keywords: dog food; extruded; fiber; palatability, sensory analysis

1. INTRODUCTION

Pet foods are manufactured with a myriad of ingredients. Mimicking human food trends, these ingredients are often of novel origin. Sensory properties of pet food products may be influenced by types of ingredients used in the formulation, added palatants, and processing factors used. Protein, grain, and fiber sources can influence the appearance, aroma, flavor and texture of extruded dog food. To understand the sensory characteristics of products manufactured with fibers of different source, eight extruded products were evaluated in this study.

Fibers are structural carbohydrates, mainly originating from plant cell walls. The energy available from fibers is limited, and because of this, fibers may be good ingredients in reduced energy diets [1]. In addition it has been found that fibers in combination with proteins help regulate satiety levels in dogs [2]. Furthermore, fibers may help regulate the digestive process and the glycemic response in dogs and cats [3, 4]. Depending on the fiber type and consumed amount, different effects on nutrient digestibility and fecal formation can be promoted [4].

Previous studies have described dietary fiber effects on sensory properties of different types of foods. The different sources and amounts of dietary fibers have an effect on the texture of extruded products due to the interaction with starch [5]. The increase of insoluble fiber (mainly wheat bran) was shown to increase the hardness of extruded cereals because of reduced expansion volumes and increased density [5]. Dietary fibers enrichment in bread showed a reduction of loaf volume, with the crumbs having higher firmness and a darker appearance [6]. Other studies on bread showed that a higher content of dietary fibers resulted in smaller cells in the crumb, which in turn resulted in a more dense appearance [7]. Popov-Ralić *et al.* [8] observed that different fiber content in cookies mostly impacted the appearance of the products, in particular the color.

Most common sources of fiber in pet foods are beet pulp and cellulose. Corn fiber, fruit fibers, rice bran, and whole grains are some of the other fiber sources available for use in pet foods [3]. Full-fat rice bran was tested for palatability and digestibility in pet foods by Pacheco *et al.* [9]. These authors found that rice bran could be used in pet foods, but at no more than a 20% inclusion rate. Sa *et al.* [10] studied an enzyme treatment effect on dog foods manufactured with wheat bran. These authors found that the enzymes did not have an effect on digestibility, but wheat bran addition resulted in a larger amount of fecal matter being produced.

Understanding pet food palatability issues is not an easy task as the test animals lack the necessary linguistic capabilities. Descriptive sensory analysis by trained human panelists may provide insight into pet food palatability [11]. Descriptive sensory analysis will not tell us how the food tastes for the target species, such as dogs or cats, but will further our understanding regarding the sensory properties of the products. Sensory studies on pet foods have found that dry dog foods are generally complex products that vary in appearance, aroma, flavor, and texture [12,13]. A study that compared baked and extruded dog foods found that the pet food textures resulting from these cooking methods were significantly different [14]. However, so far no studies have been found that compare the sensory properties of pet foods to palatability or animal liking of the foods. The primary hypothesis of this study is that fiber source has an effect on sensory characteristics such as flavor and texture properties and palatability of pet foods.

The objectives of this study were to determine the following characteristics for dry dog foods formulated with a significant proportion of dietary fiber: (a) the effect of fiber on the sensory characteristics; (b) differences between coated and uncoated kibbles for aroma and flavor characteristics; (c) palatability of selected coated treatments; and (d) potential associations between palatability and sensory flavor and texture attributes.

2. EXPERIMENTAL SECTION

2.1. Diet Formulation

A basal diet containing maize and poultry by-product meal was formulated for adult dog maintenance according to the European Pet Food Industry Federation nutritional guidelines for complete and complementary pet food for cats and dogs [15]. Different types of fiber, sourced from Dilumix (Leme, Sao Paulo, Brazil), were added to this basal diet to create eight treatments as described in Table 1: control, with no fiber addition (CO); guava fruit fiber (67% insoluble dietary fiber, less than 1% soluble fiber) at the inclusion levels of 3%, 6%, and 12% (GF3, GF6, and GF12, respectively); sugarcane fiber (90% insoluble dietary fiber, less than 1% soluble fiber) with large particle size (SC1) and small particle size (SC2; Vit2be Fiber; inclusion level 9% for both treatments); and wheat bran fiber (32% insoluble dietary fiber, 1.5% soluble fiber) with large particle size (WB1) and small particle size (WB2). Ingredients were previously analyzed and the diets containing 12% guava fiber (GF12), sugarcane fiber (SC1 and SC2) and wheat bran (WB1 and WB2) were balanced to have 16% of total dietary fiber, which is a level typically used in commercial high-fiber dog diets. All fibers were added by replacing corn.

2.2. Grinding and Mixing

The ingredients, with the exception of fiber sources, were weighed, mixed, and ground using a hammer mill fitted with a screen size of 0.8 mm (Sistema Tigre de Mistura e Moagem, Tigre, Sao Paulo, Brazil). Fiber sources were provided already ground to desired particle sizes by the supplier (Dilumix, Leme, SP, Brazil): guava fiber—213 μm; large sugarcane fiber—395 μm; small sugar cane fiber—197 μm; large wheat bran—345 μm; small wheat bran—143 μm. These sizes were determined using laser diffraction particle size analysis [16]. The other ingredients and fiber sources were then mixed, compounding the final diet.

Table 1. Treatment ingredients and nutritional composition, % *.

Ingredients, %	CO	GF3	GF6	GF12	SC1	SC2	WB1	WB2
Corn grain	57.9	54.6	51.2	44.6	47.5	47.5	30.4	30.4
Chicken byproduct meal	31.3	31.6	31.8	32.3	32.6	32.6	26.1	26.1
Chicken Fat	7.0	7.0	7.3	7.4	7.2	7.2	7.6	7.6
Guava Fiber	0.0	3.0	6.0	12.0	0.0	0.0	0.0	0.0
Sugar Cane Fiber	0.0	0.0	0.0	0.0	9.0	9.0	0.0	0.0
Wheat Bran	0.0	0.0	0.0	0.0	0.0	0.0	32.0	32.0
Fish oil	0.15	0.15	0.15	0.15	0.15	0.15	0.15	0.15
Palatant	2	2	2	2	2	2	2	2
NaCl	0.5	0.5	0.5	0.5	0.5	0.5	0.65	0.65
KCl	0.5	0.5	0.5	0.5	0.5	0.5	0.5	0.5
Vitamin and Mineral mix	0.3	0.3	0.3	0.3	0.3	0.3	0.3	0.3
Choline Chloride	0.2	0.2	0.2	0.2	0.2	0.2	0.2	0.2
Mold inhibitor agent	0.1	0.1	0.1	0.1	0.1	0.1	0.1	0.1
Antioxidant agent	0.04	0.04	0.04	0.04	0.04	0.04	0.04	0.04
Nutritional Composition in the final product (on DM-basis)								
Crude Protein	29.1	28.8	28.4	28.5	29.4	29.5	28.0	28.2
Crude Fat	15.3	15.9	15.2	14.6	15.0	14.7	14.5	15.5
Ash	6.4	6.0	7.0	6.1	5.8	6.2	6.5	6.4
Crude Fiber	1.9	3.0	3.9	4.7	4.6	4.2	3.1	3.7
Dietary fiber	8.0	9.9	11.9	15.7	16.1	16.3	16.9	16.7
Starch	40.2	38.7	35.7	34.6	35.6	34.8	32.4	32.4
Moisture	5.9	6.8	6.0	7.3	5.4	5.3	5.6	5.5

* CO-Control, GF3—3% guava fiber, GF6—6% guava fiber, GF12—12% guava fiber, SC1—sugar cane fiber large grind, SC2—sugar cane fiber small grind, WB1—wheat bran fiber, large grind, WB2—wheat bran fiber, small grind.

2.3. Extrusion

The diets were extruded in a single screw extruder (MEX 250, Mazoni, Campinas, Brazil), with a processing capacity of 250 kg/h. Each food was processed separately on two different days for replicates. The processing conditions were not changed for any treatment in order to isolate the influence of fiber. Four samples were collected per diet each day and pooled. Pooled samples from both days were combined into one batch per diet for the sensory and palatability studies. For each diet, suitable amounts of samples were selected for coating for palatability and sensory tests from the same consolidated batch as the uncoated samples. A pre-conditioner was used to treat the diets with steam and water prior to extrusion. The pre-conditioner residence time was approximately 3.5 min, and downspout temperature was 82.2–85.7 °C. The extruder screw speed was 465 rpm, and the die open area was 15.9 mm^2/ton/h. Extruder die temperature ranged between 118.7–130.3 °C and die pressure between 52.45–70.6 bar. After extrusion, the kibbles were dried in a forced air dryer at 105 °C for 30 minutes. The dried kibbles were coated in a tumble system, receiving first the poultry fat and then the commercial palatant D'TECH 6L (SPF, Descalvado, São Paulo, Brazil) at 2% of the ingredients. The fish oil in the formulation was mixed with dry ingredients before extrusion.

2.4. Descriptive Sensory Analysis

2.4.1. Panelists

Five highly trained panelists from the Sensory Analysis Center, Kansas State University (Manhattan, KS, USA) participated in this study. All of the panelists had completed 120 h of general descriptive analysis training with a variety of food products. The training included techniques and practice in attribute identification, terminology development, and intensity scoring. Each of the panelists had more than 1000 h of testing experience with a variety of food products. For this project, the panelists received further orientation on dried dog food using samples that may or may not be included in the study. Panels of similar size and training have been reported in other recent research [8,9,10].

2.4.2. Sample Presentation and Evaluation

The panelists evaluated both uncoated (no poultry fat and no palatant added) and coated (poultry fat and palatant added) kibbles to determine the effect of coating and palatant on kibble flavor.

Evaluation of uncoated kibbles (n = 8: CO, SC1, SC2, WB1, WB2, GF3, GF6, and GF12) was conducted during seven 1.5 h sessions. Evaluation of coated kibbles (n = 5: CO, SC1, SC2, WB1, and WB2) was conducted during five 1.5 h sessions.

Each sample was served in a ~100 mL plastic cup for appearance, texture, and flavor evaluation, and in a medium snifter covered with a watch glass for the evaluation of aroma. The amount of product in the snifter was 3 g. Samples were prepared 30 min prior to the testing and were coded with three-digit random numbers.

All of the uncoated samples were evaluated in a randomized order in duplicate for appearance, texture, flavor, and aroma using attributes selected from a lexicon developed for this product category by Di Donfrancesco *et al.* [12] and used by Koppel *et al.* [13]. Barnyard, brothy, toasted, grain, vitamin, stale, egg, oxidized oil, cardboard, liver, fish, metallic, and dusty/earthy aroma and flavor were evaluated in all samples. In addition sour, salty, sweet, and bitter taste and aftertaste and barnyard, vitamin, stale, oxidized oil, cardboard, liver, fish, and metallic aftertaste attributes were evaluated. Brown color intensity, porous, grainy, flecks (yes/no), and fibrous appearance characteristics and cohesiveness of mass, fracturability, hardness, initial crispness, fibrous and gritty texture attributes were evaluated. A total of two 1.5 h sessions were held for orientation purposes and five 1.5 h sessions for the evaluation phase of the samples.

The coated samples (n = 5) were evaluated in a randomized order in duplicate for aroma, flavor, and appearance attributes only. The appearance attributes evaluated were: brown, porous, grainy, fibrous, and flecks. The aroma attributes were: barnyard, brothy, toasted, grain, vitamin, egg, sour aromatics, oxidized oil, cardboard, dusty/earthy, iron, and hay-like. The flavor attributes were: barnyard, broth, toasted, grain, vitamin, egg, sour, salt, bitter, sweet, oxidized oil, cardboard, liver, fish, dusty/earthy, metallic, iron, and hay-like. A total of two 1.5 h sessions were held for orientation purposes and three 1.5 h sessions for the evaluation phase of the coated samples.

For the evaluation, a numeric scale of 0–15 with 0.5 increments where 0 represented none and 15 extremely high was applied to each attribute to provide a measure of intensity. Each panelist individually assigned intensities to the attributes present in the sample according to their perception of the appearance, aroma, flavor, and texture references included in the lexicon. The panelists were provided with a definition sheet with the list of attributes and their definitions as well as reference materials for each attribute according to Di Donfrancesco *et al.* [12].

The panelists were asked to chew one kibble for flavor and texture evaluation. The panelists were instructed to expectorate samples after evaluation. Panelists were provided with apple slices, unsalted crackers, purified water, and toothbrushes for palate cleansing in between the evaluations. The testing room was at 21 ± 1 °C and 55% ± 5% relative humidity.

2.5. Palatability Testing

2.5.1 Palatability Testing Procedure

The tests were performed in Panelis, Diana Group (Descalvado, São Paulo, Brazil). Palatability was measured for the five fiber treatments (CO, SC1, SC2, WB1, and WB2) and with the coated kibbles only, using the two-pan method on two meals in one day, each test with 38 dogs in individual kennels [17]. The combinations of the samples were CO × SC1, CO × WB1, SC1 × SC2, and WB1 × WB2. In the morning after a 12 h fast the dogs received two pans, each containing one of the experimental foods, and were allowed to eat for 30 min. The position of the food pans was alternated at the evening meal. The amount of food offered in each pan surpassed the consumption capacity of the animal to ensure there would be leftovers to measure. After 30 min the pans were removed, the remains weighed and consumption rate was calculated (Equation (1)). Due to the large differences in body weights the results were calculated as relative consumption of each diet, and the mean intake of the two meals for each dog was compared.

$$\text{Relative consumption (\%)} = \frac{\text{Food A consumption}}{\text{Food A consumption} + \text{Food B consumption}} \times 100 \quad (1)$$

2.6. Data Analysis

Significant differences ($p < 0.05$) among the uncoated kibble treatments and coated kibble treatments sensory properties were determined using SAS Glimmix procedure and Fisher's protected Least Significant Difference (LSD) (Version 9.3, SAS Institute Inc., Cary, NC, USA). Principal Components Analysis was conducted for the significantly different appearance, texture, aroma, and flavor attributes of Control, WB1, WB2, SC1, and SC2 using XLStat version 2014.1.08 (Addinsoft, New York, NY, USA). Dog palatability data were statistically evaluated using Analysis of Variance (ANOVA), using the SAS software (Version 9.3, SAS Institute Inc.).

3. RESULTS AND DISCUSSION

3.1. Descriptive Sensory Analysis

3.1.1 Uncoated Treatments

The main contributing attributes to aroma for the uncoated treatments were barnyard, toasted, grainy, stale, and cardboard [18]. These attributes were higher in intensity than other attributes, but were not necessarily different among treatments. The main contributing attributes to flavor and taste for these dog foods were barnyard, toasted, grainy, vitamin, stale, sour, salty, bitter, dusty/earthy, oxidized oil, cardboard, liver, and metallic. The highest intensity was noted for bitter taste and aftertaste (averages 6.70–8.90).

The uncoated treatments were not significantly different ($p > 0.05$) in brown color, cohesiveness of mass, hardness, vitamin aftertaste, sour aftertaste, salty aftertaste, cardboard aftertaste, liver aftertaste, metallic aftertaste, barnyard flavor, broth flavor, grain flavor, vitamin flavor, sour taste, salty taste, sweet taste, cardboard flavor, liver flavor, fish flavor, metallic flavor, iron flavor, barnyard aroma, broth aroma, toasted aroma, grainy aroma, vitamin aroma, stale aroma, eggy aroma, cardboard aroma, liver aroma, fishy aroma, or iron aroma.

Appearance attributes porous, grainy, and fibrous were significantly different ($p < 0.05$) among the uncoated treatments (Table 2). The two wheat bran fiber diets, WB1 and WB2, had the highest scores for porous attribute, while the small particle size sugar cane (SC2) has the lowest. Sugar cane large particle size (SC1) had the highest fibrous (more than three point-scales from the second highest score) and grainy appearance level while sugar cane small particle size (SC2) had the lowest level for grainy attribute.

Off-notes oxidized oil and dusty/earthy were significantly different for aroma and flavor among the uncoated treatments. Samples GF6 and GF12 showed the highest oxidized oil aroma and flavor level. Stale, egg, and bitter were significantly different in flavor. Samples GF12 and SC2 were the most bitter while GF3 was the least bitter sample. The samples varied in barnyard, stale, bitter, sweet, oxidized oil, fish, and iron aftertaste. Sample GF12 had the highest bitter aftertaste and was the highest in stale, barnyard, sweet, oxidized oil, and fish aftertaste.

Fracturability, initial crispness, fibrous, and gritty attributes differentiated the uncoated treatments in texture. The largest difference in texture was found in the fibrous attribute. Sample SC1 was more than five points higher than the second highest score among the treatments. For other texture attributes the treatments showed smaller differences.

Table 2. Average texture, appearance, aroma, flavor, and aftertaste attributes intensity scores for uncoated treatments. Significantly different attributes showed only ($p < 0.05$).

Attribute	Treatment							
	CO	GF3	GF6	GF12	SC1	SC2	WB1	WB2
Fracturability	7.03 ^{a}	6.60 ^{ab}	6.73 ^{ab}	6.37 ^{bc}	5.67 ^{d}	5.63 ^{d}	7.17 ^{a}	5.90 ^{cd}
Initial Crispness	10.47 ^{a}	10.47 ^{a}	10.43 ^{a}	9.33 ^{b}	8.87 ^{bc}	8.73 ^{c}	10.33 ^{a}	9.17 ^{bc}
Fibrous	1.63 ^{de}	3.03 ^{bc}	0.70 ^{e}	2.07 ^{cd}	9.20 ^{a}	4.07 ^{b}	1.93 ^{cde}	3.10 ^{bc}
Gritty	4.23 ^{ab}	4.23 ^{ab}	4.87 ^{a}	4.03 ^{b}	3.37 ^{c}	4.13 ^{b}	3.97 ^{bc}	3.87 ^{bc}
Porous Appearance	5.00 ^{bc}	5.27 ^{abc}	6.13 ^{a}	5.80 ^{ab}	4.60 ^{c}	2.93 ^{d}	6.17 ^{a}	6.23 ^{a}
Grainy Appearance	2.00 ^{bcd}	2.10 ^{bc}	2.60 ^{ab}	1.93 ^{cd}	2.90 ^{a}	1.47 ^{d}	2.13 ^{bc}	1.90 ^{cd}
Fibrous Appearance	1.10 ^{c}	1.00 ^{c}	0.60 ^{c}	1.07 ^{c}	6.17 ^{a}	2.40 ^{b}	0.80 ^{c}	1.30 ^{c}
Oxidized Oil Aroma	2.13 ^{ab}	1.87 ^{b}	2.37 ^{a}	2.37 ^{a}	1.77 ^{b}	2.00 ^{ab}	1.83 ^{b}	1.73 ^{b}
Dusty/Earthy Aroma	2.30 ^{bc}	2.23 ^{bc}	2.50 ^{ab}	2.60 ^{ab}	2.77 ^{a}	2.50 ^{ab}	2.07 ^{c}	2.43 ^{abc}
Stale Flavor	2.77 ^{b}	2.73 ^{b}	3.03 ^{ab}	3.07 ^{ab}	3.00 ^{ab}	3.23 ^{a}	2.83 ^{b}	2.80 ^{b}
Eggy Flavor	1.10 ^{a}	1.33 ^{a}	1.23 ^{a}	1.17 ^{a}	0.40 ^{b}	0.93 ^{a}	1.40 ^{a}	1.00 ^{a}
Bitter Taste	7.57 ^{b}	6.70 ^{c}	8.07 ^{ab}	8.33 ^{a}	7.93 ^{ab}	8.30 ^{a}	7.37 ^{bc}	7.93 ^{ab}
Dusty/Earthy Flavor	2.13 ^{c}	2.23 ^{c}	2.53 ^{abc}	2.73 ^{ab}	2.77 ^{a}	2.27 ^{c}	2.33 ^{bc}	2.43 ^{abc}
Oxidized Oil Flavor	2.24 ^{b}	2.23 ^{b}	2.80 ^{a}	2.80 ^{a}	2.20 ^{b}	2.20 ^{b}	2.20 ^{b}	2.23 ^{b}
Barnyard aftertaste	2.87 ^{d}	3.17 ^{bcd}	3.47 ^{ab}	3.67 ^{a}	3.43 ^{ab}	3.00 ^{cd}	3.20 ^{bcd}	3.27 ^{bc}
Stale aftertaste	2.67 ^{c}	3.00 ^{ab}	3.10 ^{ab}	3.17 ^{a}	2.83 ^{bc}	3.00 ^{ab}	2.67 ^{c}	2.90 ^{abc}
Bitter aftertaste	8.07 ^{bcd}	7.93 ^{cd}	8.70 ^{ab}	8.90 ^{a}	8.47 ^{abcd}	8.63 ^{ab}	7.87 ^{d}	8.53 ^{abc}
Sweet aftertaste	0.20 ^{ab}	0.03 ^{bc}	0.00 ^{c}	0.27 ^{a}	0.03 ^{bc}	0.00 ^{c}	0.13 ^{abc}	0.00 ^{c}
Oxidized oil aftertaste	2.20 ^{b}	2.33 ^{b}	2.57 ^{ab}	2.90 ^{a}	2.33 ^{b}	2.30 ^{b}	2.23 ^{b}	2.27 ^{b}
Fish aftertaste	1.27 ^{ab}	1.03 ^{b}	1.80 ^{a}	1.80 ^{a}	1.40 ^{ab}	1.10 ^{b}	1.77 ^{a}	1.27 ^{ab}
Iron aftertaste	1.77 ^{c}	2.33 ^{abc}	2.37 ^{abc}	2.87 ^{a}	2.30 ^{abc}	2.43 ^{ab}	2.17 ^{bc}	1.93 ^{bc}

Least significant differences are shown with superscript letters following the average intensity scores. Letters that are the same for a treatment attribute in a row are not significantly different ($p > 0.05$).

The results indicate that fiber source as well as fiber amount and particle size have an influence on the sensory properties of extruded dog food. For aroma and flavor, the overall differences were not too large across the treatments. Bitter taste and stale and iron aftertaste were examples of flavor attributes that differed among treatments. Sample GF12 (together with sample SC1) showed the highest scores for most of the aftertastes (including off-notes) and the highest bitterness. GF12 had the least particle size and also one of highest levels of total dietary fiber among all treatments, which might be factors contributing to highest intensity scores for various attributes described above. On the other hand, GF3 had the least dietary fiber level, which was probably the reason for its lowest bitter score. Martin *et al.* found that dietary fibers enriched breads were characterized by a higher bitterness together with higher cereal and toasted notes [6]. Sugar cane fibers with large particle size (sample SC1) seemed to influence fibrous texture more than any other type of fibers. Sugar cane samples also showed the lowest fracturability and initial crispness (texture) levels, possibly because the length/size of sugar cane fibers was higher than other fibers, leading to a strengthening effect. This was probably also the reason for the above observation related to fibrous texture and was reflected in appearance as well. SC1 showed the highest scores across the sample set for fibrous and grainy appearance attributes. Treatments with sugar cane fiber (SC1 and SC2) also had the least porous appearance. Interestingly, among all fiber sources, wheat bran appeared to have the least impact on most sensory attributes. WB2 showed lower fracturability and initial crispness than WB1. Smaller particle size has been reported previously to result in lower levels of hardness in the case of extruded cereals [5].

3.1.2. Coated Treatments

The main contributing attributes to aroma of the coated treatments were barnyard, brothy, toasted, grainy, dusty/earthy, cardboard, and hay-like [18]. The main contributing attributes to flavor and taste of these dog foods were barnyard, brothy, toasted, grainy, vitamin, sour, salty, bitter, dusty/earthy, cardboard, and hay-like. The highest intensity was noted for bitter taste (averages 5.94–6.56). Commercial pet foods are typically coated with a palatant; in general, commercial dry dog foods exhibit sensory characteristics such as barnyard, brothy, brown, grain, soy, vitamin, oxidized oil, cardboard, and stale [12]. These attributes were similar to the main flavor and taste attributes detected in experimental dry dog foods in this study.

After coating with poultry fat and a palatant there were few significant differences ($p < 0.05$) found among the treatments (Table 3). Brown and porous appearances were the attributes that varied most among the treatments. Wheat bran treatments WB1 and WB2 were significantly darker brown than sugar cane treatments. The control sample was the lightest brown in appearance. Porous appearance was more apparent for the control and WB2 samples. Toasted aroma was found to be less intense in sample SC2, while dusty/earthy aroma was less intense in sample WB1. Sample WB1 was also less intense in cardboard flavor.

Table 3. Average appearance, aroma, and flavor attribute intensity scores for coated treatments. Significantly different attributes showed only ($p < 0.05$).

Attribute	Treatment				
	CO	SC1	SC2	WB1	WB2
Brown Appearance	7.19 [c]	7.69 [ab]	7.50 [bc]	7.94 [a]	7.94 [a]
Porous Appearance	2.44 [a]	1.88 [b]	1.81 [b]	1.25 [c]	2.38 [a]
Toasted Aroma	3.38 [a]	3.00 [a]	2.56 [b]	3.13 [a]	3.19 [a]
Dusty/earthy Aroma	2.94 [a]	2.94 [a]	2.88 [a]	2.44 [b]	3.06 [a]
Cardboard Flavor	3.19 [a]	3.13 [a]	2.81 [a]	2.31 [b]	2.81 [a]

3.1.3. Coating Effect on Sensory Properties

Poultry fat and palatant effects on appearance, aroma, and flavor attributes among the uncoated and coated treatments were noticed. The coated treatments were less porous in appearance, albeit more intense in brown color. The aromatics of the treatments changed after coating with the poultry fat and palatant (Figure 1). Specifically, stale, liver, and fish aromatics were present in the uncoated kibbles, but were not detected in the coated kibbles. Similarly, sour aromatics and hay-like aroma was detected only in the coated kibbles. Brothy, grainy, and toasted attributes increased in intensity and oxidized oil and iron aromatics reduced in intensity after coating with poultry fat and palatant. The flavor of the treatments changed after coating with the poultry fat and palatant (Figure 2). Fish flavor was not detected in the coated kibbles, while hay-like flavor was detected in the coated kibbles. Brothy, toasted, and grainy flavor increased after coating the kibbles. Stale, liver, oxidized oil, metallic, and iron flavor and bitter taste decreased after coating the kibbles. It was apparent that coating had a masking effect on appearance attributes such as porosity, aroma attributes such as stale, and flavor attributes such as iron and bitter taste.

3.2. Palatability

According to the palatability testing results some of the treatments were more palatable than others (Table 4). For example the CO treatment was eaten more than SC2 or WB2, thus implying that fiber addition decreased the palatability. This is not a surprising result, given the negative attributes associated with fiber in general, including hardness and bitterness. A study conducted by Sa *et al.* [10] looked at using wheat bran in dog foods. These authors found that the dogs in the experiment actually consumed the negative control sample less than the wheat bran treatments, however, this was not a preference test. Sample WB1 was eaten more than WB2, and SC1 was eaten more than SC2. These results indicated that both fiber addition and the size of the fiber particles might have an influence when determining whether the diet is palatable or not. A full set of comparisons among the treatments would be needed in order to determine an order of preference.

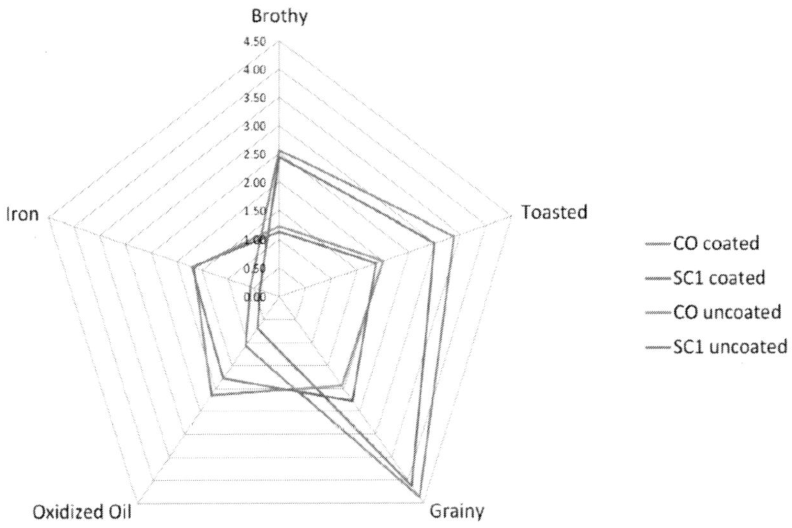

Figure 1. Poultry fat and palatant effect on kibble aromatics. Only the Control (CO) sample and Sugar Cane large particles (SC1) samples are shown.

Table 4. Palatability testing results.

Comparison	Treatment	First Choice (%)	Food intake (%)
CONTROL *versus* SC2	Control	70 **	80 ***
	SC2	30	20
CONTROL *versus* WB2 *	Control	75 **	88 ***
	WB2	25	12
SC1 *versus* SC2	SC1	87 ***	79 ***
	SC2	13	21
WB1 *versus* WB2	WB1	24	55
	WB2	76 **	45

* This test was not validated due to underconsumption of the treatments by the dogs. ** difference between groups ($p < 0.05$). *** difference between groups ($p < 0.01$).

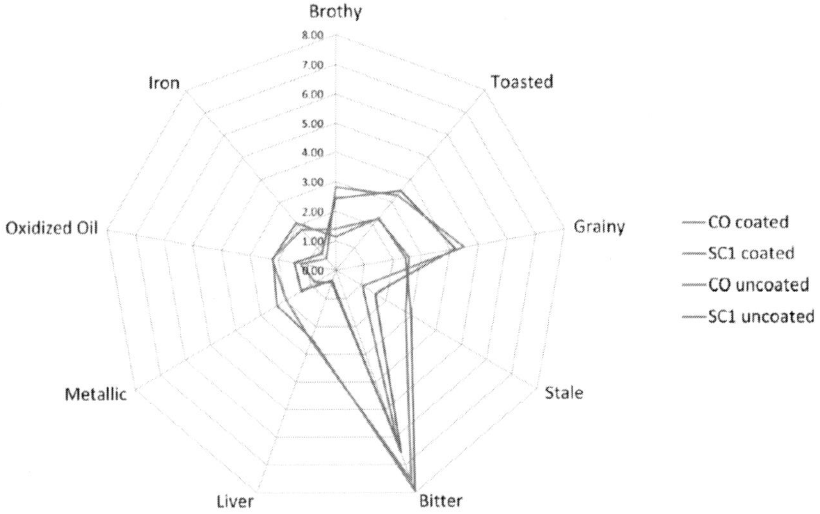

Figure 2. Poultry fat and palatant effect on kibble flavor. Only the Control (CO) sample and Sugar Cane large particles (SC1) samples are shown.

3.3. Potential Effect of Sensory Properties on Palatability

Further inspection of the treatments sensory properties and palatability testing results indicated that the texture properties of CO and WB1 were similar in fracturability, initial crispness, and fibrous attributes. The CO sample was preferred over both SC2 and WB2. The WB1 sample was only compared to WB2 sample, and the results showed a somewhat higher, but not significantly higher intake for WB1 sample (Table 4). These results indicate that texture properties may have an influence on dry pet food palatability, and higher intensity of attributes such as fracturability and initial crispness might lead to greater palatability.

Most variability was detected in the fibrous texture attribute. This was probably caused by the different fiber sources used in the formulation. The CO and WB1 samples were similarly low in this attribute, while WB2 and SC2 were higher, and SC1 the highest (9.20, Table 2). The fibrous attribute was defined as "the perception of fibers and filaments in the product after three to five chews". Most dogs do not chew the food too extensively, but the fibrous texture is likely to have an impact on the mouthfeel of the food. Based on the results from palatability testing, it seems that fibrous mouthfeel may have an effect on pet food preference. When considering that SC1 was preferred over SC2, however, and that SC1 was highest in fibrous characteristics, the role of this attribute in determining palatability should be studied further.

There were only three aroma and flavor attributes that were significantly different among the coated treatments (toasted and dusty/earthy aroma and cardboard flavor). Toasted aromatics were the highest in the CO sample, and lowest in SC2 sample. As the CO was eaten more in the comparisons, and SC2 was eaten less in the comparisons, this attribute may be important in determining palatability of dry pet food. The texture effect was expected to be more prominent in these treatments, as the palatant amount and formulation was kept constant for each of the treatments. Nevertheless these differences in aroma and flavor were likely to be the result of the ingredient differences and may have influenced palatability as well.

In a comparison between WB1 and WB2, WB2 was most often the first choice for dogs. This did not result in this sample being the most eaten, though. Treatment WB2 showed a higher dusty/earthy aromatics level, and this could be a factor that attracted the dogs before they actually started eating. A flavor note that was different between WB1 and WB2 was cardboard, and WB2 had a higher intensity of this flavor. This may be one of the reasons for preference of WB1 once dogs started eating the treatment.

A Principal Components Analysis (PCA) was conducted in order to visualize the main differences between the treatments and to relate these to dog preference testing results. A total of 78.54% of variability among the treatments was explained by the two first Principal Components (Figure 3). According to the PCA analysis in the comparison of treatments WB1 and WB2, porous appearance and cardboard flavor and dusty/earthy aromatics may have influenced palatability. The comparison between CO and WB2 or SC2 resulted in the CO treatment being more palatable. This may have been caused by higher toasted aromatics, grittiness, initial crispness, or fracturability. The comparison between SC1 and SC2 resulted in the SC1 treatment being more palatable. This may have been caused by the more fibrous texture of treatment SC1.

There are some limitations to this research. The first and most obvious is the fact that human senses are different from a dog's senses. This study was not trying to claim that the senses of a human and those of a dog are similar. Rather the study was trying to understand if there were any sensory attributes that the human panelists could distinguish in dog foods that may help indicate direction of preference in a palatability test. While the results from this study may not necessarily fully answer that question, at least as far as fiber sources in pet foods are concerned, some indication of the reasons behind higher liking were detected.

The second limitation lies in the palatability test and the amount of information collected with this test. Typically, characteristics like first choice and food intake are collected. During an eating situation, dogs, just like humans, are likely to indicate their liking with other bodily cues, such as wagging the tail, increased heart rate or respiration rate, pupil dilation, licking of the mouth, or other indications. A study conducted by van den Bos *et al.* [19] concluded that cats might indicate their liking or disliking of foods with some behavioral, facial or tongue movements. Further research is needed to understand what cues may indicate liking in dogs. In addition, the two-bowl testing situations are somewhat forced in their nature. The dogs have two choices of food to choose from, and there is no guarantee that either of the foods is more palatable. Since the dogs are likely to be hungry, they will probably eat either or both of the foods, but this may not necessarily show that either of the foods is liked more.

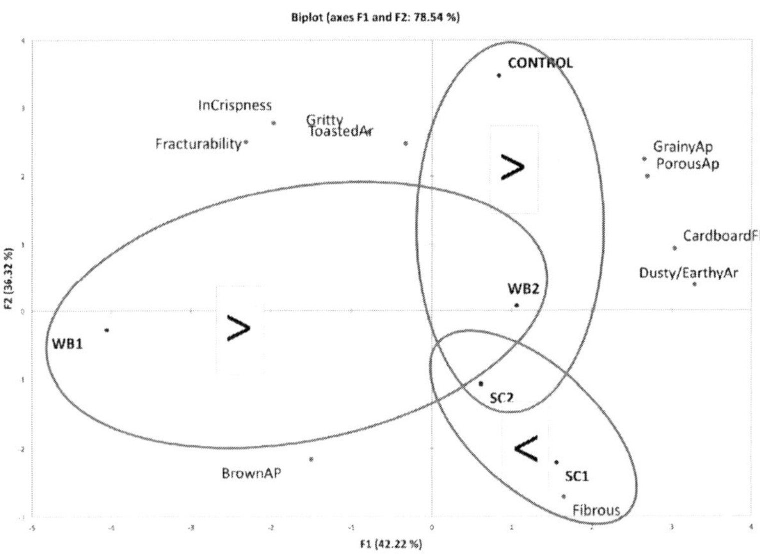

Figure 3. Principal Components Analysis of treatments for the significantly different appearance (Ap) texture, aroma (Ar), and flavor (Fl) attributes. Texture attributes were included based on the uncoated treatment descriptive sensory analysis results. In Crispness—Initial crispness. The < and > symbols indicate the direction of the palatability test for the comparisons of samples within the ovals.

Furthermore, the use of kennel dogs may provide different results than the use of actual pets that live with their owners. This has been shown by Smith *et al.* [20] in a comparison of laboratory and pet dogs in their preferences for dog foods that varied in flavor. In the current study, the focus was on texture attributes, as the flavor variation was kept at a minimum by using the same palatant on all treatments. A comparison of different textures of dog foods has been conducted by Kitchell *et al.* [21], however this comparison was among moist, canned, and dry foods. A preference of canned and moist foods over dry foods was found in that study. No studies were found that compared dog preference for different dry food textures.

4. CONCLUSIONS

The results of this study indicated that fibers (type or quality, particle size, and amount) have an influence on sensory characteristics of extruded dog food. The texture of treatments was most influenced by the different fibers, and sugar cane fibers with large particle size influenced the fibrousness of treatments more than other fiber sources. Moreover, the higher length of sugar cane fibers, causing a strengthening effect, could have played a role in the lowest score in fracturability and initial crispness. Fibers seemed to influence aroma and flavor properties; the treatment with the highest amount of guava fibers was the most bitter and had most aftertastes, including off-notes such as stale and oxidized oil. The coating process showed to be fundamental in reducing or eliminating off-notes both in flavor and aroma, and also altering the appearance. Moreover it increased notes such as brothy, grainy, and toasted in the samples and even added new characteristics not present in the uncoated samples such as sour taste and hay-like aroma and flavor.

Palatability testing results indicated that the control treatment was preferred over treatments containing small particle size of both sugar cane and wheat bran fibers. Product hardness, bitter taste and other off-notes associated with fibers might have been contributing factors. When the same fiber sources where compared, a preference for large particle size was shown for the food intake. Wheat bran fibers with small particle size showed to be preferred as first choice but not in regard to food intake.

Results showed that texture characteristics such as fracturability, initial crispness, and grittiness, together with aromas such as toasted may play a main role in increasing the palatability. Moreover, different results among first choice and food intake for a particular source of fiber may be helpful to indicate what kind of aromas may be more attractive for dogs, and what kind of flavors may cause them prefer a treatment that was perhaps not chosen in the first moment.

Further future studies on the subject may clarify the influence of different fibers on palatability further and give a better understanding of the relationship between sensory properties of extruded dog food and acceptance by dogs.

ACKNOWLEDGMENTS

The authors thank Dilumix (Leme, SP, Brazil) for processing and donating the fiber sources, and for partially financially supporting the experiment, and Panelis (Diana Group, Descalvado, São Paulo, Brazil) for their technical contribution and donation of the palatability tests with dogs. The authors would also like to thank Jeff Wilson from USDA-ARS Grain Marketing and Production Research Center for the particle size analysis.

REFERENCES

1. Mcnamara, J.P. Chapter 3: Glucose and Fatty Acids: Providers of Body Structure and Function. In *Principles of Companion Animal Nutrition*, 2nd ed.; Pearson Education, Inc.: Upper Saddle River, NJ, USA, 2014; pp. 27–47.
2. Weber, M.; Bissot, T.; Servet, E.; Sergheraert, R.; Biourge, V.; German, A.J. A high-protein, high-fiber diet designed for weight loss improves satiety in dogs. *J. Vet. Int. Med.* 2007, *21*, 1203–1208.
3. De Godoy, M.R.C.; Kerr, K.R.; Fahey, G.C., Jr. Alternative dietary fiber sources in companion animal nutrition. *Nutrients* 2013, *5*, 3099–3117.
4. Campbell, K.L.; Campbell, J.R. Chapter 9: Feeding and Nutrition of Dogs and Cats. In *Companion Animals. Their Biology, Care, Health, and Management*, 2nd ed.; Pearson Education Inc.: Upper Saddle River, NJ, USA, 2009; pp. 253–299.
5. Robin, F.; Schuchmann, P.H.; Palzer, S. Dietary fiber in extruded cereals: Limitations and opportunities. *Trends Food Sci. Tech.* 2012, *28*, 23–32.
6. Martin, C.; Chiron, H.; Issanchou, S. Impact of dietary fiber enrichment on the sensory characteristics and acceptance of French baguette. *J. Food Qual.* 2013, *36*, 324–333.
7. Wang, J.; Rosell, C.M.; Benedito de Barber, C. Effect of the addition of different fibers on wheat dough performance and bread quality. *Food Chem.* 2002, *79*, 221–226.
8. Popov-Ralić, J.V.; Mastilović, J.S.; Laličić,-Petronijević, J.G.; Kevrešan, Z.S.; Demin, M.A. Sensory and color properties of dietary cookies with different fiber sources during 180 days of storage. *Hem. Ind.* 2013, *67*, 123–134.
9. Pacheco, G.F.E.; Marcolla, C.S.; Machado, G.S.; Kessler, A.M.; Trevizan, L. Effect of full-fat rice bran on palatability and digestibility of diets supplemented with enzymes in adult dogs. *J. Anim. Sci.* 2014, *92*, 4598–4606.
10. Sa, F.C.; Vasconcellos, R.S.; Brunetto, M.A.; Filho, F.O.R.; Gomes, M.O.S.; Carciofi, A.C. Enzyme use in kibble diets formulated with wheat bran for dogs: Effects on processing and digestibility. *J. Anim. Physiol. Anim. Nutr.* 2013, *97*, 51–59.
11. Koppel, K. Sensory analysis of pet foods. *J. Sci. Food Agric.* 2014, *94*, 2148–2153.
12. Di Donfrancesco, B.; Koppel, K.; Chambers, E., IV. An initial lexicon for sensory properties of dry dog food. *J. Sens. Stud.* 2012, *27*, 498–510.
13. Koppel, K.; Adhikari, K.; di Donfrancesco, B. Volatile compounds in dry dog foods and their influence on sensory aromatic profile. *Molecules* 2013, *18*, 2646–2662.

14. Koppel, K.; Gibson, M.; Alavi, S.; Aldrich, G. The effects of cooking process and meat inclusion on pet food flavor and texture characteristics. *Animals* 2014, *4*, 254–271.
15. FEDIAF—European Pet Food Industry Federation. Nutritional Guidelines for Complete and Complementary Pet Food for Cats and Dogs. 2011. Available online: http://www.fediaf.org/press-area/press-releases/news-detail/artikel/nutritional-guidelines-cats-and-dogs/ (accessed on 26 November 2014).
16. Boac, J.M.; Maghirang, R.G.; Casada, M.E.; Wilson, J.D.; Jung, Y.S. Size distribution and rate of dust generated during grain elevator handling. *Appl. Eng. Agric.* 2009, *25*, 533–541.
17. Griffin, R.W. Section IV: Palatability. In *Petfood Technology*, 1st ed.; Kvamme, J.L., Phillips, T.D., Eds.; Watt Publishing Co.: Mt Morris, IL, USA, 2003; pp. 176–193.
18. Koppel, K.; The Sensory Analysis Center, Department of Human Nutrition, College of Human Ecology, Kansas State University, Manhattan, KS, USA. Unpublished data. 2014.
19. Van den Bos, R.; Meijer, M.K.; Spruijt, B.M. Taste reactivity patterns in domestic cats (*Felis silvestris catus*). *Appl. Anim. Behav. Sci.* 2000, *69*, 149–168.
20. Smith, S.L.; Kronfeld, D.S.; Banta, C.A. Owners' perception of food flavor preferences of pet dogs in relation to measured preferences of laboratory dogs. *Appl. Anim. Ethol.* 1983, *10*, 75–87.
21. Kitchell, R.L.; Franti, C.E.; Sprague, R.H. Palatability of commercial dog foods. *Anat. Histol. Embryol.* 1975, *4*, 371.

CHAPTER 19

Managing Mastitis in Heifers: An Initial Step in Improving Dairy Herd Health

Stephen C. Nickerson

University of Georgia,, United States

1. INTRODUCTION

Replacement heifers are critical to dairy herd productivity because they represent the future milking and breeding stock of all dairy operations. The goal should be to provide an environment for heifers to develop full lactation potential at the desired age with minimal expense. Animal health and well-being play vital roles in achieving this potential, and one disease that can influence future productivity is heifer mastitis caused by Staphylococcus aureus, the coagulase-negative staphylococci (CNS), and the environmental streptococci (Streptococcus uberis, Streptococcus dysgalactiae).

Unfortunately, most producers regard young dairy heifers as uninfected, and the presence of mastitis is not observed until freshening or until the first clinical flare-up in early lactation. Thus, animals may carry intramammary infections (IMI) for a year or more before they are diagnosed with mastitis (Boddie et al., 1987). The greatest development of milk-producing tissue in the udder occurs during the first pregnancy, so it is important to protect the heifer mammary gland from pathogenic microorganisms to ensure maximum milk production during the first and future lactations. In the USA, Louisiana researchers found that if bred heifers infected with Staph. aureus were left untreated, they produced 10%

less milk in early lactation than those receiving therapy (Owens et al., 1991; Trinidad et al., 1990c). Likewise, research in New Zealand demonstrated that Staph. aureus mastitis in heifers resulted in significant production losses during the first lactation, which carried over into the subsequent lactation because of damage to milk-producing tissues of the udder (Woolford et al., 1983).

2. PREVALENCE OF MASTITIS IN UNBRED AND PRIMIGRAVID HEIFERS

A greater focus on heifer mastitis began in the mid 1980s after several dairy producers in Louisiana complained to university researchers that a large percentage of their heifers were freshening with clinical mastitis. Subsequent study of breeding age animals in a research herd revealed that IMI may be diagnosed as early as 6 months of age, and infections persisted throughout pregnancy and into lactation (Boddie et al., 1987). Other studies demonstrated that greater than 90% of breeding age and bred heifers (12 - 24 months of age) may be infected (Trinidad et al., 1990b). Most of the infections were shown to be caused by the CNS (Staphylococcus chromogenes and Staphylococcus hyicus) followed by Staph. aureus (20%). Mixed isolates of CNS and Streptococcus species were also found.

The prevalence of mastitis in 10 unbred Jersey heifers (10 - 12 months of age) in Louisiana was initially evaluated in 1985 and monitored over a 1-year period, which covered breeding age and gestation (Boddie et al., 1987). Samplings from the teat skin, teat canal, and mammary secretion were performed bimonthly and continued through time of calving. For bacteriologic analysis, a total of 388 samples from teat skin, 388 from teat canal keratin, and 216 from secretions were examined; not all quarters contained sufficient volumes of secretions.

For teat skin, staphylococci were isolated on mannitol salt agar, and the predominant colony type of a particular isolate based on color, rough/smooth, size, hemolytic pattern, and appearance was collected, sub-plated for culture on blood agar, and identified to the species level (Boddie et al., 1987). Staphylococcus xylosus (20.9%) and Staph. chromogenes (14.9%) were the predominant flora isolated from teat skin, followed by Staph. warneri (6.7%), Staph. sciuri (6.2%), Staph. aureus (2.8%), Staph. hyicus (1.3%), and Staph. simulans (1.3%). The most prevalent bacteria found in teat canals were Staph. chromogenes (41.0%), followed by Staph. hyicus (16.8%), Staph. aureus (10.0%), Staph. xylosus (1.0%), Staph. warneri (0.8%), and Staph. sciuri (0.5%).

In mammary secretions, Staph. chromogenes (49.5%) was the predominant organism, followed by Staph. hyicus (21.3%), Staph. aureus (13.0%), Streptococcus uberis (1.4), and Staph. xylosus (0.9%). The CNS species in teat canals and in secretions from a particular mammary quarter were present at the first sampling, and the same CNS species were isolated from teat canals and secretions from the same individual quarters with each successive sampling of the trial. Also, there appeared to be a correlation between colonization of the teat canal and IMI. For example, the major CNS species colonizing the teat canal (Staph. chromogenes (41.0%), Staph. hyicus (16.8%), Staph. aureus (10.0%), and Staph. xylosus (1.0%)) were also the predominant organisms causing IMI (Staph. chromogenes (49.5%), Staph. hyicus (21.3%), Staph. aureus (13.0%), and Staph. xylosus (0.9%)). Thus, teat canals infected with Staph. chromogenes (41.0%) were positively correlated with secretions (49.5%) infected with these bacteria. The overall correlation between teat canal colonization and IMI was 82.2% across these 4 CNS species.

This initial study (Boddie et al., 1987) indicated that teat skin, teat canals, and mammary secretions of heifers are colonized with CNS as well as Staph. aureus at an early age, and that infections may persist for up to 1 year. Species identification demonstrated that nearly all isolates from the same quarter throughout the study were the same biovariant based on the API Staph-Ident System as previously modified (Watts et al., 1984), which supports the contention that isolates from each quarter over time were from persistent infections and not from new IMI occurring between sampling periods. Although the percentage of Staph. aureus isolates was lower than CNS, presence of this major pathogen demonstrated that it colonized teat canals at a much earlier age than documented previously (Rendel & Sundberg, 1962).

In a subsequent herd survey (Trinidad et al., 1990b), the prevalence of mastitis in breeding age and pregnant heifers was determined in 4 commercial dairies. Teat canal keratin and secretion samples were collected from 116 Jersey heifers, and results revealed that teat canal colonizations were present in 93.1% of heifers and 70.7% of quarters. Staph. aureus was isolated from teat canal keratin samples of 31% of heifers and 12.3% of quarters. Other organisms isolated from keratin and percentage frequencies were Staph. chromogenes (42.9%), Staph. hyicus (25.2%), other staphylococcal species (5.7%), Strep. dysgalactiae (0.6%), Strep. species (3.1%), and mixed isolates containing staphylococci and streptococci (5.7%). In this 4-herd survey, IMI were found in 96.9% of heifers and in 74.6% of quarters. Twenty-nine percent of heifers and 15.1% of quarters showed clinical symptoms of mastitis as evidenced by clots, flakes, and blood. Staph. aureus was isolated from 14.7% of quarters. This microorganism was also isolated from 25% of quarters with clinical symptoms. Staph. aureus causes severe damage to mammary tissue

(Trinidad et al., 1990a), and infections are very difficult to eliminate in lactating cows. Other organisms isolated from secretions and percentage frequencies were Staph. chromogenes (43.1%), Staph. hyicus (24.3%), other staphylococcal spp. (3.6%), Strep. dysgalactiae (0.4%), Strep. spp. (3.3%), Nocardia species (0.4%), and mixed isolates containing staphylococci and streptococci (5.1%).

These initial studies on heifer mastitis were performed in Louisiana, USA, where a warm and humid climate with a long fly season may be conducive to IMI in these young dairy animals. Thus, a subsequent national trial was carried out using 1583 breeding age heifers from 28 herds to determine the prevalence of heifer mastitis in the states of California, Washington, and Vermont as well as in Louisiana (Fox et al., 1995). The majority of quarter infections were caused by the CNS (mean = 27.1%). The mean prevalence of Staph. aureus among the 4 sites was 2.9%, but was highest in Louisiana at 10.1% during the spring season. The overall IMI prevalence was greatest in Louisiana, and the highest frequency was during the 3rd trimester of gestation just prior to parturition, which was due to an increase in IMI cause d by Staph. aureus, CNS, and environmental streptococci. Likewise, following parturition, the greatest prevalence of IMI was in Louisiana, which had the greatest percentage of heifers with IMI caused by the environmental streptococci and Staph. aureus, and the 2nd greatest prevalence of CNS among the 4 sites. For breeding age and bred heifers, the stage of pregnancy significantly affected IMI prevalence, which was highest for heifers in the last trimester of pregnancy, especially CNS IMI. This study by Fox et al. (1995) showed that site location had a significant effect on prevalence of mastitis, and that Louisiana had the greatest prevalence of IMI. This was postulated to be due to the warm and humid climate, as well as to a prolonged horn fly season in Louisiana that exposed udders and teats to an increased bacterial load that management practices failed to control.

3. MAMMARY LEUKOCYTE RESPONSE TO INTRAMAMMARY INFECTION

In lactating cows, the milk somatic cell count (SCC) composed of leukocytes (macrophages, lymphocytes, neutrophils) and a small percentage of mammary epithelial cells, is considered an important parameter for assessing mammary health status (e.g. inflammation); milk yield decreases as SCC and incidence of mastitis increase. Thus, SCC in breeding age and pregnant heifer mammary gland secretions have been analyzed to measure the degree of inflammation and potential reductions in future milk yield. In a study by Boddie et al. (1987), the mean arithmetic SCC of quarters from unbred heifers infected with Staph. chromogenes, Staph. hyicus, and Staph. aureus were 7.8, 8.5, and 9.2 x 10^6/ml, respectively, whereas the mean SCC of uninfected quarters was 3.5 x 10^6. The mean SCC of heifer secretions collected on the day of freshening were 3.2 x 10^6/ml and 1.6 x 10^6/ml for quarters infected by staphylococci and uninfected quarters, respectively. The mean SCC during the first 3 months of lactation in quarters infected with Staph. chromogenes, Staph. hyicus, and Staph. aureus were 168, 193, and 578 x 10^3/ml, respectively, while the SCC of uninfected quarters was 39 x 10^3/ml. Thus, this study found SCC approaching 200 x 10^3 for quarters infected with CNS during the first 3 months of lactation, and, based on previous studies (Jones et al., 1984; Kirk, 1984), SCC in this range are associated with milk loss. Approximately 13% of quarter secretions sampled prepartum contained Staph. aureus, and after freshening, the SCC of these quarters averaged 578 x 10^3/ml, a cell count that has been associated with a loss of >4.4 lb (2.0 kg) of milk/day (Kirk, 1984).

Other studies have also demonstrated elevated SCC in heifer mammary glands infected with mastitis-causing bacteria. For example, in a study by Trinidad et al. (1990b), average arithmetic SCC in secretions from 325 quarter samples regardless of infection status was 11.7 x 10^6/ml. Infected (n = 240) and uninfected (n = 85) quarters had secretion SCC of 13.6 x 10^6/ml and 5.7 x 10^6/ml, respectively. Of the staphylococci, Staph. aureus-infected quarters had the highest secretion SCC/ml (17.3 x 10^6), followed by Staph. chromogenes (12.8 x 10^6), and Staph. hyicus (12.4 x 10^6). The mean SCC for non-agalactiae streptococci was 15.5 x 10^6/ml.

The volume of mammary secretion is very low in breeding-age animals; thus, SCC become concentrated, resulting in high SCC even in uninfected quarters. However, SCC approach 20 x 10^6/ml in quarters infected with Staph. aureus, and over 13.6 x 10^6/ml in those infected with the CNS and Strep. species. Such elevated SCC over a long period of time suggests that mammary tissue in affected quarters are in a state of chronic inflammation, which could adversely affect development of milk-producing tissues and negatively affect future milk yield. In response to infection, neutrophils become the major leukocyte type that infiltrates mammary tissue from the vascular system. The function of the neutrophil influx via chemotaxis is to phagocytose and kill mastitis-causing bacteria. But there is also evidence that this defense response can impair and disrupt mammary function (Akers & Thompson 1987; Capuco et al., 1986; Zhao & Lacasse, 2008). For example, in addition to the damage caused by bacterial toxins, the migration of leukocytes, namely neutrophils, across the mammary epithelial surfaces is thought to cause mechanical damage and/or chemical damage (via release of reactive oxygen species) to the milk secretory cells as well as to the ductal cells of the mammary gland.

4. HISTOLOGICAL RESPONSE OF THE MAMMARY GLAND TO IMI

Initially, the mammary glands of dairy heifers were studied to determine histological responses to teat canal colonization with Staph. chromogenes and Staph. hyicus (Boddie et al. (1987). Two unbred heifers were slaughtered, one at 8 months and the other at 18 months of age, and examination of mammary tissues from both heifers demonstrated a leukocyte reaction to the colonization of the teat canal. Cross-sections through the mid-teat canal demonstrated cocci colonizing keratinized cells of the canal lumen, and sections of distal teat cisternal tissues demonstrated heavy leukocyte infiltration with lymphocytes and plasma cells at Fürstenberg's rosette compared with uninfected tissues.

Subsequently, 7 unbred heifers, 14 - 26 months of age, were studied to evaluate the effects of IMI on leukocyte infiltration and characteristics of secretory tissue in developing mammary glands (Trinidad et al., 1990a). Histologic observations of tissue samples from lobes of mammary parenchyma of uninfected quarters showed that alveoli were small; the epithelial lining was composed of a single layer of cuboidal cells surrounding a small luminal space with little or no stained secretory product (See Figures 1 through 3). Interalveolar connective tissue area composed approximately half of the observed lobes, and a few infiltrating leukocytes, mainly lymphocytes, were observed.

Infected tissues, particularly those with Staph. aureus IMI, exhibited large amounts of interalveolar connective tissues and reductions in epithelial and luminal areas (See Figures 4-8). Such areas also exhibited leukocytic infiltration, particularly lymphocytes and neutrophils, into stromal and luminal areas. Hyperplasia of ducts and cisterns as a result of infection was also observed, and macro- and microscopic abscesses were found in the parenchyma of one quarter infected with Staph. aureus. Abscesses were tubercule-like with a circular, stratified fibrosis containing numerous lymphocytes, neutrophils, plasma cells, and multinucleated giant cells.

Results of morphometric analysis on parenchymal tissue components showed that percentages of each component in uninfected quarters were very similar to percentages from quarters infected with CNS, although quarters infected with CNS exhibited significantly more stromal area. Percentages of alveolar epithelium and lumen in quarters infected with Staph. aureus were significantly lower ($P < .05$) than those in uninfected quarters and in quarters infected with CNS. Quarters infected with Staph. aureus also showed a greater percentage ($P < .05$) of interalveolar stroma than did uninfected quarters and quarters infected with CNS. Thus, Staph. aureus-infected tissue demonstrated reduced secretory activity. The greatest development of secretory tissue in young heifers occurs during the first pregnancy, and developing secretory tissues may be affected adversely by bacterial infection and inflammation, leading to deposition of connective tissue stroma instead of milk secretory tissue and a subsequent deleterious effect on future milk yield.

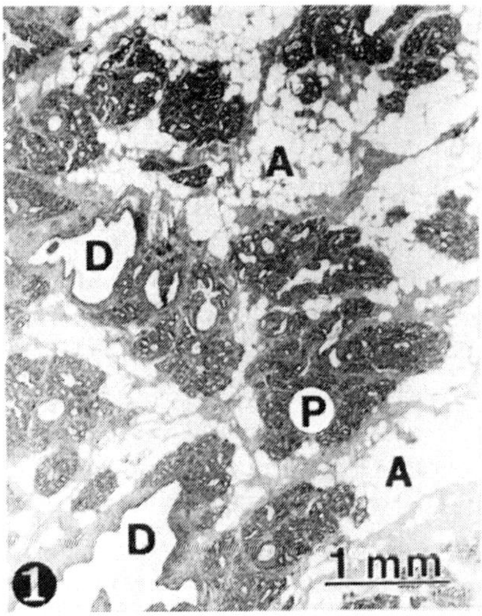

Figure 1. General view of a cross-section of a lobe of mammary tissue from an uninfected quarter exhibiting large ducts (D) and undeveloped lobules of parenchyma (P) among adipose tissue stroma. (A). x18.

Leukocyte infiltration into cisternal and parenchymal mammary tissues was also evaluated. Quarters infected with Staph. aureus exhibited the greatest tissue leukocytosis, followed by quarters infected with CNS and uninfected quarters. Leukocyte infiltration in gland cistern and secretory tissue for infected quarters was significantly higher (P <.05) than that for uninfected quarters. Leukocytosis into teat cistern tissue was similar for uninfected quarters and those infected with CNS, but significantly lower (P <.05) than quarters with Staph. aureus IMI. None of the uninfected quarters or quarters infected with CNS demonstrated marked leukocyte infiltration. However, marked leukocyte infiltration, particularly lymphocytes, into cisternal and parenchymal areas was commonly observed in quarters that were infected with Staph. aureus. The majority of leukocytes observed within alveolar lumina were neutrophils.

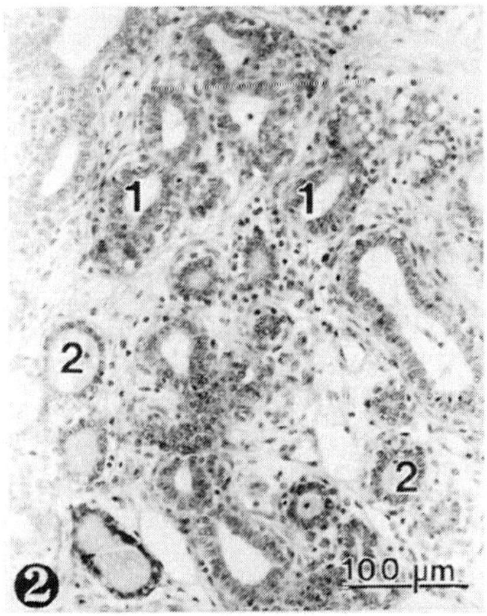

Figure 2. Portion of mammary parenchymal tissue typical of that obtained from uninfected quarters and those infected with coagulase-negative staphylococci exhibiting small alveoli with empty, ovoid lumens (1) and those with some secretions (2). x180.

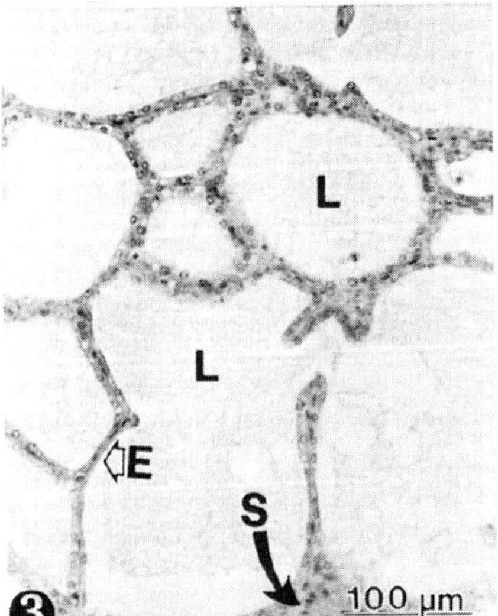

Figure 3. Portion of uninfected parenchymal tissue revealing limited stroma (S), flattened epithelium (E), and distended luminal areas (L) engorged with flocculent matter suggested active secretion. x180.

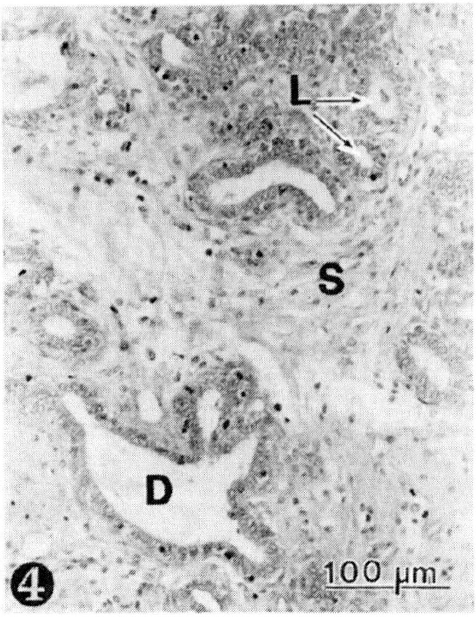

Figure 4. Parenchymal tissue from a quarter infected with Staph. aureus exhibiting a large interalveolar connective tissue stroma (S) and limited alveolar luminal areas (L). D = Duct. x180. Trinidad et al., 1990a.

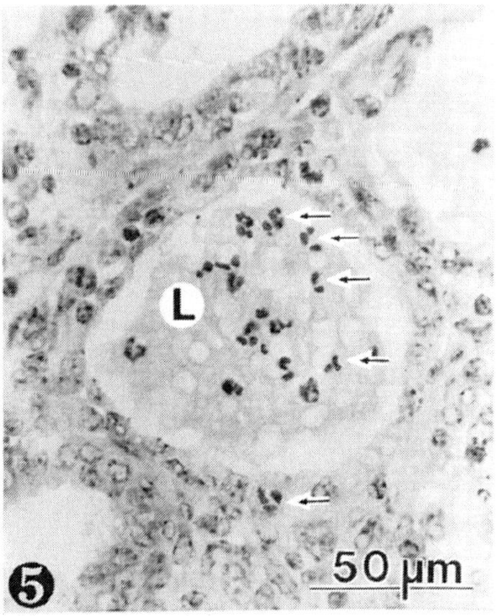

Figure 5. Parenchymal tissue from a quarter infected with Staphylococcus aureus showing numerous neutrophils (arrows) infiltrating a luminal area (L) of one alveolus. x500.

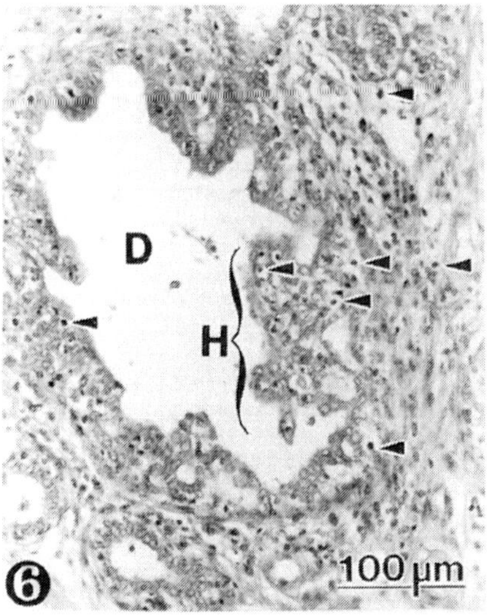

Figure 6. Extensive epithelial hyperplasia (H) was observed in ductal linings in the parenchyma from one quarter infected with Staphylococcus aureus. Lymphoid cells (arrowheads) were numerous in the epithelium as well as in the underlining connective tissue. D = duct. x180.

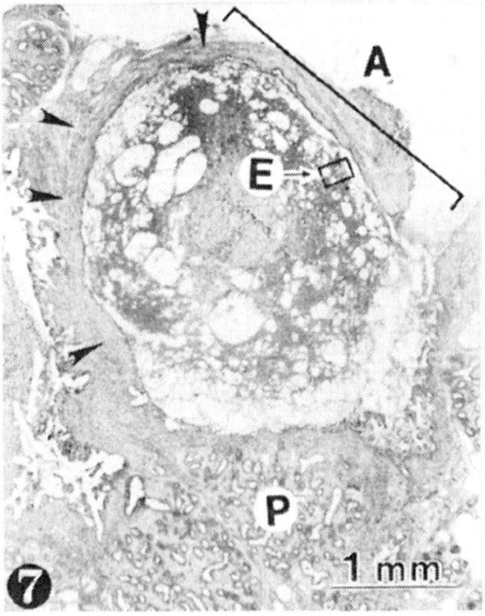

Figure 7. Abscess (A) from one quarter infected with Staphylococcus aureus exhibiting tubercle-like morphology with circular stratified fibrosis (arrows) and marked cellular infiltration. E = Portion of enlarged in Figure 8. P = Parenchyma. x18.

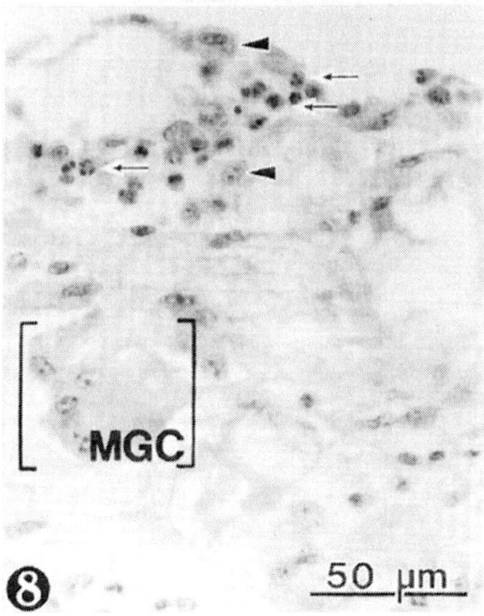

Figure 8. Magnification of an edge of the abscess shown in Figure 7. Neutrophils (arrows), macrophages (arrowheads), and multinucleated giant cells (MGC) were present in this area of the abscess. x500. Trinidad et al., 1990a.

5. EFFICACY OF NONLACTATING COW ANTIBIOTIC THERAPY

Because of the high level of infection commonly found in breeding age and pregnant heifers, especially mastitis caused by Staph. aureus, as well as the associated elevated SCC, antimicrobial therapy should be considered. The testing of various staphylococcal isolates obtained from heifers for susceptibility to antibiotics commonly incorporated into mastitis infusion tubes has shown that antibiotic resistance is usually low (Watts et al., 1995). Greater than 90% of mastitis-causing staphylococci are generally killed by the drug preparations used, based on in vitro sensitivity testing using zone diffusion analysis (Watts et al., 1995). From a practical standpoint, the administration of antibiotics by a parenteral route would be preferred; however, in the author's experience, neither subcutaneous nor intramuscular injections of drugs have been found to cure IMI in heifers because sufficient antibiotic does not pass into the mammary gland to be bactericidal. Thus, intramammary infusion is the route of choice.

In an initial study to evaluate the effectiveness of treatment, several heifers from each of the 4 commercial herds studied by Trinidad et al. (1990c) were randomly selected to receive a single intramammary treatment of a penicillin and dihydrostreptomycin product. Antimicrobial susceptibility testing demonstrated that 97% of the Staph. aureus isolates were sensitive to 12 antibiotics, including the product selected for treatment (Trinidad et al., 1990d). Teat ends were sanitized, and a dry cow antibiotic containing 1,000,000 units of penicillin and 1 gram of dihydrostreptomycin was infused into all quarters of 35 heifers using the partial insertion technique; 38 heifers served as untreated controls. Treatments were made at approximately 60 days prior to the calculated calving date.

Results showed that 97.1% of treated heifers (73.2% of quarters) were infected at the time of treatment, but, at calving, infected heifers and quarters in the treatment group were reduced to 40 and 34%, respectively. Antibiotic residues were limited to two heifers that were treated within 3 weeks of calving because estimated dates of parturition were miscalculated, but all quarters were free of antibiotic residue after 5 days; 2.9% of treated quarters had antibiotic residues at time of calving. Of the 38 untreated control heifers, 100% (71.2% of quarters) were infected at initial sampling, and at calving, mastitis in control heifers was reduced only slightly to 97.4%; percentage of infected quarters increased slightly to 77.8%. The mean SCC at calving was also reduced as a result of therapy. For treated heifers, SCC decreased significantly ($P < .001$) from 11,825 x 10^3/ml at time of treatment to 3,439 x 10^3/ml at calving. In the control group, SCC decreased from 11,047 x 10^3/ml to 5,594 x 10^3/ml ($P > .05$, not significant).

In the same study, Staph. aureus was isolated from 11 quarters of 6 treated heifers before antibiotic infusion (45.8%), but, at calving, this organism was isolated from only 1 quarter of 1 heifer (4.2%). In the control group, 18 quarters of 10 heifers were infected with Staph. aureus at time of treatment (45%). At calving, 6 of the control heifers still had Staph. aureus mastitis in 11 quarters (55%). Thus, the overall incidence of IMI was reduced 60% and that caused by Staph. aureus was reduced over 90%. From 20 to 26% of quarters with subclinical and clinical Staph. aureus mastitis are typically cured after antibiotic therapy (Dodd, 1992). In this study, 90.9% of quarters were cured; thus, therapy in heifers was highly effective in eliminating S. aureus mastitis compared with that for lactating cows.

In this study, an economic analysis was performed to justify use of the heifer treatment program (Trinidad et al., 1990c). Production data collected over the first 2 months of lactation demonstrated that Staph. aureus-infected heifers receiving nonlactating cow therapy during pregnancy produced an average of 5.5 pounds (2.5 kg) more milk per day than Staph. aureus-infected herd mates that did not receive treatment. At the milk price received at that time, the greater milk yield translated to a $42.00 increase for treated heifers, which was well worth the $5.00 cost of treatment. Other advantages include a longer productive life and higher income due to quality milk premiums.

Subsequent studies (Owens et al., 1991, 1994) with pregnant and nonpregnant heifers using a cephalosporin-based product formulated for nonlactating cows was also successful. Heifers that were either experimentally or naturally infected with Staph. aureus were infused 8 -12 weeks prepartum with one dose of 300 mg of a cephapirin benzathine product and were compared with untreated controls infected with Staph. aureus. Results demonstrated that 100% of experimentally induced IMI and 87% of naturally occurring Staph. aureus IMI were eliminated in treated heifers at calving, and cured quarters remained negative at biweekly samples collected 2 months into lactation. Quarters remaining infected at calving with Staph. aureus were treated with a lactating cow product containing 200 mg of cephapirin benzathine, but the cure rate was lower (50 - 56%). Thus, cure rates were greater when a nonlactating product was administered 8 -12 weeks prepartum than when a lactating cow product was given at time of calving.

After antibiotic infusion, SCC in infected quarters that cured decreased from 15 x 10^6/ml to 4 x 10^6/ml 1 week later, and to 700 x 10^3/ml at calving. In contrast, none of the untreated control quarters infected with Staph. aureus cured spontaneously by the time heifers calved, and SCC at calving were 5 x 10^6/ml. Treated heifers in which Staph. aureus IMI were cured yielded a mean of 36.1 lb (16.4 kg) milk/day, and untreated controls that retained Staph. aureus IMI yielded 31.9 lb (14.5 kg)/day or 11% less during the first 2 months of lactation.

Generally, spontaneous cure rates for major mastitis pathogens are low. For example, in a subsequent study on heifer mastitis (Owens et al., 1994), spontaneous cure rates for Staph. aureus and the environmental streptococci were 9 and 6%, respectively. Thus, treatment is required to cure such infected quarters in these young dairy heifers. New IMI rates in uninfected quarters receiving no therapy over the period from 8 - 10 weeks prepartum were very low for most species of staphylococci. However, new environmental streptococcal IMI were common in previously uninfected, untreated heifers. So, as part of this study (Owens et al., 1994), a trial was initiated to treat quarters found to be negative at 8 – 10 weeks prepartum with cephapirin benzathine to determine if establishment of new environmental streptococcal IMI could be prevented during this 8 – 12 week prepartum period. Results showed that prophylactic treatment of such quarters prepartum reduced new environmental streptococcal IMI at calving by 93%. Thus, in this trial, use of nonlactating cow therapy was effective in preventing new IMI as well as curing existing infections. Reasons for this high cure rate are unclear, but the relatively small secretory tissue area of heifer mammary glands compared with mature cows might allow for greater drug concentrations in the udder of the heifer. Similarly, histological studies have demonstrated less scar tissue and abscess formation in the mammary glands of heifers compared with older cows (Trinidad et al., 1990a), a condition which would allow for better drug distribution and greater contact with colonized bacteria.

6. THE OPTIMUM TREATMENT SCHEDULE FOR MAXIMIZING EFFICACY OF NONLACTATING COW THERAPY

The question arises as to when is the best time to treat bred heifers for optimizing cures against Staph. aureus mastitis. A 2-year study involving 233 Jersey heifers was designed to answer this question (Owens et al., 2001). In this trial, heifers were quarter sampled shortly after they were confirmed pregnant and at 4-week intervals thereafter. At the initial sampling, 56.5% of quarters were infected with some type of organism, and 15.4% of quarters were infected with Staph. aureus. After the initial sampling, animals were treated with a one-time infusion of 1 of 5 nonlactating cow infusion products during the first (0 - 90 days), second (91 - 180 days), or third (181 - 270 days) trimester of pregnancy. Products evaluated were: 1) a combination of 1 million units of penicillin and 1 gm dihydrostreptomycin; 2) 300 mg cephapirin benzathine; 3) a combination of 400 mg novobiocin and 200,000 units of penicillin G; 4) 300 mg tilmicosin; and 5) 250 mg cephalonium (this last product was only infused during the first trimester of gestation). Results showed that cure rates for the 5 products were high, ranging from 67 to 100%, and significantly higher than the spontaneous cure rate (25%) observed in untreated control quarters. No differences were observed among the three treatment time periods during gestation. However, fewer new Staph. aureus infections occurred after treatment in the group infused during the 3rd trimester, indicating that treatment during this time will reduce incidence of new IMI after infusion and continuing to calving. Because therapy during the first, second, or third trimester of gestation had no effect on treatment efficacy, the timing of treatment is best determined by what is most convenient for the management practices of a particular dairy. For example, heifers could be treated: 1) at time of artificial insemination; 2) during routine rectal palpation to determine pregnancy status; or 3) when moved to a new location in preparation for calving. Treatment should be administered no less than 45 days prior to expected calving date to prevent antibiotic residues at calving.

Sampimon & Sol (2005) studied the efficacy of nonlactating cow therapy (600 mg cloxacillin) administered intramammarily 8-10 weeks prepartum in 1) 5 low prevalence farms in which less than 15% of heifers had SCC of >150,000/ml at the beginning of the trial, and 2) 8 high prevalence farms in which greater than 15% of heifers had SCC of >150,000/ml at the beginning of the trial. Results showed that in high prevalence farms, treated heifers produced significantly more milk during the first 100 days of lactation compared with untreated controls; however, in low prevalence farms, treatment had no effect on production. In both groups of farms, SCC and incidence of clinical mastitis were significantly lower in heifers receiving antibiotic therapy. Authors concluded that nonlactating cow therapy was beneficial in high prevalence farms, but not in low prevalence farms.

The treatment of heifers during pregnancy with a nonlactating cow product is advantageous because: 1) the cure rate is higher than during lactation, especially against Staph. aureus; 2) there are no milk losses during therapy; 3) the risk of antibiotic residues is minimal; 4) SCC at calving is reduced; 5) new IMI with environmental streptococci is prevented; and 6) milk production is increased by approximately 10% in successfully treated animals in some herds.

7. EFFICACY OF PARENTERAL ANTIBIOTIC TREATMENT AND AN INFUSIBLE TEAT SEAL IN CURING AND PREVENTING IMI

In one study involving 1,067 pregnant heifers in 30 New Zealand herds (McDougal et al., 2005), mammary quarters were treated 1 month prepartum with 1) an infusible teat seal composed of bismuth subnitrate; 2) parenteral administration of antibiotic via injection of 5 gm of tylosin i.m. daily for 3 days; 3) teat seal plus tylosin; or 4) no treatment in order to determine if treatment reduced the prevalence of IMI and incidence of clinical mastitis postpartum. Results demonstrated that heifers treated prepartum with teat seal exhibited a reduced prevalence of IMI as well as reduced incidence of clinical mastitis. However, treatment with tylosin prepartum did not reduce prevalence of mastitis or incidence of clinical mastitis postpartum.

8. EFFICACY OF LACTATING COW PRODUCTS IN CURING IMI

Lactating cow products have been used successfully in heifers when treating infections caused by the environmental streptococci and CNS immediately prior to calving. Studies on this subject are typically performed on heifers in late gestation 2 - 3 weeks before calving. In an initial study conducted at the University of Tennessee, quarters of 115 pregnant Jersey heifers were infused one time at approximately 1 week prepartum with either 200 mg sodium cloxacillin, 200 mg cephapirin sodium, or left untreated (Oliver et al., 1992). At the time of infusion, approximately 90% of heifers were infected in one or more quarters. For heifers left untreated, 78% of animals (44.5% of quarters) remained infected at time of calving. However, only 17.6% of the heifers (4.5% of quarters) remained infected at calving if they were treated prepartum, regardless of the product used. Results demonstrated that significantly fewer antibiotic treated heifers and quarters were infected at calving compared with untreated controls. This study also examined the influence of prepartum antibiotic treatment on subsequent lactational performance, and demonstrated that heifers receiving treatment produced approximately 1,000 lb (455 kg) more milk per lactation than untreated controls.

Mastitis pathogens were isolated from 76% of untreated control quarters at 7 days before calving, from 47% of samples at 3 days after calving, and from 29% of samples at 10 days postpartum. Throughout the remainder of lactation, pathogens were isolated from 30% of control quarters. A similar percentage of samples (70%) was positive for mastitis pathogens at 7 days before calving in antibiotic-treated quarters; however, only 8% of samples obtained at 3 days after calving and 4% of samples obtained 10 days postpartum contained pathogens; throughout the remainder of lactation, mastitis pathogens were isolated from only 11% of quarters. Strep. uberis, Strep. dysgalactiae, and CNS species were isolated most frequently in both untreated controls and antibiotic-treated heifer mammary glands.

In a subsequent study reported by Oliver et al. (1997b), mastitis pathogens were isolated from 67% of samples obtained from control mammary glands (quarters to be left untreated) 14 days prior to expected calving, from 56% of samples obtained 3 days after calving, and from 36% of samples obtained 30 days postpartum; throughout the remainder of lactation, mastitis pathogens were isolated from 45% of quarters. In quarters to be treated with 200 mg cephapirin sodium, 64% were positive for mastitis pathogens prior to antibiotic treatment; however, only 16% of samples obtained at 3 days after calving and 8% of samples obtained 30 days postpartum contained pathogens. Throughout the remainder of lactation, mastitis pathogens were isolated from an average of 12% of quarters. Coagulase-negative staphylococci were isolated most frequently followed by environmental mastitis pathogens.

A follow-up study was conducted to determine if prepartum therapy with penicillin-novobiocin or pirlimycin hydrochloride was effective in reducing the percentage of infection with mastitis pathogens during early lactation (Oliver et al., 2004). Almost 73% of Holstein heifers (34.3% of quarters) were infected 14 days before expected calving. Of the quarters infected 14 days before parturition, 76% were uninfected following treatment with penicillin-novobiocin, 59% were uninfected following treatment with pirlimycin, and 26% were uninfected in the untreated control group. The majority of IMI were due to CNS (44%) and Staph. aureus (30%).

Among the Jersey heifers, 96% of animals and 71.3% of quarters were infected 14 days before calving. Of the quarters infected at 14 days before parturition, 75% were uninfected following treatment with penicillin-novobiocin, 87% were uninfected following treatment with pirlimycin, and 56% were uninfected in the untreated control group. The majority of IMI were due to CNS (61%), Strep. spp. (19%), and Staph. aureus (8%). Thus, prepartum therapy of heifer mammary glands with penicillin-novobiocin or pirlimycin hydrochloride was effective in reducing the percentage of heifers and quarters infected with mastitis pathogens during early lactation.

As a part of the University of Tennessee studies, milk production and somatic cell score data from 82 control heifers and 111 heifers treated with antibiotics before calving were evaluated (Oliver et al., 2003). Milk production (actual and 305-day) was significantly higher in heifers treated prepartum with antibiotics. Additionally, treated heifers had a significantly lower somatic cell scores than control heifers (2.04 vs. 2.63). Thus, prepartum antibiotic treatment to reduce the rate of mastitis in heifers during early lactation was economically beneficial. Treated heifers produced 1,168 lb (531kg) more actual milk than the untreated controls. Based on a milk price of $18.50/cwt this resulted in $216.24 per heifer increase in gross revenue. Considering treatment costs of $15.60 per heifer including teat hygiene ($0.10), antibiotic ($10.00), labor ($2.50), and residue testing ($3.00), the net revenue amounted to $200.64 per heifer. The researchers concluded that it would be profitable to treat heifers before calving as long as the milk price was above $0.013 per pound and as long as the increase in milk production was greater than 84 pounds (Oliver et al., 2003).

Middleton et al. (2005) also evaluated the prepartum treatment of heifers with pirlimycin hydrochloride at 10-14 days before parturition on the prevalence of IMI at calving in 2 dairy farms. Postpartum sampling revealed that heifers receiving treatment in herd A had a higher overall cure rate and higher cure rates for IMI caused by CNS and Staph. aureus as well as lower

SCC and lower prevalence of chronic IMI compared with untreated controls. The cure rate for Staph. aureus of 78% was significantly greater than that for untreated controls (8%). Treated heifers in herd B had a higher overall cure rate and higher cure rate for IMI caused by CNS compared with untreated controls, but CMT (SCC) scores and prevalence of chronic IMI were not different. In contrast to previous results of using lactating cow therapy, (Oliver et al., 2003), milk production did not differ between treatment groups. Authors concluded that routine prepartum therapy of heifers may not benefit all herds.

Subsequently, Borm et al. (2006) evaluated the efficacy of lactating cow intramammary therapy (cephapirin) administered to heifers 10-21 days prior to expected calving date in a 9-herd study involving 561 animals. Results demonstrated a 59.5% cure rate among mammary quarters that were infected prepartum and treated with antibiotic vs. untreated controls (31.7%). However, treatment did not significantly affect SCC or milk yield during the first 200 days of lactation. This observation is similar to that found by Middleton et al. (2005) who observed that although successful, treatment did not necessary reduce SCC or result in greater milk production. Authors concluded that prepartum treatment of heifers with lactating cow antibiotics may not be warranted as a universal strategy for mastitis management.

The studies on prepartum treatment with lactating cow therapy administered 7 - 21 days before calving (Borm et al., 2006; Oliver et al., 1992, 1997b, 2003, 2004) have shown treatment to be effective for quarters infected with CNS but waiting until this time to treat chronic Staph. aureus mastitis might be too late. A mammary gland that has been infected with Staph. aureus for several months to a year will not develop normally, and treatment during the immediate prepartum period would most likely be of little benefit in curing infections or salvaging mammary tissue. At this point, the tissue damage would have already been done, and affected quarters should have been treated earlier in gestation to: 1) cure existing infections; 2) reduce chronic inflammation; and 3) allow mammary tissue to develop normally during the later stages of pregnancy.

Results of these trials demonstrated that nonlactating and lactating cow antimicrobial treatment of heifers known to be at risk for developing IMI is advantageous because the cure rate is much higher than that obtained when treating infections during lactation. In addition, most studies showed that SCC are lower, there is no milk loss due to therapy, risk of antibiotic residue at calving is minimal, and future milk production is increased in heifers cured of IMI.

9. ANTIBIOTIC RESIDUES IN MILK FOLLOWING PREPARTUM LACTATING COW TREATMENT

A disadvantage of prepartum lactating cow antibiotic therapy for controlling mastitis in heifers is the potential for antibiotic residues, especially if heifers calve sooner than expected. In one study (Oliver et al., 1992), it was shown that 17% of colostrum samples from heifer mammary glands infused with cloxacillin were positive for antibiotic residues by the Bacillus stearothermophilus disc assay, the majority of which were from heifers that calved within 5 days of treatment. Only 4.5% of samples obtained at the first milking after parturition were positive for antibiotic residues if intramammary infusion of cloxacillin occurred > 7 days before parturition. All samples obtained 3 days after parturition, the time when milk would likely be marketed for human consumption, were negative for antibiotic residues. In contrast, 85% of colostrum samples and 28.2% of samples obtained 3 days after parturition were positive for cephapirin residues, and marked variability between time of antibiotic treatment and parturition with persistence of antibiotic residues was observed. Thus, antibiotic treatment of heifer mammary glands earlier in gestation may be advantageous from a residue standpoint, but the timing of antibiotic treatment and subsequent persistence in mammary secretions could impact efficacy.

Another study was conducted to determine if antibiotic treatment of heifer mammary glands earlier in the prepartum period reduced occurrence of residues in milk (Oliver et al., 1997a). A total of 82 Jersey heifers was used. Approximately half served as negative controls (n = 42) and half received an intramammary infusion of 200 mg cephapirin sodium (n = 40) 14 days before calving. Forty percent of samples from cephapirin-treated quarters were positive at the first milking after calving, but only 3.1% of samples obtained from antibiotic-treated quarters at the sixth milking (3 days) after calving were positive; 3 of the 4 positive samples were from a heifer that calved early and within 3 days of treatment.

Thus, the interval between prepartum antibiotic treatment and calving was related to persistence of residues during early lactation, and infusion of antibiotics 14 days prepartum (Oliver et al., 1992) compared with 7 days prepartum (Oliver et al., 1997a) reduced occurrence of residues in milk during early lactation. Similarly, Middleton et al. (2005) found that after heifers were treated 10-14 days prior to expected calving date with a pirlimycin lactating cow product, prevalence of IMI in early lactation was decreased without causing pirlimycin residues in milk at 3 days postpartum, even when a heifer was treated 1 day before calving (heifer calved early).

10. ROLE OF VACCINATION IN MASTITIS CONTROL

Although antimicrobial therapy is successful, the goal from a herd management perspective is to prevent new infections from occurring, and vaccination has been attempted as a prophylactic measure. Recent research has demonstrated that several experimental Staph. aureus vaccines, as well as one commercial vaccine, can increase antistaphylococcal antibody titers and reduce the new infection rate in heifers. A Staph. aureus vaccine formulated to stimulate pseudocapsule and alpha toxin antibodies was evaluated in heifers in New York (Sears et al., 1990). At 4 and 2 weeks prior to calving, heifers were given subcutaneous injections into the supramammary lymph node, and after calving, heifers were challenged with Staph. aureus. Vaccinates demonstrated a 52% reduction in new IMI. In addition, 64% of intramammary infections in control cows became chronic compared with only 12% in vaccinates.

A field study in Norway evaluated a Staph. aureus vaccine that contained two strains of whole, formalin-inactivated bacteria with pseudocapsule, alpha and beta toxoids, and mineral oil as an adjuvant (Nordhaug et al., 1994). A total of 108 pregnant heifers on 16 farms with an average Staph. aureus prevalence of 19.2% was used. Vaccinates were injected subcutaneously in the area of the supramammary lymph node with a dose of 2.5 ml at 8 and 2 weeks before calving. Results showed a 46% reduction in new IMI during the subsequent lactation. Antibody titers to Staph. aureus pseudocapsule and alpha toxin were markedly elevated in the serum of vaccinates, and these titers remained significantly higher in serum and milk during the entire lactation compared with those of unvaccinated controls.

In Argentina, a vaccine was developed based on an inactivated, encapsulated Staph. aureus strain, a crude extract of Staph. aureus exopolysaccharides, and inactivated, unencapsulated Staph. aureus and Streptococcus species in an aluminum hydroxide adjuvant (Giraudo et al., 1997). This formulation was evaluated in three groups of ten 24- to 26-month-old heifers each in a 7-month trial. The first group received an intramuscular injection of the vaccine in the neck at 8 and 4 weeks prepartum, the second group was vaccinated similarly at 1 and 5 weeks postpartum, and a third group (control) received placebo injections at 8 and 4 weeks prepartum. The research herd from which the heifers were selected had a bulk tank SCC ranging from 480,000 to 730,000, and 19% of quarters were infected with Staphylococcus aureus. This immunization program showed that the frequency of new Staph. aureus infection was reduced from 18.8% in controls to 6.7 and 6.0% for heifers vaccinated prepartum and postpartum, respectively; the protective effect was maintained for at least 6 months.

In view of more recent studies showing success of vaccines in heifers, researchers in Louisiana evaluated a commercially available Staph. aureus vaccine in young dairy animals (Nickerson et al., 1999). The vaccine was a lysed culture of polyvalent Staph. aureus somatic antigens representing 5 phage types in an aluminum hydroxide adjuvant base, including serotypes 5, 8, and 336, the most common Staph. aureus serotypes associated with clinical mastitis (Lysigin®, Boehringer Ingelheim Vetmedica, Inc., St. Joseph, MO, USA). At 6 months of age, 35 Jersey heifers were vaccinated using a 5-ml dose intramuscularly in the semimembranosus muscle of the rear leg, and 14 days later, vaccinates received a booster dose, which was repeated at 6-month intervals. Another 35 heifers served as unvaccinated controls. Results demonstrated that: (1) the number of quarters exhibiting chronic intramammary infection during pregnancy was reduced 43.1% in vaccinates compared with controls; (2) rate of new intramammary infection during pregnancy was reduced 44.8%; (3) rate of new intramammary infection at freshening was reduced 44.7%; and (4) the SCC was reduced by 50% in vaccinates compared with controls.

In a subsequent, more in depth study using the same vaccine (Lysigin®), 106 Holstein heifers from the James River Correctional Center dairy herd Goochland, VA, USA were evaluated (Nickerson et al., 2009). This herd had a 9,979-kg rolling herd average milk production with an average SCC of ~200,000/ml. Previous microbiological culture of heifer mammary secretions indicated that approximately 35% of animals were infected with Staph. aureus.

At approximately 6 to 18 months of age, heifers were processed through a restraining chute to collect aseptic quarter mammary secretion samples for microbiological. Fifty-three heifers were vaccinated using a dose of 5 ml intramuscularly that was administered as above, and the other 53 heifers served as unvaccinated controls. Fourteen days after the initial processing, the vaccinated group was again processed through the chute and boosted with Lysigin®. All animals were maintained on pasture and rotated by age group through calving. At 6-month intervals after the initiation of the trial and through time of calving, the vaccinated group was again processed through the chute for boosting.

At 2-month intervals after the trial initiation and through calving, mammary secretion samples were collected for bacteriological culture and for the determination of electronic SCC (A/SN Foss, Hillerod, Denmark). Microbiological examination of quarter samples collected from bred heifers over gestation demonstrated that 19.8% of heifers (9.4% of quarters) were infected with Staph. aureus, 68.9% of heifers (34.3% of quarters) were infected with CNS, 6.6% of heifers (2.3% of quarters) were infected with environmental streptococci, and 1% of heifers (0.3% of quarters) were infected with coliforms.

At time of calving, heifers were enrolled in the Dairy Herd Improvement Program (DHIA) and data were recorded for milk yield, percentages and actual pounds fat and protein, days in milk, and SCC. Data on vaccine efficacy were examined in terms of mean percentage reduction in rate of new Staph. aureus or CNS intramammary infections achieved among immunized heifers compared with the rate among unimmunized controls at the time of calving; differences between the percentage of heifers becoming infected among treatments was tested with the standard normal approximation.

Immunization with Lysigin® did not cause any adverse reactions at the injection site or systemically. Minimal swelling (<2.5 cm) was occasionally observed, which disappeared within 48 hours of administration. Vaccine efficacy data showed that the percentage of heifers with Staph. aureus intramammary infections at freshening was lower in vaccinates (13.3%) compared with controls (34.0%); a reduction of 60.9% ($P < 0.01$). Likewise, an examination of health records showed that the percentage of heifers that were culled or died during the trial was reduced by approximately one-third by vaccination: 16.9% in vaccinates and 24.5% in controls ($P > 0.05$). The vaccinated group also experienced a slight, insignificant reduction in mastitis caused by CNS. At freshening, intramammary infections with CNS were lower in vaccinates (64.2%) compared with controls (69.8%); a reduction of 8.1%.

Somatic cell counts in samples collected during first week of lactation irrespective of infection status were 45% lower in vaccinates compared with controls (287,317 vs. 522,345/ml). Somatic cell counts in samples collected during first week of lactation from uninfected heifers for vaccinates and controls were 66,095 and 132,754/ml, respectively; a 50.2% reduction; and for infected heifers, SCC were 441,764 and 892,176/ml, respectively; a 50.5% reduction. Somatic cell counts in samples collected during the prepartum period were highest for Staph. aureus ($6,730 \times 10^3$), followed by the environmental streptococci ($3,850 \times 10^3$), and CNS ($3,510 \times 10^3$).

An examination of the 305-day lactation milk yield for the 1st lactation of both vaccinated and unvaccinated control heifers demonstrated an approximate 8.6% increase in vaccinates vs. controls (11,217 vs. 10,332 kg, respectively) or a difference of 886 kg. On a complete lactation basis, vaccinated animals produced 839 kg more milk than controls (12,537 vs. 11,698 kg, respectively); an increase of 7.3%.

The percentage of 305-day lactation fat was higher in vaccinates than controls (3.64 vs. 3.27%, respectively); however, the percentage of 305-day lactation protein was slightly higher in controls than vaccinates (3.06 vs. 2.95, respectively). Actual 305-day kilograms of both fat and protein were higher in vaccinates than controls (fat: 408 vs. 339 kg, respectively; protein:

330 vs. 315 kg, respectively). Likewise, on a complete lactation basis, actual kilograms of both fat and protein were higher in vaccinates than controls (fat: 460 vs. 393, respectively; protein: 370 vs. 353, respectively).

An examination of the number of days in milk for the first lactation demonstrated that vaccinates persisted 13 days longer than the unvaccinated controls (349 vs. 336 days). In addition, average first lactation SCC were 11,000 cells/ml lower in vaccinates compared with controls (49,000 vs. 60,000/ml).

Results of this Virginia investigation demonstrated that vaccinating dairy heifers according to the prescribed protocol with a commercial USDA licensed Staph. aureus bacterin, Lysigin®, reduced the number of new Staph. aureus intramammary infections at time of calving by 60.9%, lowered the SCC by 50%, and decreased the culling rate by approximately one-third. In addition, overall milk yield, production of fat and protein, and number of days in milk were greater in vaccinated heifers compared with controls.

The decrease in frequency of new Staph. aureus intramammary infections at calving (60.9%) in vaccinates using Holstein heifers is higher than the 44.7% reduction observed in a Louisiana trial using the same vaccine in Jersey heifers (Nickerson et al., 1999). In both trials, SCC at calving were reduced by approximately 50%. The 60.9% efficacy found in the present trial is also higher than the 40.2% efficacy observed by Giraudo et al. (1989), the 46% efficacy observed by Nordhaug et al. (1994), and the 52% efficacy observed by Sears et al. (1990). However, it is difficult to compare among the latter three trials as the vaccine formulations were all quite different.

The question becomes: Is it economically feasible to use this commercial vaccination protocol on young dairy heifers? Based on an average of 1,864 more lb milk per vaccinated heifer, which translates to 18.64 hundredweights (cwt) of milk (1,864/100), at the current (2008) price of \$25.00/cwt, an increased income of \$466.00/heifer would be realized (18.64 cwt x \$25.00/cwt = \$466.00). If each heifer was vaccinated beginning at 6 months of age until calving, this would entail vaccinations at 1) 6 months, 2) a booster 2-weeks later, and subsequently at 3) 12 months, 4)18 months, and 5) 24 months, or a total of 5 immunizations through calving. At \$1.50/dose, this cost would total \$7.50, which when subtracted from the increased income from milk production, would yield a net income of \$458.50 per heifer (\$466.00 - \$7.50). This figure does not include the costs of labor involved in the immunization process; however, it is evident that vaccination is well worth the cost of the vaccine. Not only does it reduce new infections in first calf heifers at parturition, it may also reduce the introduction of Staph. aureus to the milking herd.

It is obvious that use of experimental and commercially available Staph. aureus vaccines can be used to prevent new infections, especially when used in heifers. Efficacy has been shown to range between 44 to 61% with reductions in SCC of 50%. This prevention strategy may represent a major control mechanism for managing Staph. aureus in the future, especially as new antigens and adjuvants are added to vaccine preparations.

11. USE OF FLY CONTROL TO MANAGE HEIFER MASTITIS

Historically, the major association between flies and intramammary infections has been with the development of summer mastitis, in which the biting fly, Hydrotoea irritans, is the proven vector (Chirico et al., 1997). Summer mastitis is an isolated seasonal problem primarily in July, August, and September in heifers and dry cows of northern Europe, and may be controlled by insecticidal sprays. In the US, fly control is used to reduce these insect pests on farm premises, and subsequently reduce animal stress, but its application as an adjunct management practice for preventing new cases of mastitis and reducing SCC has not really been considered or embraced by producers.

Surprisingly, very little research has been conducted on the relationship between mastitis in adult cows and fly control; most studies in this regard have been carried out in dairy heifers. An initial survey performed at Louisiana State University showed that prevalence of mastitis in bred heifers was significantly lower in dairy herds that used some form of fly control for their lactating cows, dry cows, and heifers compared with herds applying no fly control (Figure 9) (Nickerson et al., 1995). The greatest reductions were in numbers of Staph. aureus and the environmental streptococci, both major mastitis pathogens in adult cows associated with elevations in SCC.

Results of this survey also demonstrated that bred heifers having teats with bite lesions and scabs caused by the blood-sucking horn fly (Haematobia irritans), exhibited a 70% frequency of intramammary infection compared with a 40% frequency in heifers with normal teats. Such infections are always associated with elevated SCC in excess of 5 million/ml in these young animals. See Figure 10 below illustrating horn flies and lesions on heifer teats.

Since that first survey, researchers have proven through DNA studies that the horn fly is not only responsible for teat lesions on heifers, but is indeed a vector in the transmission of mastitis-causing bacteria, such as Staph. aureus, from heifer to heifer (Owens et al., 1998). Such mastitic heifers serve as sources of IMI for transmission to the entire lactating and nonlactating herds.

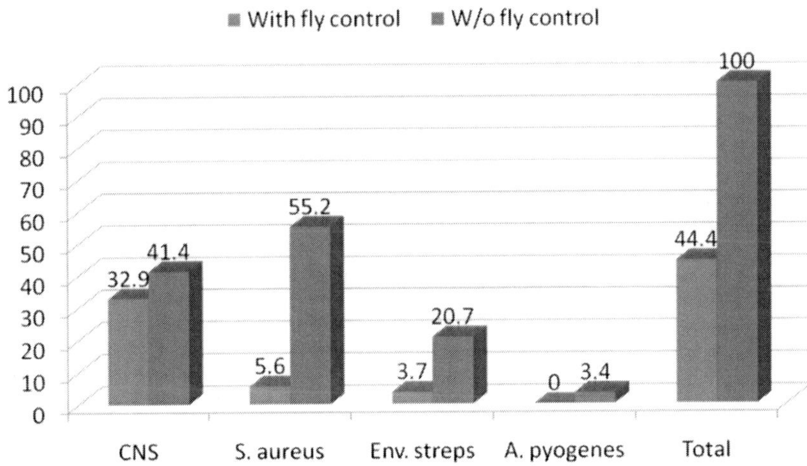

Figure 9. Prevalence of mastitis in Louisiana dairy herds with and without a fly control program.

Once it was established that the horn fly was a vector in the transmission of mastitis-causing bacteria, the next step was to develop management practices to reduce flies and lower the prevalence of intramammary infections. Insecticide-impregnated tags placed on the tail switch in close proximity to the udder during the spring and summer months were successful in reducing horn fly populations by 60% as well as the incidence of mastitis during the first 2 months after placement (Nickerson et al, 1997). In heifers with tail tags, mastitis incidence increased from 8.6 to 15% (1.7-fold increase), while in controls, incidence increased from 17.1 to 52.4% (3.1-fold increase). As observed above, infections were associated with marked elevations in SCC. However, after 2 months, tags fell off and replacing them was impractical from a management standpoint.

In a subsequent trial, the daily dietary supplementation of an insect growth regulator helped to suppress fly populations but not enough to prevent new cases of mastitis in dairy heifers (Owens et al., 2000).

Lastly, the use of an insecticidal pour-on every 2 wk for 6 wk followed by treatment with insecticidal ear tags reduced fly populations and decreased the incidence of new Staph. aureus by 83% during a 6-mo trial in heifers during the warm season in Louisiana (Owens et al., 2002). Mastitis in heifers caused by Staph. aureus was associated with SCC in excess of 10 million/ml.

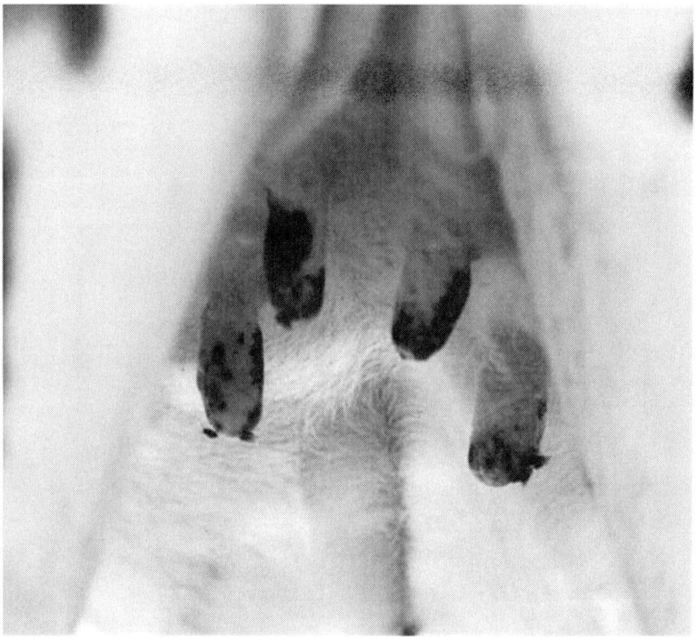

Figure 10. Udder of a 10-month-old heifer illustrating horn flies and lesions on teat ends.

These studies demonstrate that, during the warm and humid months of the year, horn flies do serve as vectors in the transmission of heifer mastitis, which is associated with elevated SCC in these young dairy animals. Although research has not been conducted to show this same association in lactating and dry adult cows, it is assumed that fly populations play some role in the elevation of mastitis and SCC observed in the hot summer months. And, with the proposed reduction in the SCC legal limit to 400,000/ml in the USA, and in light of the fact that milk buyers are imposing their own limits, some as low as 250,000/ml, it is imperative that dairymen utilize all possible means to prevent new cases of mastitis and their associated SCC. A simple fly control program can serve as an important adjunct to the basic 5-point plan of mastitis control and assist dairymen in lowering their bulk tank SCC and earn quality premiums for their product.

12. INFLUENCE OF DIETARY SUPPLEMENTATION ON MAMMARY HEALTH

Another management tool to reduce the level of infection and SCC when heifers calve as well as throughout lactation is through dietary supplementation. Diet plays a role in udder resistance to infection because certain nutrients affect various mammary resistance mechanisms, namely: (1) leukocyte function, (2) antibody transport, and (3) mammary tissue integrity. In one study, heifers received selenium (0.3 ppm/day) and vitamin E (50 to 100 ppm/day) supplementation starting 60 days prepartum (Hogan et al., 1993). A selenium booster injection (50 mg) was administered 21 days prior to freshening, and the dietary supplementation was continued throughout lactation. Dietary supplementation reduced staphylococcal and coliform infections at calving by 42%. Although rate of new infection during lactation did not differ from unsupplemented controls, the duration of infection caused by organisms other than Corynebacterium bovis was reduced 40 to 50% in supplemented heifers. Clinical mastitis in supplemented heifers was reduced 57% in early lactation and 3.2% throughout lactation, and the mean SCC was lower. Thus, vitamin E and selenium improved udder health of heifers, and the effect of dietary supplementation was most evident at calving and in early lactation.

In a more recent study, dairy heifers were fed a daily supplement beginning at 5 months of age containing an immunostimulant composed of B-complex vitamins and yeast extract (Eubanks et al., 2011). Compared with unsupplemented control animals, those supplemented with the immunostimulant exhibited greater leukocyte expression of L-selectin and interleukin-8 cell surface receptors, suggesting the capability for a greater immune response to bacterial infection. Preliminary results also showed a lower incidence of Staph. aureus mastitis in supplemented heifers.

13. CONCLUSIONS

Prevalence of mastitis in unbred, breeding age, and pregnant dairy heifers is higher than formerly realized. Infected mammary quarters, especially those with Staph. aureus IMI exhibit reduced secretory potential and marked leukocyte infiltration and accompanying inflammation. Both nonlactating and lactating commercial antibiotic infusion products have been used successfully to cure existing infections and reduce SCC, and nonlactating therapy prevents new IMI with environmental streptococci. However, the goal is to prevent new infections from occurring in these young dairy animals through management strategies aimed at vaccination, fly control, and dietary supplementation. Results of experimental and commercial vaccine trials illustrate that immunization will reduce Staph. aureus mastitis in heifers at calving between approximately 45 and 60%, with reductions in SCC of 50%. Likewise, a fly control program for heifers will decrease incidence of Staph. aureus mastitis by up to 83%. Lastly, dietary supplementation to boost the immune systems of heifers has been shown to reduce incidence of mastitis at calving as well as to lower SCC. As global milk quality standards become more stringent, management practices based on curing existing infections and preventing new IMI in heifers will ensure that these young dairy animals enter the milking herd free of mastitis with low SCC. Such practices should be considered for incorporation into dairy herd health programs in herds suffering from a high prevalence of heifer mastitis, especially that caused by Staph. aureus.

REFERENCES

1. Akers, R.M. & Thompson, W. (1987). Effect of induced leucocyte migration on mammary cell morphology and milk component biosynthesis. *Journal of Dairy Science,* Vol.70, pp. 1685-1695
2. Boddie, R.L., Nickerson, S.C., Owens, W.E. & Watts, J.L. (1987). Udder microflora in nonlactating heifers. *Agri-Practice,*Vol. 8, pp. 22-25
3. Borm, A.A., Fox, L.K., Leslie, K.E., Hogan, J.S., Andrew, S.M., Mayes, K.M., Oliver, S.P., Schukken, Y.H., Hancock, D. D., C.T. Gaskins, Owens, W.E. & Norman, C. (2006). Effects of prepartum intramammary antibiotic therapy on udder health, milk, production, and reproductive performance in dairy heifers. *Journal of Dairy Science,* Vol.89, pp. 2090-2098
4. Capuco, A.V., Paape, M.J. & Nickerson, S.C. (1986). In vitro study of polymorphonuclear leukocyte damage to mammary tissue of lactating cows. *American Journal of Veterinary Research,* Vol.47, pp. 633-668
5. Chirico, J., Jonsson, R., Kjellberg, S. & Thomas, G. (1997). Summer mastitis experimentally induced by *Hydrotaea irritans* exposed to bacteria. *Medical Veterinary Entomology,* Vol.11, pp. 187-192
6. Dodd, F.H. (1992). Role of therapy in mastitis control. In: *Large Dairy Herd Management.* Van Horn, H.H. & Wilcox,C.J. (ed). pp. 448-456, American Dairy Science Association. Champaign, IL
7. Eubanks, V.E., Forsberg, N.E., Wang, Y.Q., Zanzalari, K., Chapman, J., Hurley, D.J., Kautz, F.M., Ely, L.O. & Nickerson, S.C. (2011). Effects of OmniGen-AF on enhancing immunity in dairy heifers vaccinated with a *Staphylococcus aureus* bacterin. *Journal of Dairy Science,* Vol.94, E-supplement 1, pp. 702-703. Abstr
8. Fox, L.K., Chester, S.T., Hallberg, J., Nickerson, S.C., Pankey, J.W. & Weaver, L.D. (1995). Survey of intramammary infections in dairy heifers at breeding age and first parturition. *Journal of Dairy Science,* Vol.78, pp. 1619-1628
9. Giraudo, J.A., Calzolari, A., Rampone, H., Rampone, A., Giraudo, T., Bogni, C., Larriestra, A. & Nagel, R. (1997). Field trials of a vaccine against bovine mastitis. 1. Evaluation in heifers. *Journal of Dairy Science* Vol.80, pp. 845-853
10. Hogan, J.S., Weiss, W. P. & K. L. Smith. (1993). Role of vitamin E and selenium in the host defense responses to mastitis. *Journal of Dairy Science,* Vol.76, pp. 2795-2802
11. Jones, G., Pearson, R.E., Clabaugh, G. & Heald, C.E. (1984). Relationships between somatic cell counts and milk production. *Journal of Dairy Science,* Vol.67, pp. 1823-1829
12. Kirk, J.H. (1984). Programmable calculator program for linear somatic cell scores to estimate mastitis yield losses. *Journal of Dairy Science,* Vol.67, pp. 441-447

13. McDougal, S., Parker, K., Compton, C. & Heuer, C. (2005). Reducing subclinical and clinical mastitis in dairy heifers by precalving infusion of a teat sealant and/or parenteral Managing Mastitis in Heifers: An Initial Step in Improving Dairy Herd Health 357 antibiotic therapy. *Proceedings of the 4th International Dairy Federation International Mastitis Conference*, Maastricht, The Netherlands
14. Middleton, J.R., Timms, L.L., Bader, R., Lakritz, J., Luby, C.D. & Steevens, B.J. (2005). Effect of prepartum intramammary treatment with pirlimycin hydrochloride on prevalence of early first-lactation mastitis in dairy heifers. *Journal of the American Veterinary Association,* Vol.27, pp. 1969-1974
15. Nickerson, S.C., Owens, W.E. & Boddie, R.L. (1995). Mastitis in dairy heifers: initial studies on prevalence and control. *Journal of Dairy Science,* Vol.78, pp. 1607-1618
16. Nickerson, S.C., Owens, W.E., DeRouen, S.M., Boddie, R.L. & Tomita, G.M. (1997). Use of insecticide-impregnated tail tags may reduce incidence of mastitis in beef cows. *The Louisiana Cattleman*
17. Nickerson, S.C., Owens, W.E., Tomita, G.M. & Widel, P.W. (1999). Vaccinating dairy heifers with a *Staphylococcus aureus* bacterin reduces mastitis at calving. *Large Animal Practice,* Vol.20, pp. 16-20
18. Nickerson, S. C., Ely, L. O., Hovingh, E. P. & Widel, P. W. (2009). Immunizing dairy heifers can reduce prevalence of *Staphylococcus aureus* and reduce herd somatic cell counts. In: *Dairy Cattle Mastitis and Milking Management,* DAIReXNET.<http:// www.extension.org/ pages/Dairy_Cattle_Mastitis_and_Milking_Management>
19. Nordhaug, M.L., Neese, L.L., Norcross, N.L & Gudding, R. (1994). A field trial with an experimental vaccine against *Staphylococcus aureus* mastitis in cattle. 1. Clinical parameters. *Journal of Dairy Science* Vol.77, pp. 1267-1275
20. Oliver, S.P., Lewis, M.J., Gillespie, B.E. & Dowlen, H.H. (1992). Influence of prepartum antibiotic therapy on intramammary infections in primigravid heifers during early lactation. *Journal of Dairy Science,* Vol.75, pp. 406-414
21. Oliver, S.P., Lewis, J., Gillespie, B.E. & Dowlen, H.H. (1997a). Antibiotic residues and prevalence of mastitis pathogen isolation in heifers during early lactation following prepartum antibiotic therapy. *Journal of Veterinary Medicine Series B* Vol.44, pp. 213-220
22. Oliver, S.P., Lewis, J., Gillespie, B.E. & Dowlen, H.H. (1997b). Mastitis pathogen isolation in heifers throughout lactation following prepartum antibiotic treatment. In: *Proceedings of the National Mastitis Council.* pp. 254-255
23. Oliver, S.P., Lewis, M.J., Gillespie, B.E., Dowlen, H.H., Janicke, E.C. & Roberts, R.K. (2003). Milk production, milk quality and economic benefit associated with prepartum antibiotic treatment of heifers. *Journal of Dairy Science,* Vol.86, pp. 1187-1193
24. Oliver, S.P., Ivey, S.J., Gillespie, B.E., Lewis, M.J., Johnson, D.L., Lamar, K.C., Moorehead, H., Dowlen, H.H., Chester, S.T. & Hallberg, J.W. (2004). Influence of prepartum intramammary infusion of pirlimycin hydrochloride or penicillin- novobiocin on mastitis in heifers during early lactation. *Journal of Dairy Science,* Vol.87, pp. 1727-1731
25. Owens, W.E., Nickerson, S.C. & Boddie, R.L. (2000). The effect of methoprene on horn fly numbers and mastitis in dairy heifers. *Large Animal Practice,* Vol.21, pp. 26-28
26. Owens, W.E., Nickerson, S.C., Boddie, R.L., Tomita, G.M. & Ray, C.H. (2001). Prevalence of mastitis in dairy heifers and effectiveness of antibiotic therapy. *Journal of Dairy Science,* Vol.84, pp. 814-817
27. Owens, W.E., Nickerson, S.C., Washburn, P.J. & Ray, C.H. (1991). Efficacy of a cephapirin dry cow product for treatment of experimentally induced *Staphylococcus aureus* mastitis in heifers. *Journal of Dairy Science,* Vol.74, pp. 3376-3382 358 Milk Production – An Up-to-Date Overview of Animal Nutrition, Management and Health
28. Owens, W.E., Nickerson, S.C., Washburn, P.J. & Ray, C.H. (1994). Prepartum antibiotic therapy with a cephapirin dry cow product against naturally occurring intramammary infections in heifers. *Veterinary Medicine Series B,* Vol.41, pp. 90-100
29. Owens, W.E., Oliver, S.P., Gillespie, B.E., Ray, C.H. & Nickerson, S.C. (1998). Role of horn flies *(Haematobia irritans)* in *Staphylococcus aureus*-induced mastitis in dairy heifers. *American Journal of Veterinary Research,* Vol.59, pp.1122-1124

30. Owens, W.E., Nickerson, S.C. & Ray C.H. (2002). Effect of a pour-on and fly tag insecticide combination in controlling horn flies and *Staphylococcus aureus* mastitis in dairy heifers. *2002 Louisiana Dairy Report.* Baton Rouge, Louisiana, USA., pp. 39-42
31. Rendel, J. & Sundberg, T. (1962). Factors influencing the type and incidence of mastitis in Swedish dairy cattle. *Acta Veterinary Scandinavia.* Vol.3, pp. 13-21
32. Sampimon, O. & Sol, J. (2005). Efficacy of a cloxacillin dry cow product for treatment of heifers on farms with a low and a high prevalence of heifer mastitis in The Netherlands. *Proceedings of the 4th International Dairy Federation International Mastitis Conference,* Maastricht, The Netherlands
33. Sears, P.M., Norcross, N.L., Kenny, K., Smith, B., Gonzalez, R.N. & Romano, M.N. (1990). Resistance to *Staphylococcus aureus* infections in staphylococcal vaccinated heifers. *Proceedings of the International Symposium on Bovine Mastitis,* Indianapolis, IN, USA, March, 1990
34. Trinidad, P., Nickerson, S.C. & Adkinson, R.W. (1990a). Histopathology of staphylococcal mastitis in unbred dairy heifers. *Journal of Dairy Science,* Vol.73, pp. 639-647
35. Trinidad, P., Nickerson, S.C. & Alley, T.K. (1990b). Prevalence of intramammary infections and teat-canal colonizations in unbred and primigravid heifers. *Journal of Dairy Science,* Vol.73, pp. 107-114
36. Trinidad, P., Nickerson, S.C., Alley, T.K. & Adkinson, R.W. (1990c). Efficacy of intramammary treatment in unbred and primigravid dairy heifers. *Journal of the American Veterinary Medical Association, Vol.*197, pp. 465-470
37. Trinidad, P., Nickerson, S.C. & Luther, D.G. (1990d). Antimicrobial susceptibilities of staphylococcal species isolated from mammary glands of unbred and primigravid dairy heifers. *Journal of Dairy Science,* Vol.73, pp. 357-362
38. Watts, J.L., Pankey, J.L. & Nickerson, S.C. (1984). Evaluation of the Staph-Ident and Staphase systems for the identification of staphylococci from bovine mammary infections. *Journal of Clinical Microbiology,* Vol.20, pp. 448-453
39. Watts, J.L., Salmon, S.A., Yancey, R.J., Nickerson, S.C., Weaver, L.J., Holmberg, C., Pankey, J.W. & Fox, L.K. (1995). Antimicrobial susceptibility of microorganisms isolated from the mammary glands of dairy heifers. *Journal of Dairy Science,* Vol.78, pp. 637-1648
40. Woolford, M.H., Williamson, J.H., Day, A.M. & Copeman, P.J.A. (1983). An identical twin study of milk production losses due to subclinical mastitis. *Proceedings of the Ruakura Farmers' Conference,* Ruakura, New Zealand
41. Zhao X & Lacasse P. (2008). Mammary damage during mastitis: Causes and control. *Journal of Animal Science,* Vol.86, pp. 57-68

CHAPTER 20

Metabolic Disorders in the Transition Period Indicate that the Dairy Cows' Ability to Adapt is Overstressed

Albert Sundrum

Department of Animal Nutrition and Animal Health, University of Kassel, Nordbahnhofstraße 1a, D-37213 Witzenhause, Germany

SIMPLE SUMMARY

Metabolic disorders are a key problem in the transition period of dairy cows and often appear before the onset of further health problems. Problems derive from difficulties animals have to adapt to large variations and disturbances occurring both outside and inside the organism. A lack of success in solving these issues may be due to predominant approaches in farm management and agricultural science, dealing with such disorders as merely negative side effects. Instead, a successful adaptation of animals to their living conditions should be seen as an important end in itself. Both farm management and agricultural sciences should support animals in their ability to cope with nutritional and metabolic challenges by employing a functional and result-driven approach.

ABSTRACT

Metabolic disorders are a key problem in the transition period of dairy cows and often appear before the onset of further health problems. They mainly derive from difficulties the animals have in adapting to changes and disturbances occurring both outside and inside the organisms and due to varying gaps between nutrient supply and demand. Adaptation is a functional and target-oriented process involving the whole organism and thus cannot be narrowed down to single factors. Most problems which challenge the organisms can be solved in a number of different ways. To understand the mechanisms of adaptation, the interconnectedness of variables and the nutrient flow within a metabolic network need to be considered. Metabolic disorders indicate an overstressed ability to balance input, partitioning and output variables. Dairy cows will more easily succeed in adapting and in avoiding dysfunctional processes in the transition period when the gap between nutrient and energy demands and their supply is restricted. Dairy farms vary widely in relation to the living conditions of the animals. The complexity of nutritional and metabolic processes and their large variations on various scales contradict any attempts to predict the outcome of animals' adaptation in a farm specific situation. Any attempts to reduce the prevalence of metabolic disorders and associated production diseases should rely on continuous and comprehensive monitoring with appropriate indicators on the farm level. Furthermore, low levels of disorders and diseases should be seen as a further significant goal which carries weight in addition to productivity goals. In the long run, low disease levels can only be expected when farmers realize that they can gain a competitive advantage over competitors with higher levels of disease.

Keywords: animal science; allostasis; autopoiesis; dairy cows; downward causation; farm management; living conditions; metabolic load; production diseases

1. INTRODUCTION

Dairy cows with high genetic merits live under quite heterogeneous nutritional and environmental conditions. Nevertheless, they are capable of adapting and maintaining their internal organization in the face of changes and disturbances, while simultaneously producing high milk yields. However, individual animals differ tremendously in their adaptive success. Adaptation to changes and disturbances in the environment requires regulations on different scales within the organism and follows a common purpose: to survive as long as possible [1,2]. Thus, adaptation is a functional and target-oriented process which can be assessed in the long run by measuring longevity. In the short run, signs of metabolic and fertility disorders as well as subclinical and clinical diseases indicate varying degrees of adaptability. When disorders occur, it is an illustration that animals have difficulties in coping with external and internal conditions, endangering their own capacity to survive [3]. As farm animals cannot escape their living conditions, their ability to cope depends on the extent their environment has an impact on them and on their level of responsiveness in order to maintain or re-establish a homeostatic state. Any stimulus (external or internal) that challenges homeostasis can be viewed as a stressor [4]. Changes in biological function occur as the animal attempts to respond to stressors and associated challenges.

Over the past decades, the focus of the dairy industry has been on maximizing milk yield, thereby creating a "nutrient highway" for nutrients from the daily feed and body reserves to travel directly to the udder to sustain milk production [5]. Indeed, farmers are not only interested in high yielding but also in healthy, long (re-)producing animals. From the farmers' perspective, production diseases are also a cost issue to which farmers can react differently [6]. For example, production diseases could be considered as undesired side effects of the production process, and factored in as associated failure costs. Alternatively, farmers could make use of investments and their work force to optimize living conditions and take preventive measures to reduce the occurrence of production disease and keep failure costs to a minimum. However, when striving for this alternative, farmers often have to face an associated increase in preventive costs.

The transition phase, defined as the period between three weeks before to three weeks after parturition [7], is the most challenging and critical period in relation to the dairy cow's health status during the lactation cycle. Major physiological, nutritional, metabolic, and immunological changes occur within this time frame as the production cycle of the cow shifts from a gestational nonlactating state to the onset of copious milk synthesis and secretion [8,9]. Cows must adjust metabolically to the dramatic increase in energy and nutrient requirements needed to ensure milk production in the ensuing lactation. Gaps between nutrient demand and supply in the first weeks of the transition period can coincide with substantial variations in the diets' nutrient content and in the daily dry matter intake (DMI), requiring comprehensive adaptation and regulation of the metabolism. Disturbances of one or multiple metabolic processes, related to the regulation of a certain metabolite in the body fluids, are known as metabolic disorders [10], and are a manifestation of the cow's inability to cope with metabolic demands [11]. Clinical diseases most related to nutritional management are ketosis, milk fever, retained placenta, metritis, displaced abomasum, and lameness [12]. The colloquial associations among various metabolic stresses and their relationships to other diseases, particular infectious and inflammatory diseases of early lactation are interrelated and have become a central focus of the interest in metabolic diseases of dairy cattle [13].

Little is known about the limitations of dairy cows in their ability to adapt to and deal with variation and disturbances in their living conditions and the gap between nutrient supply and nutrient demand. Questions arising include: how to assess what is decisive for adaptation in the specific farm contexts. Making trade-offs between various nutritional and physiological traits requires further in-depth reflections on the background, the regulatory processes involved in metabolic pathways and the possible consequences. The complexity of the issue makes it necessary to consider it from different perspectives. The main objective of this paper is to reflect on the nutritional and physiological process of adaptation to the living conditions and on the prerequisites for a successful adaptation which avoids metabolic disorders in the transition period.

2. HOW DOES ADAPTATION WORK?

To survive, dairy cows need to adapt to changing living conditions, challenged by varying gaps between individual needs and supply and by environmental changes or sudden and unforeseen disturbances. Stress response may be either non-adaptive, resulting in increased morbidity and mortality or adaptive, promoting overall fitness, emphasizing the ambivalent nature of stress. Not are only the stress factors manifold; the possible reactions of the animals to the same or to different stressors vary greatly. The adaptive responses of an individual animal may be specific to a given stressor, or general and independent of the type of stressor. Thus, adaptive success depends not only on the level and type of threat and on the responsiveness of the organism but primarily on the interactions between both.

Different terms are in use to describe the outcome of reactions in face of changes and disturbances. In this context, "robustness" can be seen as the property of being strong and healthy in constitution. The term means a kind of resistance to disturbance and irritation which prevents changes in a system's composition occurring. Thus, robustness implies endurance but not necessarily adaptation which is a special manner of being tolerant to challenges by actively monitoring disturbances and compensating for their tendencies. The term "resilience" derives from materials' sciences, meaning the power or ability to return to the original form, position, *etc.*, after being bent, compressed, or stretched. In contrast to "robustness", changes within the system's composition that occurred due to disturbances but were only temporary. In ecology, the term refers to "the ability of an ecosystem to return to its original state after being disturbed" [14]. With respect to agricultural systems, "resilience thinking focuses on enabling a system to cope with unexpected change and disturbance" [15]. In the case of dairy cows, this term might be used in describing the ability of an animal to regain a state within a normal physiological range and/or to recover from disorders and diseases. The term "adaptation" is used in biology in relation to how living beings adjust to their environments. In the face of changes and disturbances, reactions of organisms lead to changes in the composition of their system which remain in force and are more or less permanent. According to Di Paolo [16], adaptivity is a system's capacity to regulate, according to the circumstances, its states and its relation to the environment with the result that, if the states are sufficiently close to the boundary of viability, (1) tendencies are distinguished and acted upon depending on whether the states will approach or recede from the boundary and, as a consequence; (2) tendencies of the first kind are moved closer to or transformed into tendencies of the second and so future states are prevented from reaching the boundary with an outward velocity. The specific rate of function for each organ, describing the specific range where it is able to operate, is specified by the regulatory system according to both environmental and internal conditions. A process becomes malfunctional when—due to the fact that its particular qualities limit its range of modulation—it is unable to do what the regulatory (sub)system "tells" it to do within the framework of a specific regime of self-maintenance. Diseases and disorders can be understood as negative bodily occurrences where a part of the organism fails to appropriately perform one of its biological functions [17]. Biological functions are interpreted as specific causal effects of a part that contributes to a complex web of mutual interactions, which, in turn, maintain the organization and, consequently, the part itself. Generally speaking, a well-adapted system is a set of interacting and interdependent entities, forming an integrated whole that together are able to respond to environmental changes or changes in the interacting parts.

Biological flexibility in the face of environmental demands is a fundamental prerequisite in regulatory physiological systems [18]. Moreover, regulation over a wide range of physiological and developmental conditions is required. Different concepts and terms have been coined in the past which refer to the necessary changes in regulatory processes. It is beyond the scope of this review to describe these concepts comprehensively. In short: In theoretical biology, an enduring tradition has placed heavy emphasis on the idea that biological systems employ what could be referred to as "self-determination" [19]. That is the capacity of a system's constitutive organization to contribute to the determination and maintenance of its own conditions of existence through the effects of its activity. Following Bernard's seminal work [20,21] during the first half of the 20th century self-determination was initially investigated as "homeostasis" [22] and mathematically expressed in terms of feedback loops by cybernetics [23,24]. For Cannon [25], homeostatic regulation reflects "a condition, which may vary but which is relatively constant". A common example to explain homeostasis is the ability of the organism to maintain a glucose concentration in the blood within a certain range despite considerable changes in supply and demand of glucose and in turnover rates ("stability through constancy"). To keep the glucose concentration in the blood within a physiological range, changes in the organism have to take place, regulated primarily by neuroendocrine and endocrine secretions. The need to maintain the glucose concentration within a certain range is crucial also in the context of metabolic disorders and will be discussed in further detail below. Homeostasis is traditionally seen as belonging to short-term physiological adaptations. Selye [26] was searching for a concept beyond homeostasis which also described long-term adaptations, which he called "heterostasis" and defined it as "the establishment of a new steady state by exogenous stimulation of adaptive mechanisms through the development and maintenance of dormant defensive tissue reactions". Waddington [27] suggested that homeostasis should be interpreted as "homeorhesis" which means stability of dynamics rather than stability of state. A crucial step in the theoretical development in understanding biological self-determination is the account put forward by Piaget [2]. His core idea was to integrate two inherent dimensions of biological systems into a single coherent picture: thermodynamic openness and organizational closure. In Piaget's view, biological systems are self-determining because they are organized and they are organized because they realize closure. Organizational closure is the basis of many subsequent accounts of biological self-determination [28]. One of the best-known theories is the concept of "autopoiesis" [29]. In the context of dairy production, Bauman and Currie [30] refer to homeorhesis regarding the regulatory mechanisms of nutrient partitioning and physiological processes in different stages of life history, e.g., a lactating mammal is in a different homeostatic state from a pregnant female. A similar concept ("rheostasis") involves the changes in regulation of state during circadian rhythms [31]. Knight *et al.* [32] used the terms "metabolic load" and "metabolic stress". Metabolic load was defined as "the burden imposed by the synthesis and secretion of milk", whereas metabolic stress was defined as "that amount of metabolic load which cannot be sustained, such that some energetic processes, including those that maintain general health, must be downregulated".

The term "allostasis" was introduced by Sterling and Eyer [33] to refer to changing regulatory systems ("stability through change"). Allostasis can be considered as the process of maximizing fitness in the face of environmental change as well as unpredictable challenges. Regulatory mechanisms must change in order to maintain or achieve a state appropriate for the time of day or year and in response to disturbances. McEwen [34] developed the concept of allostasis further and coined the term "allostatic load" to describe potential permanent overburdening of homeostatic processes, e.g., by diseases. One can imagine allostatic load increasing due to the rising energetic cost of fueling regulatory processes. Accordingly, allostatic load is the sum of the energy required to maintain basic homeostasis and to acclimate to changing environmental conditions. Here, energy is assumed to be the sum of an individual's requirements (including protein, essential fatty acids, vitamins, minerals, and so on) in any given state. McEwen and Wingfield [35] extended this idea further. They see allostasis as the process of adjusting morphology, physiology, and behavior to cope with allostatic load, while allostatic state is the hormonal state required to regulate predictable changes in an individual's life cycle as well as unpredictable disturbances. Energetic demands on the individual can be further exacerbated by injury, disease, or social stress that could add cumulative allostatic load before all of the other energetic costs associated with the life cycle are considered.

Changes may occur as variation within a normal range. Predictable changes over time often emerge in an individual's life cycle. However, unpredictable events, including many potential stressors, requiring immediate physiological and behavioral adjustments may also need to be dealt with. Disturbances can occur slowly or abruptly, e.g., as change in diet, climatic conditions (heat stress), pathogen pressure, competition with other individuals that may reduce access to trophic resources, injuries, diseases, and other challenges. Effects may be cumulative over hours or days and weeks, and are disruptive or lead to abnormal activity. Additionally, current infection and disease, social status, *etc.*, may influence how an individual animal goes about its routines and responds to unpredictable disturbances. In the case of dairy cows, challenges to their adaptation can be differentiated as follows:

- A group of animals or the dairy herd is exposed to the same feeding and housing conditions, although individual cows can vary considerably in their needs, due to differences in live weight, milk yield, age, or social rank. The bigger the individual gap between existing and optimum living conditions, the more cows are challenged to adapt.

- Unforeseen and abrupt changes in the animal's environment, e.g., from chemical and physical factors generally concern the whole group of animals. However, individual animals may vary in their capability to react to these changes for various reasons.
- Cows within a herd face individual challenges due to disturbances from ranking conflicts or due to handicaps caused by clinical and subclinical diseases or injuries.
- Cows which simultaneously undergo changes in the environment and within the organism face the biggest challenges; *i.e.*, in the transition period. Considerable morphological, physiological, hormonal, and behavioral changes occur which serve in part to prepare the body for the period leading up to birth and for calving itself and also to get the cow's metabolism ready for impending lactation. These adaptations consist of mid-pregnancy anabolism followed by a significant shift to catabolism in late pregnancy and a dramatic catabolic shift in early lactation. At the same time, cows are offered a different diet than in their dry period and are often moved to different living conditions.

This paper focusses on the possible implications of metabolic stress caused by changes and disturbance in nutrition on a metabolic level. The current knowledge about the relationship between metabolic disorders and production diseases is outlined in brief before the above issue is further explored with reference to variation on different process levels.

3. METABOLIC DISORDERS AND PRODUCTION DISEASES

Desirable outcomes in farm management are cows that are successful in adapting metabolically to challenges in the transition period with minimal to no disease events, reduced avoidable culls, and efficient productive and reproductive performance. However, the reality often differs greatly. Negative side effects of the production process are obvious. They have been intensively discussed since the first Intl. Conference on Production Diseases in Farm Animals in the year 1968 which has occurred periodically ever since [36]. Traditionally the term "production disease" was regarded as encompassing the significant metabolic disorders of dairy cows (hypocalcaemia, hypomagnesaemia, and ketosis), but has been broadened in the course of time to include conditions such as a retained placenta, displacement of the abomasum, metritis, and laminitis [11]. With the increase of milk production, diagnostic method sensitivity as well as recognition of subclinical disease also increased. Dairy farms continue to be plagued by high prevalence of cow disorders, adversely affecting productivity, reproduction, and animal health and welfare [37,38]. Much intensive research has been conducted to address nutritional requirements, physiological adaptations, and metabolic associations with periparturient disease of cows in the transition stage. According to LeBlanc [39], dairy production is challenged by the fact that 30% to 50% of dairy cows are affected by some form of metabolic or infectious disease around the time of calving. Compiled periparturient disease incidence rates are presented in Table 1. According to Van Saun and Sniffen [40], most in the dairy industry still believe that there are tremendous opportunities to improve transition cow health and reproductive performance without compromising milk production, yet the solution is not clearly evident.

Table 1. Compiled periparturient prevalence of metabolic disorders from various published studies according to Van Saun and Sniffen [40].

Disease	Median Incidence Risk (%)	Range of Incidence Risk (%)
Hypocalcemia	6.5	0.3–22
Subclinical hypocalcemia	2.2	8–54
Retained fetal membranes	8.6	1.3–39.2
Metritis	10.1	2–37
Subclinical metritis	53	37–74
Ketosis	4.8	1.3–18.3
Subclinical ketosis	43	26–55
Lameness	7.0	1.8–30
Clinical mastitis	14.2	1.7–54.6
Subclinical mastitis	30	15–60

In contrast to the comprehensive data sets on the performance traits of dairy cows, the monitoring of metabolic disorders is often restricted to data from single scientific investigations. Unlike clinical diseases, subclinical metabolic disorders are even more difficult to detect; they require additional diagnostic tests, and obtaining a complete picture is not always possible. This refers not only to the severity of subclinical disorders but also to impacts which metabolic disorders might have on the development of infectious diseases and fertility disorders or mortality and culling rates, respectively. In the past, comparably few studies have been carried out to investigate relationships across a broad spectrum of health problems in relation to milk production. For example, in the case of subclinical ketosis, the reported prevalence for hyperketonemia in the first two months of lactation ranged widely from 8.9% to 34% in various studies [11,12]. Comparing current incidence rates of disease with those from past years is not helpful as records are not necessarily reliable, and cows who produce more milk than their herdmates are not automatically at increased risk of developing disorders [43]. According to Mulligan and Doherty [11] the hypothesis that high yielding cows automatically have higher levels of production diseases is likely to be as false as the hypothesis that lower yielding cows suffer from lower levels of production disease.

In early lactation, cows are in a stage of negative energy balance (NEB) caused by a rapid increase in the demand for nutrients to support milk production which exceed the increase in food intake. This results in lower blood glucose and mobilization of body reserves to provide additional energy. These processes are accompanied by elevated blood concentrations of non-esterified fatty acids (NEFA) and ß-hydroxybutyrate (BHBA) and decreased levels of calcium and phosphorus. According to Drackley [7], many health disorders in dairy cattle are attributed to the process of uncontrolled lipid mobilization in response to excessive NEB in early lactation. All of these metabolic changes increase the risk for ketosis, hepatic lipidosis, hypocalcemia, and infectious diseases, such as mastitis and metritis [44,45,46,47]. However, there is considerable variation in the plasma concentration of substrates during the day (diurnal variation) and over a period of days in early lactation. For example, plasma BHBA concentration of cows with clinical signs of metabolic disorders, generally used as an indicator for ketosis, differs widely as does the maximum plasma concentration measured in cows that did not develop clinical ketosis in the first six lactation weeks [48]. Changes in the concentrations of metabolites and hormones during the postpartum period differ remarkably amongst animals kept under similar and highly standardized conditions on a research farm, indicating that the ability to cope with metabolic stress varies considerably between individual cows [49].

Sordillo and Raphael [9] addressed the possible connections between fat mobilization and dysfunctional inflammation responses that may contribute to increased morbidity and mortality in the transition phase. While an efficient inflammatory response eliminates the invading pathogen, restores immune homeostasis, and returns tissues back to normal function and morphology, many aspects of the bovine immune system are compromised around the time of calving, especially on the level of inflammatory responses [50]. The authors conclude that the delicate balance between a sufficient inflammatory response needed for optimal pathogen clearance and the prompt return to immune homeostasis is often lost during the transition period. Immune suppression in the periparturient dairy cow is a commonly observed phenomenon and has been linked to poor metabolic status and negative energy balance [51,52]. In cows suffering severe NEB, inflammation immune genes are upregulated [53], whereas genes involved in the acquired immune responses are downregulated [54]. The mechanisms of impairment of immune defense in the transition period have been described in further detail with regard to the mammary gland [55] and the uterus [56,57]. Cows with severe metritis ate less than healthy cows in the two to three weeks preceding the clinical signs of metritis [58]. Lower feed intake is associated with increased circulating concentrations of non-esterified fatty acids (NEFA) which may directly [59] or indirectly [60] inhibit neutrophil function. Due to both high metabolic demands and pathogen challenges, cattle routinely experience substantial oxidative stress in early lactation too [61], which also contributes to a pro-inflammatory state that impairs immune defense [62]. The authors highlight that a major underlying factor in the development of transition cow disorders is metabolic stress, which occurs when cows fail to adapt physiologically to an increase in nutrient requirements needed for parturition and the onset of milk synthesis and secretion. The combined effects of altered nutrient metabolism, dysfunctional inflammatory responses, and oxidative stress can form destructive feedback loops that exacerbate metabolic stress and cause health disorders.

The role of inflammation-associated pathways in adaptation processes is not yet fully understood. The findings by Farney *et al.* [63] suggest that inflammation induced insulin resistance is, in some cases, an adaptive rather than pathological, phenomenon. Some degree of inflammation might even be required for successful adaptation. On the other hand, increased concentrations of analytes associated with stress and inflammation during the periparturient period are associated with lower milk yield and compromised reproductive performance later in lactation [64]. An underlying factor of dysfunctional inflammatory responses is the overproduction of reactive oxygen species, leading to oxidative stress [65]. While free radicals are essential for physiological processes, imbalanced or excessive free radical production plays a key role in the pathogenesis of diseases. Oxidative stress has been identified as a link between nutrient metabolism and inflammation during the transition period [62].

A close relationship between NEB and fertility disorders has been revealed and reviewed in many studies [5,66,67,68]. Increased locomotive problems are also associated with longer and more extreme periods of negative energy balance [69]. As a consequence, individual production diseases of the dairy cow should not be considered in isolation. For example, ketosis, fatty liver, retained placenta, hypocalcaemia, metritis, and displaced abomasum may be all interrelated

[11]. Primary causes and the confounding factors which contribute to the development of metabolic disorders are manifold and vary considerably between farms. These factors are not only related to the feeding regime but are complicated by various management issues. Irrespective of average milk yields, some farms do well whilst others fail quite markedly in reducing clinical and subclinical problems. Vastly different nutrition and management programs produce equally good or equally poor success in relation to metabolic disorders [7]. However, little is known about the relationships between factors like body condition, feed intake, and postpartum health and performance, each varying so widely among individual cows.

4. CHALLENGES DUE TO VARYING LIVING CONDITIONS

Dairy cows live in environments that change over time. Changes that affect their self-determination and self-preservation are particularly relevant. In the first place, reactions of animals to changes in the environment are thus self-referential [70]. Cows with different and varying needs have to adapt to more or less identical living conditions. Total mixed rations are generally formulated according to the average performance of a herd or a group of farm animals. As such, rations are not adequately adapted to the requirements of individual cows within a feeding group or herd. Group size, grouping strategy, and group feeding behavior have a considerable impact on the competition between animals for feed and space and thus on feed intake [71]. Competition is perceived differently by the animals, depending to a high degree on the social rank they hold. Cows in uncomfortable conditions are less likely to use their potential for adaptation.

Living conditions are quite complex, encompassing various nutritional, hygienic, and housing spheres as well as social environment. Each sphere includes a multitude of influencing factors, differing considerably in their impact on the animals. Furthermore, they change in varying degrees over time. While climatic conditions can change quite quickly, the dimensions of space allowance, lying areas, and bunk space are more or less given conditions in a stable. Also feeding regimes on a farm might appear steady, but nevertheless can evolve more or less extended modifications in the nutrient and energy supply. Diets are composed of various ingredients, which differ not only between proportion of carbohydrates, proteins, fat, minerals, and trace elements but vary considerably in the solubility of the different carbohydrate and protein fractions and correspondingly in the availability for the animals within the digestive tract [72]. Charges of roughage or concentrate as part of the total ration differ widely in their composition. Furthermore, the proportion of single components can vary substantially due to imprecision in allocation or in mixing procedure. A comprehensive overview of the countless nutritional factors on the farm level that cause variation between and within dairy farms has recently been provided by Oetzel [12].

The amount of daily feed intake is a resultant of interactions between the composition of the diet itself, the environment in which a diet is offered and various intrinsic processes [73,74]. Feed and energy intake can change dramatically in response to changes in diet composition or metabolic state. Such changes are poorly predicted by traditional models of feed intake regulation [75]. Variation in feed intake does not only refer to the daily dry matter intake (DMI). The same DMI per day can be achieved by different frequencies and durations of eating time and meal sizes, respectively. Thus, it is not surprising that a large inter- and intra-individual variation of feed intake is observed in farm practice [76]. Even if a balanced diet following the defined average intake for a given feeding group is provided, a large variation in DMI within a group results.

When provided with free access to high nutrient dense feed, dry cows do not regulate and restrict their intake to meet their requirements and thus will over-consume energy. Dann [77] recorded intakes by dry cows that were up to 60% in excess of their requirements when fed ad libitum, with a large variation between cows. Lactating cows, provided with a suitable ration will readily consume in excess of 3.5% of their body weight as feed DM daily. This equates to between 22 and 25 kg DM/day, yet in many situations, intakes often fail to achieve even 20 kg DM/day [74]. Feed intake is depressed when highly fibrous, bulky food is fed or diets are deficient in essential components such as protein or feed access is restricted by time, bunk space, or competition, and when environmental temperatures and humidity increase and create heat stress [71,78,79]. Bunk space is most critical in farms where feed availability is limited or where overcrowding occurs. Fresh cows seem to be more sensitive to behavioral competition and intimidation, which results in altered feeding behaviors where feed bunk space is inadequate [79]. As the social hierarchy is more frequently altered, observational studies suggest a decline in feed intake, with observed greater disease prevalence [80]. The reasons for the large variation in DMI, even under standardized living conditions, are manifold, reliant on influencing factors both outside and inside the organism. Interactions between the different influencing factors create a virtually unlimited variety of combinations. What makes it nearly impossible to assess the extent of the impacts of feeding, social, and housing conditions is the fact that impacts do not emerge separately from one other. Living conditions have impacts and are challenging the animals as a whole. On the other hand, a reduction in feed intake is generally the first observed sign in cows developing metabolic disorders, particularly clinical ketosis [81]. Thus, the question always remains open, whether the feed intakes are primarily caused by impending disease, or whether the feeding patterns are induced and then result in increased disease risk?

5. VARIATION IN METABOLISM

Digestion by ruminants is the net result of a sequence of processes that occur in different segments of the gastrointestinal tract (GIT). These processes are characterized by degradation, passage, and absorption; they compete directly with one other. The sequence includes fermentation of dietary components by microbes in the rumen, acid hydrolysis, and degradation by the host animal's enzymes in the abomasum and small intestine, and secondary fermentation in the cecum and large intestine. The passage rate determines the time that feed is retained in various compartments of the GIT for digestive action and affects the nature of end-products absorbed [82]. First of all, the feed reaches the rumen, which is a large fermentation chamber, containing a complex microbial ecosystem that works in a dynamic, symbiotic relationship with the host to convert feed into energy and protein. This microbial ecosystem consists of bacteria, protozoa, archaea, fungi, and bacteriophages and is highly responsive to dietary changes [83,84]. The diverse synergies and antagonism of ruminal microbes allow ruminants to efficiently use a range of feeds. Individual cows seem to have a rumen ecosystem comprising a core rumen microbiome that adapts to different feed substrates [85]. Thus, the composition of the rumen microbiome reflects the diet, feed additives, health, age, and seasonal conditions [86,87].

To understand the dynamics of ruminal fermentation, it is necessary to understand the interactions between the rate of passage of feed through the rumen and the rate of digestion of feed in the rumen. The competing rates of fermentation and passage from the rumen dictate how much of a fermentable fraction is fermented in the rumen and how much passes undigested to the abomasum. Passage from the rumen involves mixing, disruption, and comminution of various components. The rumen operates as a partially stirring, continuous-flow reactor and different sub-compartments with different flow characteristics may be distinguished: liquid, escaping particles, and retained particles [88]. In contrast to the extensive amount of data available on substrate degradation rates, the availability of data on ruminal retention times is relatively scarce. Production and absorption of volatile fatty acids (VFA) constitute dynamic processes resulting in large postprandial variations in ruminal concentrations. Close relationships exist between production rates, the ruminal concentrations and the fluxes of VFA across portal drained viscera. The fractional ruminal VFA absorption rates increase with chain length and decrease with pH [89,90]. Diets higher in non-fiber carbohydrates (NFC) content promote the development of ruminal papillae for adequate absorption of volatile fatty acids produced during ruminal fermentation [91].

Effects of pH on ruminal VFA concentrations and kinetics of VFA absorption by the rumen wall differ considerably [89]. While removal of volatile fatty acids (VFA) from the rumen by absorption through the rumen wall are major processes that influence ruminal pH, VFA and lactic acid can build up in the rumen and reduce ruminal pH. In addition, the buffering capacity of rumen fluid is variable and is generally assumed to depend primarily on bicarbonate. Bicarbonate dependent absorption is not just a primary absorption pathway of VFA but bicarbonate can also be secreted at a capacity equal to that from saliva, thus removing protons from the rumen by neutralization. In addition, the inherent buffering capacity of the diet is involved in pH regulation, largely explained by the cation exchange capacity of feedstuffs. Low ruminal pH for prolonged periods each day can affect feed intake, microbial metabolism, and feed digestion, and has also been related to inflammation, diarrhea, and milk fat depression [92]. In face of the various impact factors, empirical models to predict ruminal pH have so far had limited success. In a recent study [93], large differences in reticulum rumen pH development were observed postpartum between individual animals when assessed continuously with an indwelling and wireless monitoring system. Considerable differences occurred despite the fact that all cows were kept under the same conditions and received the same diet before and after parturition. The results indicate that rumen pH-values are not exclusively influenced by the diet and it is thus hard to predict using experimental models; at least not on the level of the individual animal.

Fermentation rates of NDF decrease from relatively acidic to almost neutral pH [94]. The rate of fermentation of water-soluble carbohydrates and starch is, among others, affected by the amount of ruminal degradation of dietary protein [95]. Increases in the amount of ruminal degradation of dietary protein relative to readily available carbohydrates result in reduced amounts of carbohydrate stored as glycogen by the microbes, which means that the carbohydrate from the diet is fermented soon rather than stored. Another effect of increasing ruminal degradation of dietary protein relative to available carbohydrate is an increase in yield of microbial protein per unit of carbohydrate [96]. The competing rate of fermentation of starch has been reported to increase as the amount of starch in the diet increases [97]. The quantity of amino acids available to dairy cows is, among others, dependent on the ruminal degradation of dietary protein, the capacity of the rumen to use non-protein nitrogen sources for the synthesis of microbial protein, the degree to and efficiency with which ruminal degradation of dietary protein can be synthesized into microbial protein, the variation in value of different proteins to the animal that escape ruminal degradation, and endogenous protein losses [78]. Thus, there is considerable day to day variation in the efficiency of the conversion of metabolizable protein (MP) to net protein and estimates of microbial nitrogen outflow from the rumen [98]. To sum up, substantial day to day variations in digestion and fermentation processes in dairy cows cause considerable variations in the relative quantities and supply of essential nutritional elements from the intermediate metabolic processes. This supply variation represents a real challenge for the metabolism in the face of demands for milk production and maintenance.

6. DISCREPANCY BETWEEN NUTRIENT DEMAND AND SUPPLY

The average milk yield per cow has increased considerably over the last decades, primarily as the result of genetic selection. Additionally, increase in production has been supported by the improvement in environmental factors such as feed composition, feeding strategies, housing conditions, and farm management. Breeding programs have become quite successful because of the comparatively high accuracy of breeding value estimation, the moderate to high heritability of most production traits, and the use of large and fast databases containing production records of many animals and their genetic relationships [38]. The authors emphasize that increase in milk production is twice as heritable as feed intake, leading to the consequences that extent and duration of negative energy balance (NEB) has considerably increased over the last decades. During lactation, an increase in milk yield with a peak between the fourth and sixth week postpartum is functionally based on a marked increase of cell differentiation and tissue hypertrophy in the udder [99]. The number of vital mammary epithelial cells control the initial conditions for the amount of milk produced and the amount of glucose needed for the production and secretion of lactose [100]. Treating cows with bovine somatotropin could help to improve persistence of milk production because it prevents, among other things, an increase of the enzyme plasmin in the milk and thus the involution of udder epithelia. This emphasizes that the power drop in milk performance is an important programmed, endocrine and autocrine regulated process [101].

Mammary epithelial cells have an incredible level of organization and a remarkable ability to convert circulating nutrients into milk components. Cows with a high genetic performance capacity for milk production are characterized by the ability to perform intensive gluconeogenesis and partitioning of the glucose into the udder. According to Bauman *et al.* [102], the productivity of this biological factory is extensive and in terms of the use of nutrients and energy, the cow should be viewed as an "appendage to the mammary gland" rather than *vice versa*. To produce milk, the udder requires a corresponding amount of nutrients, particularly glucose, from the intermediate metabolism which correlates greatly with the amount of milk secreted. Hepatic glucose production is essential for the ultimate production of lactose which controls milk volume. The demand for glucose varies in connection with the intra- and inter-individual variation of daily milk yields. In contrast, fat is the most variable milk component, predominantly affected by nutrition. Being the major energy component in milk, milk fat percentage and fat yield can be reduced by 50% or more with little or no change in the milk yield or other milk components [102]. Milk fat is composed of fatty acids of varying chain length and degree of saturation. These are derived in almost equal manner from the uptake of circulating preformed fatty acids and the synthesis of new (de novo) fatty acids in the mammary gland. Gardner *et al.* [103] compared starch and fat supplemented rations fed to multiparous cows and noted a 4 kg/day improvement in milk yield for those fed enhanced fat levels. Fat supplement reduced milk protein content by 8% whilst milk fat levels were unaffected, but overall milk fat and protein output was increased by 9%. Computing their energy balances showed that the fat supplemented diet increased tissue mobilization.

In a study by van Knegsel *et al.* [104] cows were fed on an isocaloric and isonitrogenous basis either a mainly glucogenic (by-pass starch) or a mainly lipogenic diet. Results showed that the glucogenic diet stimulated energy partitioning towards body reserves in early lactation. Cows fed the glucogenic diet had lower NEB and reduced fat mobilization, which led to milk fat depression and less energy partitioned to milk. Thus, energy partitioning between milk and body tissue can be altered considerably by diets that differ in lipogenic and glucogenic nutrient content. High yielding cows fed with a diet enriched with crude protein (19.9%) produced an extra 5 kg milk/day of similar fat but lower protein contents in comparison to a diet with 16.7% crude protein, while ME intakes showed no treatment differences [105]. The author concluded that the extra production of milk and milk solids in high protein diets had been achieved by increased mobilization or reduced depletion of body condition. Due to a sudden increase of nutrient requirements for milk production postpartum at a time when dry matter intake and nutrient supply lag behind, nearly every high yielding cow faces the challenge of negative energy balance. NEB describes the portion of the total energy requirements of the dairy cow for maintenance and milk performance which is not covered by nutrient supply from feed intake [106]. This portion has to be provided by the organism in making use of depot nutrients to gain a balance between output and input levels. Increases in genetic merit for milk yield go together with increases in feed intake, but the latter does not fully compensate for the extra energy demands during early lactation, resulting in a more or less extended negative energy balance and increased mobilization of body reserves [107,108].

Calculated energy balance is typically mostly negative within the first 12 days postpartum [106]. Differences among cows in nadir and total energy deficits are large. Over the course of 122 lactations, mean values of total energy deficits during early lactation amounted to 1451 MJ NE_L with a standard deviation of 1062 MJ NE_L. The postpartum interval to nadir of the estimated energy balance averaged to 48 ± 29 days. First lactation cows showed a smaller energy deficit in early lactation than older cows. Moreover, cows differ considerably with regard to the partitioning of energy between different physiological systems. Thus, cows with similar energy intakes and expenditures via the milk may actually experience differences in the burden of NEB. This does not only refer to genetic make-up (e.g., high v. low genetic merit) or the stage of lactation, but varies greatly between individuals of the same genotype or in the same stage of lactation [109]. This makes predicting the dynamic of nutrient partitioning in a generalizable way (*i.e.*, across physiological stages and genotypes) more challenging.

Furthermore, emotional stress affects NEB via feed intake and glucose and free fatty acid metabolism, mainly via the sympathetic nervous system, and, to a lesser extent, via HPA axis [110,111]. Glucose is the main substrate covering energy expenditure during activities which contribute to acute emotional stress, whereas fatty acids are the main substrate which covers energy expenditure during prolonged demands. Inflammation postpartum, coupled with an increased cellular metabolism, accelerates immune gene expression while mitochondrial uncoupling further increases energy requirements and exacerbates negative energy status [112]. While numerous studies revealed that NEB can be responsible for various health disorders such as ketosis and infectious diseases [51], health problems (e.g., digestive or locomotive problems) can also be a trigger for NEB and may affect the NEB negatively in early lactating cows.

7. METABOLIC ADAPTATION

For a successful transition from late pregnancy into lactation, a cow needs to carefully coordinate metabolism across multiple tissue layers to provide sufficient nutrients and energy to support productive needs. Daily requirements for glucose, amino acids, fatty acids, and calcium for an early lactation cow are more than 2.7, 2.0, 4.5, and 6.8 times greater, respectively, than those needed for pregnancy [8]. These differences represent changes in nutrient requirements over a short period of only one to two weeks, highlighting the tremendous metabolic alterations necessary to adequately support lactation. This means that it is understandable that a cow with an inadequate ability to adapt metabolically to this transition can be the root of most postpartum diseases. Imbalance in energy and nutrient supply of high yielding cows in early lactation is inescapable. The intensity of the imbalance is influenced by both, the level of milk yield and the degree of endogenous energy provision. Levels required from the nutrient stores to compensate deficits can be theoretically calculated as the discrepancy between the total requirements and the resources deriving from daily feed intake. Generally, increase in genetic merit for milk yield is the most pronounced determinant of increased body condition loss and allostatic load [109,113]. In a study by Sutter and Beever [114], tissue mobilized over the first eight weeks of lactation was estimated to support the production of 300 L milk, from a total recorded production of 1820 L milk. Total milk production in week one was 33.8 L/day, of which 10.8 L/day (32%) were estimated to be derived from mobilized tissue. High-yielding dairy cows have been genetically selected to partition even more energy reserves into milk production with the effect that reliance on body reserves has dramatically increased [115]. Even in the case of higher dietary intake, the increased input will primarily result in greater milk production, while a similar energy imbalance remains, with no beneficial effects on body condition and reserves at all [116].

First and foremost, the mobilization requires well-functioning regulatory capacities that enable an efficient exploitation of resources, orchestrating a release of nutrients matching the requirements to a high degree and dealing with possible bottlenecks in metabolic pathways. The goal of regulation is not to preserve the constancy of the internal milieu; rather, it is to continually adjust the milieu to promote survival. The question is how the organism assesses and deals with the differences on a short- and long-term basis respectively; allowing the mobilization process to become more efficient. Furthermore, there is a question of whether the organism can rely on the regulatory opportunities available to it to prevent exhaustion due to overwhelming demands. The process of adaptation to NEB involves metabolic interrelationships between the energy storage site in the adipose tissue and skeletal muscle, the fuel processing site in the liver and the energy use sites in the mammary gland, placenta, fetus, and other tissues of the body. An increased competition for resources arises within the organism in the light of limited availability of resources. Health problems occur when these interrelationships and the adaptation process do not function appropriately to allow the cow to get through the period of NEB in early lactation.

Adaptation consists in the first place of physiological shifts in the use and conservation of body fuels: carbohydrates, amino acids and fats. The major mechanism involved is shifting the body's fuel supply away from glucose and toward the use of lipid energy sources. Lactation has an obligatory requirement for glucose that cannot be met by other fuels. Substrate-mediated events initiated by a decline in blood glucose concentration result in mobilization of non-esterified fatty acids (NEFA), which in turn results in several actions that stabilize or increase blood glucose. Besides NEFA, glycerol and proinflammatory cytokines, (TNF, IL-6) are also released [117]. Neuroendocrine and endocrine mediators are essential in the ability to adapt to NEB because fuel-mediated adaptation is not sensitive enough to function independently. Thus, endocrine actions primarily serve to control the rate of fuel metabolism. Additionally, the presence and number of specific receptors as well as their affinity play a decisive role in the mechanisms of metabolic regulation [118]. Initiation and persistence of a nutritional imbalance in early lactation in favor of the provision of external and internal nutrients for milk production is managed by simultaneous changes in the concentration of insulin, glucagon and growth hormone [119]. Low molar quotients of hormone concentrations ensure a maximum nutrient flux and nutrient uptake towards the udder. Growth hormone (GH) increases before calving [120]. High GH concentrations not only stimulate milk production but also provoke liver gluconeogenesis and lipolysis in adipocytes. Despite increased gluconeogenesis, blood glucose levels are low because of the drain of glucose to the udder. The mobilized NEFAs serve as an alternative energy source for other tissues to preserve glucose, which is used preferentially by the mammary gland to form lactose [121]. Hypoinsulinaemia promotes gluconeogenesis in the liver and acts as a massive lipolytic trigger. The udder benefits because it does not need insulin to facilitate glucose uptake into cells by the glucose-transport molecules [122]. Furthermore, low insulin concentrations uncouple the growth hormone (GH)-insulin like growth factor (IGF-I) axis in the liver because of down regulation of GH 1A receptors [123]. The resulting high blood NEFA and GH concentrations antagonize insulin action and create a further state of peripheral insulin resistance in liver, muscles, and adipose tissues [124]. In this way even more glucose is conserved to be available for lactose synthesis. Fat has been described as the "master regulator in the development of systemic insulin resistance" with good reasons [125].

Problems can occur in several steps in the adaptation process, particularly involving the adipose tissue and the liver. In the dairy cow, adipose tissue lipid accumulates during pregnancy, and catabolism begins prior to parturition and increases dramatically afterward. The duration and magnitudes of these adaptations depend on milk energy secretion, net energy intake, genotype, and endocrine environment [126]. Adipose tissue lipid synthesis is decreased and lipolysis is increased in early lactation. The magnitude and duration of these adaptations are increased in animals either consuming less energy or producing more milk. Adipose tissue is more sensitive to catecholamines in early and in midlactation and in animals with higher production. Adaptive responses of adipose tissue turn out to be inappropriate because of alterations in adipose sensitivity [9]. As adipose sensitivity increases, so does the NEFA response to a given stimulus such as catecholamine. When adipose tissue is highly sensitive, antilipolytic feedback responses may be blunted, releasing adipose tissue from metabolic and endocrine controls and allowing the rate of NEFA mobilization to go unchecked [124]. Different mechanisms lead to increased adipose sensitivity in obese cows, resulting in an interruption of homeostatic mechanisms regulating adipose mobilization. For example, high genetic merit in dairy cows is associated with increased adipose sensitivity [127,128] and a decline in circulating insulin levels [66]. During NEB, the GH-IGF axis uncouples due to a downregulation in liver GHR and this is associated with a reduction in circulating IGF-I and elevated GH concentrations, providing an endocrine environment that promotes the direct action of GH on lipolysis and gluconeogenesis in early lactation [129]. The adipose levels in a cow are also a critical factor affecting adipose sensitivity. Studies conducted by Rukkwamsuk *et al.* [130,131] revealed that cows on a high-energy diet ad libitum throughout the dry period had adipose tissue that was less responsive *in vitro* to the antilipolytic effects of glucose and ketone bodies compared with restricted-fed controls. This reduces the effects of fuel-induced, autoregulatory negative feedback signals on lipolysis, allowing increased NEFA mobilization in obese animals. In response to NEB, over conditioned cows mobilize more body fat than thin cows and subsequently are prone to develop metabolic disorders [132]. Obese cows have less effective insulin feedback regulation in the adaptive response to NEB and greater adipose sensitivity and NEFA mobilization compared to leaner cows [133]. This can result in inappropriately high blood NEFA concentrations that overwhelm the capacity of the liver for fuel processing and interconversion.

Recent findings revealed the effect of dietary level of energy during the dry period on feed intake, energy balance, milk production, and composition [134]. Cows being fed a diet exceeding energy requirements during the whole dry period showed an increased incidence of hyperketonemia while a controlled-energy dry period feeding approach was successful in minimizing the degree of negative energy balance postpartum as well as hyperketonemia. Several studies have shown that the energy level and composition of the pre-partum diet also have an impact on lipid metabolism, liver function, and the overall health of the cow post-partum [135,136]. A transient decline in high density lipoproteins and loss of this molecule's anti-inflammatory and antioxidant actions may account for a portion of the increased oxidative stress recognized to alter immune function in transition cows [9,137]. Besides fat, protein also plays an important role in the provision of glucose. Absorbed and mobilized amino acids from the muscles are partitioned to either gluconeogenesis or ketogenesis (oxidized for energy), milk protein or body (and conceptus) proteins. Although utilization of propionate for gluconeogenesis is extensive, amino acids have the potential to increase in importance as gluconeogenetic substrates when glucose demand is substantially increased [138]. Van der Drift *et al.* [48] observed large variation in the onset and duration of periparturient fat and protein mobilization. Cows mobilized in total 35% ± 26% of back fat thickness and 18% ± 9% of longissimus muscle thickness when kept under similar conditions at an experimental farm. Ketone body production was restricted in cows that mobilized more muscle protein relative to fat tissue after parturition. The relative partitioning amongst these pools depends on the mix of nutrients entering the body, the availability or need to replenish body tissue stores, and other physiological influences on the cow (stage of lactation, phenotypic, and genetic capacity for milk production, environmental conditions, age, growth stage, previous nutrition).

The hepatic response to NEB is to produce glucose and ketone bodies and take up large amounts of NEFAs from the blood for mobilization as other forms of energy. Apart from the activity of hormones, metabolic regulatory events in the liver are directed primarily by the level of metabolites, either in excess (e.g., free fatty acids) or shortage of supply (e.g., glucose) [139]. Reynolds *et al.* [140] reported major changes in the hepatic flux of NEFA as cows progressed from a pregnant state into lactation. Measured NEFA flux at day 19 before calving was equivalent to approximately 1 mol palmitate equivalent/day and a similar level was recorded at day nine before calving. In contrast, by 11 days post-calving, NEFA flux increased to 5.5 mol palmitate/day and whilst the level declined as lactation progressed, it was still almost twice the pre-calving baseline at lactation day 83. Maladaptive responses in the liver can arise at several points in the metabolic processes. Comprehensive reviews on the pathophysiological basis of ruminant adaptation to NEB and the development of ketosis and fatty liver is provided by Herdt [141], Drackley *et al.* [47] and Ingvartsen [112]. In short: Type I ketosis occurs when the demand for glucose outstrips the capacity of the liver for gluconeogenesis. Pathways for gluconeogenesis might be at the height of stimulation, but the supply of glucose precursors is insufficient to permit maximal glucose production. Thus, glucose production is limited by substrate supply in the first place.

Type II ketosis occurs when large amounts of NEFAs are delivered to the liver, but gluconeogenesis and ketogenesis are not at the height of stimulation. Mitochondrial uptake of NEFA is not as active as in type I ketosis. NEFAs not used for ketone body synthesis are esterified in the cytosol, forming triglyceride. Transport of triglycerides from the liver to other tissues requires the synthesis of very low-density lipoproteins (VLDL). Secretion may be near disabled when blood NEFA concentrations are high, leading to an accumulation of triglycerides in the liver. As the degree of fatty infiltration increases, normal functions of the liver are adversely affected. For example, the development of a fatty liver can interfere with hepatic gluconeogenesis capacity [142,143]. The ensuing hypoglycemia can lead to further adipose mobilization, creating a destructive spiral of events leading to fatty liver of increasing severity [141]. Fat infiltration also impairs the ability of the liver to detoxify ammonia to urea [144]. Ammonia decreases the ability of the liver to convert propionate to glucose [138]. Fatty liver and type II ketosis are usually seen clinically shortly after parturition, well before peak milk production. This reflects the association with adipose sensitivity [145]. Fat accumulation in the liver begins to occur in the last weeks of gestation, as adipose sensitivity increases. Factors influencing adipose sensitivity and energy balance in the late dry period are important concerns in the prevention of fatty liver and type II ketosis.

8. HIERARCHICALLY ORGANIZED AND NESTED SYSTEMS

The previous explanations followed the processes during the flux of nutrients from the environment through the digestive tract, and processes within the intermediate metabolism on different scales. As stated above, transformation processes are influenced by countless variables on each scale. Changes in perspectives from a top down to a bottom up approach and *vice versa* might help to improve our understanding of what restricts cows' metabolic capabilities to adapt to nutritional changes and to grasp metabolism with a genuinely systemic approach. Ideally metabolism should be expressed mathematically, but this is made difficult by the fact that its elements are simultaneously both variables and functions [146]. Metabolism is the inherent process of living organisms to obtain nutrients and energy from the feed ingested for self-referential purposes. Feed is made up of substrates (proteins, carbohydrates, and fats, *etc.*), degraded by enzymes in the digestive system into absorbent nutrients and fuel. The body can use the nutrients and the fuel right away, or it can store them in body tissues, such as body fat and muscles. A metabolic disorder occurs when abnormal chemical reactions in the body disrupt or impair the process of utilization. When this happens, the animal might have too much of some substrates, too little of other ones, or too much and too little simultaneously so that proper function is disturbed. A metabolic disorder can develop also when some organs become diseased or do not function properly. Thus metabolic disorders can be divided to structural and functional components. A structural component is represented by the arrangement and configuration of compartments (organs, tissues, cells). Details of a structure can vary considerably between animals, e.g., in relation to the proportional size of sub-systems and their functionality. For example, the large variation in number and form of villi in the rumen, and the condition of other semipermeable borders which have a considerable effect on the flow of nutrients between the compartments [147], illustrate structural variability. Functional components contribute to self-regulatory mechanisms, responsible for the stability of the body's internal environment [148]. Their effects can be regarded as dynamic constants, including blood metabolites, osmotic pressure, or pH amongst others. However, metabolic and external influences cause permanent or periodic changes in parameters of the internal environment. Therefore, the mode of activity and parameters of results of metabolic functional systems change consecutively. Self-regulation of functional systems is the mechanism which maintains these parameters at optimal levels.

Changes in the amounts and patterns of nutrients flowing through body compartments where these nutrients are degraded, synthesized, or stored, reflect the large day to day variation in nutrient availability the metabolism has to deal with. Metabolic disorders primarily develop when discrepancies exist between what is needed to sustain a certain balance and what is supplied from various sources, and when the organism cannot cope appropriately with these discrepancies. Discrepancies are not limited to single substrates but encompass a wide range of factors and processes, including interactions between different substrates within and between different compartments on different scales. The complex systems in living organisms operate both on a level of time and of space. The hierarchy of systems or scales is introduced because it corresponds to our subjective views of a system based usually on our various discrete experimental viewpoints. In their famous work, Oppenheim and Putnam [149] conceived of nature as being constituted by a hierarchy of objects that in turn, defined a hierarchy of distinct sciences. In agricultural and animal science, multiscale approaches inspired from theoretical physics have been developed in an essentially unidirectional (bottom-up) approach to integrate parameters at a given scale into reduced reproduction at higher scales [150]. However, lower-scale properties are also directly coupled with properties at a higher scale. The very complexity of living systems and biological functions lies partly in the presence of these bi-directional feedbacks between the upper and lower scale. An illustration of the hierarchical structure of living (autopoietic) systems is presented in Figure 1.

Based on basic biological considerations by Maturana and Varela [1] concerning the characteristics of self-organizing autopoietic systems, reworked, and reformulated by Razeto-Barry [70], the vertical structure of living systems in agriculture can be also described as a hierarchy of scales. An example is provided in Figure 2 with the basic level of cells (1); interacting with material elements, and organizing themselves in tissues and organs (2) as superordinate systems, integrated in a whole organism (3); which is part of a herd (4); while the herd is an intermediary level between the single animal and the agro-eco-system (5). Microbiologists would probably see the genome as the basic level [150], while agriculturalists might extend the hierarchy of systems to far larger scales [151]. In the current context, it is primarily important to bear in mind that each superordinate system acts as the environment for the corresponding sub-system. It is the duality of the perspectives in particular which enables a more comprehensive understanding of biological processes. Without the whole living organism, its parts—the organs and the cells—could not exist or be sustained. Thus, the living system continuously produces the components, which constitute it by itself. Furthermore, these components steadily sustain and regenerate the superordinate system [152].

From a biological-systemic perspective, an organism is an integral and functional system in continual process of exchange with the environment. It represents an entity with a functional integrity [153]. Maintenance of the entity is an active process of self-organization which requires resources. In striving for maintenance, external material is regulated by the living systems following to their own principles and purposes, while integrating and transforming the material within the system. This principle is valid for the single cell as well as for the organism and for the farm as an agro-ecosystem. For instance, cells adapt their own metabolism and their functional capacity to the availability of external resources surrounding the cells [154]. To a certain degree, cells are able to perceive the availability of nutrients directly and transform this information into modified metabolic flux rates. Thus, processes are not exclusively determined by internal dispositions (e.g., genome or proteome) and conditions (e.g., anabolic or katabolic metabolism) or by impacts of the physical environment (e.g., availability of nutrients or external stressors). Living systems react due to the way the system is configured and adapt constantly to their environment. They select the appropriate components from the environment and give them a meaning of incentives, objects of perception, and interests [152].

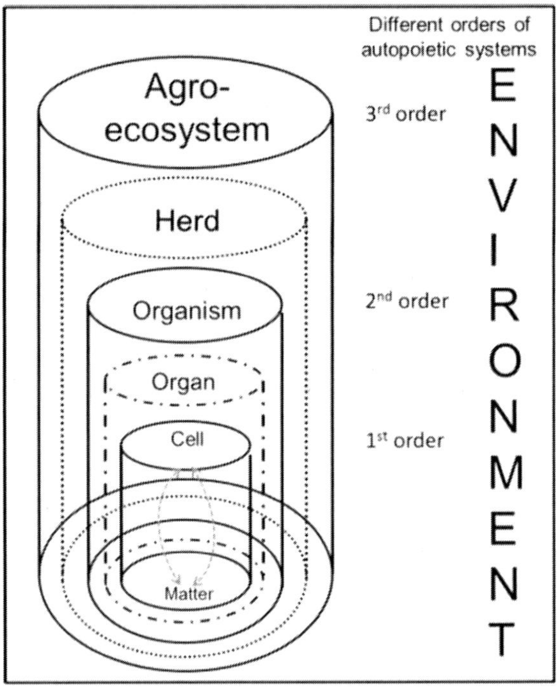

Figure 1. Hierarchical structure of autopoietic living systems of different orders within an agro-ecosystem.

Nutrients that would enable the individual animals to meet their requirements efficiently and comprehensively on a daily basis are neither available in the wild nor in livestock farming. Animals can rely on different metabolic pathways by which nutrient deficits or surplus can be compensated for an extended time period. Ability and skills to deal with nutrient imbalances are decisive if the animal is to succeed trying to adapt to inappropriate and varying supply conditions. Metabolic and reproductive disorders, diseases, and mortality indicate that the cows' ability to adapt to variations in its environment is overstressed due to disturbances in the interdependency between super- and sub-ordinated systems and their dynamic transition processes. Disorders and diseases can be understood as negative bodily occurrences where a part of the organism fails to perform one of its biological functions appropriately. This interpretation corresponds with the concept of biological malfunction, recently described by Saborido and Moreno [17]. In this concept, biological functions are interpreted as specific causal effects of a part or a trait that contributes to a complex web of mutual interactions, which, in turn, maintain the organization and, consequently, the part itself.

Change and disturbance occur both between and on all scales. Due to a large variation in composition and ingredients of: feedstuffs, diets, daily feed intake, as well as digestive and absorbent capacities, a variable supply from the intermediate metabolism meets a variable need in nutrient and energy requirements in the various compartments of the body. The gap between demand and supply has to be covered by regulatory processes which for their part consist of influencing factors with respect to input, partitioning, and output variables. Dairy cows differ considerably in their adaptability to nutritional changes and metabolic disturbances, particularly to NEB. Last but not least, this is evident through large differences in prevalence and incidence of metabolic disorders despite similar environmental conditions. The reasons behind the differences in adaptability are manifold. How to evaluate the degree to which adaptation occurs in this specific situation, taking into consideration the large intra- and inter-individual variation in the nutrient flow that takes place on and between various scales, is therefore of particular interest. This is also true for various other factors involved in the adaptation process. The large individual differences in input, partitioning, and output variables indicate a variability of metabolic processes on various scales that cannot be narrowed to a single factor but which needs to encompass the interconnectedness of variables and the nutrient flow. Added to this is the fact that most problems which challenge organisms have a number of equivalent solutions, which can be thought of as existing in a vast neutral space [155]. A neutral space is a collection of equivalent solutions to the same biological problem, embodied in the biological systems ensuring an organism's survival. This is the case for the multiple levels of biological organization. Understanding the structure of neutral spaces is critical to understanding the robustness of biological systems and to what extent the robustness itself can evolve.

A limited availability of nutrients and energy provokes severe competition between different tissues in the need for nutrients to sustain their various functions. Limitations require partitioning, and partitioning requires prioritization in guiding the nutrient flows, and ensuring that the demands of other cells, tissues, and organs within the organism are not completely neglected [30]. Thus, there is a need to prevent ruinous competition between sub-systems, to avoid being swamped by unwanted side reactions which clog the system or parasitic reactions by single organs to the expense of other organs which may cause the whole organism to collapse. Nutrient prioritization in early lactation to favor milk production over fertility is a reasonable strategy in biology [156]. As nutrition becomes scare, the lactating dam will preferentially invest the limited resources in the survival of living offspring rather than gambling on the oocyte that is yet to be ovulated, fertilized, and cared for during an entire gestation. Selection of high milk yields takes advantage of the genetically programmed readiness of the dairy cow to enter into a negative energy balance at the onset of lactation and to mobilize resources from its body tissues. However, selection has advanced into dimensions that are far beyond the initial intention to ensure nutrient supply to the off-spring via milk. Thus, there is no basic conflict between milk production to safeguard the off-spring on the one hand and self-preservation of the dam on the other. Life-threatening conflicts occur, when the gap between demand and supply gets to the stage where metabolic regulations are at risk of failing to mobilize the bodily resources needed to compensate for the deficits between nutrient output and intake. In case of failed regulation, the maternal ability to nourish the off-spring can be completely lost due to the death of the dam.

Adaptation depends *inter alia* on the degree to which the individual requirements of cows are met through the structure and organization of the farm system as the individual peculiarities of dairy cows mean that they differ considerably both in their general requirements and respecting their specific performance within the lactation course. To fulfil these demands, the farm system itself is in need of appropriate resources (high quality feed, investments, labor capacity, knowledge, *etc.*). What is needed on individual farms cannot be generalized but also depends on the specific farm system. At the same time, farmers faced with low milk prices are striving and fighting for the sustainability of their own farm system in a global market where all dairy farmers produce more or less the same commodity, thereby placing high pressure on production costs.

9. CHALLENGED FARM MANAGEMENT

Intense selection for milk production has resulted in an immense priority for the high-producing dairy cow to partition energy to milk, at the cost of body reserves. This has resulted in an excessive negative energy balance (NEB) and high prevalence of metabolic disorders and poor reproductive performance. Thus, milk production, animal health, and reproductive performance in high-producing dairy cows generate conflicting interests. Traits in relation to cow health have only recently been included in the selection criteria on the basis that considerable genetic diversity with respect to health and reproductive performance exists in the current population [67]. However, functional traits have a comparably low heritability thus questioning whether such approaches will bring the desired improvements in the foreseeable future. In the meantime, animals will continue to suffer from production diseases while milk yields will further increase and linked with the negative side effects. The preparedness of the dairy cow to enter into a negative energy balance in early lactation while mobilizing nutrients from body tissues is an essential precondition for a high milk performance in the total lactation period. Apparently, dairy cows do not possess an effective feedback loop, e.g., to reduce milk yield and depletion of body reserves to a degree that prevents metabolic disorders and does not compromise the survival of the cow itself. Thus, it is the responsibility of the farm management with their various organizational skills and resources to avoid overstretching the gap between demand and supply. The crucial question is: when is the border which endangers the ability of individual animals to sustain themselves crossed? As this paper's concept explains, such thresholds can neither be generalized nor predicted for the individual animal.

As milk yield increases, the management challenge of meeting a cow's dietary nutrient needs becomes greater. This implies that increases in milk yield achieved through genetic selection requires improving the entire availability of nutritional resources. However, not all high yielding cows are capable of achieving an adequate energy intake, especially under grass-based management systems [157]. While energy requirements have increased more rapidly than dry matter intake (DMI), the nutrient density of diets has increased simultaneously, enhancing the risk of subacute ruminal acidosis [158]. Although advancements have been made in feeding practice to minimize the risk of metabolic diseases, the periparturient period continues to present some of the greatest challenges in the animals' health [107]. Nutritional approaches vary enormously between farms and in many instances are failing to meet the more exacting demands on modern dairy cows [159]. Too many farmers are still concerned about feed costs per ton when the real focus should be on feed cost per liter milk produced. Larger herds usually result in more cows managed per available staff member. Declining availability of qualified dairy staff has added concerns that cows are receiving less individual attention, which is likely to have contributed to the high level of metabolic and fertility disorders. Dry cows are still often considered as second-class citizens or nonworking cows and thus receive little attention and poor-quality forage. Separation of dry cows into an independent group or into sub-groups is not readily practiced. Dry cows are often housed in a low-producing group [67].

Given that nearly all cows experience reduced feed intake and loss in body condition postpartum, there is a large variation between cows in the extent and duration of negative energy balance. Because the energy balance calculation requires knowledge of feed intake and digestibility, milk composition, and weight; in farm practice, energy balance values can often not be calculated due to lack of data on these components. Correspondingly, farm management is often eager to follow general recommendations instead of trying to adjust feeding regimes to the degree of NEB and the variation between cows, as this is easier. In the light of highly variable initial and boundary conditions in farm practice and limited availability of resources in dairy management, it is more than questionable whether "best practice" is still a realistic and valid concept. There is much evidence that specific measures operating under certain conditions are not necessarily effective, when compared with others in achieving the same purpose. Feeding higher-concentrate diets in close-up groups to meet increasing energy needs, is often practiced but recognized as decreasing the feed intake due to high availability of propionic acids which provokes satiety [73]. This situation can be counterproductive at a time when cows are accelerating their milk output, which accelerates lipid mobilization and results in associated negative consequences. On the other hand, excessive intakes of energy relative to requirements in late lactation and dry period is almost certainly expected to increase the risk of "fat cow" or "fatty liver" syndrome [77] and increase the risk of insulin resistance in cows and to elicit changes similar to obesity and type II diabetes [160]. Besides the availability of energy, the relevance of the dietary provision of metabolizable protein has been emphasized, based on strong indirect evidence backing the importance of the maintenance of maternal protein stores on long-term health, productivity, and reproduction [119].

With the continued challenges to cows navigating the transition period, some researchers have considered altering the dry period to minimize metabolic changes. Shortening or even skipping the dry period improves dry matter intake peripartum, reduces milk production in early lactation, improves energy balance, and reduces the number of days postpartum till resumption of ovarian activity [161,162]. A recent meta-analysis of 24 studies manipulating dry period length showed lower milk production, improved energy balance, and decreased risk of ketosis, but no difference in other diseases or fertility [163]. In a recent study by Chen *et al.* [164], cows with a 0-d dry period had an improved metabolic status in early lactation, indicated by lower plasma concentrations of non-esterified fatty acids, greater plasma concentrations of glucose, insulin-like growth factor-I, insulin and lower mRNA expression of pyruvate carboxylase in the liver, compared with cows with a 30-d or 60-d dry period. However, adoption of shorter dry periods has been limited because of the milk loss concerns.

A way to reduce the imbalance between nutrient supply and demand in early lactation is not only to improve the first but also to temporarily decrease the latter. While most dairy cows are milked twice daily, it is not uncommon in intensive dairying systems to increase milking frequency up to four times daily to increase milk production. Reducing milking frequency is much less common. However, the overall energy balance of cows during early lactation is improved with one-daily milking [165]. In dairying systems where an emphasis is placed on animal health rather than on milk production per cow, this practice might fit well under certain infrastructural farm conditions. Depending on the stage of lactation, breed, and parity the reduction in milk yield losses varies considerably and can amount to approximately 22% [166]. This figure has been revealed as an average milk loss across 30 different international short-term studies. Production expressed as milk solids (*i.e.*, combined fat and protein yields) is also lower with once-daily milking, despite the fact that milk fat and protein content of the milk increase [52,167]. The increase in fat and protein content is not sufficient to compensate for the decrease in milk volume. However, it may provide a tool to manage the metabolism and energy balance of cows during early lactation better [168], e.g., at highest negative energy balance period. Results from a study by Carbonneau *et al.* [169] revealed that blood concentrations of glucose were higher and concentrations of non-esterified fatty acids and ß-hydroxybutyrate were lower in partially milked cows compared to control cows during the treatment period. In other studies too, cows milked once a day showed an improvement in their metabolic profile [52] and immune function [59]. In contrast, cows subjected to an increased milking frequency were 1.4 times more likely to be classified as subclinically ketotic as the control cows [170].

One-sided selection for high milk yield is likely to produce cows that are increasingly more vulnerable to disease and poor fertility [171]. The authors concluded from different studies that negative genetic correlations between production and fitness traits increase in less favorable environments and that selection for increased yield has decreased the adaptability of modern cows. Keeping such animals in living conditions where farmers have increasingly less time available per cow demands even higher management skills. In situations of inferior management, unfavorable genetic correlations between milk yield and metabolic disorders are expected to be stronger than in herds with superior management. In a study allowing discrimination between the roles of genotype (G), environment (E) (e.g., feed caloric density and milking frequency), and GxE interactions, the effects of genetic merit and milking frequency were significant only in the groups that were fed rations with high caloric density [109]. However, signs of severe negative energy balance, poor protein balance, and low body condition scores were not concentrated in the highest producing cows. A lower feed caloric density and extra milking had strong unfavorable effects on both energy and protein balance, and emphasize the possible effect of mismanagement on animal health risks. Increased milking frequency increased milk yield, although only significantly when the cows received high caloric density rations. Milking the cows three times a day instead of twice significantly decreased energy balance. Thus, environmental factors like feeding management and milking frequency strongly influence the cows' energy balances. High genetic merit cows seem more prone to allocate resources toward high fat output in times of nutrient-deficient diets and toward high protein output with high nutrient diets. High genetic merit for milk yield seems intrinsically connected with the allocation of resources from maintenance toward milk. To prevent compromise of animal health status; demands on the provision of appropriate feed and feeding conditions; farmers' time and management skills are clearly increasing. In a direct comparison of low- and high-yielding genetic lines of cattle, at relatively similar levels of dry matter intake, there is no evidence of a difference in the fractional recovery of DE from intake energy or ME from DE [172]. Similarly, no differences were found in the efficiency of ME use for milk energy. Thus, increase in productivity due to higher milk performance per cow is based primarily on the proportional reduction in nutrient resources needed to cover the demands for maintenance. In terms of maintenance requirement, the high yielding cow is more efficient as maintenance cost per unit milk is diluted [173].

Against an industry-held view that milk output is the single most important aspect in an efficient dairy business, it is contended that producers need to refocus on factors which also affect profitability, mainly: nutrient efficiency, cow health, fertility, and longevity [6,174,175]. In animal nutrition, the most efficient use of nutrient resources is generally obtained when each animal is supplied with energy and nutrients according to their individual demands [40]. Both exceeding supply before calving and under-nutrition postpartum not only reduces efficiency but requires additional resources of regulation to maintain the internal processes within a certain range of homeostatic conditions [9,176]. Feeding farm animals according to their current requirements regarding nutrients and energy provides the best orientation for the farmer to gain a high efficiency in the use of nutrients while simultaneously reducing the risks of metabolic disturbances. However, to meet such a goal requires additional resources, amongst others: high quality feed, good size and type of stalls, suitable amount of eating space available to the cows or ventilation systems in place, and last but not least knowledge and skills to reduce possible gaps and compensate for deficits [12]. Thus market conditions which do not provide enough income for farmers and do not fully cover the expenditure during the production process threaten not only the sustainability of the farm systems but also that of the dairy cows and their functional integrity of organs and tissues through downward causation.

While farmers have to consider changes in the availability of resources, they can no longer ignore consumer concerns about animal health and welfare, and food safety. Increasing disadvantages of the intensification process and conflicts with the interests of other stakeholders call for a critical assessment of the sustainability of dairy production [177]. What is sustainable for agricultural systems is subject not only to system-inherent and self-referential demands but also to changes in the market. Thus, farm managers are not only agents of control but components of the agricultural system itself. The fact that market prices and consumer preferences are primarily dominated by forces beyond the farmers' control makes the issue even more complex and somewhat confusing. In providing the essentials for remaining both competitive and sustainable, the farm management is restricted to only a few options; either to further reduce production costs per product unit or to obtain a higher selling price. As long as qualitative traits such as the production of "healthy" food from healthy cows are not honored by premium prices, farm management has to focus on how to improve the overall efficiency of production processes [178]. This cannot be achieved solely by addressing the milk yield of dairy cows. In fact, farmers have to take into account the cost-benefit relationships on various scales, including failure costs caused by metabolic disorders and production diseases as well as the preventive costs needed to reduce these. Expenditures for preventive measures can be seen as additional input to reduce monetary losses of disorders and diseases [6]. The higher the preventive costs, the lower the failure costs and *vice versa* (see Figure 2). If no preventive and control measures are taken, the losses due to disease are at their maximum (Lmax). With maximum prevention, the failure costs due to disease will be at a minimal (Lmin). Because the relationships between prevention and failure is not linear, there is an optimal level of control (Lopt). To gain profit from reduced prevalence of production diseases, the farm management has to strive for an optimum between failure and preventive costs. Due to the various and highly varying factors involved in the development of productivity and in the development of disorders, and due to the large differences in the availability of resources mentioned above, such a balance is different for each farm system. Furthermore, the balance varies in relation to changes in the environment of the farms, especially regarding changes in food and feed markets.

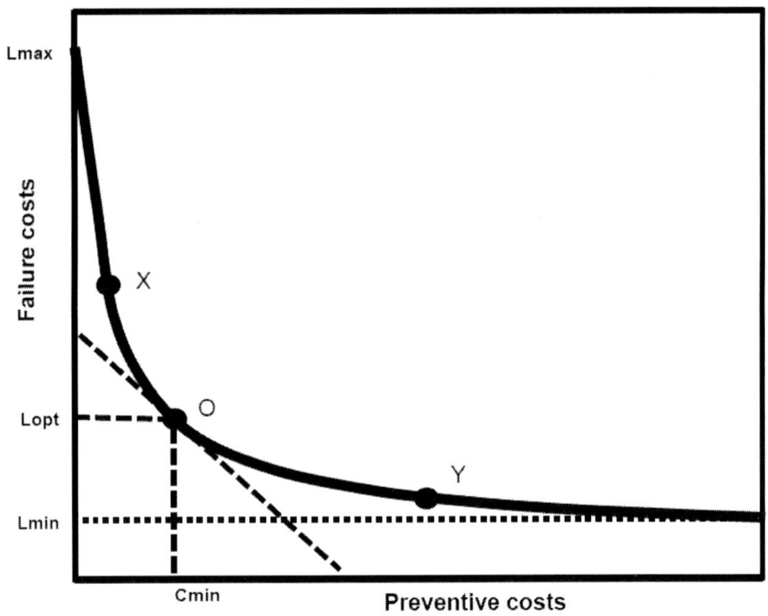

Figure 2. Schematic representation of the relation between failure and preventive costs [6].

It is in the nature of much-circulated general recommendations that they only marginally address the context in which the biological processes take place. Correspondingly, they may appear counterproductive, especially when they distract the farm management from looking for farm specific solutions. The latter should rely on valid data which unfortunately are often not available to the desired quality, probably because data acquisition itself is very time consuming and reflecting upon data needs much mental energy [179]. Moreover, changes in perspectives are required to implement and assess bilateral adaptation processes, on the one hand by the animals towards their living conditions and on the other by the farm management in relation to the specific demands and limitations in the adaptation capacities of dairy cows, to the benefit of both animals and farmers.

10. CHALLENGED AGRICULTURAL SCIENCE

Agricultural research and animal science as a sub-discipline deal with agricultural systems and with socio-technical and socio-economic circumstances as well as the superordinate systems which influence them. Criteria for assessing the quality of agricultural research are shaped on a problem-oriented systems research basis [180]. This consists of the close connection between basic and applied research, the multi- and interdisciplinary character of agricultural research and the orientation of research towards the solving of issues relevant to society. Regarding the high prevalences of metabolic disorders and associated production diseases in dairy production, agricultural science so far cannot claim to have been able to reduce health problems. Despite tremendous scientific efforts in the past, the situation has not improved adequately. Animal scientists might at least claim that without technical and scientific progress in gaining more insights and knowledge about the pathophysiological processes, the situation might have become even worse. However, lack of success in reducing production diseases does not meet goals set by agricultural scientists and there is thus an impetus to reflect on whether the previous approaches in dealing with health problems are still appropriate to the questions on hand. In the first instance, a reductionist approach has been employed, accumulating large quantities of detailed information with little attempt to incorporate it into a broad view of the whole subject. Thus, scientific efforts have provided a wealth of knowledge but to a large extend, this information has not been integrated into coherent biological models. Members of animal science in particular have not done well to integrate output productivity and negative side effects with respect to animal welfare [181].

In the case of adaptation to metabolic challenges too, there is a predominant tendency to reduce the differences between animals' ability to cope with nutritional and metabolic challenges to single factors, with a priority to differences in the genome (e.g., [47,182,183]). Based on the methodological advances made in the human genome project, and the need to interpret findings in a more biologically meaningful fashion, "system biology" has emerged as a new approach [184]. In the meantime this has been applied to many species, including dairy cows. The pace at which transcriptomics has expanded has forced the discipline of genomics to embrace the notion of holism as far as animal function is concerned [185]. The new approach focusses on the gene/protein expression (proteomics) on the cell/tissue level and the associated changes in biological functions. It tries to combine omics-generated data with mathematics and computer science to gain knowledge of biological behavior of a cell in either a bottom-up or top-down direction [186]. Recent developments in technology have paved the way for new modelling approaches at various levels of complexity [115,187,188]. One of the main objectives is to be able to predict the ability of an animal to respond to the nutritional limitations that arise from the environment in which it is placed. The perspective is that of incorporating increasingly detailed descriptions of genotype via genomics, and of the downstream metabolic machinery via gene expression, proteomic and metabolomic information [115]. As indicated by the authors, this remains a daunting task; not only due to serious methodological hurdles. The authors also acknowledge the danger that even if all the omic information were available, it might be still too complex for the minds of scientists. Whilst generating biological information about dairy cattle through the application of genome-enabled tools is no longer a bottleneck, the biggest challenge is to interpret findings in a more biologically meaningful fashion and eventually to communicate biological knowledge systematically by linking the genome to the whole organism [188]. The authors admit that the use of the omics' tools alone cannot provide a full picture of the biological system and can thus be misleading when interpreting biological outcomes, while scientists are still struggling to interpret the results from any one of the omics' technologies in a holistic and unbiased way.

While many genome-wide association studies have frequently shown that they do not explain the majority of the variation in whole-animal phenotypes, it has become clear that the relationship between the genome and the phenome is best characterized in terms of causal interdependency [150]. Genes are often found to possess multiple functions, which are sometimes critically dependent on the context and the specificity of enhancers for promotor sequences is often surprisingly loose [189]. Thus, even for fixed genotypes and environments, a large variability of phenotypes can occur across a population [190]. According to McNamara [181], "systems biology" means different things to different people. At minimum, it is the recognition that each piece of the system has a specific function related to the outcome of the entire system, not just the subsystem in which the molecule acts. For Cornish-Bowden *et al.* [146], the term "system biology" is little more than a euphemism for gathering even more details on an even larger scale, and not, as it should be, the study of biological systems as systems rather than as a collection of components. The more complex the level at which one seeks to explain a living system, the greater the need to examine the network of interactions.

In opposition to the omics' approach, it might be further remarked that in search for general principles and generalizable regularities, the mathematical approach is faced with the tremendous variation that occurs on the nutrient pathways from nutrient intake to utilization and secretion via milk. The variation is increased further by the interconnectedness of the variables on different scales, organized by various forward and feedback loops. As origin, scales of time and space, control, and functional significance of fluctuations in biological systems are largely unknown, it is questionable whether the genomic approach will be able to achieve biologically meaningful aggregated phenotypic descriptions in a foreseeable time period, let alone provide a valid base for decision-making processes on a farm level.

It has become increasingly clear that understanding nutrient partitioning is central in grasping the complexity of adaptation to shortages and imbalances in nutrient supply in relation to requirements, and thus of the ability of dairy cows to cope, especially when challenged with abrupt and unforeseeable changes. This refers not only to the nutrient partitioning within the animals but also to the level of the farm. In looking for approaches to gain knowledge in predicting nutrient partitioning in dairy cows, animal scientists often recognize the central role of the genotype. This however, leads to the question, how a valid prediction regarding the reactions of animals with a high variability in their medical history towards more or less unpredictable challenges can be obtained. Although the degree of deficits might be comparable between cows, for instance in the extent of NEB, there are large differences in the possible reasons behind the deficits and thus in the reactions and adaptation processes of the individual animals as outlined above. In the light of metabolic disorders and production diseases, affecting more than every other cow in early lactation to a varying degree, expectations concerning success in gaining predictions on the capacities to adapt are—to formulate it cautiously—astonishing, at least from the perspective of animal nutrition and animal health. Probabilities for successful adaptation can be at best predicted on the population level. This however, does not necessarily mean that what might be expected on the population level is valid for the corresponding farm level where adaption takes place in reality.

To look beyond narrowly molecular perspectives for answers to biological questions, it is important for metabolic adaptation not just to look at the total sum of the individual parts, but how they interact, and especially if they interact successfully in preventing the animals from suffering metabolic disorders and associated production diseases. A widespread view is that causation operates in an exclusively bottom-up way, from the microscopic to the macroscopic. Adherents of such a view often express perplexity at the idea that causation might even run in the reverse direction [150]. In tracking the flows of causal influence that might overstress the ability of an animal to adapt and increase the risk of metabolic disorders, it should be evident that the main causalities do not occur on the cell or genome level and thus cannot be solved there. In the first place, difficulties to adapt are due to extended and dynamic gaps between nutrient requirements and nutrient supply that already begin with the nutrient partitioning on the farm or even a larger scale. The central role of the whole organism, the genome being only a minor part, is to deal with the challenges deriving from the gaps between synchronized and tight causal coupling among the subsystems on the different scales. Correspondingly, it is the organism as a whole that succeeds or fails in adapting to nutritional change and disturbance.

Biological systems as a whole embody solutions to important biological challenges. Complex systems conveniently provide a conceptual framework and effective tools to solve problems and to decouple emergent and immergent features from molecules to organisms and back. The latter is also described as downward causation meaning the necessity for all processes at the lower level of hierarchy to act in conformity to the laws of the higher level. This means that some macro-level constraints are expected to cascade back onto micro-levels, the macro-level being itself an emergence [191]. Accordingly, pathological symptoms indicating disturbed adaptation to nutritional and metabolic changes cannot be reduced to a single factor but result from the complex interactions between various processes at different scales, thus expressing the present pathophysiology state of the whole organism.

Biological systems are goal-driven and above all they strive to survive and sustain. Facing a restricted availability of resources, while being confronted with extended change and disturbance, some animals are more successful than others at coping. However, animals that show competitive advantages in a specific unbalanced situation do not necessarily have an advantage in a quite different nutritional situation and might not show better results in the long run with respect to lifetime efficiency. The fact that metabolic disorders and production diseases have been occurring for a long time might suggest looking for solutions in the long run, e.g., breeding measures. However, metabolic disorders and production diseases are an instant problem, causing suffering, and thus require shot-term reactions and counteractive measures. Metabolic problems that derive from inappropriate and unbalanced nutrient supplies cannot be solved by selections between different genotypes but can only be reduced by the farm management in the specific context and environmental conditions. There is a significant body of knowledge on how to provide optimal farm management and prevention of bovine metabolic disorders and production diseases. For instance, tremendous efforts have been made in animal nutrition to assess and predict nutrient requirements of cows in their various life stages and following different animal and environment related variables. However, knowledge on how to adapt requirements to each unique case does not provide information on how to deal with the intra- and inter-individual variability in both nutrient demands and nutrient supply within a herd, and how to support endangered animals in coping when facing a large gap between demand and supply.

In the light of high prevalences and incidences of metabolic disorders and production diseases, it is obvious that many farm systems do not provide adequate solutions to the apparent problems; they thus overstretch the adaptation capacities of their animals. Farm systems themselves are facing fierce competition and have to adapt to change and disturbance; amongst others in feed and food markets. Farmers thus often lack the resources in terms of quality feed, labor time, investments, know how, *etc.*, needed to improve the status of animal health and welfare. Partitioning of the available resources in the most efficient way is not only a challenge for the animal but also for farm management. It can be expected that farmers would be willing to invest more in animal health and welfare if they could expect a competitive advantage. A large variation between farms regarding failure and preventive costs indicate that there is room for improvement on many farms meaning an appropriate balance which benefits a farm's profitability.

The task of agricultural and animal science should be to identify the most effective preventive measures and/or therapeutic options for farm animals but also the most cost-effective and economic approach in the corresponding complex farm system. In the first place, trade-offs have to be addressed on the scale where they emerge (principle of subsidiarity), following a bi-directional approach which simultaneously reflects upward and downward causation. Applied agricultural and animal science should support the farmer in striving for the goal to find an economic balance between productivity and animal health and welfare to the benefit of the animals, the farmers, and the prosperity of the common good. This can only be obtained by the development of more comprehensive concepts and evidence based solutions which simultaneously consider a larger number of causal relationships to discover regularities beyond the scope of individual scientific disciplines.

11. CONCLUSIONS

The pathophysiology of metabolic disorders and postpartum diseases is quite complex, interrelated with many processes within or outside the intermediary metabolism, they are also multifactorial and often driven by several interconnected risk factors. In this way, it is difficult for both scientists and the farm management to grasp the complexity and make the appropriate decisions to improve the adaptation capacity of dairy cows, particularly in the transition period. Metabolic processes in early lactation are dominated by a pull effect originating from the high glucose demand of the epithelial cells in the udder and by a large variation on the various process levels. The capacity of the dairy cows to cope with unbalanced situations between requirements and supply is limited and often exhausted. Especially in the transition period, dairy cows face varying initial conditions (e.g., body condition, medical history) and different boundary conditions (e.g., feeding and housing conditions) which interact with physiological regularities and thus result in highly variable outcomes. As a result, the interconnectedness of numerous factors on different scales create a large variation in adaptation capacities and in the occurrence of metabolic disorders and production diseases both on the animal and dairy farm levels.

It is concluded that the fundamental principles of metabolism are based on network properties which are able to contribute to the adaptability of the organism and not on the properties of single components. A metabolic network is a functional union of the different organs and tissues involved, providing a responsiveness to metabolic change and disturbance as a functional unit or as a whole organism, respectively. An understanding of symbiosis implies that the inter-tissue competition becomes an insufficient explanation for metabolic function but must incorporate a wide range of very intimate modes of cooperation. In a new perspective, the object of biological investigation becomes the process that pervades multiple levels of organization. Cells, and tissues must be understood as sub-systems within the framework of organs, and these sub-systems are embedded within broader organismal systems of interconnected and communicating structures. This represents a cooperative aspect of life and carries much importance in relation to fitness and reproduction.

The ability to cope with suboptimal living conditions and a large gap between nutrient supply and demand cannot be left to the dairy cow itself. It lies in the first place under the responsibility of the farm management to regulate nutrient and energy input as well as output to prevent the exhaustion of this adaptation capacity. On the other hand, dairy farming is a business that people engage in for financial gains.

Dairy production has capitalized on the metabolic drive of the dairy cow to ensure provision of milk to the calf by selecting for increasingly higher milk production. This strategy has been very successful. Unfortunately, the goals of economic efficiency and optimal dairy cow health are often in conflict. Metabolic disorders in early lactation indicate an overstressed ability of the animals to adapt to living conditions that do not appropriately provide the specific nutrient and energy requirements. Dairy cows will succeed more easily in adaptation and in avoiding dysfunctional processes in the transition period when the gap between nutrient and energy demands and supply is restricted. The fact that on average, one in two cows succumbs to health problems during the transition period highlights the fragility of many production systems. From these previous considerations, it seems obvious that animals within a herd will be better able to adapt in order to survive, if appropriate resources and living conditions are offered which meet the individual requirements at the different stages to a high degree and if nutritional and other disturbances are reduced to a minimum. Any attempts to reduce prevalences of metabolic disorders and associated production diseases should rely on a continuous and comprehensive monitoring of appropriate indicators on the farm level. If low prevalences of metabolic disorders and production diseases would go along with a competitive advantage, it would support the farm management in providing appropriate living conditions. As a first step, low prevalences for the most relevant production disease should be established as a separate production goal, continuously assessed by appropriate criteria and monitored as an emergent health output of the farm practice. Furthermore, those farms able to combine a low prevalence of production diseases with a high productive output should be rewarded by premium prices. In the light of a lack of success in a strong reduction in the prevalence of metabolic disorders and production diseases; the deficiencies in reductionist approaches on the farm level are becoming increasingly apparent, challenging the disciplines of agricultural and animal science. Thus, scientists should ask themselves whether the focus on single aspects has contributed to promoting the tendency of oversimplification and of widely ignoring the context in which adaptation processes take place. One of the central questions should be to what degree each of the subsystems contributes to the benefit of the superordinate systems and *vice versa*.

ACKNOWLEDGMENTS

The author thanks Hoischen-Taubner, K. Müller, and F. Brand for their support and the very fruitful and comprehensive discussions we shared on this issue of metabolic disorders.

REFERENCES

1. Maturana, H.R.; Varela, F.J. *Autopoiesis and Cognition: The Realization of the Living*; Springer Netherlands: Dordrecht, The Netherlands, 1980.
2. Piaget, J. *Biology and Knowledge: An Essay on the Relations between Organic Regulations and Cognitive Processes*; Edinburgh University Press; University of Chicago Press: Edinburgh, UK, 1971.
3. Broom, D.M. Assessing the welfare of modified or treated animals. *Livest. Prod. Sci.* **1993**, *36*, 39–54.
4. Wingfield, J.C.; Romero, L.M. Adrenocortical responses to stress and their modulation in free-living vertebrates. In *Handbook of Physiology, Section 7, Coping with the Environment: Neural and Endocrine Mechanisms*; McEwen, B.S., Ed.; Oxford University Press: Oxford, UK, 2000; Volume 4, pp. 211–236.
5. Leroy, J.L.; Vanholder, T.; van Knegsel, A.T.M.; Garcia-Ispierto, I.; Bols, P.E.J. Nutrient prioritization in dairy cows early postpartum: Mismatch between metabolism and fertility? *Reprod. Domest. Anim.* **2008**, *43* (Suppl. 2), 96–103.
6. Hogeveen, H. Costs of production diseases. In Proceedings of the 27th World Buiatrics Congress, Cairns, Australia, 27 July–1 August 2014; pp. 36–42.
7. Drackley, J.K. Biology of dairy cows during the transition period: The final frontier? *J. Dairy Sci.* **1999**, *82*, 2259–2273.
8. Bell, A.W. Regulation of organic nutrient metabolism during transition from late pregnancy to early lactation. *J. Anim. Sci.* **1995**, *73*, 2804–2819.
9. Sordillo, L.M.; Raphael, W. Significance of metabolic stress, lipid mobilization, and inflammation on transition cow disorders. *Vet. Clin. North Am. Food Anim. Pract.* **2013**, *29*, 267–278.
10. Ametaj, B.N. Metabolic disorders of dairy cattle. In *Encyclopedia of Life Support Systems (EOLSS): Veterinary Science*; Developed Under the Auspices of the UNESCO, Eolss Publishers: Paris, France, 2014.
11. Mulligan, F.J.; Doherty, M.L. Production diseases of the transition cow. *Vet. J.* **2008**, *176*, 3–9.
12. Oetzel, G.R. Undertaking nutritional diagnostic investigations. *Vet. Clin. North Am. Food Anim. Pract.* **2014**, *30*, 765–788.
13. Herdt, T.H. Metabolic diseases of dairy cattle. *Vet. Clin. North Am. Food Anim. Pract.* **2013**, *29*.
14. Brand, F.S.; Jax, K. Focusing the meaning(s) of resilience: Resilience as a descriptive concept and a boundary object. *Ecol. Soc.* **2007**, *12*, 23.
15. Darnhofer, I. Strategies of family farms to strengthen their resilience. *Environ. Policy Gov.* **2010**, *20*, 212–222.
16. Di Paolo, E. Autopoiesis, adaptivity, teleology, agency. *Phenomenol. Cogn. Sci.* **2005**, *4*, 429–452.
17. Saborido, C.; Moreno, A. Biological pathology from an organizational perspective. *Theor. Med. Bioeth.* **2015**, *36*, 83–95.
18. Wingfield, J.C. Allostatic load and life cycles: Implications for neuroendocrine control mechanisms. In *Allostasis, Homeostasis, and the Costs of Physiological Adaptation*; Schulkin, J., Ed.; Cambridge University Press: Cambridge, UK, 2004; pp. 302–342.

19. Montévil, M.; Mossio, M. Biological organisation as closure of constraints. *J. Theor. Biol.* **2015**, *372*, 179–191.
20. Bernard, C. *Introduction à L'étude de la Médecine Expérimentale*; Baillière: Paris, France, 1865.
21. Bernard, C. *Leçons sur les Phénomènes de la vie Communs aux Animaux et aux Végétaux*; J.-B. Baillière et fils: Paris, France, 1878.
22. Cannon, W.B. Organization for physiological homeostasis. *Physiol. Rev.* **1929**, *9*, 399–431.
23. Ashby, W.R. *An Introduction to Cybernetics*, 2nd ed.; Chapman & Hall: London, UK, 1956.
24. Wiener, N. *Cybernetics: Or Control and Communication in the Animal and the Machine*, 2nd ed.; M.I.T. Press: Cambridge, MA, USA, 1948.
25. Cannon, W.B. *The Wisdom of the Body*; WW Norton & Co.: New York, NY, USA, 1932.
26. Selye, H. Homeostasis and heterostasis. *Perspect. Biol. Med.* **1973**, *16*, 441–445.
27. Waddington, C.H. The basic ideas of biology. In *Towards a Theoretical Biology: Prolegomena*; Waddington, C.H., Ed.; Edinburgh University Press: Edinburgh, UK; Aldine Publ. Co.: Chicago, IL, USA, 1968; pp. 238–253.
28. Letelier, J.-C.; Cárdenas, M.L.; Cornish-Bowden, A. From l'homme machine to metabolic closure: Steps towards understanding life. *J. Theor. Biol.* **2011**, *286*, 100–113.
29. Varela, F.G.; Maturana, H.R.; Uribe, R. Autopoiesis: The organization of living systems, its characterization and a model. *Biosystems* **1974**, *5*, 187–196.
30. Bauman, D.E.; Currie, B.W. Partitioning of nutrients during pregnancy and lactation: A review of mechanisms involving homeostasis and homeorhesis. *J. Dairy Sci.* **1980**, *63*, 1514–1529.
31. Mrosovsky, N. *Rheostasis: The Physiology of Change*; Oxford University Press: New York, NY, USA, 1990.
32. Knight, C.H.; Beever, D.E.; Sorensen, A. Metabolic loads to be expected from different genotypes under different systems: Metabolic stress in dairy cows. *Br. Soc. Anim. Sci. Occas. Publ.* **1999**, 27–36.
33. Sterling, P.; Eyer, J. Allostasis: A new paradigm to explain arousal pathology. In *Handbook of Life Stress, Cognition and Health*; Fisher, S., Reason, J., Eds.; John Wiley & Sons: Hoboken, NJ, USA, 1988; pp. 629–649.
34. McEwen, B.S. Stress, adaptation, and disease: Allostasis and allostatic load. *Ann. N. Y. Acad. Sci.* **1998**, *840*, 33–44.
35. McEwen, B.S.; Wingfield, J.C. The concept of allostasis in biology and biomedicine. *Horm. Behav.* **2003**, *43*, 2–15.
36. Herdt, T.H. The history and influence of the ICPD. In *Production Diseases in Farm Animals: 12th International Conference*; Joshi, N.P., Herdt, T.H., Eds.; Wageningen Academic Publishers: Wageningen, The Netherlands, 2006; pp. 19–22.
37. Knaus, W. Dairy cows trapped between performance demands and adaptability. *J. Sci. Food Agric.* **2009**, *89*, 1107–1114.
38. Rauw, W.; Kanis, E.; Noordhuizen-Stassen, E.; Grommers, F. Undesirable side effects of selection for high production efficiency in farm animals: A review. *Livest. Prod. Sci.* **1998**, *56*, 15–33.
39. LeBlanc, S. Managing critical periods—transition dairy cows. In *Book of Abstracts, 15th Conference on Production Diseases in Farm Animals*; Dalin, G., Ed.; Swedish University of Agricultural Science (SLU): Uppsala, Sweden, 2013; pp. 62–65.
40. Van Saun, R.J.; Sniffen, C.J. Transition cow nutrition and feeding management for disease prevention. *Vet. Clin. North Am. Food Anim. Pract.* **2014**, *30*, 689–719.
41. Dohoo, I.R.; Wayne Martin, S. Disease, production and culling in Holstein-Friesian cows: IV. Effects of disease on production. *Prev. Vet. Med.* **1984**, *2*, 755–770.
42. Andersson, L.; Emanuelson, U. An epidemiological study of hyperketonaemia in Swedish dairy cows; Determinants and the relation to fertility. *Prev. Vet. Med.* **1985**, *3*, 449–462.
43. Erb, H.N. Interrelationships among production and clinical disease in dairy cattle: A review. *Can. Vet. J.* **1987**, *28*, 326–329.
44. Melendez, P.; Marin, M.P.; Robles, J.; Rios, C.; Duchens, M.; Archbald, L. Relationship between serum nonesterified fatty acids at calving and the incidence of periparturient diseases in Holstein dairy cows. *Theriogenology* **2009**, *72*, 826–833.

45. Ospina, P.A.; Nydam, D.V.; Stokol, T.; Overton, T.R. Evaluation of nonesterified fatty acids and beta-hydroxybutyrate in transition dairy cattle in the northeastern United States: Critical thresholds for prediction of clinical diseases. *J. Dairy Sci.* **2010**, *93*, 546–554.
46. Goff, J.P.; Horst, R.L. Physiological changes at parturition and their relationship to metabolic disorders. *J. Dairy Sci.***1997**, *80*, 1260–1268.
47. Drackley, J.K.; Dann, H.M.; Douglas, G.N.; Guretzky, N.A.J.; Litherland, N.B.; Underwood, J.P.; Loor, J.J. Physiological and pathological adaptations in dairy cows that may increase susceptibility to periparturient diseases and disorders. *Italian J. Animal Sci.* **2005**.
48. Van der Drift, S.G.A.; Jorritsma, R.; Schonewille, J.T.; Knijn, H.M.; Stegeman, J.A. Routine detection of hyperketonemia in dairy cows using Fourier transform infrared spectroscopy analysis of β-hydroxybutyrate and acetone in milk in combination with test-day information. *J. Dairy Sci.* **2012**, *95*, 4886–4898.
49. Kessel, S.; Stroehl, M.; Meyer, H.H.D.; Hiss, S.; Sauerwein, H.; Schwarz, F.J.; Bruckmaier, R.M. Individual variability in physiological adaptation to metabolic stress during early lactation in dairy cows kept under equal conditions. *J. Anim. Sci.* **2008**, *86*, 2903–2912.
50. Aitken, S.; Corl, C.; Sordillo, L. Immunopathology of mastitis: Insights into disease recognition and resolution: Journal of Mammary Gland Biology and Neoplasia. *J. Mammary Gland Biol. Neoplasia* **2011**, *16*, 291–304.
51. Esposito, G.; Irons, P.C.; Webb, E.C.; Chapwanya, A. Interactions between negative energy balance, metabolic diseases, uterine health and immune response in transition dairy cows. *Anim. Reprod. Sci.* **2014**, *144*, 60–71.
52. Loiselle, M.C.; Ster, C.; Talbot, B.G.; Zhao, X.; Wagner, G.F.; Boisclair, Y.R.; Lacasse, P. Impact of postpartum milking frequency on the immune system and the blood metabolite concentration of dairy cows. *J. Dairy Sci.* **2009**, *92*, 1900–1912.
53. Wathes, D.C.; Cheng, Z.; Chowdhury, W.; Fenwick, M.A.; Fitzpatrick, R.; Morris, D.G.; Patton, J.; Murphy, J.J. Negative energy balance alters global gene expression and immune responses in the uterus of postpartum dairy cows. *Physiol. Genomics* **2009**, *39*, 1–13.
54. Moyes, K.M.; Drackley, J.K.; Morin, D.E.; Rodriguez-Zas, S.L.; Everts, R.E.; Lewin, H.A.; Loor, J.J. Mammary gene expression profiles during an intramammary challenge reveal potential mechanisms linking negative energy balance with impaired immune response. *Physiol. Genomics* **2010**, *41*, 161–170.
55. Burvenich, C.; Bannerman, D.D.; Lippolis, J.D.; Peelman, L.; Nonnecke, B.J.; Kehrli, M.E.; Paape, M.J. Cumulative physiological events influence the inflammatory response of the bovine udder to *Escherichia coli* infections during the transition period. *J. Dairy Sci.* **2007**, *90* (Suppl. 1), E39–E54.
56. LeBlanc, S.J.; Osawa, T.; Dubuc, J. Reproductive tract defense and disease in postpartum dairy cows. *Theriogenology***2011**, *76*, 1610–1618.
57. Sheldon, I.M.; Cronin, J.; Goetze, L.; Donofrio, G.; Schuberth, H.-J. Defining postpartum uterine disease and the mechanisms of infection and immunity in the female reproductive tract in cattle. *Biol. Reprod.* **2009**, *81*, 1025–1032.
58. Huzzey, J.M.; Veira, D.M.; Weary, D.M.; von Keyserlingk, M.A.G. Prepartum behavior and dry matter intake identify dairy cows at risk for metritis. *J. Dairy Sci.* **2007**, *90*, 3220–3233.
59. Ster, C.; Loiselle, M.-C.; Lacasse, P. Effect of postcalving serum nonesterified fatty acids concentration on the functionality of bovine immune cells. *J. Dairy Sci.* **2012**, *95*, 708–717.
60. Hammon, D.S.; Evjen, I.M.; Dhiman, T.R.; Goff, J.P.; Walters, J.L. Neutrophil function and energy status in Holstein cows with uterine health disorders. *Vet. Immunol. Immunopathol.* **2006**, *113*, 21–29.
61. Sordillo, L.M.; Contreras, G.A.; Aitken, S.L. Metabolic factors affecting the inflammatory response of periparturient dairy cows. *Anim. Health Res. Rev.* **2009**, *10*, 53–63.
62. Sordillo, L.M.; Mavangira, V. The nexus between nutrient metabolism, oxidative stress and inflammation in transition cows. *Anim. Prod. Sci.* **2014**, *54*, 1204–1214.
63. Farney, J.K.; Mamedova, L.K.; Coetzee, J.F.; KuKanich, B.; Sordillo, L.M.; Stoakes, S.K.; Minton, J.E.; Hollis, L.C.; Bradford, B.J. Anti-inflammatory salicylate treatment alters the

metabolic adaptations to lactation in dairy cattle. *Am. J. Physiol. Regul. Integr. Comp. Physiol.* **2013**, *305*, R110–R117.
64. Huzzey, J.M.; Mann, S.; Nydam, D.V.; Grant, R.J.; Overton, T.R. Associations of peripartum markers of stress and inflammation with milk yield and reproductive performance in Holstein dairy cows. *Prev. Vet. Med.* **2015**, *120*, 291–297.
65. Abuelo, A.; Hernández, J.; Benedito, J.L.; Castillo, C. The importance of the oxidative status of dairy cattle in the periparturient period: Revisiting antioxidant supplementation. *J. Anim. Physiol. Anim. Nutr. (Berl.)* **2015**.
66. Taylor, V.; Beever, D.; Wathes, D.; Kebreab, E.; Mills, J. Physiological adaptions to milk production that affect the fertility of high yielding dairy cows. In *Dairying: Using Science to Meet Consumers' Needs*; University of Reading: Reading, UK, 2004; pp. 37–71.
67. Beever, D.E. The impact of controlled nutrition during the dry period on dairy cow health, fertility and performance. *Anim. Reprod. Sci.* **2006**, *96*, 212–226.
68. Robinson, J.J.; Ashworth, C.J.; Rooke, J.A.; Mitchell, L.M.; McEvoy, T.G. Nutrition and fertility in ruminant livestock. *Anim. Feed Sci. Technol.* **2006**, *126*, 259–276.
69. Collard, B.L.; Boettcher, P.J.; Dekkers, J.; Petitclerc, D.; Schaeffer, L.R. Relationships between energy balance and health traits of dairy cattle in early lactation. *J. Dairy Sci.* **2000**, *83*, 2683–2690.
70. Razeto-Barry, P. Autopoiesis 40 years later. A review and a reformulation. *Orig. Life Evol. Biosph.* **2012**, *42*, 543–567.
71. Grant, R.J.; Albright, J.L. Feeding behavior and management factors during the transition period in dairy cattle. *J. Anim. Sci.* **1995**, *73*, 2791–2803.
72. Tylutki, T.P.; Fox, D.G.; Durbal, V.M.; Tedeschi, L.O.; Russell, J.B.; van Amburgh, M.E.; Overton, T.R.; Chase, L.E.; Pell, A.N. Cornell Net Carbohydrate and Protein System: A model for precision feeding of dairy cattle. *Anim. Feed Sci. Technol.* **2008**, *143*, 174–202.
73. Allen, M.S.; Piantoni, P. Metabolic control of feed intake: Implications for metabolic disease of fresh cows. *Vet. Clin. North Am. Food Anim. Pract.* **2013**, *29*, 279–297.
74. Allen, M.S. Effects of diet on short-term regulation of feed intake by lactating dairy cattle. *J. Dairy Sci.* **2000**, *83*, 1598–1624.
75. Allen, M.S.; Bradford, B.J.; Oba, M. The hepatic oxidation theory of the control of feed intake and its application to ruminants. *J. Anim. Sci.* **2009**, *87*, 3317–3334.
76. Meyer, U.; Horstmann, K.; Kaske, M.; Flachowsky, G. *Inter- and Intra-Individual Variation of Feed Intake and Metabolic Parameters of Dairy Cows Related to Energy Supply*; European Association for Animal Production: Roma, Italy, 2007; Volume 124, p. 571.
77. Dann, H.M. Dietary Energy Restriction during Late Gestation in Multiparous Holstein Cows. Ph.D. Thesis, University of Illinois, Champaign, IL, USA, 2004.
78. Lean, I.J.; Golder, H.M.; Hall, M.B. Feeding, evaluating, and controlling rumen function. *Vet. Clin. North Am. Food Anim. Pract.* **2014**, *30*, 539–575.
79. Proudfoot, K.L.; Veira, D.M.; Weary, D.M.; von Keyserlingk, M.A.G. Competition at the feed bunk changes the feeding, standing, and social behavior of transition dairy cows. *J. Dairy Sci.* **2009**, *92*, 3116–3123.
80. Cook, N.B.; Bennett, T.B.; Nordlund, K.V. Effect of free stall surface on daily activity patterns in dairy cows with relevance to lameness prevalence. *J. Dairy Sci.* **2004**, *87*, 2912–2922.
81. Kuhla, B. *Central Regulation of Feed Intake in Early Lactation*; Leibniz Institute: Jena, Germany, 2012.
82. Merchen, N.R.; Elizalde, J.C.; Drackley, J.K. Current perspective on assessing site of digestion in ruminants. *J. Anim. Sci.* **1997**, *75*, 2223–2234.
83. Kamra, D.N. Rumen microbial ecosystem. *Curr. Sci.* **2005**, *89*, 124–135.
84. Tajima, K.; Aminov, R.I.; Nagamine, T.; Matsui, H.; Nakamura, M.; Benno, Y. Diet-dependent shifts in the bacterial population of the rumen revealed with real-time PCR. *Appl. Environ. Microbiol.* **2001**, *67*, 2766–2774.
85. Brown, M.S.; Krehbiel, C.R.; Galyean, M.L.; Remmenga, M.D.; Peters, J.P.; Hibbard, B.; Robinson, J.; Moseley, W.M. Evaluation of models of acute and subacute acidosis on dry matter intake, ruminal fermentation, blood chemistry, and endocrine profiles of beef steers. *J. Anim. Sci.* **2000**, *78*, 3155–3168.

86. Pers-Kamczyc, E.; Zmora, P.; Cieślak, A.; Szumacher-Strabel, M. Development of nucleic acid based techniques and possibilities of their application to rumen microbial ecology research. *J. Anim. Feed Sci.* **2011**, *20*, 315–337.
87. Stewart, C.S.; Flint, H.J.; Bryant, M.P. The rumen bacteria. In *The Rumen Microbial Ecosystem*; Hobson, P.N., Stewart, C.S., Eds.; Springer: Dordrecht, The Netherlands, 1997; pp. 10–72.
88. Dijkstra, J.; Kebreab, E.; Mills, J.A.N.; Pellikaan, W.F.; López, S.; Bannink, A.; France, J. Predicting the profile of nutrients available for absorption: From nutrient requirement to animal response and environmental impact. *Animal* **2007**, *1*, 99–111.
89. Nozière, P.; Hoch, T. Modelling fluxes of volatile fatty acids from rumen to portal blood. In *Nutrient Digestion and Utilization in Farm Animals: Modelling Approaches*; Kebreab, E., Nozière, P., Hoch, T., Eds.; CABI Publishing: Wallingford, CT, USA, 2006; pp. 40–47.
90. Dijkstra, J.; Boer, H.; van Bruchem, J.; Bruining, M.; Tamminga, S. Absorption of volatile fatty acids from the rumen of lactating dairy cows as influenced by volatile fatty acid concentration, pH and rumen liquid volume. *Br. J. Nutr.* **1993**, *69*, 385–396.
91. Rabelo, E.; Rezende, R.L.; Bertics, S.J.; Grummer, R.R. Effects of transition diets varying in dietary energy density on lactation performance and ruminal parameters of dairy cows. *J. Dairy Sci.* **2003**, *86*, 916–925.
92. Dijkstra, J.; Ellis, J.L.; Kebreab, E.; Strathe, A.B.; López, S.; France, J.; Bannink, A. Ruminal pH regulation and nutritional consequences of low pH. *Anim. Feed Sci. Technol.* **2012**, *172*, 22–33.
93. Stein, S.K.; Sundrum, A. Detection of SARA in the transition period of dairy cows. *J. Dairy Sci.* **2015**. submitted.
94. Bossen, D.; Mertens, D.R.; Weisbjerg, M.R. Influence of fermentation methods on neutral detergent fiber degradation parameters. *J. Dairy Sci.* **2008**, *91*, 1464–1476.
95. Hall, M.B. Dietary starch source and protein degradability in diets containing sucrose: Effects on ruminal measures and proposed mechanism for degradable protein effects. *J. Dairy Sci.* **2013**, *96*, 7093–7109.
96. Argyle, J.L.; Baldwin, R.L. Effects of amino acids and peptides on rumen microbial growth yields. *J. Dairy Sci.* **1989**, *72*, 2017–2027.
97. Oba, M.; Allen, M.S. Effects of corn grain conservation method on feeding behavior and productivity of lactating dairy cows at two dietary starch concentrations. *J. Dairy Sci.* **2003**, *86*, 174–183.
98. Firkins, J.L.; Allen, M.S.; Oldick, B.S.; St-Pierre, N.R. Modeling ruminal digestibility of carbohydrates and microbial protein flow to the duodenum. *J. Dairy Sci.* **1998**, *81*, 3350–3369.
99. Erdman, R.A.; Varner, M. Fixed yield responses to increased milking frequency. *J. Dairy Sci.* **1995**, *78*, 1199–1203.
100. Stefanon, B.; Colitti, M.; Gabai, G.; Knight, C.H.; Wilde, C.J. Mammary apoptosis and lactation persistency in dairy animals. *J. Dairy Res.* **2002**, *69*, 37–52.
101. Murney, R.; Stelwagen, K.; Wheeler, T.T.; Margerison, J.K.; Singh, K. The effects of milking frequency in early lactation on milk yield, mammary cell turnover, and secretory activity in grazing dairy cows. *J. Dairy Sci.* **2015**, *98*, 305–311.
102. Bauman, D.E.; Mather, I.H.; Wall, R.J.; Lock, A.L. Major advances associated with the biosynthesis of milk. *J. Dairy Sci.* **2006**, *89*, 1235–1243.
103. Gardner, N.H.; Reynolds, C.K.; Phipps, R.H.; Jones, A.K.; Beever, D.E. Effects of different diet supplements in the pre- and post-partum period on reproductive performance in the dairy cow. In *Fertility in the High Producing Dairy Cow*; Occasional Publication No. 26; British Society of Animal Science: London, UK, 2001; pp. 313–322.
104. Van Knegsel, A.; van den Brand, H.; Dijkstra, J.; van Straalen, W.M.; Jorritsma, R.; Tamminga, S.; Kemp, B. Effect of glucogenic *vs.* lipogenic diets on energy balance, blood metabolites, and reproduction in primiparous and multiparous dairy cows in early lactation. *J. Dairy Sci.* **2007**, *90*, 3397–3409.
105. Hattan, A.J. *Energy Utilisation in High Yielding Dairy Cows*; University of Reading: Reading, UK, 2003.

106. De Vries, M.J.; van der Beek, S.; Kaal-Lansbergen, L.; Ouweltjes, W.; Wilmink, J.B. Modeling of energy balance in early lactation and the effect of energy deficits in early lactation on first detected estrus postpartum in dairy cows. *J. Dairy Sci.* **1999**, *82*, 1927–1934.
107. Eastridge, M.L. Major advances in applied dairy cattle nutrition. *J. Dairy Sci.* **2006**, *89*, 1311–1323.
108. Van Arendonk, J.A.; Nieuwhof, G.J.; Vos, H.; Korver, S. Genetic aspects of feed intake and efficiency in lactating dairy heifers. *Livest. Prod. Sci.* **1991**, *29*, 263–275.
109. Beerda, B.; Ouweltjes, W.; Šebek, L.B.; Windig, J.J.; Veerkamp, R.F. Effects of genotype by environment interactions on milk yield, energy balance, and protein balance. *J. Dairy Sci.* **2007**, *90*, 219–228.
110. Rhoads, M.L.; Rhoads, R.P.; VanBaale, M.J.; Collier, R.J.; Sanders, S.R.; Weber, W.J.; Crooker, B.A.; Baumgard, L.H. Effects of heat stress and plane of nutrition on lactating Holstein cows: I. Production, metabolism, and aspects of circulating somatotropin. *J. Dairy Sci.* **2009**, *92*, 1986–1997.
111. Steffens, A.B.; de Boer, S.F. Impact of stress on animal intermediate metabolism. In *Stress in Physiology of Animals*; Balm, P.H., Ed.; Sheffield Academic Press: Sheffield, UK, 1999; pp. 108–129.
112. Ingvartsen, K.L. Feeding- and management-related diseases in the transition cow: Physiological adaptations around calving and strategies to reduce feeding-related diseases. *Anim. Feed Sci. Technol.* **2006**, *126*, 175–213.
113. Veerkamp, R.F.; Brotherstone, S. Genetic correlations between linear type traits, food intake, live weight and condition score in Holstein Friesian dairy cattle. *Anim. Sci.* **1997**, *64*, 385–392.
114. Sutter, F.; Beever, D.E. Energy and nitrogen metabolism in Holstein-Friesian cows during early lactation. *Anim. Sci.***2000**, *70*, 503–514.
115. Friggens, N.C.; Brun-Lafleur, L.; Faverdin, P.; Sauvant, D.; Martin, O. Advances in predicting nutrient partitioning in the dairy cow: Recognizing the central role of genotype and its expression through time. *Animal* **2013**, *7* (Suppl. 1), 89–101.
116. Patton, J.; Kenny, D.A.; Mee, J.F.; O'Mara, F.P.; Wathes, D.C.; Cook, M.; Murphy, J.J. Effect of milking frequency and diet on milk production, energy balance, and reproduction in dairy cows. *J. Dairy Sci.* **2006**, *89*, 1478–1487.
117. Bradford, B.J. Immunity and inflammation in transition cows. In Proceedings of AABP Annual Meeting, St. Louis, MO, USA, 22–24 September 2011; pp. 48–52.
118. Boutinaud, M.; Ben Chedly, M.H.; Delamaire, E.; Guinard-Flament, J. Milking and feed restriction regulate transcripts of mammary epithelial cells purified from milk. *J. Dairy Sci.* **2008**, *91*, 988–998.
119. Bell, A.W.; Burhans, W.S.; Overton, T.R. Protein nutrition in late pregnancy, maternal protein reserves and lactation performance in dairy cows. *Proc. Nutr. Soc.* **2000**, *59*, 119–126.
120. Lucy, M.C. Mechanisms linking the somatotrophic axis with insulin: Lesson from the postpartum dairy cow. *Proc. N. Z. Soc. Anim. Prod.* **2004**, *64*, 19–23.
121. Vernon, R.G. Nutrient partitioning, lipid metabolism and relevant imbalances. In *Recent Developments and Perspectives in Bovine Medicine*; Kasle, M., Scholz, H., Holtershinken, M., Eds.; Clinik fur Rinderkrankheiten: Hanover, Germany, 2002.
122. Zhao, F.-Q.; Dixon, W.T.; Kennelly, J.J. Localization and gene expression of glucose transporters in bovine mammary gland. *Comp. Biochem. Physiol. B Biochem. Mol. Biol.* **1996**, *115*, 127–134.
123. Butler, S.T.; Marr, A.L.; Pelton, S.H.; Radcliff, R.P.; Lucy, M.C.; Butler, W.R. Insulin restores GH responsiveness during lactation-induced negative energy balance in dairy cattle: Effects on expression of IGF-I and GH receptor 1A. *J. Endocrinol.* **2003**, *176*, 205–217.
124. Lucy, M.C. Fertility in high-producing dairy cows: Reasons for decline and corrective strategies for sustainable improvement. *Soc. Reprod. Fertil. Suppl.* **2007**, *64*, 237–254.
125. Osborn, O.; Olefsky, J.M. The cellular and signaling networks linking the immune system and metabolism in disease. *Nat. Med.* **2012**, *18*, 363–374.
126. McNamara, J.P. Regulation of adipose tissue metabolism in support of lactation. *J. Dairy Sci.* **1991**, *74*, 706–719.

127. McNamara, J.P.; Harrison, J.H.; Kincaid, R.L.; Waltner, S.S. Lipid metabolism in adipose tissue of cows fed high fat diets during lactation. *J. Dairy Sci.* **1995**, *78*, 2782–2796.
128. Smith, T.H.; McNamara, J.P. Regulation of bovine adipose tissue metabolism during lactation. 6. Cellularity and hormone-sensitive lipase activity as affected by genetic merit and energy intake. *J. Dairy Sci.* **1990**, *73*, 772–783.
129. Lucy, M.C.; Jiang, H.; Kobayashi, Y. Changes in the somatotrophic axis associated with the initiation of lactation. *J. Dairy Sci.* **2001**, *84*, E113–E119.
130. Rukkwamsuk, T.; Wensing, T.; Geelen, M.J. Effect of overfeeding during the dry period on regulation of adipose tissue metabolism in dairy cows during the periparturient period. *J. Dairy Sci.* **1998**, *81*, 2904–2911.
131. Rukkwamsuk, T.; Wensing, T.; Geelen, M.J. Effect of overfeeding during the dry period on the rate of esterification in adipose tissue of dairy cows during the periparturient period. *J. Dairy Sci.* **1999**, *82*, 1164–1169.
132. Locher, L.; Häussler, S.; Laubenthal, L.; Singh, S.P.; Winkler, J.; Kinoshita, A.; Kenéz, Á.; Rehage, J.; Huber, K.; Sauerwein, H. Effect of increasing body condition on key regulators of fat metabolism in subcutaneous adipose tissue depot and circulation of non-lactating dairy cows. *J. Dairy Sci.* **2015**, *98*, 1057–1068.
133. Katamoto, H.; Yukawa, T.; Shimada, Y. Lipogenic and lipolytic activities in isolated adipocytes from cattle with fat necrosis. *Res. Vet. Sci.* **1996**, *61*, 214–217.
134. Mann, S.; Yepes, F.A.L.; Overton, T.R.; Wakshlag, J.J.; Lock, A.L.; Ryan, C.M.; Nydam, D.V. Dry period plane of energy: Effects on feed intake, energy balance, milk production, and composition in transition dairy cows. *J. Dairy Sci.* **2015**, *98*, 3366–3382.
135. Andersen, J.B.; Ridder, C.; Larsen, T. Priming the cow for mobilization in the periparturient period: Effects of supplementing the dry cow with saturated fat or linseed. *J. Dairy Sci.* **2008**, *91*, 1029–1043.
136. Litherland, N.B.; da Silva, D.N.L.; Hansen, W.P.; Davis, L.; Emanuele, S.; Blalock, H. Effects of prepartum controlled-energy wheat straw and grass hay diets supplemented with starch or sugar on periparturient dairy cow performance and lipid metabolism. *J. Dairy Sci.* **2013**, *96*, 3050–3063.
137. Newman, A.; Mann, S.; Nydam, D.V.; Overton, T.R.; Behling-Kelly, E. Impact of dietary plane of energy during the dry period on lipoprotein parameters in the transition period in dairy cattle. *J. Anim. Physiol. Anim. Nutr. (Berl.)* **2015**.
138. Overton, T.R.; Drackley, J.K.; Ottemann-Abbamonte, C.J.; Beaulieu, A.D.; Emmert, L.S.; Clark, J.H. Substrate utilization for hepatic gluconeogenesis is altered by increased glucose demand in ruminants. *J. Anim. Sci.* **1999**, *77*, 1940–1951.
139. Kreipe, L.; Vernay, M.C.M.B.; Oppliger, A.; Wellnitz, O.; Bruckmaier, R.M.; van Dorland, H.A. Induced hypoglycemia for 48 hours indicates differential glucose and insulin effects on liver metabolism in dairy cows. *J. Dairy Sci.* **2011**, *94*, 5435–5448.
140. Reynolds, C.K.; Aikman, P.C.; Lupoli, B.; Humphries, D.J.; Beever, D.E. Splanchnic metabolism of dairy cows during the transition from late gestation through early lactation. *J. Dairy Sci.* **2003**, *86*, 1201–1217.
141. Herdt, T.H. Ruminant adaptation to negative energy balance. Influences on the etiology of ketosis and fatty liver. *Vet. Clin. North Am. Food Anim. Pract.* **2000**, *16*, 215–230.
142. Hammon, H.M.; Stürmer, G.; Schneider, F.; Tuchscherer, A.; Blum, H.; Engelhard, T.; Genzel, A.; Staufenbiel, R.; Kanitz, W. Performance and metabolic and endocrine changes with emphasis on glucose metabolism in high-yielding dairy cows with high and low fat content in liver after calving. *J. Dairy Sci.* **2009**, *92*, 1554–1566.
143. Cadórniga-Valiño, C.; Grummer, R.R.; Armentano, L.E.; Donkin, S.S.; Bertics, S.J. Effects of fatty acids and hormones on fatty acid metabolism and gluconeogenesis in bovine hepatocytes. *J. Dairy Sci.* **1997**, *80*, 646–656.
144. Strang, B.D.; Bertics, S.J.; Grummer, R.R.; Armentano, L.E. Effect of long-chain fatty acids on triglyceride accumulation, gluconeogenesis, and ureagenesis in bovine hepatocytes. *J. Dairy Sci.* **1998**, *81*, 728–739.
145. Vazquez-Añon, M.; Bertics, S.; Luck, M.; Grummer, R.R.; Pinheiro, J. Peripartum liver triglyceride and plasma metabolites in dairy cows. *J. Dairy Sci.* **1994**, *77*, 1521–1528.

146. Cornish-Bowden, A.; Cárdenas, M.L.; Letelier, J.-C.; Soto-Andrade, J.; Abarzúa, F.G. Understanding the parts in terms of the whole. *Biol. Cell* **2004**, *96*, 713–717.
147. Bannink, A.; France, J.; Lopez, S.; Gerrits, W.; Kebreab, E.; Tamminga, S.; Dijkstra, J. Modelling the implications of feeding strategy on rumen fermentation and functioning of the rumen wall. *Anim. Feed Sci. Technol.* **2008**, *143*, 3–26.
148. Sudakov, K. The theory of functional systems: General postulates and principles of dynamic organization. *Integr. Physiol. Behav. Sci.* **1997**, *32*, 392–414.
149. Oppenheim, P.; Putnam, H. The unity of science as a working hypothesis. In *Concepts, Theories, and the Mind-Body Problem*; Feigl, H., Ed.; University of Minnesota Press: Minneapolis, MI, USA, 1958; Volume 2, pp. 3–36.
150. Powell, A.; Dupré, J. From molecules to systems: The importance of looking both ways. *Stud. Hist. Philos. Biol. Biomed. Sci.* **2009**, *40*, 54–64.
151. Conway, G.R. The properties of agroecosystems. *Agric. Syst.* **1987**, *24*, 95–117.
152. Varela, F.J. Patterns of life: Intertwining identity and cognition. *Brain Cogn.* **1997**, *34*, 72–87.
153. Thompson, P.B. The varieties of sustainability in livestock farming. *EAAP Publ.* **1997**, *89*, 5–15.
154. Fox, C.J.; Hammerman, P.S.; Thompson, C.B. Fuel feeds function: Energy metabolism and the T-cell response. *Nat. Rev. Immunol.* **2005**, *5*, 844–852.
155. Wagner, A. *Robustness and Evolvability in Living Systems*; Princeton University Press: Princeton, NJ, USA, 2005.
156. Lucy, M.C. Mechanisms linking nutrition and reproduction in postpartum cows. *Reprod. Suppl.* **2003**, *61*, 415–427.
157. Buckley, F.; Dillon, P.; Rath, M.; Veerkamp, R.F. The relationship between genetic merit for yield and live weight, condition score, and energy balance of spring calving Holstein Friesian dairy cows on grass based systems of milk production. *J. Dairy Sci.* **2000**, *83*, 1878–1886.
158. Krause, K.M.; Oetzel, G.R. Understanding and preventing subacute ruminal acidosis in dairy herds: A review. *Anim. Feed Sci. Technol.* **2006**, *126*, 215–236.
159. Haerle, C.; Sundrum, A. Animal health on the farm level 2. Communication: nutrient supply on bavarian dairy farms. *Züchtungskunde* **2013**, *85*, 396–412.
160. Holtenius, K.; Agenäs, S.; Delavaud, C.; Chilliard, Y. Effects of feeding intensity during the dry period. 2. Metabolic and hormonal responses. *J. Dairy Sci.* **2003**, *86*, 883–891.
161. Rastani, R.R.; Grummer, R.R.; Bertics, S.J.; Gümen, A.; Wiltbank, M.C.; Mashek, D.G.; Schwab, M.C. Reducing dry period length to simplify feeding transition cows: Milk production, energy balance, and metabolic profiles. *J. Dairy Sci.* **2005**, *88*, 1004–1014.
162. Gümen, A.; Rastani, R.R.; Grummer, R.R.; Wiltbank, M.C. Reduced dry periods and varying prepartum diets alter postpartum ovulation and reproductive measures. *J. Dairy Sci.* **2005**, *88*, 2401–2411.
163. Van Knegsel, A.; van der Drift, S.G.A.; Cermáková, J.; Kemp, B. Effects of shortening the dry period of dairy cows on milk production, energy balance, health, and fertility: A systematic review. *Vet. J.* **2013**, *198*, 707–713.
164. Chen, J.; Gross, J.J.; van Dorland, H.A.; Remmelink, G.J.; Bruckmaier, R.M.; Kemp, B.; van Knegsel, A.T.M. Effects of dry period length and dietary energy source on metabolic status and hepatic gene expression of dairy cows in early lactation. *J. Dairy Sci.* **2015**, *98*, 1033–1045.
165. McNamara, S.; Murphy, J.J.; O'Mara, F.P.; Rath, M.; Mee, J.F. Effect of milking frequency in early lactation on energy metabolism, milk production and reproductive performance of dairy cows. *Livest. Sci.* **2008**, *117*, 70–78.
166. Phyn, C.; Kay, J.K.; Rius, A.G.; Morgan, S.R.; Roach, C.S.; Grala, T.M.; Roche, J.R. Review: Impact of short—Term alterations to milking frequency in early lactation. In Proceedings of 4th Australian Dairy Science Symposium; Caxton Press: Caldwell, NJ, USA, 2010.
167. Clark, D.A.; Phyn, C.; Tong, M.J.; Collis, S.J.; Dalley, D.E. A systems comparison of once-*versus* twice-daily milking of pastured dairy cows. *J. Dairy Sci.* **2006**, *89*, 1854–1862.

168. Stelwagen, K.; Phyn, C.V.C.; Davis, S.R.; Guinard-Flament, J.; Pomiès, D.; Roche, J.R.; Kay, J.K. Invited review: Reduced milking frequency: Milk production and management implications. *J. Dairy Sci.* **2013**, *96*, 3401–3413.
169. Carbonneau, E.; de Passillé, A.M.; Rushen, J.; Talbot, B.G.; Lacasse, P. The effect of incomplete milking or nursing on milk production, blood metabolites, and immune functions of dairy cows. *J. Dairy Sci.* **2012**, *95*, 6503–6512.
170. Soberon, F.; Ryan, C.M.; Nydam, D.V.; Galton, D.M.; Overton, T.R. The effects of increased milking frequency during early lactation on milk yield and milk composition on commercial dairy farms. *J. Dairy Sci.* **2011**, *94*, 4398–4405.
171. Oltenacu, P.A.; Algers, B. Selection for increased production and the welfare of dairy cows: Are new breeding goals needed? *Ambio* **2005**, *34*, 311–315.
172. Gordon, F.J.; Patterson, D.C.; Yan, T.; Porter, M.G.; Mayne, C.S.; Unsworth, E.F. The influence of genetic index for milk production on the response to complete diet feeding and the utilization of energy and nitrogen. *Anim. Sci.* **1995**, *61*, 199–210.
173. Tyrrell, H.F.; Brown, A.C.; Reynolds, P.J.; Haaland, G.L.; Bauman, D.E.; Peel, C.J.; Steinhour, W.D. Effect of bovine somatotropin on metabolism of lactating dairy cows: Energy and nitrogen utilization as determined by respiration calorimetry. *J. Nutr.* **1988**, *118*, 1024–1030.
174. Fourichon, C.; Seegers, H.; Beaudeau, F.; Verfaille, L.; Bareille, N. Health-control costs in dairy farming systems in western France. *Livest. Prod. Sci.* **2001**, *68*, 141–156.
175. Lawson, L.G.; Agger, J.F.; Lund, M.; Coelli, T. Lameness, metabolic and digestive disorders, and technical efficiency in Danish dairy herds: A stochastic frontier production function approach. *Livest. Prod. Sci.* **2004**, *91*, 157–172.
176. Lean, I.J.; van Saun, R.; DeGaris, P.J. Energy and protein nutrition management of transition dairy cows: Metabolic diseases of dairy cattle. *Vet. Clin. North Am. Food Anim. Pract.* **2013**, *29*, 337–366.
177. Von Keyserlingk, M.; Martin, N.P.; Kebreab, E.; Knowlton, K.F.; Grant, R.J.; Stephenson, M.; Sniffen, C.J.; Harner, J.P.; Wright, A.D.; Smith, S.I. Invited review: Sustainability of the US dairy industry. *J. Dairy Sci.* **2013**, *96*, 5405–5425.
178. Sundrum, A. "Healthy food" from healthy cows. In *Organic Farming and Food Production*; Konvalina, P., Ed.; InTech: Rijeka, Croatia, 2012; pp. 95–120.
179. Kahneman, D. *Thinking, Fast and Slow*; Macmillan: New York, NY, USA, 2011.
180. Deutsche Forschungsgemeinschaft. *Future Perspectives of Agricultural Science and Research: Perspektiven der Agrarwissenschaftlichen Forschung*; Wiley-VCH: Bonn, Germany, 2005.
181. McNamara, J.P. Ruminant Nutrition Symposium: A systems approach to integrating genetics, nutrition, and metabolic efficiency in dairy cattle. *J. Anim. Sci.* **2012**, *90*, 1846–1854.
182. Friggens, N.C.; Newbold, J.R. Towards a biological basis for predicting nutrient partitioning: The dairy cow as an example. *Animal* **2007**, *1*, 87–97.
183. Khan, M.J.; Hosseini, A.; Burrell, S.; Rocco, S.M.; McNamara, J.P.; Loor, J.J. Change in subcutaneous adipose tissue metabolism and gene network expression during the transition period in dairy cows, including differences due to sire genetic merit. *J. Dairy Sci.* **2013**, *96*, 2171–2182.
184. Ideker, T.; Galitski, T.; Hood, L. A new approach to decoding life: Systems biology. *Annu. Rev. Genomics Hum. Genet.***2001**, *2*, 343–372.
185. Loor, J.J.; Bionaz, M.; Drackley, J.K. Systems physiology in dairy cattle: Nutritional genomics and beyond. *Annu. Rev. Anim. Biosci.* **2013**, *1*, 365–392.
186. Van Dien, S.; Schilling, C.H. Bringing metabolomics data into the forefront of systems biology. *Mol. Syst. Biol.* **2006**, *2*.
187. Bannink, A.; France, J.; Dijkstra, J. Modeling approaches to link various levels of organization in animal physiology. In *Systems Biology and Livestock Science*; Te Pas, M.F.W., Bannink, A., Woelders, H., Eds.; Wiley-Blackwell: Ames, IA, USA, 2011; pp. 191–214.
188. Loor, J.J.; Bionaz, M.; Invermizzi, G. Systems biology and animal nutrition: Insights from the dairy cow during growth and the lactation cycle. In *Systems Biology and Livestock*

Science; Te Pas, M.F.W., Bannink, A., Woelders, H., Eds.; Wiley-Blackwell: Ames, IA, USA, 2011; pp. 215–246.

189. Hampf, M.; Gossen, M. Promoter crosstalk effects on gene expression. *J. Mol. Biol.* **2007**, *365*, 911–920.
190. Kaern, M.; Elston, T.C.; Blake, W.J.; Collins, J.J. Stochasticity in gene expression: From theories to phenotypes. *Nat. Rev. Genet.* **2005**, *6*, 451–464.
191. Mossio, M.; Bich, L.; Moreno, A. Emergence, closure and inter-level causation in biological systems. *Erkenntnis* **2013**, *78*, 153–178.

INDEX

S

T

U

V

W

Y